Current Topics in Behavioral Neurosciences

Series Editors:
Mark Geyer, La Jolla, CA, USA
Bart Ellenbroek, Hamburg, Germany
Charles Marsden, Nottingham, UK

About this series

Current Topics in Behavioral Neurosciences provides critical and comprehensive discussions of the most significant areas of behavioral neuroscience research, written by leading international authorities. Each volume offers an informative and contemporary account of its subject, making it an unrivalled reference source. Titles in this series are available in both print and electronic formats.

With the development of new methodologies for brain imaging, genetic and genomic analyses, molecular engineering of mutant animals, novel routes for drug delivery, and sophisticated cross-species behavioral assessments, it is now possible to study behavior relevant to psychiatric and neurological diseases and disorders on the physiological level. The *Behavioral Neurosciences* series focuses on "translational medicine" and cutting-edge technologies. Preclinical and clinical trials for the development of new diagnostics and therapeutics as well as prevention efforts are covered whenever possible.

Dave Kendall • Stephen Alexander
Editors

Behavioral Neurobiology of the Endocannabinoid System

Springer

Editors
Prof. David A. Kendall
Dr. Stephen Alexander
University of Nottingham
Queen's Medical Centre
School of Biomedical Sciences
Nottingham, NG7 2UH
United Kingdom
dave.kendall@nottingham.ac.uk
steve.alexander@nottingham.ac.uk

ISSN 1866-3370 e-ISSN 1866-3389
ISBN 978-3-540-88954-0 e-ISBN 978-3-540-88955-7
DOI: 10.1007/978-3-540-88955-7
Springer Heidelberg Dordrecht London New York

Library of Congress Control Number: 2009928098

Cover illustration: Artistic representation of oscillatory synchrony and timing of neurons in networks by Gyorgy Buzsaki

Cover design: WMXDesign GmbH, Heidelberg, Germany

Printed on acid-free paper

Springer is part of Springer Science+Business Media (www.springer.com)

Preface

It is abundantly clear that the endocannabinoid signalling system is ubiquitously expressed throughout the animal kingdom from simple invertebrates to man, and that it is represented in almost every cell type.

The components of the system in relation to the synthesis and metabolism of the endocannabinoids and their biological targets are beginning to be unravelled, but the degree of complexity involved is enormous. This is particularly true for endocannabinoid signalling in the central nervous system, which is the subject of this volume.

In the book, a variety of authors, who are all very active researchers in the field, provide current accounts of the roles of the endocannabinoid system in normal brain physiology in relation to the neurobiology of essential behaviors and in a number of central disease states.

The first part provides a background and "tool box", detailing what is known about the endocannabinoids themselves and their target receptors and how they influence synaptic activity. It goes on to describe the genetic and pharmacological methods available for investigating the system.

The second part describes endocannabinoid roles in key systems controlling appetite, pain, memory and learning, stress responses and reproduction.

The final group of chapters reviews the current state of knowledge surrounding the function of the endocannabinoid system in depression, drug addiction, schizophrenia, feeding disorders and Tourette's syndrome.

Given the enormous amount of information available and the rate of progress in research, it is impossible for the volume to be totally comprehensive, but we trust that it will provide an excellent background to researchers wanting to expand their area of interest and to newcomers to the field.

Nottingham, United Kingdom

David Kendall
Stephen Alexander

Contents

Contributors

Stephen P.H. Alexander
School of Biomedical Sciences and Institute of Neuroscience, University of Nottingham Medical School, Queens Medical Centre, Nottingham NG7 2UH, UK; steve.alexander@nottingham.ac.uk

Bradley E. Alger
Departments of Physiology and Psychiatry, University of Maryland School of Medicine, 655 W. Baltimore St./BRB 5-025, Baltimore, MD 21201, USA; balger@umaryland.edu

Cristina Cervino
Endocrinology Unit and Center of Applied Biomedical Research (C.R.B.A.), Department of Internal Medicine and Gastroenterology, S.Orsola-Malpighi Hospital, Alma Mater Studiorum University of Bologna, 40138 Bologna, Italy

Victoria Chapman
School of Biomedical Sciences, University of Nottingham, Nottingham NG7 2UH, UK; victoria.chapman@nottingham.ac.uk

Sudhansu K. Dey
Departments of Pharmacology, Pediatrics and Cell & Developmental Biology, Division of Reproductive and Developmental Biology, Vanderbilt University Medical Center, Nashville, TN 37232, USA; sk.dey@vanderbilt.edu

Steven R. Goldberg
Preclinical Pharmacology Section, Behavioral Neuroscience Research Branch, Intramural Research Program, National Institute on Drug Abuse, National Institutes of Health, Department of Health and Human Services, Baltimore, MD 21224, USA; sgoldber@intra.nida.nih.gov

Boris B. Gorzalka
Department of Psychology, University of British Columbia, 2136 West Mall, Vancouver, British Columbia V6T1Z4, Canada; bgorzalka@psych.ubc.ca

Matthew N. Hill
Department of Psychology, University of British Columbia, 2136 West Mall, Vancouver, British Columbia V6T1Z4, Canada

Cecilia J. Hillard
Department of Pharmacology and Toxicology, Medical College of Wisconsin, 8701 Watertown Plank Road, Milwaukee, WI 53226, USA; chillard@mcw.edu

Maulik Jhaveri
School of Biomedical Sciences, University of Nottingham, Nottingham NG7 2UH, UK

Zuzana Justinova
Preclinical Pharmacology Section, Behavioral Neuroscience Research Branch, Intramural Research Program, National Institute on Drug Abuse, National Institutes of Health, Department of Health and Human Services, Baltimore, MD 21224, USA and Maryland Psychiatric Research Center, Department of Psychiatry, University of Maryland School of Medicine, Baltimore, MD 21228, USA

István Katona
Institute of Experimental Medicine, Hungarian Academy of Sciences, H-1083 Budapest, Szigony utca 43. Hungary; katona@koki.hu

David A Kendall
School of Biomedical Sciences and Institute of Neuroscience, University of Nottingham Medical School, Queens Medical Centre, Nottingham NG7 2UH, UK; dave.kendall@nottingham.ac.uk

Tim C Kirkham
School of Psychology, University of Liverpool, Eleanor Rathbone Building, Bedford Street South, Liverpool L69 7ZA, England; t.c.kirkham@liverpool.ac.uk

Pauline Lafenêtre
Group Molecular Mechanisms of Behavioural Adaptation, INSERM U862 Neuro-Centre Magendie Université Bordeaux 2, 146, rue Léo Saignat, 33077 Bordeaux, France

F. Markus Leweke
Department of Psychiatry and Psychotherapy, University of Cologne, 50924 Cologne, Germany; leweke@ecnp.net

Alessio Lodola
Dipartimento Farmaceutico, Università degli Studi di Parma, viale G. P. Usberti 27/A Campus Universitario, I-43100 Parma, Italy; alessio.lodola@unipr.it

Beat Lutz
Department of Physiological Chemistry, Johannes Gutenberg University Mainz, Duesbergweg 6, D-55099 Mainz, Germany; blutz@uni-mainz.de

Ken Mackie
Gill Center for Biomolecular Science, Department of Physiology and Brain Science, Indiana University, 1101 E 10th St., Bloomington, IN 47405, USA; kmackie@indiana.edu

Giovanni Marsicano
Group Molecular Mechanisms of Behavioural Adaptation, INSERM U862 Neuro-Centre Magendie Université Bordeaux 2, 146, rue Léo Saignat, 33077 Bordeaux, France; marsicano@bordeaux.inserm.fr

Krisztina Monory
Department of Physiological Chemistry, Johannes Gutenberg University Mainz, Duesbergweg 6, D-55099 Mainz, Germany; monory@uni-mainz.de

Marco Mor
Dipartimento Farmaceutico, Università degli Studi di Parma, viale G. P. Usberti 27/A Campus Universitario, I-43100 Parma, Italy; marco.mor@unipr.it

Kirsten R Müller-Vahl
Department of Clinical Psychiatry and Psychotherapy, Medical School Hannover, Carl-Neuberg-Str.1, D-30625 Hannover, Germany; mueller-vahl.kirsten@mh-hannover.de

Uberto Pagotto
Endocrinology Unit and Center of Applied Biomedical Research (C.R.B.A.), Department of Internal Medicine and Gastroenterology, S.Orsola-Malpighi Hospital, Alma Mater Studiorum University of Bologna, 40138 Bologna, Italy; uberto.pagotto@unibo.it

Leigh V. Panlilio
Preclinical Pharmacology Section, Behavioral Neuroscience Research Branch, Intramural Research Program, National Institute on Drug Abuse, National Institutes of Health, Department of Health and Human Services, Baltimore, MD 21224, USA

Renato Pasquali
Endocrinology Unit and Center of Applied Biomedical Research (C.R.B.A.), Department of Internal Medicine and Gastroenterology, S.Orsola-Malpighi Hospital, Alma Mater Studiorum University of Bologna, 40138 Bologna, Italy

Sachin Patel
Department of Psychiatry, Vanderbilt University, Nashville, TN USA

Devi Rani Sagar
School of Biomedical Sciences, University of Nottingham, Nottingham NG7 2UH, UK; Devi.sagar@nottingham.ac.uk

Ken Soderstrom
Department of Pharmacology and Toxicology, Brody School of Medicine, East Carolina University, Greenville, NC 27834, USA; soderstromk@ecu.edu

Xiaofei Sun
Departments of Pharmacology, Division of Reproductive and Developmental Biology, Vanderbilt University Medical Center, Nashville, TN 37232, USA

Valentina Vicennati
Endocrinology Unit and Center of Applied Biomedical Research (C.R.B.A.), Department of Internal Medicine and Gastroenterology, S.Orsola-Malpighi Hospital, Alma Mater Studiorum University of Bologna, 40138 Bologna, Italy

Betty Yao
Neuroscience Research, Abbott Laboratories, 100 Abbot Park Road, Dept. R47W, Bldg AP9A, Abbott Park, IL 60048, USA; Betty.yao@abbott.com

Part I
Biochemistry, Pharmacology, Genetics & Chemistry

The Life Cycle of the Endocannabinoids: Formation and Inactivation

Stephen P.H. Alexander and David A. Kendall

Contents

Abstract In this chapter, we summarise the current thinking about the nature of endocannabinoids. In describing the life cycle of these agents, we highlight the synthetic and catabolic enzymes suggested to be involved. For each of these, we provide a systematic analysis of information on sequence, subcellular and cellular distribution, as well as physiological and pharmacological substrates, enhancers and inhibitors, together with brief descriptions of the impact of manipulating enzyme levels through genetic mechanisms (dealt with in more detail in the chapter "Genetic Models of the Endocannabinoid System" by Monory and Lutz, this

S.P.H. Alexander (✉) and D.A. Kendall
School of Biomedical Sciences and Institute of Neuroscience, University of Nottingham Medical School, Queens Medical Centre, Nottingham, NG7 2UH, UK
e-mail: steve.alexander@nottingham.ac.uk, dave.kendall@nottingham.ac.uk

D. Kendall and S. Alexander (eds.), *Behavioral Neurobiology of the Endocannabinoid System*, Current Topics in Behavioral Neurosciences 1,
DOI: 10.1007/978-3-540-88955-7_1, © Springer-Verlag Berlin Heidelberg 2009

volume). In addition, we describe experiments investigating the stimulation of endocannabinoid synthesis and release in intact cell systems.

Keywords 2-arachidonoylglycerol • Anandamide • Diacylglcyerol lipase • Endocannabinoid turnover • Fatty acid amide hydrolase • N-acylphosphatidylethanolamine phospholipase D

Abbreviations

2AG	2-Arachidonoylglycerol
2AG-3P	2-Arachidonoylglycerol-3-phosphate
AEA	Anandamide, *N*-arachidonoylethanolamine
COX	Cyclooxygenase
DAG	Diacylglycerol
DGL	Diacylglycerol lipase
DSI	Depolarization-evoked suppression of inhibition
ECB	Endocannabinoid
EET	Epoxyeicosatrienoic acid
Epac	Exchange protein activated by cyclic AMP
FAAH	Fatty acid amide hydrolase
LOX	Lipoxygenase
LPI	Lysophosphatidylinositol
LPLC	Lysophospholipase C
LPLD	Lysophospholipase D
lysoNAPE	Lyso-*N*-acylphosphatidylethanolamine
MAFP	Methylarachidonylfluorophosphonate
MGL	Monoacylglycerol lipase
NAAA	*N*-Acylethanolamine acid amidase
NAE	*N*-Acylethanolamine
NAPE	*N*-Acylphosphatidylethanolamine
ODA	Oleamide
OEA	*N*-Oleoylethanolamine
PE	Phosphatidylethanolamine
PEA	*N*-Palmitoylethanolamine
PIP_2	Phosphatidylinositol-4,5-bisphosphate
PLA_1	Phospholipase A_1
PLA_2	Phospholipase A_2
PLB	Phospholipase B
PLC	Phospholipase C
PLD	Phospholipase D
SEA	N-Stearoylethanolamine
THL	Tetrahydrolipstatin

1 Cannabinoid Signalling in the CNS

The widely accepted phenomenon of synaptic plasticity highlights the fact that the efficiency of synaptic transmission can alter, dependent on the local environment. Two relevant aspects of synaptic plasticity involving cannabinoid receptors are depolarization-evoked suppression of excitation and inhibition (Gerdeman and Lovinger 2003; Diana and Marty 2004). Although dealt with in more detail in the chapter "Endocannabinoid Signaling in Neural Plasticity" by Brad Alger in this volume, in brief, these phenomena are proposed to result from transmitter-mediated post-synaptic depolarization of neurones leading to an elevation of intracellular calcium ions, resulting in the generation of a retrograde messenger which acts on the presynaptic neurone to alter neurotransmitter release. The involvement of the CB_1 cannabinoid receptor has been identified through the use of the relatively selective antagonists, rimonabant (Wilson and Nicoll 2001) and AM251 (Kreitzer and Regehr 2001), as well as animal models with disruption of the gene encoding CB_1 cannabinoid receptors (Varma et al. 2001; Wilson et al. 2001; Kim et al. 2002; Yoshida et al. 2002). The majority of evidence favours the involvement of ester endocannabinoids (ECBs) in mediating these retrograde effects. In contrast, the amide ECBs have been proposed to act in an anterograde fashion, subserving a more conventional neurotransmitter-like role (Egertova et al. 2008).

From these studies and previous investigations (Di Marzo et al. 1994), the hypothesis has emerged that ECBs are made "on demand" as a result of heightened neuronal activity. In this chapter, we will look at synthetic pathways which appear to be consistent with this hypothesis and more recent developments, which suggest alternative strategies for endocannabinoid biosynthesis, potentially of more relevance to the pathophysiological state. In addition, we will examine metabolic pathways which inactivate, or potentially transform, ECBs.

2 What Are Endocannabinoids?

Anandamide (*N*-arachidonoylethanolamine, AEA) is the archetypal ECB described first by Raphael Mechoulam, Roger Pertwee and colleagues in 1992 (Devane et al. 1992). It was identified in the classical fashion by screening solvent extracts of brain in a cannabinoid receptor radioligand binding assay, with subsequent determination of structure by GC-MS and re-synthesis. *N*-Palmitoylethanolamine (PEA) was identified in the same experiments, but was not considered to be an ECB since only one cannabinoid receptor (CB_1) had been identified at the time and PEA had negligible affinity for this. The situation is much more complex now with a variety of putative ECB receptors of the G-protein-coupled, ion channel and nuclear receptor families proposed (see the chapter "Endocannabinoid Receptor Pharmacology" by Mackie and Yao, this volume) along with chemically related agents having affinities for one or more of these. As a note of caution, it is not always clear whether endogenous levels of some of these agents in different tissues are sufficient to activate cognate receptors allowing them to be labelled as true ECBs (Oka et al. 2003). 2-Arachidonoylglycerol (2AG) has been suggested to be the most biologically important ECB, as it occurs in greater concentrations in tissues, and shows greater efficacy at these targets, than AEA (Sugiura et al. 1997, 1999; Sugiura and Waku 2000, 2002). Although AEA and 2AG are considered the principal ECBs, the range of endogenous agents active at cannabinoid receptors is certainly not limited to these two (Hanus et al. 1993). As a pair, they are closely structurally related in that they are both based on the polyunsaturated fatty acid arachidonate. They are both hydrophobic entities, with partition coefficients (XlogP values, indices of hydrophobicity) of 5.5 and 5.4, respectively. In comparison, conventional neurotransmitters like dopamine, glutamate and GABA have XlogP values of 0.9, −3.3 and −0.7, and are considerably more hydrophilic, partitioning readily into aqueous solutions. This hydrophobicity is also considerably more marked than that of prostaglandin E_2 (2.8) and more similar to leukotriene A_4 (5.0). In comparison to arachidonic acid (6.5), however, the endocannabinoids AEA and 2AG are less hydrophobic. Similarly, the precursor molecules 1-stearoyl-2-arachidonoylglycerol (14.3) and *N*-arachidonoyl-1-stearoyl-2-arachidonoylglycerolphosphoethanolamine (XlogP likely in excess of 20) are likely only to be found dissolved in membranes. 1-Stearoyl-2-arachidonoylglycerol and 1-palmitoyl-2-oleoylglycerol, in particular, partition into enriched domains of membranes (Basanez et al. 1996; Jimenez-Monreal et al. 1998). This hydrophobicity has a marked influence on the life cycle of ECBs. For example, it has been hypothesised that AEA is able to merge into the phospholipid bilayer as an extended conformation with the ethanolamine headgroup protruding, and access the receptor binding site by lateral diffusion without leaving the plane of the membrane (Tian et al. 2005), suggesting a role as an autocrine messenger without the need for a specific release mechanism.

Despite sharing similar structural features, the turnovers of AEA and 2AG follow parallel pathways with little overlap in selectivity. A convenient division of the ECBs is into ester or amide derivatives (see the chapter "Pharmacological Tools in Endocannabinoid Neurobiology" by Mor and Lodola, this volume).

3 Ester Endocannabinoids

Given the levels of 2AG in rodent brain (in our assays of rat brain, between 10 and 30 nmol g^{-1}), and the relative ability of isolated astrocytes and neurones to generate 2AG, it has been suggested that astrocytes are the major source of 2AG in the brain (Walter et al. 2004).

3.1 Synthesis of Ester Endocannabinoids

The "classical" pathway for 2AG synthesis is through the sequential activation of phospholipase C and diacylglycerol lipase (DGL) enzymes (Figs. 1 and 2). The intermediate in 2AG synthesis through this pathway, diacylglycerol (DAG), is more

Phosphatidylinositol-4,5-bisphosphate

Inositol-1,4,5-trisphosphate

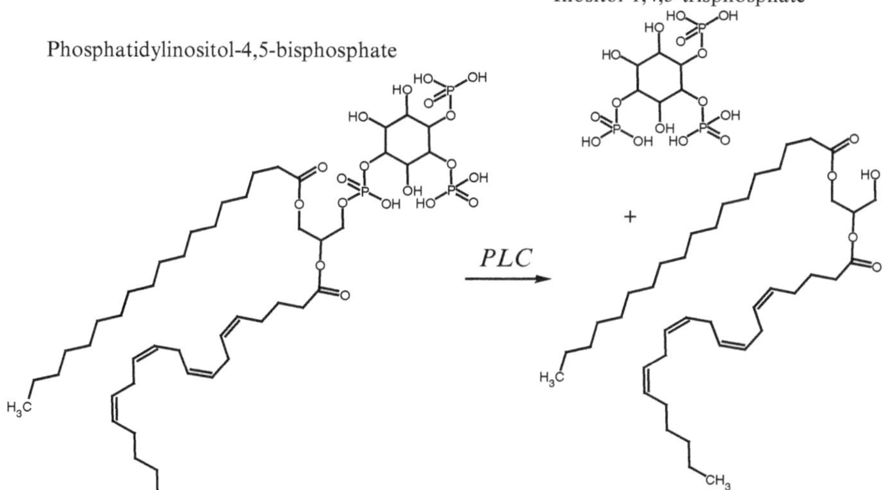

sn-1-Stearoyl-2-arachidonoylglycerol

Fig. 1 Reaction scheme for phospholipase C action. Phosphatidylinositol-4,5-bisphosphate, accumulated primarily in plasma membranes, is cleaved to form the water-soluble second messenger inositol 1,4,5-trisphosphate, which causes calcium release from intracellular stores. The co-product, diacylglycerol, shown here as 1-stearoyl-2-arachidonoylglycerol, remains inserted in the plasma membrane and can activate protein kinase C, be recycled to form phospholipids or be hydrolysed to generate monoacylglycerols

sn-1-Stearoyl-2-arachidonoylglycerol

Stearic acid

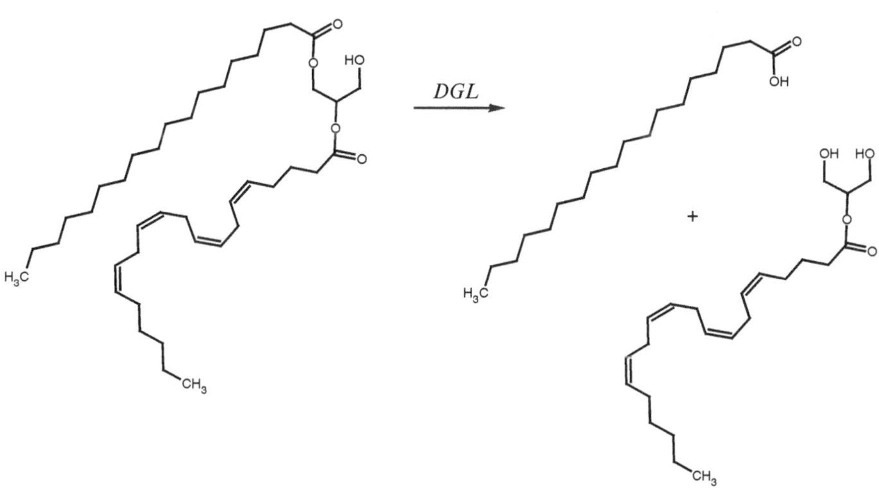

Sn-2-Arachidonoylglycerol

Fig. 2 Reaction scheme for diacylglycerol lipase action. Membrane-associated diacylglycerol, shown here as 1-stearoyl-2-arachidonoylglycerol, is hydrolysed to generate free fatty acid and 2-acylglycerol, both of which are much more water soluble than the parent and may have multiple cellular actions

widely associated as a second messenger in phosphoinositide turnover, activating protein kinase C. Enzymes competing for DAG include DAG kinase (seven isoforms), which is able to generate phosphatidic acid, and DAG acyltransferases (two isoforms prominent in adipose tissue), which generate triacylglycerols.

DGL is a membrane-associated enzyme generated as two separate gene products, DGLα and DGLβ (Table 1). The β isoform (694 aa) is a truncated paralogue of the α isoform (1,004 aa), although both show similar topology, with a short intracellular N-terminus and four transmembrane domains in the first 10% of the molecule. The remainder of the protein encompasses the active site and putative

Table 1 Molecular parameters of ECB-related enzymes

Enzyme	Gene name/Ensembl ID	Size	Species orthologues identity (homology)	Genetic variation
DGLα	*DAGLA*/ ENSG00000134780	1,042 aa/ 115 kDa	h/r 97% (99%) h/m 97% (97%)	Intronless; four non-synonymous SNPs: 595 G/A, 735 G/T, 889 C/T, 945 C/G
DGLβ	*DAGLB*/ ENSG00000164535	672 aa/ 74 kDa 34% (52%) identity to DGLα	h/r 78% (88%) h/m 79% (88%)	Intronless; three non-synonymous SNPs: 456 G/C, 517 G/A, 664 T/C
MGL	*MGLL*/ ENSG00000074416	303 aa/ 33 kDa	h/r 83% (92%) h/m 84% (93%)	Eight exons generating two isoforms of 303 and 273 aa; two non-synonymous SNPs: 202 C/T, 288 A/C
NAPE-PLD	*NAPEPLD*/ ENSG00000161048	393 aa/ 46 kDa	h/r 90% (95%) h/m 89% (94%)	Six exons; four non-synonymous SNPs: 152 A/C, 207 C/G, 380 T/C, 389 C/T
FAAH1	*FAAH*/ ENSG00000117480	579 aa/ 63 kDa	h/r 82% (91%) h/m 84% (91%)	15 exons; four non-synonymous SNPs: 129 C/A, 208 G/A, 370 A/G, 504 G/A
FAAH2	*FAAH2*/ ENSG00000165591	532 aa/ 58 kDa		11 exons; one non-synonymous SNP: 293 G/T
NAAA	*NAAA*/ ENSG00000138744	331 aa/ 36 kDa	h/r 79% (88%) h/m 78% (87%)	11 exons generating three isoforms; three non-synonymous SNPs: 107 G/T, 151 C/G, 334 A/G
COX2	*PTGS2*/ ENSG00000073756	587 aa/ 68 kDa	h/r 84% (91%) h/m 86% (93%)	10 exons; six non-synonymous SNPs: 1 C/T, 228 C/T, 428 G/C, 488 T/C, 511 A/G, 587 C/T

sites for regulation. Analysis of the protein sequence suggests two consensus sequences in the cytoplasmic C-terminus of the DGLα isoform for serine/threonine phosphorylation, one of which (S-727) is a potential target for both protein kinases A and C.

3.1.1 Regulation of Phospholipase C Activity

Although PLC appears capable of hydrolysing a variety of phosphoinositides in vitro, phosphatidylinositol-4,5-bisphosphate (PIP_2) appears to be the physiological substrate (Fig. 1). This substrate and one of the products (DAG) are sufficiently hydrophobic to be retained in the plasma membrane, while the second product of PLC action, inositol-1,4,5-trisphosphate, is much more hydrophilic (XlogP of –7) and so can migrate away from the membrane. Currently, 13 isoforms of PLC have been identified, which are widely distributed in the body (Suh et al. 2008). Within the cell, PLC-β isoforms (β1-β4) are membrane-associated and activated by G-proteins of the Gq family, while PLC-γ isoforms (γ1, γ2) are recruited to membranes by activation by tyrosine kinase-linked receptors of the growth factor family. PLC-δ isoforms (δ1, δ3, δ4) associate with PIP_2 in the plasma membrane and are activated by elevated concentrations of intracellular calcium ions leading to the view that PLC-δ is a calcium amplifier. PLC-ε1 is activated by the low molecular weight G-proteins Ras and Rho, as well as the exchange protein activated by cyclic AMP (Epac). Much less is known about the regulation of the ζ1, η1 and η2 isoforms (Suh et al. 2008). Gene expression of all of these isoforms appears abundant in CNS tissues, with the exception of the ζ1 isoform, which appears to have a crucial role in oocyte fertilisation. Clearly, therefore, the apparent potential for regulation of this route of ECB synthesis is huge.

U73122 is an aminosteroid which has been used to inhibit PLC activity, although its activity has not been assessed against all 13 isoforms. It has been shown to inhibit 2AG synthesis in a macrophage cell line (Berdyshev et al. 2001), 3T3 mouse fibroblasts (Parrish and Nichols 2006) and rat brain synaptosomes (Oka et al. 2007a), as well as inhibiting DSI in the hippocampus (Edwards et al. 2006). However, U73122 has also been shown to interfere with 2AG-evoked regulation of excitability in rat microglial cells (Carrier et al. 2004) or rat hippocampal slices (Hashimotodani et al. 2008). Furthermore, using mice in which genes encoding three of the isoforms of phospholipase C (PLCδ1, PLCδ3 and PLCδ4) were disrupted failed to alter cannabinoid-induced DSI responses (Hashimotodani et al. 2008).

The role of phospholipase C in 2AG generation in the CNS is, therefore, inconclusive.

3.1.2 Regulation of DGL Activity

DGL (Table 1) hydrolyses DAG to generate monoacylglycerol and free fatty acid (Fig. 2) with some selectivity for the *sn*-1 position (Bisogno et al. 2003). The substrate specificity is not well understood, but a dually monounsaturated DAG appeared better hydrolysed than a mixed monounsaturated/saturated or monounsaturated/polyunsaturated DAG (Bisogno et al. 2003). DGL action, therefore, takes a predominantly membrane-associated substrate and generates two products, both of which are much more able to migrate away from the membrane. The recombinant enzymes are activated by calcium at supra-physiological concentrations of 100 μM or above, albeit to levels less than those evoked by glutathione (Bisogno et al. 2003). Whether these modulations are replicated with the enzyme in situ awaits further investigation.

Tetrahydrolipstatin (THL, also known as orlistat), an agent used to target pancreatic lipase in the treatment of obesity, was also found to inhibit the recombinant enzymes with IC_{50} values of 60–100 nM (Bisogno et al. 2003), although the activity in bovine aorta was more sensitive by an order of magnitude (Lee et al. 1995). THL is ineffective at 25 μM against MGL or FAAH activities, but does show inhibition of NAPE-PLD (IC_{50} of 10 μM) and triacylglycerol lipase (IC_{50} of 10 μM) (Szabo et al. 2006). Intriguingly, it also shows some occupancy of cannabinoid receptors (CB_1 IC_{50} of 4 μM vs. CB_2 IC_{50}>25 μM) (Szabo et al. 2006). RHC80267 shows low potency inhibition of DGL in platelets with an IC_{50} of 1–4 μM, with some selectivity vs. other enzymes expressed (no inhibition at 100 μM against phospholipase C or phospholipase A_2 activities (Sutherland and Amin 1982) or MGL (Rindlisbacher et al. 1987)). A recent investigation of 'activity-based protein profiling' of mouse brain using fluorophosphonate probes indicated that these two agents interfered with multiple serine hydrolases (Hoover et al. 2008), including FAAH and ABHD12 (see below). Intriguingly, the two isoforms of DGL were not identified using this methodology, suggesting either low abundance in this tissue, or reduced activity against the fluorophosphonate substrate. Despite this apparent lack of selectivity, it was noted that very few enzyme activities were inhibited by both THL and RHC80267, leading the authors to suggest the use of both agents to identify the role of DGL in biological processes.

These inhibitors have been used to identify the essential role of DGL in 2AG accumulation in the action of the Ca^{2+} ionophores ionomycin in neuroblastoma cells (Bisogno et al. 1999; Szabo et al. 2006) and A23187 in RTMGL1 rat microglial cells (Carrier et al. 2004) as well as ATP in astrocytes (Walter et al. 2004).

Currently, there are no published reports of genetic interference with DGL expression.

3.1.3 Alternative Pathways of DAG and 2AG Synthesis

Although the best established route of 2AG biosynthesis described above involves a two-step process utilising sequential activities of PLC (Fig. 1) and DGL (Fig. 2) activities, at least three further routes are possible using phosphatidylinositol, phosphatidylcholine or phosphatidylserine as starting points (Fig. 3).

A Ca^{2+}-independent phospholipase A_1 (PLA_1) activity in rat brain hydrolyses phosphatidylinositol to generate LPI, lysophosphatidylinositol (Kobayashi et al.

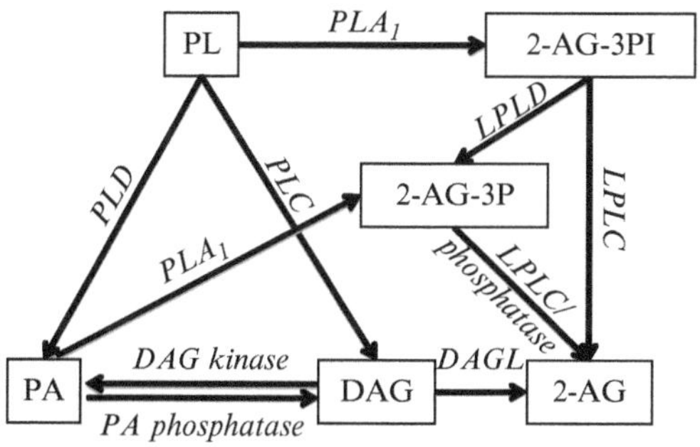

Fig. 3 Alternative routes of 2AG formation. Aside from the canonical route of 2AG formation through PLC/DGL action, diacylglycerol can also be formed via phosphatidic acid generated by PLD action. Additionally, PLA_1 activity can generate a lysophospholipid, which may be used to generate 2AG directly through a LPLC activity, or indirectly through a LPLD/LPLC sequence. Phospholipase A_1 hydrolysis of phosphatidic acid can also allow generation of 2AG, through the intermediate 2AG-3-phosphate, which can then be hydrolysed by a phosphatase/ lysophospholipase C activity

1996), which has recently been suggested to be the endogenous ligand for a cannabinoid-related receptor, GPR55 (Oka et al. 2007b, 2009). A PLA_1 activity able to hydrolyse phosphatidylinositol in cytosolic and microsomal fractions of rat brain has been described, which was less active than a PLC-like phosphodiesterase activity (Hirasawa et al. 1981). Later reports described a PLA_1 activity found in the soluble fraction of brains, which exhibited some selectivity for phosphatidylinositol over other phospholipid substrates (Ueda et al. 1993a, b).

In molecular terms, three isoforms of PLA_1 have been identified (Aoki et al. 2007). PS-PLA_1 (also known as PLA1A, ENSG00000144837) is a soluble enzyme released by activated platelets (Sato et al. 1997), which hydrolyses phosphatidylserine to produce lysophosphatidylserine and a fatty acid. Two further, membrane-associated PLA_1 activities have been identified (mPA-PLA1α, LIPH or PLA1B, ENSG00000163898 and mPA-PLA1β, LIPI or PLA1C, ENSG00000188992), which appear to hydrolyse preferentially phosphatidic acid, giving rise to lysophosphatidic acid and a fatty acid (Hiramatsu et al. 2003). Other lipase activities, such as hepatic lipase (LIPC, ENSG00000166035) and endothelial lipase (LIPG, ENSG00000101670), have also been reported to exhibit phospholipase A_1 activity when presented with phosphatidylcholine as a substrate (Gillett et al. 1993; Jaye et al. 1999).

Following PLA_1 degradation of phospholipid, a lysophospholipase C (LPLC) activity of rat brain, with some selectivity for LPI, is able to generate 2AG (Tsutsumi et al. 1994). This enzyme, although not precisely identified at the molecular level, appears to be an integral membrane protein (Tsutsumi et al. 1995).

Two further alternative routes of 2AG synthesis, independent of the phosphatidylinositol/PLC pathway, involve phospholipase D (PLD) activity, which favours phosphatidylcholine as a substrate, generating phosphatidic acid. In mouse N18TG2 neuroblastoma cells stimulated by the calcium ionophore ionomycin, this appears to be the major synthetic route (Bisogno et al. 1999), with sequential formation of phosphatidic acid, DAG and then 2AG. The conversion of phosphatidic acid to DAG is catalysed by phosphatidic acid phosphatases or lipid phosphate phosphatases (Brindley 2004).

The phosphatidic acid phosphatase can be inhibited by high concentrations (100 μM) of the β-adrenoceptor antagonist propranolol, which has allowed identification of the involvement of this enzyme in 2AG biosynthesis in cultured neuroblastoma (Bisogno et al. 1999) and microglial (Carrier et al. 2004) cells. It has, however, not been widely applied to investigate mechanisms of ECB biosynthesis, presumably because of "non-specific" effects, for example, directly interfering with electrophysiological recordings (Hashimotodani et al. 2008) due to its local anaesthetic-like action.

A further alternative pathway for 2AG synthesis involves the generation of 2-arachidonoylglycerol-3-phosphate (2AG-3P), a lysophosphatidic acid (Nakane et al. 2002). This may theoretically be generated from phosphatidic acid by phospholipase A_1 or from lysophospholipids by lysophospholipase D (LPLD). A phosphatidic acid-hydrolysing PLA_1 activity was identified in porcine platelet membranes (Inoue and Okuyama 1984), and subsequently in rat liver (Kucera et al. 1988) and bovine brain (Higgs and Glomset 1994), leading to cloning of the enzyme from bovine testis (DDHD1, ENSG00000100523) (Higgs et al. 1998).

To date, a single isoform of LPLD has been identified at the molecular level. This is autotaxin (ENPP2, ENSG00000136960), a membrane-associated enzyme initially characterised as an ecto-nucleotide pyrophosphatase/phosphodiesterase. The primary physiological role of this enzyme, however, is thought to be the regulation of levels of lysophosphatidic acid, which it produces from lysophosphatidylcholine (Goding et al. 2003). It remains to be determined whether this entity is able to regulate ECB production, however; the fact that it contains extracellular enzymatic activity allows some speculation about a particular signalling role.

Although the enzymatic pathway involved in 2AG-3P synthesis has not been unequivocally defined, levels in rat brain of 2AG-3P (530 pmol g^{-1}) were lower than those of 2AG (37,000 pmol g^{-1} (Artmann et al. 2008)), suggesting either lower rates of 2AG-3P synthesis or higher rates of 2AG-3P dephosphorylation. The rapid conversion of 2AG-3P to 2AG (70% in 2 min) by rat brain homogenate (Nakane et al. 2002) suggests that this may be a feasible route for 2AG synthesis in vivo.

3.2 *Hydrolysis of Ester Endocannabinoids*

FAAH appears to be the primary enzyme involved in amide ECB hydrolysis (see Sect. 4.3 below). Initial characterization of cell-free preparations from cells expressing recombinant FAAH (Goparaju et al. 1998) or endogenously expressing FAAH

(Di Marzo et al. 1998) suggested that 2AG might also be hydrolysed through this route. However, tissues from mice with disruption of the gene encoding FAAH show an unchanged ability to hydrolyse 2AG, suggesting FAAH plays only a minor role in turnover of ester ECBs (Lichtman et al. 2002). The identification of monoacylglycerol lipase (MGL) as a serine hydrolase enzyme capable of hydrolysing ester ECBs in vitro drew attention to an alternative route of ECL turnover (Dinh et al. 2002). Although this enzyme was shown to have a central role in lipid turnover over 30 years ago (Tornqvist and Belfrage 1976), it appears to have an additional important role in the regulation of ester ECBs. It is generally thought to be a cytosolic enzyme, with the primary sequence consistent with a lack of predicted transmembrane domains. Experimentally, however, both soluble and membrane-associated activities are observed (Dinh et al. 2002; Saario et al. 2004; Vandevoorde et al. 2005), with some pharmacological evidence to indicate minor differences between the two (Vandevoorde et al. 2005; Duncan et al. 2008).

Primary sequence analysis indicated the possibility for phosphorylation of MGL and, presumably, regulation of activity by protein kinases, in particular calcium/calmodulin kinase II and cyclic AMP- and cyclic GMP-dependent protein kinases (Dinh et al. 2002). Although this has not been investigated directly, this suggests the possibility that 2AG hydrolysis can be regulated by intracellular levels of calcium and cyclic nucleotides. Immunostaining analysis suggested predominant expression of MGL in nerve fibres and cell bodies of brain regions rich in CB_1 cannabinoid receptors (Dinh et al. 2002).

The substrate specificity of MGL showed hydrolytic activity towards 2AG, but not AEA (Dinh et al. 2002), but with little specificity between acylglycerols (Ghafouri et al. 2004; Vandevoorde et al. 2005).

The available MGL inhibitors described to date have little selectivity. Fluorophosphonate analogues, such as methylarachidonylfluorophosphonate (MAFP), are potent inhibitors of MGL activity in the nanomolar range (Saario et al. 2004; Duncan et al. 2008). However, they exhibit similar activity at FAAH (De Petrocellis et al. 1997), and are thus unhelpful in defining a role of MGL in intact tissues. URB754, on the other hand, appeared at first to have selectivity for MGL over FAAH (Makara et al. 2005). Subsequently, a retraction was published indicating that a contaminant of the preparation was found to be responsible (Makara et al. 2007). URB602 was initially described as a non-competitive selective inhibitor of MGL activity (Hohmann et al. 2005). However, this compound has been reported not to show selectivity over FAAH (Vandevoorde et al. 2007; Duncan et al. 2008). OMDM169, a recently reported analogue of the DAGL inhibitor THL, shows submicromolar potency at MGL activity and enhances levels of 2AG, but not AEA, in ionomycin-stimulated N18TG2 neuroblastoma cells, but still is only tenfold selective over FAAH (Bisogno et al. 2009). JZL184, a carbamate analogue, on the other hand, appears to be almost 1,000-fold selective for MGL over FAAH (Long et al. 2009). Intraperitoneal administration of this agent caused an elevation of 2AG, but not AEA, in KCl-perfused microdialysate of mouse nucleus accumbens.

Although a knockout mouse with the gene encoding MGL has not yet been described, siRNA silencing of the enzyme in HeLa human cervical carcinoma cells causes an elevation of cellular 2AG levels equivalent to those obtained in the presence of MAFP (Dinh et al. 2004).

A functional, activity-based protein profiling approach to studying the enzymes in rat brain responsible for 2AG hydrolysis indicated MGL accounted for the vast majority of activity. Two further, poorly characterised enzymes, abhd6 and abhd12, were identified as contributing up to 15% of 2AG hydrolysis, but evidence for their physiological significance is currently lacking.

4 Amide Endocannabinoids

The canonical pathway of AEA formation in neural tissues is thought to be via a two-step reaction – a transacylase-phosphodiesterase pathway. The intermediate involved is a low abundance phospholipid, which acts as a precursor for *N*-acylethanolamides (NAEs), including AEA. It is generally considered that the formation of this precursor, rather than its metabolism, is the rate-determining step in AEA synthesis.

4.1 Synthesis of NAPEs

N-Acylphosphatidylethanolamines (NAPEs) were described in plants about 40 years ago (Dawson et al. 1969), and observed to be mobilised during seed germination and to accumulate during stress. More recently, they were identified as precursors of the ethanolamide ECBs (Di Marzo et al. 1994). They are synthesised through the action of an acyltransferase (E.C. 2.3.1.-), which catalyses the lysophospholipase A_1-style hydrolysis of a fatty acid from the sn-1 position of phosphatidylcholine and transfers it to the amine of phosphatidylethanolamine (PE) (Fig. 4). In mammalian systems, this activity was initially identified in dog heart and reported to be calcium-dependent (Natarajan et al. 1982; Reddy et al. 1983). In mouse cerebral cortical neurones, NAPE formation was also enhanced substantially in the presence of the calcium ionophore A23187 (Hansen et al. 1995). Intriguingly, although the adenylyl cyclase activator forskolin and the Gs-coupled receptor agonist vasoactive intestinal polypeptide both failed to enhance NAPE accumulation in cultured neurones, they potentiated the stimulatory effects of the calcium ionophore ionomycin (Cadas et al. 1996a). The protein kinase inhibitor, H89, was able to prevent this potentiation, indicating a role for phosphorylation of a key enzyme in this process. In a comparison of cell types, NAPE synthesis appeared restricted to cultured neurons rather than astrocytes (Cadas et al. 1996a). In cultured neurons, the use of exogenous PLD activities indicated that approximately half of cellular NAPEs were available for hydrolysis, indicating a likely accumulation in the plasma membrane (Cadas et al. 1996b). A molecular correlate for this calcium-dependent transferase has yet to be identified.

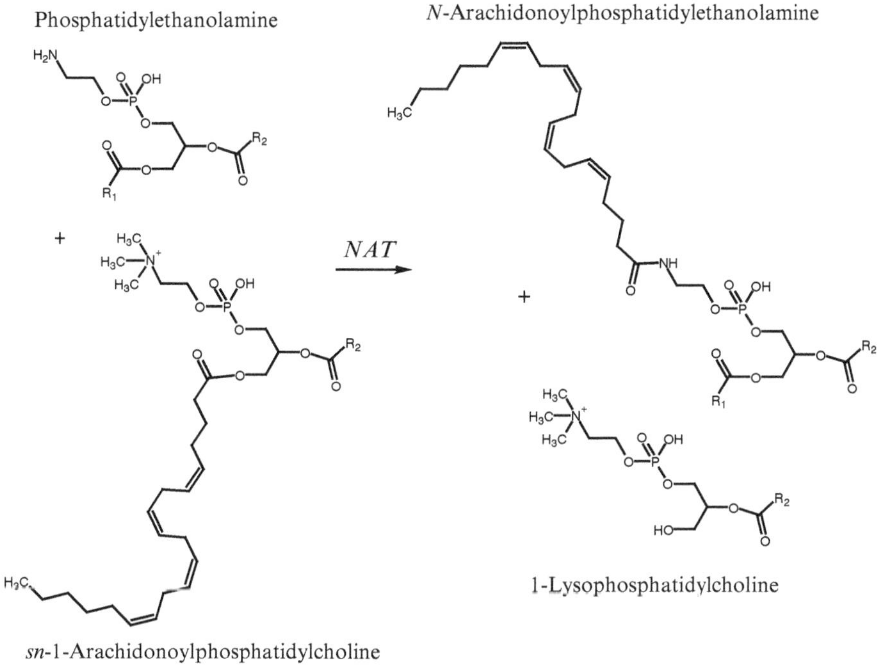

Fig. 4 Reaction scheme for *N*-acyltransferase action. The two phospholipids, phosphatidylethanolamine and phosphatidylcholine are co-substrates for *N*-acyltransferase activity, where the *sn*-1 fatty acid (depicted here as arachidonic acid) from the phosphatidylcholine is transferred to the amine of the phosphatidylethanolamine

In contrast, a Ca^{2+}-independent PE *N*-acyltransferase has recently been described (Jin et al. 2007). This predominantly cytosolic activity appears to be identical to a protein termed rat lecithin-retinol acyltransferase-like protein 1 (RLP-1, ENSG00000168004). Given that this enzyme is highly expressed in testis and pancreas, with much lower levels expressed in brain, it seems unlikely that it contributes significantly to ECB precursor formation in neural tissues.

Very recently, NAPEs have been described to have functions beyond acting simply as precursors for ECBs. Reportedly, NAPEs are synthesised in the gut, prompted by fat ingestion, and released into the circulation where they appear to have a hormonal function. Administration of exogenous NAPE led to reduced food intake which was independent of CB_1 receptors (Gillum et al. 2008).

4.2 Synthesis of Amide Endocannabinoids

As the (perhaps inappropriately considered) archetypal ECB, AEA synthesis has received the most attention. In the chemistry lab, AEA can be synthesised as a simple condensation product of arachidonic acid and ethanolamine, but in vivo, generation of the ethanolamide ECBs is thought to occur mainly as a result of hydrolysis of a minor membrane phospholipid, *N*-arachidonoylphosphatidylethanolamine (Di Marzo et al. 1994). This is a substrate for a phospholipase D-type activity (NAPE-PLD, ENSG00000161048, Table 1) which can produce a wide range of endogenous fatty acid ethanolamides, including AEA (Okamoto et al. 2004).

4.2.1 Pharmacological and Biochemical Manipulation of NAPE-PLD Activity

An early report of crude preparations of rat heart homogenates identified a membrane-associated NAPE-PLD activity (Fig. 5) capable of hydrolysing NAPEs to generate phosphatidic acid and diacylglycerols (Schmid et al. 1983). With the inhibition of phosphatidic acid phosphatase activity, the production of diacylglycerol was inhibited indicating that the latter was produced in a two-step process. In addition to the phosphatidic acid, NAEs were produced apparently in equimolar quantities, and in the absence of synthesis of *N*-acylethanolamine phosphates (indicating the lack of involvement of PLC). Similar levels of lyso-*N*-acylphosphatidylethanolamine (lysoNAPE) to phosphatidic acid were observed, indicating

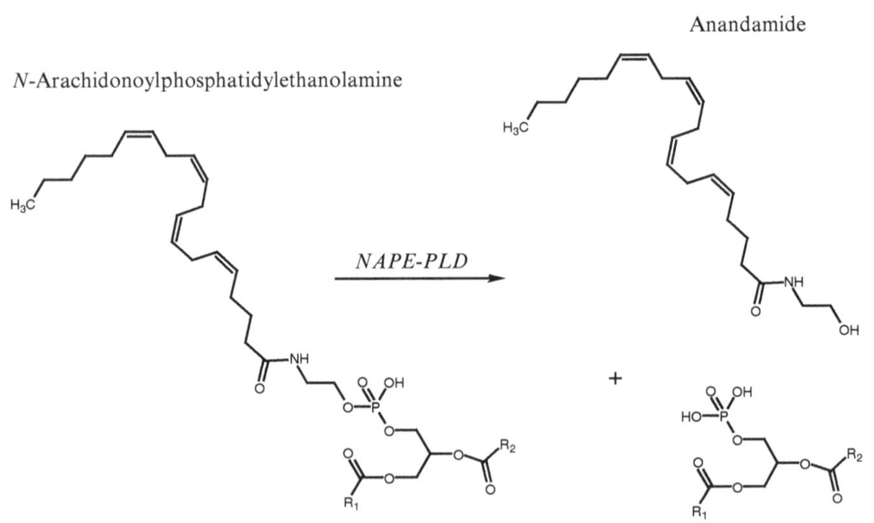

Fig. 5 Reaction scheme for NAPE-PLD action. The action of a selective phospholipase D activity allows cleavage of *N*-arachidonoylphosphatidylethanolamine to generate anandamide and phosphatidic acid. The latter is highly hydrophobic (dipalmitoylphosphatidic acid has an XlogP value of 12.9) and so stays associated with the membrane, while the anandamide can more readily move into the aqueous milieu

activity of phospholipase A_1 and/or A_2 in this preparation. In this crude preparation, supplementation with calcium or magnesium ions at concentrations up to 5 mM was without effect (Schmid et al. 1983). The same crude preparations were also able to hydrolyse LNAPE and an ether analogue of NAPE with activities only slightly less than those with NAPE itself, although it is uncertain whether these activities reside in the NAPE-PLD activity or are present in parallel enzymes (Schmid et al. 1983).

NAPE-PLD activity from rat brain microsomes was observed to generate AEA at a slightly lower rate compared to other shorter chain, more saturated NAEs (Sugiura et al. 1996). In the presence of calcium ions, the generation of these latter shorter chain, more saturated NAEs appeared enhanced, while AEA production was unchanged, although the mechanism for this selective action has not been elucidated. A further stimulus for NAPE-PLD activity is the presence of polyamines. Spermine, spermidine and putrescine were able to replace calcium ions or detergent (see below) as enhancers of NAPE-PLD activity at concentrations within the physiological range (Liu et al. 2002), although whether polyamine levels are a physiological influence on AEA levels has not been identified.

Intriguingly, addition of the non-ionic detergent Triton X-100 (up to 0.2%) led to a doubling of NAPE-PLD activity, while the same concentrations of an alternative non-ionic detergent, Tween 20, inhibited activity to 4% of the level in control preparations (Schmid et al. 1983). Ionic detergents, such as SDS or taurodeoxycholate, also inhibited NAPE-PLD activity in these preparations. It may be that these influences are more physical than biochemical, with the possibility that Triton X-100 allows a particular conformation of enzyme:substrate interaction to occur, which the other detergents are unable to facilitate. Recently, it was noted that solubilisation of NAPE-PLD from the membrane by detergents revealed a greater sensitivity to divalent cations, including calcium (Wang et al. 2008a), leading to the suggestion that a membrane component was able to substitute for calcium. A heat-stable membrane fraction was able to enhance enzyme activity, which was later suggested to be the phospholipid PE. It was surmised that membrane components, including PE, were able to maintain activity of NAPE-PLD in a tonically active state, implying that formation of the NAPE precursor was the rate-determining state in amide ECB synthesis (Wang et al. 2008a).

Although this is dealt with by Monory and Lutz in the chapter "Genetic Models of the Endocannabinoid System" in this volume, it is pertinent to consider briefly the impact of genetic manipulation of NAPE-PLD activity. Cloning of the gene encoding this enzyme allowed identification of a primary sequence distinct from classical phospholipase D activities, with characteristics of a metallo-β-lactamase family (Okamoto et al. 2004), including the obligate incorporation of a zinc atom (Wang et al. 2006). Subsequently, it was observed that disruption of the gene encoding NAPE-PLD leads to increased levels of many forms of the precursor NAPE and decreased levels of the cognate NAEs (Leung et al. 2006). In particular, OEA, PEA and SEA levels were reduced, while their cognate precursors were enhanced. In comparison, AEA and DHEA, as well as their precursors, were unaltered. This has been taken as evidence for alternative pathways for synthesis of NAEs, particularly AEA (see below).

Using viral over-expression of NAPE-PLD activity in HeLa cells, it was observed that cellular levels of OEA and PEA were increased, without altering AEA levels (Fu et al. 2008). This suggests either that AEA synthesis can be selectively driven by synthesis of the precursor NAPE, or that AEA synthesis does not involve NAPE-PLD activity.

4.2.2 Alternative Pathways of Amide ECB Generation

In NAPE-PLD knockout mice, lower brain levels of saturated *N*-acylethanolamines were detected but concentrations of polyunsaturated NAEs, including AEA, were essentially unchanged (Leung et al. 2006), indicating the existence of more than one synthetic pathway (Fig. 6). Indeed, a further three routes for AEA synthesis have been proposed, although their roles in the physiological generation of AEA in neural preparations is unclear.

Studies by Natarajan et al. (1984) provided in vitro evidence for multi-step enzymatic activities capable of producing NAEs from NAPEs. This involved the hydrolysis of one or both acyl chains from NAPEs followed by cleavage of the phosphodiester bond of the resulting lysoNAPE or glycerophospho (GP)-NAE, respectively. A secreted PLA_2 has been shown to catalyse the deacylation of

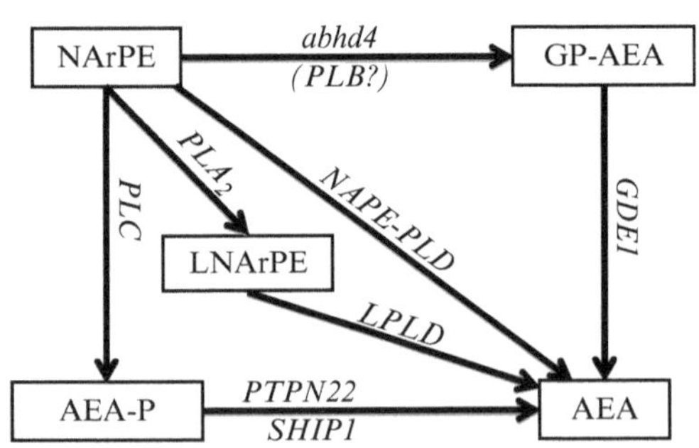

Fig. 6 Alternative routes of AEA formation. Aside from the canonical pathway of NAPE hydrolysis to form AEA, through NAPE-PLD action, phospholipase C action can generate anandamide-phosphate, which can subsequently be hydrolysed by at least two phosphatases, PTPN22 and SHIP1, to generate AEA. Phospholipase A_2 hydrolyses NAPE to generate a lysoNAPE, which can then be hydrolysed by a lysophospholipase D activity to generate AEA. A third route, utilising a phospholipase B-like action of abhd4 generates glycerophospho-AEA. A membrane-associated glycerophosphodiesterase, GDE1, is able to then generate AEA

NAPE to yield lysoNAPE in vitro (Sun et al. 2004), although this enzyme was primarily expressed in the gut, with little expression in the brain. This suggests the existence of additional enzymes responsible for the calcium-independent NAPE hydrolase activity detected in NAPE-PLD knockout mouse brain.

The lysoNAPE evolved following a PLA_2-mediated hydrolysis of NAPE can itself be hydrolysed by a LPLD activity which produces NAEs and lysophosphatidic acid (LPA). This activity was, however, found to be enriched in brain and testis (Sun et al. 2004). Given the profound biological actions of LPA, it is interesting to speculate on the dual functions of products of this enzyme.

More recently, Simon and Cravatt (2006) identified a novel enzyme αβ-hydrolase 4 (abhd4, ENSG00000100439) as a lysophospholipase/phospholipase B that selectively hydrolyzes NAPEs and lysoNAPEs to yield GP-NAE. This enzyme is indeed present in the brain and probably represents the NAPE-PLD-independent route for NAE biosynthesis observed in both NAPE-PLD-knockout and wild-type mice. Currently, very little is known about the distribution of abhd4 between neuronal and glial populations, as well as its subcellular location. The enzyme shows little selectivity between acyl groups, generating PEA at an equal rate to AEA, and is inhibited by fluorophosphonates, with 5 μM MAFP proving an effective inhibitor (Simon and Cravatt 2006). As yet, no selective inhibitors have been described. However, since NAPE-PLD is insensitive to MAFP up to 100 μM (Petersen and Hansen 1999), it is possible that this agent allows some discrimination of the two routes of AEA synthesis.

GDE1 (ENSG00000006007) is an integral membrane protein which has been identified as a glycerophosphodiesterase (Zheng et al. 2000). Initial characterization indicated an interaction with RGS16, a regulator of G-protein signalling, implying the possibility that enzyme activity might be modulated by cell-surface receptors. Indeed, in a recombinant system, the enzyme was able to hydrolyse glycerophosphoinositol preferentially (compared to glycerophosphocholine) and this activity was enhanced by isoprenaline and reduced by phenylephrine (Zheng et al. 2003). Recently, the substrate profile of this enzyme was extended to include GP-NAE, including a glycerophospho derivative of AEA (Simon and Cravatt 2008). Since the enzyme activity is stimulated by magnesium ions and inhibited by calcium ions, chelation of these allowed accumulation of several GP-NAEs in a rat brain membrane fraction, including saturated, mono-unsaturated and polyunsaturated fatty acid derivatives. Analysis of multiple recombinant glycerophosphodiesterase activities suggested identity with GDE1. Taken together, these data suggest a role for abhd4 and GDE1 in calcium-independent generation of AEA (and other NAEs) in neural tissue.

Another route that has been identified is the PLC hydrolysis of NAPE and the consequent production of acylethanolamine-*O*-phosphates, which may subsequently be hydrolysed by the phosphatases, PTPN22 or SHIP1 (Liu et al. 2008). Intriguingly, the PLC route was suggested to react more rapidly (<10 min) than the PLB route (≥ 1 h). Whether the PLC activity which is able to hydrolysis NAPE is a member of the conventional phosphoinositide-specific PLC activities described earlier is as yet unknown.

The physiogical role of PTPN22 and SHIP1 in ECB turnover in neural tissues is, as yet, almost completely unexplored.

4.3 Hydrolysis of Amide Endocannabinoids

The best characterised and investigated pathway of ECB turnover is the hydrolysis of amide ECBs. This is partly because of the relative ease of assay and synthesis of substrates, but also because inhibitors of ECB hydrolysis show some promise as therapeutic agents.

4.3.1 FAAH1 Activity

FAAH was cloned from rat tissues on the basis of identifying the enzyme responsible for hydrolysis of oleamide, an ECB-related fatty acid amide (Cravatt et al. 1996). Expression of human and rat enzymes in recombinant systems indicated intracellular location of both enzymes, although the two appeared to have distinct patterns. The rat enzyme appeared to associate with Golgi and ER membranes, predominantly in perinuclear regions, while the human enzyme appeared more associated with cytoskeletal elements (Cravatt et al. 1996). In neural tissues, FAAH-like immunoreactivity is associated primarily with neurons, in a pattern extensively (although not completely) complementary to the expression of CB_1 cannabinoid receptors (Egertová et al. 1998; Tsou et al. 1998).

Hydrolysis rates of AEA were greater than those of oleamide, OEA and PEA in mouse brain and liver, but were diminished by ~99% in both tissues in mice in which the *faah* gene was disrupted (Lichtman et al. 2002) indicating the predominant role for FAAH in the hydrolysis of AEA, at least in "normal" neural tissues. It is, therefore, easy to understand the focus on development of FAAH inhibitors as therapeutic alternatives to receptor agonists. Intriguingly, there is a possibility that endogenous inhibitors of FAAH are able to regulate ECB turnover. Thus, *N*-arachidonoyl amino acids, such as *N*-arachidonoylglycine and *N*-arachidonoylalanine show species-dependent inhibition of FAAH activity (Grazia Cascio et al. 2004), although whether these are physiological regulators of ECB hydrolysis is unknown. Of potentially more direct influence is the observation that FAAH has a wide substrate profile, such that many endogenous fatty acid amides, including OEA and PEA, but not limited to NAEs, are also substrates for the enzyme and are present in quantities up to 100 times those of AEA. This observation led to the hypothesis that these compounds act as "entourage" compounds. That is, although they have no direct activity at CB_1 or CB_2 cannabinoid receptors, they are able to slow the hydrolysis of AEA through competition for FAAH activity sufficiently so that they can indirectly influence cannabinoid activity. The issue is complicated further by studies of the putative ECB-like receptor GPR119, which suggested that it was activated by OEA, PEA and SEA (Overton et al. 2006). Whether there is a convergence between GPR119 and conventional cannabinoid receptors awaits further investigation.

Synthetic inhibitors of FAAH abound and can be divided into two broad groups. One group is based around mimicking endogenous ligands, while the second is structurally unrelated compounds. Although α-keto ethyl esters and trifluoromethylketone analogues of AEA were effective FAAH inhibitors, the overlap in pharmacophore meant that activity at CB_1 receptors and other eicosanoid-metabolising enzymes reduced their applicability (Koutek et al. 1994). Assessment of a number of carbamate analogues identified an irreversible inhibitor with nanomolar potency, URB597 (Kathuria et al. 2003). Although this compound has some "off-target" activity (Zhang et al. 2007), including activation of TRPA1 channels (Niforatos et al. 2007), the profile of its action in vitro and in vivo is consistent with a predominant action to elevate NAEs. Although systemic administration of URB597 has been demonstrated to increase levels of AEA, OEA and PEA in rat CNS tissues (Gobbi et al. 2005; Russo et al. 2007). Moise et al. (2008) reported that it elevated brain levels of OEA and PEA but not AEA in the hamster brain, indicating the possibility of species-selective effects of the inhibitor on multiple enzyme activities.

The observation that some, but not all, non-steroidal anti-inflammatory drugs, previously thought to exert their therapeutic effects through inhibition of cyclooxygenase activity, were also able to inhibit FAAH activity at relevant concentrations (Fowler et al. 1997) raised the possibility that some of the therapeutic effects of these agents might be mediated through cannabinoid receptors.

4.3.2 FAAH2 Activity

A second isoform of FAAH, FAAH2 (ENSG00000165591), has a limited species distribution in mammals, being found in man, other primates, elephants and rabbits, but not mice, rats, pigs, dogs, sheep or cows (Wei et al. 2006). Although the subcellular distribution of this isoform has not been precisely identified, it was predicted to be membrane-associated with the active site oriented towards the luminal side of the membrane. Whether FAAH2 regulates ECB levels in the extracellular medium or in subcellular organelles is, as yet, unknown. Although FAAH2 appears to hydrolyse the conventional fatty acid ethanolamine ECB-like compounds, the activity against AEA, OEA and PEA is greatly reduced, while ODA hydrolysis is similar to that evoked by FAAH1 (Wei et al. 2006). Unlike FAAH1, FAAH2 appears not to be expressed in brain or small intestine, but in contrast to FAAH1, shows low expression in heart, muscle and ovary (Wei et al. 2006). Both isoforms show high expression in kidney, liver, lung and prostate. In comparison with FAAH1, there appear to be no inhibitors of FAAH2 with greater than 100-fold selectivity (Wei et al. 2006), although both URB597 and OL135 show more than tenfold selectivity. In contrast, JNJ1661010 appears to be 100-fold selective for FAAH1 (Karbarz et al. 2009). It appears unlikely, therefore, that FAAH2 is a major regulator of AEA levels in human neural tissues.

4.3.3 NAAA Activity

N-Acylethanolamine acid amidase (NAAA, ENSG00000138744) activity is a lysosomal enzyme, with an acid pH optimum, and structural similarity to acid ceramidase. The mature enzyme is glycosylated and requires proteolysis for activation (Wang et al. 2008b). It appears to hydrolyse preferentially PEA compared to AEA in cell-free systems (Ueda et al. 1999). In intact cells, however, NAAA appeared capable of hydrolysing a variety of fatty acid ethanolamides, including AEA (Sun et al. 2005), but not 2AG (Tsuboi et al. 2005). The enzyme is expressed to relatively high levels in lung, spleen and large intestine, but in contrast to FAAH activity, is less well expressed in liver, testis and brain (Tsuboi et al. 2005). Under normal circumstances, therefore, it appears unlikely to contribute significantly to AEA turnover in neural tissues.

In counterpoint to FAAH activities, NAAA is not inhibited by MAFP concentrations up to 10^{-5} M (Ueda et al. 1999). The enzyme is also insensitive to URB597, but can be inhibited by a retroamide, *N*-cyclohexylcarbonylpentadecylamine, in the micromolar range (Tsuboi et al. 2004). As yet, genetic disruption of this enzyme has not been reported.

5 Other Routes of ECB Transformation

Other than FAAH, NAAA and MGL, the most prominent route of ECB inactivation appears to be through oxidative metabolism. Intriguingly, there is the possibility that this is not simply an inactivation, but rather a transformation to metabolites, which may themselves be active, albeit not only at canonical cannabinoid receptors.

Aside from oxidative metabolism, a further form of transformation of NAEs was identified using tissue from $FAAH^{-/-}$ mice (Mulder and Cravatt 2006). *O*-Phosphorylcholine derivatives of NAEs (PC-NAEs) were identified in the brain and/or spinal cord of $FAAH^{-/-}$ mice, but not wild-type mice. Intriguingly, although AEA levels were elevated in $FAAH^{-/-}$ mice, there were no detectable levels of PC-AEA, although PC-PEA and PC-OEA were detectable. Whether these metabolites are generated in other species or under pathological conditions or indeed whether they have biological activity in their own right is unknown. The mechanism of PC-NAE formation is also unknown; however, an enzyme activity has been identified which is capable of hydrolysing PC-NAEs to generate *O*-phosphorylcholine and NAE (Mulder and Cravatt 2006). This is ENPP6 (ENSG00000164303), a membrane-associated member of the nucleotide pyrophosphate/phosphatase family (Sakagami et al. 2005), which is expressed highly in human (although not mouse) brain. ENPP6 exhibits LPLC activity with some selectivity for lysophosphatidylcholine over any other lysophospholipid, including LPI (Sakagami et al. 2005). Although PC-NAEs were poor substrates for FAAH activity, they were efficiently hydrolysed by recombinant ENPP6 (Mulder and Cravatt 2006).

5.1 Oxidative Metabolism of ECBs

5.1.1 Cyclooxygenase Activity

Cyclooxygenase (COX) activities are membrane-bound enzymes responsible for the production of prostanoids (prostaglandins and thromboxanes) from arachidonic acid. Of the two isoforms, COX-1 is generally held to be constitutively expressed and responsible for the "house-keeping" roles of prostanoids, while COX-2 is generally inducible, although it is constitutively expressed in some tissues (e.g. spinal cord) and thought to be responsible for the inflammatory, pyrexic and hyperalgesic prostanoids. Given that the two major ECBs, AEA and 2AG, are arachidonate derivatives, it is, in retrospect, not too surprising that COX metabolises ECBs to produce prostanoid-like molecules. Intriguingly, AEA and 2AG appear to be poor substrates for COX-1, but are readily metabolised by COX-2 (Yu et al. 1997; Kozak et al. 2000). Perhaps more intriguing is the observation that ECBs, through the CB_1 receptor, are able to induce COX-2 expression in the cerebral microvasculature (Chen et al. 2005), indicating the possibility of diversion of ECBs through alternative metabolic routes following repeated administration.

The products of COX-2 oxidative metabolism of ECBs are biologically active (Sang et al. 2006, 2007; Hu et al. 2008) and so COX-2 metabolism represents transformation of ECBs rather than inactivation. The prostanoid ethanolamides and glyceryl esters appear not to be active at conventional cannabinoid or prostanoid receptors, however, but rather through separate targets, as yet undefined at the molecular level (Fowler 2007; Woodward et al. 2008). The major route of prostaglandin inactivation, via 15-hydroxyprostaglandin dehydrogenase, appears to be less effective for oxidation of COX-2 metabolites of ECBs (Kozak et al. 2001). In parallel, FAAH or MGL hydrolysis of the COX-2 metabolites of AEA or 2AG, respectively, was reduced in comparison to the untransformed parent ECB (Vila et al. 2007). It appears, therefore, that whilst the ECBs themselves are transient species, COX-2 metabolism is able to generate derivatives which are far more long-lasting.

Numerous cellular and tissue preparations have been shown to be able to metabolise administered ECBs through the COX-2 pathway (Kim and Alger 2004; Patsos et al. 2005; Ahn et al. 2007; Ho and Randall 2007; Rockwell et al. 2008; Jhaveri et al. 2008; Bajo et al. 2009); however, definitive evidence for COX-2 metabolism of endogenous ECBs is currently lacking. Intriguingly, however, the observation that typical antibody-based assays for prostanoids fails to distinguish prostaglandins from prostamides suggests that there is more to be elucidated from the COX-2 metabolism of ECBs (Glass et al. 2005). The picture is further obscured by the observation that many, but not all, non-steroidal anti-inflammatory drugs, previously thought to target COX activities selectively, are also able to inhibit FAAH activity at pharmacologically relevant concentrations (Fowler et al. 1997, 1999, 2003). The possibility exists, therefore, that clinical efficacy of some of these agents may be due to a combination of preventing the accumulation of inflammatory prostanoids, as well as promoting the accumulation of anti-inflammatory ECBs (Jhaveri et al. 2008).

5.1.2 Lipoxygenase Activity

Mammalian lipoxygenases (LOXs) are bound to membranes inside the cell, including the nuclear membrane, and generate hydroperoxides of unsaturated fatty acids (typically of the 1Z,5Z pentadiene structure) by introducing molecular oxygen at the points of unsaturation. For arachidonate, 5-, 12- and 15-hydroperoxidation generates hydroperoxyeicosatetraenoic acids (HPETEs). 5-LOX metabolism of arachidonate, primarily in white blood cells, generates leukotrienes, while sequential oxidative metabolism of arachidonate by 5- and 15-LOXs generates lipoxins. Although a sequence of metabolic steps allows AEA to be metabolised to 12-oxygenated species in splenocytes, this appears to be due to FAAH-mediated hydrolysis of AEA generating arachidonate, thereafter metabolised by 12-LOX (Bobrov et al. 2000). Both 12- and 15-LOX, but not 5-LOX, appear capable of metabolising AEA and 2AG (Hampson et al. 1995; Ueda et al. 1995; Edgemond et al. 1998; Moody et al. 2001; Kozak et al. 2002). Indeed, incubation of AEA with plant-derived 5-LOX generates a 12-hydroperoxide derivative (Van Zadelhoff et al. 1998).

In the brain, the majority of LOX activity appears to reside in the pineal gland (Nishiyama et al. 1993; Hada et al. 1994), through which enzyme activity a product consistent with 12-hydroxyAEA was identified (Hampson et al. 1995). Rat brain homogenates are able to generate both 12-hydroxyAEA and 15-hydroxyAEA (Veldhuis et al. 2003). Human platelets are able to convert AEA to 12(S)-hydroxyAEA, while both 12(S)-hydroxyAEA and 15(S)-hydroxyAEA appeared following incubation of AEA with human polymorphonuclear lymphocytes (Edgemond et al. 1998).

As with COX activity, LOX metabolism represents transformation, rather than inactivation, of ECBs, since LOX products are active at CB_1 and CB_2 receptors (Edgemond et al. 1998), as well as TRPV1 (Craib et al. 2001) and PPAR (Kozak et al. 2002) receptors.

5.1.3 Cytochrome P450s and Epoxygenase Activity

Arachidonic acid is subject to an additional form of oxidative metabolism, in which 5,6-epoxyeicosatrienoic acid (EET), 8,9-EET, 11,12-EET or 14,15-EET may be formed. These epoxides are thought to be important signalling molecules in the vascular system and are metabolised by epoxide hydrolases to form diols. Human liver microsomes, containing multiple cytochrome P450 activities, were found to catalyse epoxide formation at all four unsaturations of AEA (Snider et al. 2007). One isoform of cytochrome P450, 4X1, generates 14,15-EET ethanolamide from

AEA (Stark et al. 2008), while a second, 2D6, is not only able to generate all four epoxides from AEA, but also oxidises them further to generate diol analogues of AEA (Snider et al. 2008).

One isoform of cytochrome P450, 4F2, oxidises the ω-carbon of the fatty acid chain to produce *N*-20-hydroxyarachidonoylethanolamine.

The potential for oxidation at the terminal alcohol of AEA has recently been demonstrated, with alcohol dehydrogenase metabolism generating *N*-arachidonoyl-glycine (Aneetha et al. 2009), which may prove to be the endogenous ligand for the putative ECB-like receptor GPR18 (Kohno et al. 2006).

6 Stimulation of ECB Synthesis and Release

It is widely thought that AEA and 2AG (which are very lipophilic compounds) are produced from their precursor membrane phosphoglycerides via Ca^{2+}-sensitive biosynthetic pathways, activated on demand, rather than being pre-synthesised and stored in secretory vesicles awaiting exocytosis, as is the case for many neuro-transmitters. Hence, it is likely that ECB agents act largely as local (autocrine/paracrine) mediators rather than conventional hormones. ECBs can, however, be detected in the plasma, although their tissues of origin are not clear, and a longer range hormonal action should not be completely disregarded. AEA and 2AG are regarded as retrograde mediators in the brain where post-synaptic depolarisation leads to the elevation of intracellular Ca^{2+} from intracellular stores, entry through receptor/voltage-operated Ca^{2+} channels (or both). This is assumed to stimulate Ca^{2+}-sensitive enzymes such as NAPE-PLD catalysing the biosynthesis of AEA and, particularly, 2AG. The released ECB mediators retrogradely traverse the synapse to activate presynaptic CB_1 receptors resulting in inhibition of voltage-activated calcium channels, activation of K^+ channels and inhibition of neurotrans-mitter release.

However, in spite of the massive research effort expended on the ECBs in recent years, there have been remarkably few direct studies of stimulated ECB synthesis and release, particularly in native cells and tissues. The chemical nature of the ECBs probably explains the difficulty in measuring extracellular concentrations, their high lipophilicity suggesting that they are unlikely to exist alone in the aqueous extracellular medium for any length of time. However, elegant electro-physiological studies by Brad Alger (see the chapter "Endocannabinoid Signaling in Neural Plasticity" by Alger, this volume) have unambiguously shown that de-polarisation and agonist-mediated Ca^{2+} mobilisation stimulates release of ECBs, indicated by depolarisation-induced suppression of inhibition and excitation in the CA1 region of the hippocampus (Kim et al. 2002). More direct studies (Bisogno et al. 1997) showed that stimulation of mouse neuroblastoma cells with the Ca^{2+} ionophore, ionomycin, caused the synthesis, release and subsequent degradation of 2AG. Stella and Piomelli (2001) also demonstrated Ca^{2+} mobilising receptor-mediated enhancement of 2AG, OEA and PEA, but not AEA, in rat cortical neurones. These authors (Stella et al. 1997) had previously reported that high frequency stimulation of hippocampal slices had increased the synthesis of 2AG but not AEA.

Despite the weight of evidence favouring the Ca^{2+}-mediated formation of 2AG but less so of AEA, Di Marzo's group have proposed that AEA acts as a kind of intracellular "volume switch" for Ca^{2+}; van der Stelt et al. (2005) reported that in dorsal root ganglion (DRG) cells, purinoceptor or muscarinic cholinergic receptor activation leads to AEA synthesis which acts on TRPV1 channels, gating extracellular Ca^{2+} allowing more AEA synthesis, thus providing a feed-forward mechanism. This potentially vicious cycle is suggested to be interrupted by released AEA acting on extracellular facing CB_1 receptors and inhibiting TRPV1 function, as demonstrated by Millns et al. (2001). In more recent studies, Vellani et al. (2008) again suggested a central role for the TRPV1 channel in the control of DRG and, by extension, sensory nerve activity. They showed that, in addition to activation by AEA, TRPV1 channels were activated and/or sensitised by stimulating protein kinases A or C leading to enhanced AEA but not 2AG or PEA levels. This indicates that, in addition to Ca^{2+} mobilisation, the generation of other second messengers following receptor activation has the potential to modulate ECB synthesis and release.

In our own studies of ECB synthesis and release in rat cerebral cortical slices in vitro, we have found little evidence for Ca^{2+}-mobilising stimuli elevating the

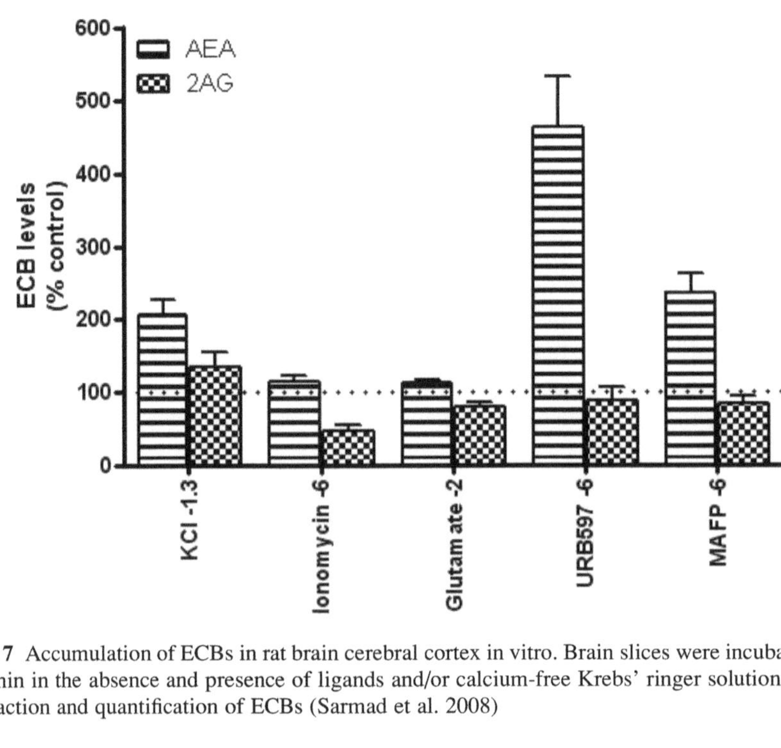

Fig. 7 Accumulation of ECBs in rat brain cerebral cortex in vitro. Brain slices were incubated for 30 min in the absence and presence of ligands and/or calcium-free Krebs' ringer solution before extraction and quantification of ECBs (Sarmad et al. 2008)

levels of ECBs but it is notable that the FAAH inhibitor URB597 (see above) causes a robust increase in tissue levels and accompanying release, suggesting that, in this preparation, there is a high on-going turnover of ECBs independent of calcium ions (Fig. 7).

There have been a few attempts to monitor in vivo ECB release using microdialysis coupled with LC/MS analysis. Béquet et al. (2007) reported that, in the rat hypothalamus, local depolarisation following high K^+ or glutamate perfusion enhanced AEA and 2AG release independently of Ca^{2+}. Their experiments supported a release-modulating role for CB_1 receptors in that the antagonist rimonabant enhanced, while the CB agonist WIN55212-2 reduced, AEA release, although, intriguingly, the same treatments induced opposite changes in 2AG. The mechanisms underlying the control of release clearly require further investigation. At the present time, even basic questions such as whether the release process is an active, energy-dependent mechanism or simply a passive flow down concentration gradients remain unanswered.

7 Conclusion

Despite a massive research effort over the last two decades, there is still a plethora of questions to be addressed concerning the ECB system. There has been a probably unwarranted concentration on AEA, given its archetypal status, and the challenge now is to clarify the roles of the many related fatty acids and their interactions, not only with CB_1 and CB_2 receptors but with the ever-growing family of G-protein-coupled, nuclear and ion channel receptors responsive to ECBs. The immense complexity of the synthetic and metabolic pathways followed by the ECBs provides a great challenge but also an opportunity for the development of selective therapeutic agents to tackle some of the diseases involving the ECB system which are described in later chapters in this volume.

References

Ahn DK, Choi HS, Yeo SP et al. (2007) Blockade of central cyclooxygenase (COX) pathways enhances the cannabinoid-induced antinociceptive effects on inflammatory temporomandibular joint (TMJ) nociception. Pain 132:23–32

Aneetha H, O'Dell DK, Tan B et al. (2009) Alcohol dehydrogenase-catalyzed in vitro oxidation of anandamide to *N*-arachidonoyl glycine, a lipid mediator: Synthesis of *N*-acyl glycinals. Bioorg Med Chem Lett 19:237–241

Aoki J, Inoue A, Makide K et al. (2007) Structure and function of extracellular phospholipase A_1 belonging to the pancreatic lipase gene family. Biochimie 89:197–204

Artmann A, Petersen G, Hellgren LI et al. (2008) Influence of dietary fatty acids on endocannabinoid and *N*-acylethanolamine levels in rat brain, liver and small intestine. Biochim Biophys Acta-Mol Cell Biol L 1781:200–212

Bajo M, Roberto M, Schweitzer P (2009) Differential alteration of hippocampal excitatory synaptic transmission by cannabinoid ligands. J Neurosci Res 87:766–775
Basanez G, Nieva JL, Goni FM et al. (1996) Origin of the lag period in the phospholipase C cleavage of phospholipids in membranes. Concomitant vesicle aggregation and enzyme activation. Biochemistry 35:15183–15187
Bequet F, Uzabiaga F, Desbazeille M et al. (2007) CB1 receptor-mediated control of the release of endocannabinoids (as assessed by microdialysis coupled with LC/MS) in the rat hypothalamus. Eur J Neurosci 26:3458–3464
Berdyshev EV, Schmid PC, Krebsbach RJ et al. (2001) Activation of PAF receptors results in enhanced synthesis of 2-arachidonoylglycerol (2-AG) in immune cells. FASEB J 15:2171–2178
Bisogno T, Maurelli S, Melck D et al. (1997) Biosynthesis, uptake, and degradation of anandamide and palmitoylethanolamide in leukocytes. J Biol Chem 272:3315–3323
Bisogno T, Melck D, De Petrocellis L et al. (1999) Phosphatidic acid as the biosynthetic precursor of the endocannabinoid 2-arachidonoylglycerol in intact mouse neuroblastoma cells stimulated with ionomycin. J Neurochem 72:2113–2119
Bisogno T, Howell F, Williams G et al. (2003) Cloning of the first sn1-DAG lipases points to the spatial and temporal regulation of endocannabinoid signaling in the brain. J Cell Biol 163:463–468
Bisogno T, Ortar G, Petrosino S et al. (2009) Development of a potent inhibitor of 2-arachidonoylglycerol hydrolysis with antinociceptive activity in vivo. Biochim Biophys Acta 1791:53–60
Bobrov MY, Shevchenko VP, Yudushkin IA et al. (2000) Hydrolysis of anandamide and eicosapentaenoic acid ethanolamide in mouse splenocytes. Biochemistry (Mosc) 65:615–619
Brindley DN (2004) Lipid phosphate phosphatases and related proteins: Signaling functions in development, cell division, and cancer. J Cell Biochem 92:900–912
Cadas H, Gaillet S, Beltramo M et al. (1996a) Biosynthesis of an endogenous cannabinoid precursor in neurons and its control by calcium and cAMP. J Neurosci 16:3934–3942
Cadas H, Schinelli S, Piomelli D (1996b) Membrane localization of *N*-acylphosphatidylethanolamine in central neurons: studies with exogenous phospholipases. J Lipid Mediat Cell Signal 14:63–70
Carrier EJ, Kearn CS, Barkmeier AJ et al. (2004) Cultured rat microglial cells synthesize the endocannabinoid 2-arachidonylglycerol, which increases proliferation via a CB2 receptor-dependent mechanism. Mol Pharmacol 65:999–1007
Chen P, Hu S, Yao J et al. (2005) Induction of cyclooxygenase-2 by anandamide in cerebral microvascular endothelium. Microvasc Res 69:28–35
Craib SJ, Ellington HC, Pertwee RG et al. (2001) A possible role of lipoxygenase in the activation of vanilloid receptors by anandamide in the guinea-pig bronchus. Br J Pharmacol 134:30–37
Cravatt BF, Giang DK, Mayfield SP et al. (1996) Molecular characterization of an enzyme that degrades neuromodulatory fatty-acid amides. Nature 384:83–87
Dawson RMC, Clarke N, Quarles RH (1969) *N*-acylphosphatidylethanolamine, a phospholipid that is rapidly metabolized during the early germination of pea seeds. Biochem J 114:265–267
De Petrocellis L, Melck D, Ueda N et al. (1997) Novel inhibitors of brain, neuronal, and basophilic anandamide amidohydrolase. Biochem Biophys Res Commun 231:82–88
Devane WA, Hanus L, Breuer A et al. (1992) Isolation and structure of a brain constituent that binds to the cannabinoid receptor. Science 258:1946–1949
Di Marzo V, Fontana A, Cadas H et al. (1994) Formation and inactivation of endogenous cannabinoid anandamide in central neurons. Nature 372:686–691
Di Marzo V, Bisogno T, Sugiura T et al. (1998) The novel endogenous cannabinoid 2-arachidonoylglycerol is inactivated by neuronal- and basophil-like cells: connections with anandamide. Biochem J 331:15–19
Diana MA, Marty A (2004) Endocannabinoid-mediated short-term synaptic plasticity: depolarization-induced suppression of inhibition (DSI) and depolarization-induced suppression of excitation (DSE). Br J Pharmacol 142:9–19

Dinh TP, Carpenter D, Leslie FM et al. (2002) Brain monoglyceride lipase participating in endocannabinoid inactivation. Proc Natl Acad Sci USA 99:10819–10824

Dinh TP, Kathuria S, Piomelli D (2004) RNA interference suggests a primary role for monoacylglycerol lipase in the degradation of the endocannabinoid 2-arachidonoylglycerol. Mol Pharmacol 66:1260–1264

Duncan M, Thomas AD, Cluny NL et al. (2008) Distribution and function of monoacylglycerol lipase in the gastrointestinal tract. Am J Physiol – Gastrointest Liver Physiol 295:G1255–G1265

Edgemond WS, Hillard CJ, Falck JR et al. (1998) Human platelets and polymorphonuclear leukocytes synthesize oxygenated derivatives of arachidonylethanolamide (anandamide): their affinities for cannabinoid receptors and pathways of inactivation. Mol Pharmacol 54:180–188

Edwards DA, Kim J, Alger BE (2006) Multiple mechanisms of endocannabinoid response initiation in hippocampus. J Neurophysiol 95:67–75

Egertova M, Simon GM, Cravatt BF et al. (2008) Localization of *N*-acyl phosphatidylethanolamine phospholipase D (NAPE-PLD) expression in mouse brain: A new perspective on *N*-acylethanolamines as neural signaling molecules. J Comp Neurol 506:604–615

Egertová M, Giang DK, Cravatt BF et al. (1998) A new perspective on cannabinoid signalling: complementary localization of fatty acid amide hydrolase and the CB1 receptor in rat brain. Proc R Soc Lond B Biol Sci 265:2081–2085

Fowler CJ (2007) The contribution of cyclooxygenase-2 to endocannabinoid metabolism and action. Br J Pharmacol 152:594–601

Fowler CJ, Tiger G, Stenstrom A (1997) Ibuprofen inhibits rat brain deamidation of anandamide at pharmacologically relevant concentrations. Mode of inhibition and structure–activity relationship. J Pharmacol Exp Ther 283:729–734

Fowler CJ, Janson U, Johnson RM et al. (1999) Inhibition of anandamide hydrolysis by the enantiomers of ibuprofen, ketorolac, and flurbiprofen. Arch Biochem Biophys 362:191–196

Fowler CJ, Holt S, Tiger G (2003) Acidic nonsteroidal anti-inflammatory drugs inhibit rat brain fatty acid amide hydrolase in a pH-dependent manner. J Enzyme Inhib Med Chem 18:55–58

Fu J, Kim J, Oveisi F et al. (2008) Targeted enhancement of oleoylethanolamide production in proximal small intestine induces across-meal satiety in rats. Am J Physiol Regul Integr Comp Physiol 295:R45–R50

Gerdeman GL, Lovinger DM (2003) Emerging roles for endocannabinoids in long-term synaptic plasticity. Br J Pharmacol 140(5):781–789

Ghafouri N, Tiger G, Razdan RK et al. (2004) Inhibition of monoacylglycerol lipase and fatty acid amide hydrolase by analogues of 2-arachidonoylglycerol. Br J Pharmacol 143:774–784

Gillett MP, Vieira EM, Dimenstein R (1993) The phospholipase activities present in preheparin mouse plasma are inhibited by antiserum to hepatic lipase. Int J Biochem 25:449–453

Gillum MP, Zhang D, Zhang XM et al. (2008) *N*-acylphosphatidylethanolamine, a gut-derived circulating factor induced by fat ingestion, inhibits food intake. Cell 135:813–824

Glass M, Hong JW, Sato TA et al. (2005) Misidentification of prostamides as prostaglandins. J Lipid Res 46:1364–1368

Gobbi G, Bambico FR, Mangieri R et al. (2005) Antidepressant-like activity and modulation of brain monoaminergic transmission by blockade of anandamide hydrolysis. Proc Natl Acad Sci USA 102:18620–18625

Goding JW, Grobben B, Slegers H (2003) Physiological and pathophysiological functions of the ecto-nucleotide pyrophosphatase/phosphodiesterase family. Biochim Biophys Acta – Mol Basis Dis 1638:1–19

Goparaju SK, Ueda N, Yamaguchi H et al. (1998) Anandamide amidohydrolase reacting with 2-arachidonoylglycerol, another cannabinoid receptor ligand. FEBS Lett 422:69–73

Grazia Cascio M, Minassi A, Ligresti A et al. (2004) A structure–activity relationship study on *N*-arachidonoyl-amino acids as possible endogenous inhibitors of fatty acid amide hydrolase. Biochem Biophys Res Commun 314:192–196

Hada T, Hagiya H, Suzuki H et al. (1994) Arachidonate 12-lipoxygenase of rat pineal glands: catalytic properties and primary structure deduced from its cDNA. Biochim Biophys Acta 1211:221–228

Hampson AJ, Hill WAG, Zanphillips M et al. (1995) Anandamide hydroxylation by brain lipoxygenase: metabolite structures and potencies at the cannabinoid receptor. Biochim Biophys Acta – Lip Lip Met 1259:173–179

Hansen HS, Lauritzen L, Strand AM et al. (1995) Glutamate stimulates the formation of *N*-acylphosphatidylethanolamine and *N*-acylethanolamine in cortical neurons in culture. Biochim Biophys Acta – Lip Lip Met 1258:303–308

Hanus L, Gopher A, Almog S et al. (1993) Two new unsaturated fatty acid ethanolamides in brain that bind to the cannabinoid receptor. J Med Chem 36:3032–3034

Hashimotodani Y, Ohno-Shosaku T, Maejima T et al. (2008) Pharmacological evidence for the involvement of diacylglycerol lipase in depolarization-induced endocanabinoid release. Neuropharmacol 54:58–67

Higgs HN, Glomset JA (1994) Identification of a phosphatidic acid-preferring phospholipase A1 from bovine brain and testis. Proc Natl Acad Sci USA 91:9574–9578

Higgs HN, Han MH, Johnson GE et al. (1998) Cloning of a phosphatidic acid-preferring phospholipase A1 from bovine testis. J Biol Chem 273:5468–5477

Hiramatsu T, Sonoda H, Takanezawa Y et al. (2003) Biochemical and molecular characterization of two phosphatidic acid-selective phospholipase A_1s, mPA-PLA$_1\alpha$ and mPA-PLA$_1\beta$. J Biol Chem 278:49438–49447

Hirasawa K, Irvine RF, Dawson RMC (1981) The catabolism of phosphatidylinositol by an EDTA-insensitive phospholipase A1 and calcium-dependent phosphatidylinositol phosphodiesterase in rat brain. Eur J Biochem 120:53–58

Ho W-SV, Randall MD (2007) Endothelium-dependent metabolism by endocannabinoid hydrolases and cyclooxygenases limits vasorelaxation to anandamide and 2-arachidonoylglycerol. Br J Pharmacol 150:641–651

Hohmann AG, Suplita RL, Bolton NM et al. (2005) An endocannabinoid mechanism for stress-induced analgesia. Nature 435:1108–1112

Hoover HS, Blankman JL, Niessen S et al. (2008) Selectivity of inhibitors of endocannabinoid biosynthesis evaluated by activity-based protein profiling. Bioorg Med Chem Lett 18:5838–5841

Hu SSJ, Bradshaw HB, Chen JSC et al. (2008) Prostaglandin E-2 glycerol ester, an endogenous COX-2 metabolite of 2-arachidonoylglycerol, induces hyperalgesia and modulates NFκB activity. Br J Pharmacol 153:1538–1549

Inoue M, Okuyama H (1984) Phospholipase A1 acting on phosphatidic acid in porcine platelet membranes. J Biol Chem 259:5083–5086

Jaye M, Lynch KJ, Krawiec J et al. (1999) A novel endothelial-derived lipase that modulates HDL metabolism. Nat Genet 21:424–428

Jhaveri MD, Richardson D, Robinson I et al. (2008) Inhibition of fatty acid amide hydrolase and cycloxygenase-2 increases levels of endocannabinoids and produces analgesia via peroxisome proliferator-activated receptor-alpha in a model of inflammatory pain. Neuropharmacol 55:85–93

Jimenez-Monreal AM, Villalain J, Aranda FJ et al. (1998) The phase behavior of aqueous dispersions of unsaturated mixtures of diacylglycerols and phospholipids. Biochim Biophys Acta 1373:209–219

Jin XH, Okamoto Y, Morishita J et al. (2007) Discovery and characterization of a Ca^{2+}-independent phosphatidylethanolamine *N*-acyltransferase generating the anandamide precursor and its congeners. J Biol Chem 282:3614–3623

Karbarz MJ, Luo L, Chang L et al. (2009) Biochemical and biological properties of 4-(3-phenyl-[1, 2, 4] thiadiazol-5-yl)-piperazine-1-carboxylic acid phenylamide, a mechanism-based inhibitor of fatty acid amide hydrolase. Anesth Analg 108:316–329

Kathuria S, Gaetani S, Fegley D et al. (2003) Modulation of anxiety through blockade of anandamide hydrolysis. Nat Med 9:76–81

Kim J, Alger BE (2004) Inhibition of cyclooxygenase-2 potentiates retrograde endocannabinoid effects in hippocampus. Nat Neurosci 7:697–698
Kim J, Isokawa M, Ledent C et al. (2002) Activation of muscarinic acetylcholine receptors enhances the release of endogenous cannabinoids in the hippocampus. J Neurosci 22:10182–10191
Kobayashi T, Kishimoto M, Okuyama H (1996) Phospholipases involved in lysophosphatidylinositol metabolism in rat brain. J Lipid Mediat Cell Signal 14:33–37
Kohno M, Hasegawa H, Inoue A et al. (2006) Identification of *N*-arachidonylglycine as the endogenous ligand for orphan G-protein-coupled receptor GPR18. Biochem Biophys Res Commun 347:827–832
Koutek B, Prestwich GD, Howlett AC et al. (1994) Inhibitors of arachidonoyl ethanolamide hydrolysis. J Biol Chem 269:22937–22940
Kozak KR, Rowlinson SW, Marnett LJ (2000) Oxygenation of the endocannabinoid, 2-arachidonylglycerol, to glyceryl prostaglandins by cyclooxygenase-2. J Biol Chem 275:33744–33749
Kozak KR, Crews BC, Ray JL et al. (2001) Metabolism of prostaglandin glycerol esters and prostaglandin ethanolamides *in vitro* and *in vivo*. J Biol Chem 276:36993–36998
Kozak KR, Gupta RA, Moody JS et al. (2002) 15-lipoxygenase metabolism of 2-arachidonylglycerol: Generation of a PPARα agonist. J Biol Chem 277:23278–23286
Kreitzer AC, Regehr WG (2001) Retrograde inhibition of presynaptic calcium influx by endogenous cannabinoids at excitatory synapses onto Purkinje cells. Neuron 29:717–727
Kucera GL, Sisson PJ, Thomas MJ et al. (1988) On the substrate specificity of rat liver phospholipase A1. J Biol Chem 263:1920–1928
Lee MW, Kraemer FB, Severson DL (1995) Characterization of a partially purified diacylglycerol lipase from bovine aorta. Biochim Biophys Acta 1254:311–318
Leung D, Saghatelian A, Simon GM et al. (2006) Inactivation of *N*-acyl phosphatidylethanolamine phospholipase D reveals multiple mechanisms for the biosynthesis of endocannabinoids. Biochemistry 45:4720–4726
Lichtman AH, Hawkins EG, Griffin G et al. (2002) Pharmacological activity of fatty acid amides is regulated, but not mediated, by fatty acid amide hydrolase in vivo. J Pharmacol Exp Ther 302:73–79
Liu Q, Tonai T, Ueda N (2002) Activation of *N*-acylethanolamine-releasing phospholipase D by polyamines. Chem Phys Lipids 115:77–84
Liu J, Wang L, Harvey-White J et al. (2008) Multiple pathways involved in the biosynthesis of anandamide. Neuropharmacol 54:1–7
Long JZ, Li W, Booker L et al. (2009) Selective blockade of 2-arachidonoylglycerol hydrolysis produces cannabinoid behavioral effects. Nat Chem Biol 5:37–44
Makara JK, Mor M, Fegley D et al. (2005) Selective inhibition of 2-AG hydrolysis enhances endocannabinoid signaling in hippocampus. Nat Neurosci 8:1139–1141
Makara JK, Mor M, Fegley D et al. (2007) Selective inhibition of 2-AG hydrolysis enhances endocannabinoid signaling in hippocampus. Nat Neurosci 10(1):134
Millns PJ, Chapman V, Kendall DA (2001) Cannabinoid inhibition of the capsaicin-induced calcium response in rat dorsal root ganglion neurones. Br J Pharmacol 132:969–971
Moise AM, Eisenstein SA, Astarita G et al. (2008) An endocannabinoid signaling system modulates anxiety-like behavior in male Syrian hamsters. Psychopharmacol 200:333–346
Moody JS, Kozak KR, Ji C et al. (2001) Selective oxygenation of the endocannabinoid 2-arachidonylglycerol by leukocyte-type 12-lipoxygenase. Biochemistry 40:861–866
Mulder AM, Cravatt BF (2006) Endocannabinoid metabolism in the absence of fatty acid amide hydrolase (FAAH): discovery of phosphorylcholine derivatives of *N*-acyl ethanolamines. Biochemistry 45:11267–11277
Nakane S, Oka S, Arai S et al. (2002) 2-Arachidonoyl-sn-glycero-3-phosphate, an arachidonic acid-containing lysophosphatidic acid: occurrence and rapid enzymatic conversion to 2-arachidonoyl-sn-glycerol, a cannabinoid receptor ligand, in rat brain. Arch Biochem Biophys 402:51–58
Natarajan V, Reddy PV, Schmid PC et al. (1982) *N*-Acylation of ethanolamine phospholipids in canine myocardium. Biochim Biophys Acta 712:342–355

Natarajan V, Schmid PC, Reddy PV et al. (1984) Catabolism of *N*-acylethanolamine phospholipids by dog brain preparations. J Neurochem 42:1613–1619

Niforatos W, Zhang X-F, Lake MR et al. (2007) Activation of TRPA1 channels by the fatty acid amide hydrolase inhibitor 3'-carbamoylbiphenyl-3-yl cyclohexylcarbamate (URB597). Mol Pharmacol 71:1209–1216

Nishiyama M, Watanabe T, Ueda N et al. (1993) Arachidonate 12-lipoxygenase is localized in neurons, glial cells, and endothelial cells of the canine brain. J Histochem Cytochem 41:111–117

Oka S, Tsuchie A, Tokumura A et al. (2003) Ether-linked analogue of 2-arachidonoylglycerol (noladin ether) was not detected in the brains of various mammalian species. J Neurochem 85:1374–1381

Oka S, Arai S, Waku K et al. (2007a) Depolarization-induced rapid generation of 2-arachidonoylglycerol, an endogenous cannabinoid receptor ligand, in rat brain synaptosomes. J Biochem 141:687–697

Oka S, Nakajima K, Yamashita A et al. (2007b) Identification of GPR55 as a lysophosphatidylinositol receptor. Biochem Biophys Res Commun 362:928–934

Oka S, Toshida T, Maruyama K et al. (2009) 2-Arachidonoyl-sn-glycero-3-phosphoinositol: A Possible Natural Ligand for GPR55. J Biochem 145:13–20

Okamoto Y, Morishita J, Tsuboi K et al. (2004) Molecular characterization of a phospholipase D generating anandamide and its congeners. J Biol Chem 279:5298–5305

Overton HA, Babbs AJ, Doel SM et al. (2006) Deorphanization of a G protein-coupled receptor for oleoylethanolamide and its use in the discovery of small-molecule hypophagic agents. Cell Metab 3:167–175

Parrish JC, Nichols DE (2006) Serotonin 5-HT_{2A} receptor activation induces 2-arachidonoylglycerol release through a phospholipase C-dependent mechanism. J Neurochem 99:1164–1175

Patsos HA, Hicks DJ, Dobson RR et al. (2005) The endogenous cannabinoid, anandamide, induces cell death in colorectal carcinoma cells: a possible role for cyclooxygenase-2. Gut 54:1741–1750

Petersen G, Hansen HS (1999) *N*-acylphosphatidylethanolamine-hydrolysing phospholipase D lacks the ability to transphosphatidylate. FEBS Lett 455:41–44

Reddy PV, Natarajan V, Schmid PC et al. (1983) *N*-Acylation of dog heart ethanolamine phospholipids by transacylase activity. Biochim Biophys Acta – Lip Lip Met 750:472–480

Rindlisbacher B, Reist M, Zahler P (1987) Diacylglycerol breakdown in plasma membranes of bovine chromaffin cells is a two-step mechanism mediated by a diacylglycerol lipase and a monoacylglycerol lipase. Biochim Biophys Acta 905:349–357

Rockwell CE, Raman P, Kaplan BLF et al. (2008) A COX-2 metabolite of the endogenous cannabinoid, 2-arachidonyl glycerol, mediates suppression of IL-2 secretion in activated Jurkat T cells. Biochem Pharmacol 76:353–361

Russo R, LoVerme J, La Rana G et al. (2007) The fatty-acid amide hydrolase inhibitor URB597 (cyclohexyl carbamic acid 3'-carbamoyl-biphenyl-3-yl ester) reduces neuropathic pain after oral administration in mice. J Pharmacol Exp Ther 322:236–242

Saario SM, Savinainen JR, Laitinen JT et al. (2004) Monoglyceride lipase-like enzymatic activity is responsible for hydrolysis of 2-arachidonoylglycerol in rat cerebellar membranes. Biochem Pharmacol 67:1381–1387

Sakagami H, Aoki J, Natori Y et al. (2005) Biochemical and molecular characterization of a novel choline-specific glycerophosphodiester phosphodiesterase belonging to the nucleotide pyrophosphatase/phosphodiesterase family. J Biol Chem 280:23084–23093

Sang N, Zhang J, Chen C (2006) PGE_2 glycerol ester, a COX-2 oxidative metabolite of 2-arachidonoyl glycerol, modulates inhibitory synaptic transmission in mouse hippocampal neurons. J Physiol 572:735–745

Sang N, Zhang J, Chen C (2007) COX-2 oxidative metabolite of endocannabinoid 2-AG enhances excitatory glutamatergic synaptic transmission and induces neurotoxicity. J Neurochem 102:1966–1977

Sarmad S, Patel A, Barrett DA et al. (2008) Calcium-independent formation of endocannabinoids in rat brain slices. Proceedings of the British Pharmacological Society, p 111P

Sato T, Aoki J, Nagai Y et al. (1997) Serine phospholipid-specific phospholipase A that is secreted from activated platelets. A new member of the lipase family. J Biol Chem 272:2192–2198

Schmid PC, Reddy PV, Natarajan V et al. (1983) Metabolism of *N*-acylethanolamine phospholipids by a mammalian phosphodiesterase of the phospholipase D type. J Biol Chem 258:9302–9306

Simon GM, Cravatt BF (2006) Endocannabinoid biosynthesis proceeding through glycerophospho-*N*-acyl ethanolamine and a role for α/β hydrolase 4 in this pathway. J Biol Chem 281:26465–26472

Simon GM, Cravatt BF (2008) Anandamide biosynthesis catalyzed by the phosphodiesterase GDE1 and detection of glycerophospho-*n*-acyl ethanolamine precursors in mouse brain. J Biol Chem 283:9341–9349

Snider NT, Kornilov AM, Kent UM et al. (2007) Anandamide metabolism by human liver and kidney microsomal cytochrome P450 enzymes to form hydroxyeicosatetraenoic and epoxyeicosatrienoic acid ethanolamides. J Pharmacol Exp Ther 321:590–597

Snider NT, Sikora MJ, Sridar C et al. (2008) The endocannabinoid anandamide is a substrate for the human polymorphic cytochrome P450 2D6. J Pharmacol Exp Ther 327:538–545

Stark K, Dostalek M, Guengerich FP (2008) Expression and purification of orphan cytochrome P450 4X1 and oxidation of anandamide. FEBS J 275:3706–3717

Stella N, Piomelli D (2001) Receptor-dependent formation of endogenous cannabinoids in cortical neurons. Eur J Pharmacol 425:189–196

Stella N, Schweitzer P, Piomelli D (1997) A second endogenous cannabinoid that modulates long-term potentiation. Nature 388:773–778

Sugiura T, Waku K (2000) 2-Arachidonoylglycerol and the cannabinoid receptors. pp. 89-106.

Sugiura T, Waku K (2002) Cannabinoid receptors and their endogenous ligands. J Biochem 132:7–12

Sugiura T, Kondo S, Sukagawa A et al. (1996) Transacylase-mediated and phosphodiesterase-mediated synthesis of *N*-arachidonoylethanolamine, an endogenous cannabinoid receptor ligand, in rat brain microsomes: comparison with synthesis from free arachidonic acid and ethanolamine. Eur J Biochem 240:53–62

Sugiura T, Kodaka T, Kondo S et al. (1997) Is the cannabinoid CB1 receptor a 2-arachidonoylglycerol receptor? Structural requirements for triggering a Ca2+ transient in NG108-15 cells. J Biochem (Tokyo) 122:890–895

Sugiura T, Kodaka T, Nakane S et al. (1999) Evidence that the cannabinoid CB1 receptor is a 2-arachidonoylglycerol receptor. Structure–activity relationship of 2-arachidonoylglycerol ether-linked analogues, and related compounds. J Biol Chem 274:2794–2801

Suh PG, Park JI, Manzoli L et al. (2008) Multiple roles of phosphoinositide-specific phospholipase C isozymes. BMB Rep 41(6):415–434

Sun YX, Tsuboi K, Okamoto Y et al. (2004) Biosynthesis of anandamide and *N*-palmitoylethanolamine by sequential actions of phospholipase A_2 and lysophospholipase D. Biochem J 380:749–756

Sun YX, Tsuboi K, Zhao LY et al. (2005) Involvement of *N*-acylethanolamine-hydrolyzing acid amidase in the degradation of anandamide and other *N*-acylethanolamines in macrophages. Biochim Biophys Acta 1736:211–220

Sutherland CA, Amin D (1982) Relative activities of rat and dog platelet phospholipase A2 and diglyceride lipase. Selective inhibition of diglyceride lipase by RHC 80267. J Biol Chem 257:14006–14010

Szabo B, Urbanski MJ, Bisogno T et al. (2006) Depolarization-induced retrograde synaptic inhibition in the mouse cerebellar cortex is mediated by 2-arachidonoylglycerol. J Physiol 577:263–280

Tian XY, Guo JX, Yao FM et al. (2005) The conformation, location, and dynamic properties of the endocannabinoid ligand anandamide in a membrane bilayer. J Biol Chem 280:29788–29795

Tornqvist H, Belfrage P (1976) Purification and some properties of a monoacylglycerol-hydrolyzing enzyme of rat adipose tissue. J Biol Chem 251:813–819

Tsou K, Nogueron MI, Muthian S et al. (1998) Fatty acid amide hydrolase is located preferentially in large neurons in the rat central nervous system as revealed by immunohistochemistry. Neurosci Lett 254:137–140

Tsuboi K, Hilligsmann C, Vandevoorde S et al. (2004) *N*-cyclohexanecarbonylpentadecylamine: a selective inhibitor of the acid amidase hydrolysing *N*-acylethanolamines, as a tool to distinguish acid amidase from fatty acid amide hydrolase. Biochem J 379:99–106

Tsuboi K, Sun YX, Okamoto Y et al. (2005) Molecular characterization of *N*-acylethanolamine-hydrolyzing acid amidase, a novel member of the choloylglycine hydrolase family with structural and functional similarity to acid ceramidase. J Biol Chem 280:11082–11092

Tsutsumi T, Kobayashi T, Ueda H et al. (1994) Lysophosphoinositide-specific phospholipase C in rat brain synaptic plasma membranes. Neurochem Res 19:399–406

Tsutsumi T, Kobayashi T, Miyashita M et al. (1995) A lysophosphoinositide-specific phospholipase C distinct from other phospholipase C families in rat brain. Arch Biochem Biophys 317:331–336

Ueda H, Kobayashi T, Kishimoto M et al. (1993a) A possible pathway of phosphoinositide metabolism through EDTA-insensitive phospholipase A1 followed by lysophosphoinositide-specific phospholipase C in rat brain. J Neurochem 61:1874–1881

Ueda H, Kobayashi T, Kishimoto M et al. (1993b) The presence of Ca^{2+}-independent phospholipase A_1 highly specific for phosphatidylinositol in bovine brain. Biochem Biophys Res Commun 195:1272–1279

Ueda N, Yamamoto K, Yamamoto S et al. (1995) Lipoxygenase-catalyzed oxygenation of arachidonylethanolamide, a cannabinoid receptor agonist. Biochim Biophys Acta – Lip Lip Met 1254:127–134

Ueda N, Yamanaka K, Terasawa Y et al. (1999) An acid amidase hydrolyzing anandamide as an endogenous ligand for cannabinoid receptors. FEBS Lett 454:267–270

van der Stelt M, Trevisani M, Vellani V et al. (2005) Anandamide acts as an intracellular messenger amplifying Ca^{2+} influx via TRPV1 channels. EMBO J 24:3026–3037

Van Zadelhoff G, Veldink GA, Vliegenthart JFG (1998) With anandamide as substrate plant 5-lipoxygenases behave like 11-lipoxygenases. Biochem Biophys Res Commun 248:33–38

Vandevoorde S, Saha B, Mahadevan A et al. (2005) Influence of the degree of unsaturation of the acyl side chain upon the interaction of analogues of 1-arachidonoylglycerol with monoacylglycerol lipase and fatty acid amide hydrolase. Biochem Biophys Res Commun 337:104–109

Vandevoorde S, Jonsson KO, Labar G et al. (2007) Lack of selectivity of URB602 for 2-oleoylglycerol compared to anandamide hydrolysis in vitro. Br J Pharmacol 150:186–191

Varma N, Carlson GC, Ledent C et al. (2001) Metabotropic glutamate receptors drive the endocannabinoid system in hippocampus. J Neurosci 21:RC133

Veldhuis WB, van der Stelt M, Wadman MW et al. (2003) Neuroprotection by the endogenous cannabinoid anandamide and arvanil against *in vivo* excitotoxicity in the rat: Role of vanilloid receptors and lipoxygenases. J Neurosci 23:4127–4133

Vellani V, Petrosino S, De Petrocellis L et al. (2008) Functional lipidomics. Calcium-independent activation of endocannabinoid/endovanilloid lipid signalling in sensory neurons by protein kinases C and A and thrombin. Neuropharmacol 55:1274–1279

Vila A, Rosengarth A, Piomelli D et al. (2007) Hydrolysis of prostaglandin glycerol esters by the endocannabinoid-hydrolyzing enzymes, monoacylglycerol lipase and fatty acid amide hydrolase. Biochemistry 46:9578–9585

Walter L, Dinh T, Stella N (2004) ATP induces a rapid and pronounced increase in 2-arachidonoylglycerol production by astrocytes, a response limited by monoacylglycerol lipase. J Neurosci 24:8068–8074

Wang J, Okamoto Y, Morishita J et al. (2006) Functional analysis of the purified anandamide-generating phospholipase D as a member of the metallo-β-lactamase family. J Biol Chem 281:12325–12335

Wang J, Okamoto Y, Tsuboi K et al. (2008a) The stimulatory effect of phosphatidylethanolamine on *N*-acylphosphatidylethanolamine-hydrolyzing phospholipase D (NAPE-PLD). Neuropharmacol 54:8–15

Wang J, Zhao LY, Uyama T et al. (2008b) Amino acid residues crucial in pH regulation and proteolytic activation of *N*-acylethanolamine-hydrolyzing acid amidase. Biochim Biophys Acta 1781:710–717
Wei BQ, Mikkelsen TS, McKinney MK et al. (2006) A second fatty acid amide hydrolase with variable distribution among placental mammals. J Biol Chem 281:36569–36578
Wilson RI, Nicoll RA (2001) Endogenous cannabinoids mediate retrograde signalling at hippocampal synapses. Nature 410:588–592
Wilson RI, Kunos G, Nicoll RA (2001) Presynaptic specificity of endocannabinoid signaling in the hippocampus. Neuron 31:453–462
Woodward DF, Carling RW, Cornell CL et al. (2008) The pharmacology and therapeutic relevance of endocannabinoid derived cyclo-oxygenase (COX)-2 products. Phamacol Ther 120(1):71–80
Yoshida T, Hashimoto K, Zimmer A et al. (2002) The cannabinoid CB1 receptor mediates retrograde signals for depolarization-induced suppression of inhibition in cerebellar Purkinje cells. J Neurosci 22:1690–1697
Yu M, Ives D, Ramesha CS (1997) Synthesis of prostaglandin E_2 ethanolamide from anandamide by cyclooxygenase-2. J Biol Chem 272:21181–21186
Zhang D, Saraf A, Kolasa T et al. (2007) Fatty acid amide hydrolase inhibitors display broad selectivity and inhibit multiple carboxylesterases as off-targets. Neuropharmacol 52:1095–1105
Zheng B, Chen D, Farquhar MG (2000) MIR16, a putative membrane glycerophosphodiester phosphodiesterase, interacts with RGS16. Proc Natl Acad Sci USA 97:3999–4004
Zheng B, Berrie CP, Corda D et al. (2003) GDE1/MIR16 is a glycerophosphoinositol phosphodiesterase regulated by stimulation of G protein-coupled receptors. Proc Natl Acad Sci USA 100:1745–1750

Endocannabinoid Receptor Pharmacology

Betty Yao and Ken Mackie

Contents

B. Yao (✉) and K. Mackie
Neuroscience Research, Abbott Laboratories, 100 Abbot Park Road, Dept. R47W, Bldg AP9A, Abbott Park, IL, 60048 USA
e-mail: betty.yao@abbott.com

Department of Physiology and Brain Science, Gill Center for Biomolecular Science, Indiana University, 1101 E 10th St., Bloomington, IN, 47405 USA
e-mail: kmackie@indiana.edu

D. Kendall and S. Alexander (eds.), *Behavioral Neurobiology of the Endocannabinoid System*, Current Topics in Behavioral Neurosciences 1,
DOI: 10.1007/978-3-540-88955-7_2, © Springer-Verlag Berlin Heidelberg 2009

Abstract This chapter will review the basic pharmacology of endocannabinoid receptors. As the best-described cannabinoid receptors are G-protein-coupled receptors (GPCRs), those will be the focus of this chapter. We will start with a basic review of GPCR signaling, as these concepts are critical to understanding the function of cannabinoid receptors. Next, several features of cannabinoid receptor signaling will be presented, with an emphasis on the effectors modulated by cannabinoid receptors. Finally, we will finish with a discussion of cannabinoid receptor agonists and antagonists and future directions. The aim of this chapter is to introduce the cannabinoid receptor pharmacology that will be necessary to appreciate the intricacies of endocannabinoid signaling presented in later chapters.

Keywords Allosteric modulator • Efficacy • Potency • Protean agonism • Radioligand binding

Abbreviations

2AG	2-Arachidonoylethanolamine
AEA	Anandamide
GIRK	G-protein-coupled inwardly rectifying potassium channels
GPCR	G-protein-coupled receptor
RTK	Receptor tyrosine kinase

1 GPCR Overview

G-protein-coupled receptors (GPCRs) are membrane receptor proteins, whose primary function is to transduce extracellular stimuli (communicated as ligands) into intracellular signals. GPCRs comprise the largest protein family with 1,000–2,000 members (>1% of the mammalian genome), of which most encode receptors for odorants and pheromones. Natural ligands for GPCRs are stimuli characterized by their diversity, from photons, ions and amino acid derivatives to large protein hormones. Based on their homology with rhodopsin, a photon receptor, GPCRs are predicted to contain an extracellular N-terminus, an intracellular C-terminus and seven membrane-spanning helices (TMs), the latter giving them the designation of 7-TM receptors. Stimulation of GPCRs by a ligand induces conformational changes, an event that initiates intracellular signal transduction cascades through the interaction of GPCR intracellular domains with heterotrimeric (comprised of α and βγ subunits) G proteins (Palczewski et al. 2000) as well as via other protein–protein interactions (Sun et al. 2007).

GPCRs are classified into three main receptor families based on their structural characteristics. Family 1 is by far the largest, and contains characteristic amino acid signatures conserved across members, such as an aspartate in TM2 that has been proposed to be an important amino acid required for receptor activation, a DRY (or ERW) motif immediately C-terminal to TM3, and cysteine residue(s) C-terminal to TM7 serving as a palmitoylation site(s) that plays an important role for G protein coupling and receptor desensitization (Morello and Bouvier 1996). Based on the nature of receptor/ligand interactions, family 1 GPCRs are further divided into three subfamilies: family 1a composed of receptors for small ligands such as odorants, histamine and anandamide (AEA), family 1b for short peptides and cytokines, and family 1c for large glycoproteins and hormones. Family 2 GPCRs are receptors for large peptides such as glucagons and calcitonin, and family 3 are receptors for glutamate, GABA, pheromones, etc. Family 3 GPCRs contain unique, large N-terminal domains often described as a Venus flytrap (Bockaert and Pin 1999). Cannabinoid receptors CB_1, CB_2, and GPR55 all belong to family 1a, and have the basic characteristics of this family, the significance of which will be discussed below.

2 Receptor Pharmacology

An appreciation of the fundamentals of GPCR signaling is essential to understanding cannabinoid receptor signaling, so these concepts will be reviewed in this section. Modern receptor pharmacology is currently based on in vitro pharmacological assays and then their extension to the organism. Both native systems and recombinant receptor expression systems are used, and both come with their limitations.

2.1 GPCR Signaling

In their inactive state GPCRs are associated with quiescent heterotrimeric G proteins. The inactive G protein consists of a GDP-bound α subunit together with its β and γ subunits. Agonist binding to the receptor catalyzes the exchange of GTP for GDP on the α subunit. The binding of GTP prompts the dissociation of the α subunit from the βγ subunits and the receptor. Both the GTP-bound α subunit (G_{α}) and the βγ subunits ($G_{\beta\gamma}$, which remain together) modulate an array of signaling pathways. After a variable period of time, signal transduction is terminated by the hydrolysis of GTP to GDP catalyzed by G_{α}. GDP-bound G_{α} protein re-associates with $G_{\beta\gamma}$, as the GDP-bound form exhibits higher affinity for $G_{\beta\gamma}$ than the GTP-bound form (Selinger 2007). It should be emphasized that these processes are highly regulated by a number of other proteins and factors and the above is only the simplest description of the G protein cycle.

2.2 *Radioligand Binding*

Key characteristics of a receptor are its affinity for a ligand (a ligand being a molecule that binds with high affinity to a receptor) and the number of receptors in a cell. Both of these parameters can be determined by radioligand binding assays (saturation and competition binding assays). To initially characterize a receptor, a saturation binding assay is performed with increasing concentrations of ligand in order to determine the affinity (K_D) of the radioligand for the receptor, as well as the density of receptor sites (B_{max}) in the preparation. The K_D value (the equilibrium dissociation constant) is an intrinsic property of the radioligand at the receptor and is defined as the free ligand concentration at which 50% receptor occupancy is achieved. Radioligands that have been employed to study CB_1 and CB_2 receptor pharmacology include non-selective agonist ligands [^{3}H] CP55,940, [^{3}H] WIN55,212-2, [^{3}H] HU243 and [^{3}H] BAY387271 (Mauler et al. 2002), the CB_2 receptor-selective inverse agonist [35S] SCH225336 and the CB_1 receptor-selective inverse agonist [^{3}H] rimonabant. Although extensively used in studying cannabinoid receptor pharmacology, non-selective radioligands need to be employed with caution when assays are performed using native tissues that express both CB_1 and CB_2 receptors. Inverse agonist radioligands also have limitations, as studies have shown that although inverse agonist ligands compete efficiently with both agonist and inverse agonist radioligand, agonist ligands are less efficient in competing with an inverse agonist radioligand than with an agonist radioligand (Thomas et al. 1998).

Radioligand binding assays are usually performed in membranes prepared from either native tissues, such as the spleen for CB_2 or brain for CB_1, or recombinant cell lines heterologously expressing cannabinoid receptors. CB_1 receptor binding sites are highly abundant in brain (Govaerts et al. 2004; Mauler et al. 2002), exemplified by high B_{max} values (1–5 pmol mg^{-1} protein), comparable to the expression levels of recombinant systems (B_{max}=1–5 pmol mg^{-1}) (McAllister et al. 2002; Tao and Abood 1998). These high levels of CB_1 expression in native tissues potentially have considerable significance in CB_1 signaling, which will be discussed below. In contrast, the level of CB_2 receptor (Govaerts et al. 2004) binding sites are significantly lower in native tissues (B_{max}=697 fmol mg^{-1} in spleen; and 100–300 binding sites per splenic T cell) compared with the level of CB_1 in the brain or the levels that can be achieved when CB_2 is heterologously expressed (Tao and Abood 1998). The density of receptors impacts downstream signaling (Tao and Abood 1998). This is important to keep in mind when evaluating the results of experiments examining GPCR signaling in cells heterologously expressing high levels of GPCRs.

Only a small number of cannabinoid ligands are available in a radiolabeled form. Thus, the binding affinities of non-radiolabeled ligands are usually determined indirectly in radioligand competition binding assays, which determine their ability to compete with a radioligand at the receptor binding site. In radioligand competition binding assays, IC_{50} values, defined as the concentration of non-radiolabeled ligand displacing 50% of the bound radioligand at equilibrium for a given concentration of the radioligand used, are obtained. The dissociation constant (K_i) for a non-radiolabeled ligand can be calculated based on the Cheng–Prusoff equation $K_i = \frac{IC_{50}}{1+\frac{[L]}{K_D}}$ using the IC_{50} value experimentally measured and the radioligand's K_D and concentration ([L]) (Tao and Abood 1998). Although IC_{50} values will vary depending upon the concentrations of the radioligand used, the K_i value (like K_D) of a ligand represents an intrinsic property of the ligand–its affinity for the receptor.

Among the radioligands described above, [^{3}H] CP55,940 and [^{3}H] WIN55,212-2 are the most widely used to characterize cannabinoid receptor pharmacology. Although in general most cannabinoid receptor ligands displace both radioligands in a similar fashion in recombinant cell lines, some discrepancies of receptor binding properties have been observed for the two (radio)ligands. For example, in 2001, Breivogel et al. demonstrated that WIN55,212-2 activates a GPCR in the brain of CB_1 knockout mice with a pharmacology consistent with a non-CB_1, non-CB_2 receptor (Breivogel et al. 2001). Reyes et al. (SFN poster, 2007) reported the presence of a high affinity and saturable binding site for [^{3}H] WIN55,212-2 on HEK cell membranes. Since these cells do not express CB_1 or CB_2 receptors, this indicates that WIN55,212-2 has binding sites besides those of CB_1 and CB_2 receptors.

Binding kinetics have been performed for at the CB_2 receptor. [^{3}H] CP55,940 has demonstrated a fast on-rate (0.263 $nM^{-1}min^{-1}$) and a slower off-rate (0.041 $nM^{-1}min^{-1}$) with a calculated K_D value of 0.156 nM, consistent to those derived from saturation binding analysis. On-rates are similarly fast and off-rates similarly slow for CP55,940 and rimonabant binding to CB_1 receptors (Herkenham et al. 1991; Rinaldi-Carmona et al. 1996).

Cannabinoid ligands in general are highly lipophilic. Receptor mutation studies suggest cannabinoid ligands interact with the hydrophobic TM domains of cannabinoid receptors. Consistent with this site of interaction, it has been proposed that cannabinoid ligands approach their receptors by fast lateral diffusion within the cell membrane (Tian et al. 2005).

2.3 GTPγS Binding as a Measure of GPCR Function

Although radioligand binding assays are widely used to determine the affinities of ligands for a receptor and the number of receptors in a cell, they reveal little information about how ligands modulate receptor activity. Thus, functional receptor assays are required in order to evaluate the properties of a ligand (most fundamentally, is it an agonist, neutral antagonist or inverse agonist?) and the receptor. GTPγS binding assays are often employed to determine the receptor activation level by measuring the binding of GTPγS (a GTP analog that is resistant to hydrolysis) to the receptor/G protein complex. Active GPCRs will catalyze the exchange of GDP for GTPγS. Since the GTPγS is not hydrolyzed, it will remain associated with the G protein α subunit and, if the GTPγS is labeled with ^{35}S, GTPγS can be detected by standard radiochemical assays. Thus, the amount of ^{35}S incorporated into the G protein α subunit pool will be proportional to the number of G proteins activated. Like radioligand binding assays, [^{35}S] GTPγS assays are typically performed using membrane preparations. However, like radioligand receptor binding assays, this technique can also be adapted to tissue sections (Sim et al. 1995).

Several useful concepts relevant for receptor signaling emerge from [^{35}S] GTPγS binding experiments. The most important of these for understanding cannabinoid receptor signaling is intrinsic efficacy (Galandrin et al. 2007). The concept of intrinsic efficacy is that all agonists are not equal – some will more strongly activate receptors than others. Thus, at full receptor occupancy agonist A might stimulate signaling substantially more than agonist B (Fig. 1a). In this case, agonist B is said to have a lower intrinsic efficacy. One way of conceptualizing intrinsic efficacy is that different agonists will favor distinct receptor conformations and some of these receptor conformations will more vigorously activate G proteins than others. It is important to note that potency and efficacy are independent concepts: Efficacy is a measure of the consequence of receptor activation. Potency is a measure of the concentration of agonist required to achieve certain levels of efficacy. For example, the concentration required to achieve 50% of the full efficacy is defined as EC_{50}. It is entirely possible to have a very potent compound which is highly efficacious and vice versa. Examples of low efficacy cannabinoid receptor agonists include anandamide and Δ^9THC, while WIN55,212-2, HU210, and 2AG are high efficacy agonists (Luk et al. 2004). A low intrinsic efficacy agonist may show partial agonism; however this will depend on receptor and downstream effector density. Specifically, partial agonism will be favored by low receptor density and/or less efficient effector coupling.

An important corollary of intrinsic efficacy is that different agonists acting at the same receptor (by inducing distinct receptor conformations) may activate different repertoires of G proteins. This is known as functional selectivity, biased agonism, or agonist-induced trafficking (Fig. 1b) (Schonbrunn 2007; Urban et al. 2007). This is a very important concept with significant therapeutic ramifications. It emphasizes the principle that all agonists are not equal and different agonists (which may appear identical based on binding affinities and stimulation of GTPγS binding) may produce very different signaling, cellular, and physiological effects. Functional selectivity is relevant for both CB_1 and CB_2 signaling (Bonhaus et al. 1998; Lauckner et al. 2005; Shoemaker et al. 2005).

Another concept that emerges with GTPγS studies is that of spare receptors. Evidence for "spare receptors" in a system comes when maximal signaling is observed despite submaximal receptor occupancy. GTPγS binding can also measure

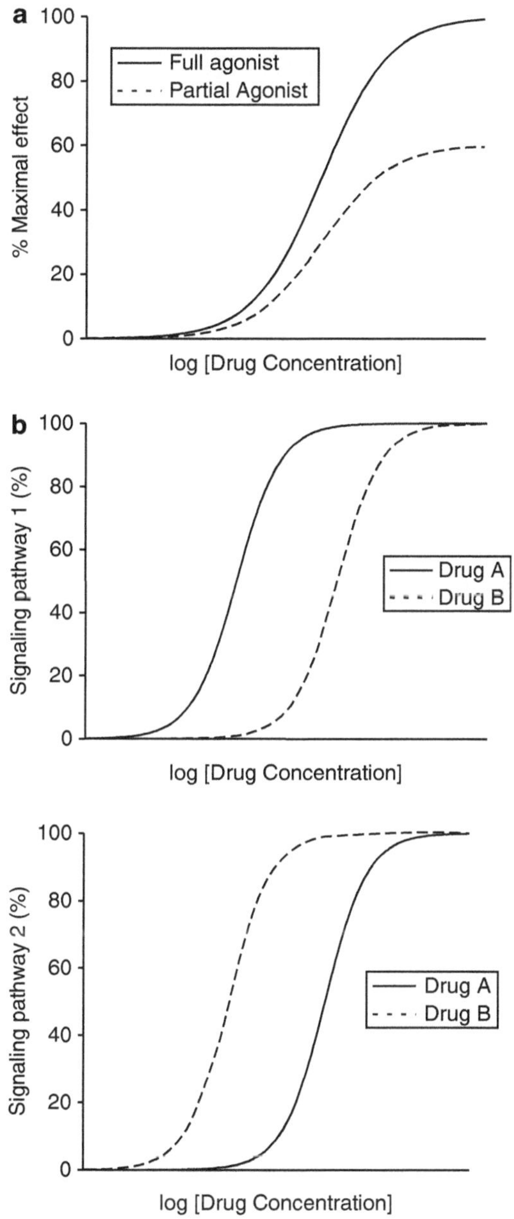

Fig. 1 *Partial agonism and functional selectivity.* (**a**) *Partial agonism.* Two hypothetical drugs have similar potencies (concentration eliciting half maximal effect). However, the partial agonist only has 60% of the efficacy of the full agonist at its maximal effective concentration. (**b**) *Drugs may have different potencies for activating different signaling pathways (functional selectivity).* In this example, for signaling pathway 1 (for example, inhibition of adenylyl cyclase) drug A is more potent. For signaling pathway 2 (for example, stimulation of MAP kinase) drug B is more potent than drug A

the efficiency of G protein activation by GPCRs. In these experiments, the number of G proteins activated by a single receptor is calculated. As an example, CB_1 receptors are inefficient in activating G proteins relative to opioid receptors (Sim et al. 1996).

While $[^{35}S]$ GTPγS binding is a useful way to assess GPCR signaling some caveats must be kept in mind. The first is that $[^{35}S]$ GTPγS binding preferentially identifies activation of the most abundant G proteins (and/or those that are most efficiently activated by the receptor). In brain, the most abundant G proteins are those of the $G_{i/o}$ class. Thus activation of other G proteins, such as $G_{q/11}$, might be overlooked in $[^{35}S]$ GTPγS binding studies. Also, the development of GTPγS binding assays requires considerable optimization, thus it can be difficult to compare results between different laboratories. In addition, GTPγS binding assays measure the first step of the signal transduction pathway, and lack the signal amplification inherent in other functional assays such as those measuring changes of cAMP levels, calcium responses, and transcriptional activity, therefore assay windows and signal-to-noise ratios are sometimes low. In addition, GTPγS binding assays give little information on the spectrum of G proteins coupling to the receptor.

3 CB_1 Receptor Gene Structure

CB_1 receptor cDNA was originally cloned from rat using a homology approach to identify orphan GPCRs (Matsuda et al. 1990). Subsequently, it has been found in all vertebrates and several vertebrates. CB_1 phylogeny is the topic of several excellent reviews (Anday and Mercier 2005; Elphick and Egertova 2005; McPartland et al. 2007).

3.1 Chromosomal Structure, Potential Alternative Splicing

The genes for human, rat, and mouse CB_1 receptors (*CNR1*) are found on chromosomes 6, 5, and 4, respectively. While the translated regions of rodent CB_1 appear to be intronless, two splice variants of human CB_1 have been described. While they may vary in their pharmacology (Ryberg et al. 2005), both variants are found in low abundance and their physiological significance remains to be elucidated (Ryberg et al. 2005; Shire et al. 1995). The potential splice donor sites present in the coding regions of human CB_1 receptors are absent in rodent CB_1 receptors (Howlett et al. 2002).

3.2 CNR1 Polymorphisms

As discussed elsewhere in this volume, substantial evidence suggests that endocannabinoids play a major role in metabolic regulation and psychiatric disorders. A logical extension of this relationship is to determine if mutations in the *CNR1* locus are associated with human diseases or responsiveness to endocannabinoid-based therapies. A number of studies examining single nucleotide polymorphisms and other *CNR1* variants have been conducted. Several of these are limited by small sample size and other methodological constraints. Because of the involvement of the endocannabinoid system in various aspects of drug dependence, several studies have searched for associations of *CNR1* polymorphisms with drug dependence. Two studies have reported associations between the intronic *CNR1* SNPs rs64,54,674 and rs8,06,368 with increased substance dependence (Ehlers et al. 2007; Zuo et al. 2007). Epidemiological and animal studies have proposed a link between schizophrenia and cannabis use. Linkage analysis studies so far have failed to find a strong link between *CNR1* SNPs and susceptibility to schizophrenia, but one report suggests that the exonic 1,359G/A SNP was associated with responsiveness to atypical antipsychotics, with an improved response in individuals with the "A" allele (Hamdani et al. 2008; Seifert et al. 2007). In addition to SNP analysis, variations in trinucleotide repeats with *CNR1* have been associated with a form of anorexia (Siegfried et al. 2004), aspects of polysubstance abuse (Hoenicka et al. 2007), and a subtype of schizophrenia (Ujike et al. 2002). Clearly, much work remains to be done to determine the contributions of these variations of *CNR1* to human disease and response to endocannabinoid system-based therapeutics, but this is an area of active research and interesting discoveries are likely to be forthcoming.

4 CB_2 Receptor Gene Structure

CB_2 receptor cDNA was originally cloned from the HL60 human promyelocytic leukemic cell line in 1992 (Munro et al. 1993). Subsequently, CB_2 receptor cDNAs have been isolated from various species including rat, mouse, zebra fish, and domestic cattle.

4.1 CB_2 Receptor Chromosomal Localization and Potential Alternative Splicing

The human CB_2 gene (*CNR2*) is located at p36.11 on chromosome 1. Other than an intron present in the 5'-untranslated region (5'-UTR), the coding sequence is intronless (Valk et al. 1997). The mouse CB_2 gene, located on chromosome 4, is also intronless in its coding region. In contrast, two variants have been reported in the literature for the rat CB_2 receptor an intronless isoform with identical length of coding sequence to the human CB_2 receptor (Griffin et al. 2000), and a longer isoform (Brown et al. 2002). The human CB_2 receptor and the short isoform of rat CB_2 receptors contain 360 amino acids, and they share 82% sequence identity and 88% sequence homology in their overall sequence. The long isoform of rat CB_2 receptors contains of a total 410 amino acids, of which the N-terminal 343 residues are identical to the short isoform. The sequence from amino acids 343–410 is unique to the long isoform, resulting from two additional splicing events – an excision of two introns of 1,239 and 143 bp respectively plus an addition of two exons encoding for 45 and 39 amino acids, respectively. The genomic DNA at the junction of 5' and 3' end of the first intron in the coding sequence of the long isoform receptor contains 5' AG/GTGA 3' and 5' CAG/A 3', respectively, consistent with the consensus sequences that often serve as splicing donor and acceptor sites.

4.2 CNR2 Polymorphisms

Three non-synonymous single nucleotide polymorphisms (SNPs) have been identified for the human CB_2 receptor: 63Q/R, 316H/Y and 342A/T. Both 63Q and 63R SNPs are prevalent with 46:54 ratio in the Caucasian population (Sipe et al. 2005). Thus far, three haploids (concomitant occurrence of more than one SNP in the same protein), 63Q/316H, 63Q/316Y and 63R/316H, have been reported in humans. Haploid 63R/316H has been shown to have a significantly high linkage to the occurrence of osteoporosis and autoimmune disease (Karsak et al. 2005).

5 Structural Characteristics of the CB_1 Receptor

CB_1 receptors belong to the family 1a of the GPCR superfamily. Remarkable features for CB_1 receptors include a relatively long (about 100 residues) amino terminus in the absence of a signal sequence (Andersson et al. 2003) and an unusually high degree of primary sequence identity across species (Fig. 2a). Considerable effort has been directed towards identifying residues important in binding CB_1 agonists and antagonists. Noteworthy residues identified include K192 (important for binding of agonists, except those of the aminoalkylindole class), as well as rimonabant (Song and Bonner 1996), Y275 and W255 (aromatic stacking, important for recognition of multiple cannabinoid ligands) (McAllister et al. 2003), F170 and F189 (interactions with the double bonds in the arachidonoyl component of endocannabinoids), and a cluster of hydrophobic amino acids in TMs 3, 5, and 6 (McAllister et al. 2003). A disulfide bond between cysteines 257 and 264 in the second extracellular loop also appears critical for receptor trafficking and activity (Fay et al. 2005; Shire et al. 1996). Several domains have been identified to be important for regulation of CB_1 receptor signaling. Regulation by phosphorylation appears to involve (residues are numbered according to rat CB_1) S317 (protein kinase C phosphorylation and uncoupling from G protein signaling) (Garcia et al. 1998) and S426 and S430 (desensitization of CB_1 activation of ERK1/2 and inwardly rectifying potassium channels) (Jin et al. 1999). The distal C-terminus appears to be involved in ligand-induced internalization of CB_1 receptors and its interactions with CRIP1a (cannabinoid receptor interacting protein 1a) and GASP1 (a protein involved in the endosomal targeting of ligand-bound GPCRs) (Hsieh et al. 1999; Martini et al. 2007; Niehaus et al. 2007).

a

```
                           1                                                                              80
        human CB1    (1) ---------MKSILDGLADTTFRTITTDLLYVGSNDIQYEDIKGDMASKLGYFPQKFPLTSFRGSPFQEKMTAGDNPQLV
          rat CB1    (1) ---------MKSILDGLADTTFRTITTDLLYVGSNDIQYEDIKGDMASKLGYFPQKFPLTSFRGSPFQEKMTAGDNSPLV
          dog CB1    (1) ---------MKSILDGLADTTFRTITTDLLYVGSNDIQYEDIKGDMASKLGYFPQKFPLTSFRGSPFQEKMTAGDNAQLV
        mouse CB1    (1) ---------MKSILDGLADTTFRTITTDLLYVGSNDIQYEDIKGDMASKLGYFPQKFPLTSFRGSPFQEKMTAGDNSPLV
      chicken CB1    (1) ---------MKSILDGLADTTFRTITTDLLYVGSNDIQYEDMKGDMASKLGYYPQKFPLSSFRGDPFQEKMTAGDDPLLS
  zebra finch CB1    (1) ---------MKSILDGLADTTFRTITTDLLYVGSNDIQYEDMKGDMASKLGYYPQKFPLSSFRGDPFQEKMTGGDDSLLS
  edible frog CB1    (1) ---------MKSVLDGLADTTFRTITTDLLYMGPNEVQYEDTKSDLS-KLGYYPQKLPLSSY-----QEKIIDGQSTLHL
         newt CB1    (1) ---------MKSILDGLADTTFRTITTDLLYMGSNDVQYEDTKGEMASKLGYFPQKLPLSSFRRDHSPDKMTIGDDNLLS
    zebrafish CB1    (1) MLFPASKSDVKSVLDGVAETTFRTITSGLQYIGSNDIGYDDHIIDGDFSKSGYPLPKPFAAYRRSSFADKVAPDEELIVK
        Consensus    (1)          MKSILDGLADTTFRTITTDLLYVGSNDIQYEDIKGDMASKLGYFPQKFPLSSFRGSPFQEKMTAGD  LLV

                           81                                                                            160
        human CB1   (72) PA--DQVNITEFYNKSLSSFKENEENIQCGENFMDIECFMVLNPSQQLAIAVLSLTLGTFTVLENLLVLCVILHSRSLRC
          rat CB1   (72) PAG-DTTNITEFYNKSLSSFKENEENIQCGENFMDMECFMILNPSQQLAIAVLSLTLGTFTVLENLLVLCVILHSRSLRC
          dog CB1   (72) PA--DQVNITEFYNKSLSSYKENEENIQCGENFMDMECFMILNPSQQLAIAVLSLTLGTFTVLENLLVLCVILHSRSLRC
        mouse CB1   (72) PAG-DTTNITEFYNKSLSSFKENEDNIQCGENFMDMECFMILNPSQQLAIAVLSLTLGTFTVLENLLVLCVILHSRSLRC
      chicken CB1   (72) IIPSDQINITEFYNKSLSTFKENEENIQCGENFMDMECFMILNPSQQLAIAVLSLTLGTFTVLENLLVLCVILHSRSLRC
  zebra finch CB1   (72) IIPSEQVNITEFYNKSLSTFKDNEENIQCGENFMDMECFMILNPSQQLAIAVLSLTLGTFTVLENLLVLCVILHSRSLRC
  edible frog CB1   (66) DS----FNATEFYNKSITTFKDGDGNIQCGNNFMDMECFMILTPSQQLVIAALSITLGTFTVLENMLVLCVIFQSRTLRC
         newt CB1   (72) FYPLDQFNVTEFFNRSVSTFKENDDNLKCGENFMDMECFMILTASQQLIIAVLSLTLGTFTVLENFLVLCVILQSRTLRC
    zebrafish CB1   (81) GLPFYPTNSSDVFGN---WSHAEDGSLQCGENFMDMECFMILTPSQQLAIAVLSLTLGTFTVLENLVVLCVILQSRTLRC
        Consensus   (81)  A  DQ NITEFYNKSLSSFKENEENIQCGENFMDMECFMILNPSQQLAIAVLSLTLGTFTVLENLLVLCVILHSRSLRC

                           161                                                                           240
        human CB1  (150) RPSYHFIGSLAVADLLGSVIFVYSFIDFHVFHRKDSRNVFLFKLGGVTASFTASVGSLFLTAIDRYISIHRPLAYKRIVT
          rat CB1  (151) RPSYHFIGSLAVADLLGSVIFVYSFVDFHVFHRKDSPNVFLFKLGGVTASFTASVGSLFLTAIDRYISIHRPLAYKRIVT
          dog CB1  (150) RPSYHFIGSLAVADLLGSVIFVYSFVDFHVFHRKDSPNVFLFKLGGVTASFTASVGSLFLTAIDRYISIHRPLAYKRIVT
        mouse CB1  (151) RPSYHFIGSLAVADLLGSVIFVYSFVDFHVFHRKDSPNVFLFKLGGVTASFTASVGSLFLTAIDRYISIHRPLAYKRIVT
      chicken CB1  (152) RPSYHFIGSLAVADLLGSVIFVYSFVDFHVFHRKDSPNVFLFKLGGVTASFTASVGSLFLTAIDRYISIHRPLAYKRIVT
  zebra finch CB1  (152) RPSYHFIGSLAVADLLGSVIFVYSFVDFHVFHRKDSPNVFLFKLGGVTASFTASVGSLFLTAIDRYISIHRPLAYKRIVT
  edible frog CB1  (142) RPSYHFIGSLAVADLLGSVIFVYSFVDFHVFHRIDSPNVFLFKLGGVTASFTASVGSLFLTAIDRYISIHRPLSYKRIVT
         newt CB1  (152) RPSYHFIGSLAVADLLGSVIFVYSFLDFHVFHRKDSSNVFLFKLGGVTASFTASVGSLFLTAIDRYISIHRPLAYKRIVT
    zebrafish CB1  (158) RPSYHFIGSLAIADLLGSVIFVYSFLDFHVFHRKDSPNVFLFKLGGVTASFTASVGSLFLTAIDRYVSIHRPLSYRRIVT
        Consensus  (161) RPSYHFIGSLAVADLLGSVIFVYSFVDFHVFHRKDSPNVFLFKLGGVTASFTASVGSLFLTAIDRYISIHRPLAYKRIVT

                           241                                                                           320
        human CB1  (230) RPKAVVAFCLMWTIAIVIAVLPLLGWNCEKLQSVCSDIFPHIDETYLMFWIGVTSVLLLFIVYAYMYILWKAHSHAVRMI
          rat CB1  (231) RPKAVVAFCLMWTIAIVIAVLPLLGWNCKKLQSVCSDIFPLIDETYLMFWIGVTSVLLLFIVYAYMYILWKAHSHAVRMI
          dog CB1  (230) RPKAVVAFCLMWTIAIVIAVLPLLGWNCKKLQSVCSDIFPLIDETYLMFWIGVTSVLLLFIVYAYMYILWKAHSHAVRMI
        mouse CB1  (231) RPKAVVAFCLMWTIAIVIAVLPLLGWNCKKLQSVCSDIFPLIDETYLMFWIGVTSVLLLFIVYAYMYILWKAHSHAVRMI
      chicken CB1  (232) RPKAVVAFCVMWTIAIVIAVLPLLGWNCKKLNSVCSDIFPLIDETYLMFWIGVTSVLLLFIVYAYMYILWKAHSHAVRML
  zebra finch CB1  (232) RPKAVVAFCVMWTIAIVIAVLPLLGWNCKKLNSVCSDIFPLIDETYLMFWIGVTSILLLFIVYAYMYILWKAHSHAVRML
  edible frog CB1  (222) RTKAVIAFCMMWTIAIVIAVLPLLGWNCKKLKSVCSDIFPLIDETYLMFWIGVTSVLLLFIVYAYMYILWKAHHHAVRML
         newt CB1  (232) RTKAVIAFCVMWTIAIIIAVLPLLGWNCKKLKSVCSDIFPLIDENYLMFWIGVTSILLLFIVYAYVYILWKAHSHAVRML
    zebrafish CB1  (238) RTKAVIAFCMMWAISIIIAVLPLLGWNCKRLNSVCSDIFPLIDENYLMFWIGVTSVLVLFIIYAYMYILWKAHHHAVRML
        Consensus  (241) RPKAVVAFCLMWTIAIVIAVLPLLGWNCKKLQSVCSDIFPLIDETYLMFWIGVTSVLLLFIVYAYMYILWKAHSHAVRML

                           321                                                                           400
        human CB1  (310) QRGTQKSIIIHTSEDGKVQVTRPDQARMDIRLAKTLVLILVVLIICWGPLLAIMVYDVFGKMNKLIKTVFAFCSMLCLLN
          rat CB1  (311) QRGTQKSIIIHTSEDGKVQVTRPDQARMDIRLAKTLVLILVVLIICWGPLLAIMVYDVFGKMNKLIKTVFAFCSMLCLLN
          dog CB1  (310) QRGTQKSIIIHTSEDGKVQVTRPDQARMDIRLAKTLVLILVVLIICWGPLLAIMVYDVFGKMNKLIKTVFAFCSMLCLLN
        mouse CB1  (311) QRGTQKSIIIHTSEDGKVQVTRPDQARMDIRLAKTLVLILVVLIICWGPLLAIMVYDVFGKMNKLIKTVFAFCSMLCLLN
      chicken CB1  (312) QRGTQKSIIIQSTEDGKVQITRPDQTRMDIRLAKTLVLILVVLIICWGPLLAIMVYDVFGKMNKLIKTVFAFCSMLCLLN
  zebra finch CB1  (312) QRGTQKSIIIQSTEDGKVQITRPDQTRMDIRLAKTLVLILVVLIICWGPLLAIMVYDVFGKMNKLIKTIFAFCSMLCLLN
  edible frog CB1  (302) QRGTQKSIIVHTSEDGKVHITRPDQTRMDIRLAKTLVLILVVLIICWGPLLAIMVYDVFGKMNKTVKTVFAFCCMLCLLN
         newt CB1  (312) QRGTQKSIIIHTSEDGKVQITRPEQTRMDIRLAKTLVLILVVLIICWGPLLAIMVYDVFGKMNNPIKTVFAFCSMLCLMD
    zebrafish CB1  (318) RRTSQKSLVVHSADGTKVQTPRPDQARMDIRLAKTLVLILVVLVICWGPLLAIMVYDLFWRMGDNIKTVFAFCSMLTLLN
        Consensus  (321) QRGTQKSIIIHTSEDGKVQITRPDQARMDIRLAKTLVLILVVLIICWGPLLAIMVYDVFGKMNKLIKTVFAFCSMLCLLN

                           401                                                                           480
        human CB1  (390) STVNPIIYALRSKDLRHAFRSMFPSCEG---TAQPLDNSMGDSDCLHKHANNAASVHRAAESCIKSTVKIAKVTMSVSTD
          rat CB1  (391) STVNPIIYALRSKDLRHAFRSMFPSCEG---TAQPLDNSMGDSDCLHKHANNTASMHRAAESCIKSTVKIAKVTMSVSTD
          dog CB1  (390) STVNPIIYALRSKDLRHAFRSMFPSCEG---TAQPLDNSMGDSDCLHKHANNAASVHRAAESCIKSTVKIAKVTMSVSTD
        mouse CB1  (391) STVNPIIYALRSKDLRHAFRSMFPSCEG---TAQPLDNSMGDSDCLHKHANNTASMHRAAESCIKSTVKIAKVTMSVSTD
      chicken CB1  (392) STVNPIIYALRSKDLRHAFRSMFPTCEG---TAQPLDNSM-ESDCQHKHANNAGNVHRAAESCIKSTVKIAKVTMSVSTD
  zebra finch CB1  (392) STVNPIIYALRSKDLRHAFRSMFPTCEG---TAQPLDNSM-ESDCQHKHANNAGNVHRAAESCIKSTVKIAKVTMSVSTD
  edible frog CB1  (382) STVNPIIYALRSKDLRSAFCSMFPNCEG---TAQPLDNSM-ESDGQNRHAHNS-NVHRAAESCIKSTVKIA---------
         newt CB1  (392) STVNPIIYALRSQDLRHAFLEQCPPCEG---TSQPLDNSM-ESDCQHRHGNNAGNVHRAAENCIKSTVKIAKVTMSVSTE
    zebrafish CB1  (398) STVNPIIYALRSKDLRRAFLAACQGCRGTSTTPLQLDNSL-ESDCHR-------NQHRAAESCVKTTVKIAKLTMSVSAE
        Consensus  (401) STVNPIIYALRSKDLRHAFRSMFPSCEG   TAQPLDNSM ESDC HKHANNAANVHRAAESCIKSTVKIAKVTMSVSTD

                           481
        human CB1  (467) TSAEAL
          rat CB1  (468) TSAEAL
          dog CB1  (467) TSAEAL
        mouse CB1  (468) TSAEAL
      chicken CB1  (468) TTAEAL
  zebra finch CB1  (468) TTAEAL
  edible frog CB1  (448) ------
         newt CB1  (468) TSGEAV
    zebrafish CB1  (470) TSAEAV
        Consensus  (481) TSAEAL
```

Fig. 2 *Alignments of CB_1 and CB_2 protein sequences from representative vertebrates.* Transmembrane domains are indicated in *gray* and conserved motifs discussed in the text are highlighted in *green*, *a*. CB_1. *b*. CB_2

b

```
                 1                                                                              80
 human CB2   (1) MEECWVTEIANGSKDGLDSNPMKDYMILSGPQKTAVAVLCTLLGLLSALENVAVLYLILSSHQLRRKPSYLFIGSLAGAD
   rat CB2S  (1) MEGCRELELTNGSNGGLEFNPMKEYMILSDAQQIAVAVLCTLMGLLSALENVAVLYLILSSQRLRRKPSYLFIGSLAGAD
   rat CB2L  (1) MAGCRELELTNGSNGGLEFNPMKEYMILSDAQQIAVAVLCTLMGLLSALENVAVLYLILSSQRLRRKPSYLFIGSLAGAD
 mouse CB2   (1) MEGCRETEVTNGSNGGLEFNPMKEYMILSSGQQIAVAVLCTLMGLLSALENMAVLYIILSSRRLRRKPSYLFISSLAGAD
   cow CB2   (1) MEICLKIEAANGSSDGLNFNPMKEYMILSGPQKIAIAVLCTLLGLLSALENLVVLYLIGSSHRLRKKPSYLFIGSLAGAD
Consensus    (1) MEGCRELELTNGSNGGLEFNPMKEYMILS AQQIAVAVLCTLMGLLSALENVAVLYLILSS RLRRKPSYLFIGSLAGAD

                 81                                                                            160
 human CB2  (81) FLASVVFACSFVNFHVFHGVDSKAVFLLKIGSVTMTFTASVGSLLLTAIDRYLCLRYPPSYKALLTRGRALVTLGIMWVL
   rat CB2S (81) FLASVIFACNFVIFHVFHGVDSRNIFLLKIGSVTMTFTASVGSLLLTAVDRYLCLCYPPTYKALVTRGRALVALGVMWVL
   rat CB2L (81) FLASVIFACNFVIFHVFHGVDSRNIFLLKIGSVTMTFTASVGSLLLTAVDRYLCLCYPPTYKALVTRGRALVALGVMWVL
 mouse CB2  (81) FLASVIFACNFVIFHVFHGVDSNAIFLLKIGSVTMTFTASVGSLLLTAVDRYLCLCYPPTYKALVTRGRALVALCVMWVL
   cow CB2  (81) FLASVVFASSFVHFHVFDGVDSKAVFLLKIGSVTLTFTASLGSLLLTAIDRYLCLRYPPTYKALLTRRRALVTLGIMWVL
Consensus   (81) FLASVIFACSFVIFHVFHGVDSKNIFLLKIGSVTMTFTASVGSLLLTAIDRYLCLCYPPTYKALVTRGRALVALGVMWVL

                 161                                                                           240
 human CB2 (161) SALVSYLPLMGWTCCPR--PCSELFPLIPNDYLLSWLLFIAFLFSGIIYTYGHVLWKAHQHVASLSGHQDR---------
   rat CB2S (161) SALISYLPLMGWTCCPS--PCSELFPLIPNDYLLGWLLFIAILFSGIIYTYGYVLWKAHQHVASLTEHLDR---------
   rat CB2L (161) SALISYLPLMGWTCCPS--PCSELFPLIPNDYLLGWLLFIAILFSGIIYTYGYVLWKAHQHVASLTEHQDR---------
 mouse CB2 (161) SALISYLPLMGWTCCPS--PCSELFPLIPNDYLLGWLLFIAILFSGIIYTYGYVLWKAHRHVATLAEHQDR---------
   cow CB2 (161) AALVSYLPLMGWTCCPR--PCSELFPLIPNDYLLGWLLFIAALFAGIIYTYAHVLWKAHQHVASLAEHRDR---------
Consensus  (161) SALISYLPLMGWTCCPS  PCSELFPLIPNDYLLGWLLFIAILFSGIIYTYGYVLWKAHQHVASLTEHQDR

                 241                                                                           320
 human CB2 (230) ----QVPGMARMRLDVRLAKTLGLVLAVLLICWFPVLALMAHSLATTLSDQVKKAFAFCSMLCLINSMVNPVIYALRSGE
   rat CB2S (230) ----QVPGIARMRLDVRLAKTLGLVMAVLLICWFPALALMGHSLVTTLSDKVKEAFAFCSMLCLVNSMVNPIIYALRSGE
   rat CB2L (230) ----QVPGIARMRLDVRLAKTLGLVMAVLLICWFPALALMGHSLVTTLSDKVKEAFAFCSMLCLVNSMINPIIYALRSGE
 mouse CB2 (230) ----QVPGIARMRLDVRLAKTLGLVLAVLLICWFPALALMGHSLVTTLSDQVKEAFAFCSMLCLVNSMVNPIIYALRSGE
   cow CB2 (230) ----HLSGIARMRLDVRLAKTLGMLLAVLFIFWFPVLALMVYSLGARLSDQVKKVFAFCSLLCLVNSMVNPIIYALRSGE
Consensus  (241)     QVPGIARMRLDVRLAKTLGLVLAVLLICWFPALALMGHSLVTTLSDQVKEAFAFCSMLCLVNSMVNPIIYALRSGE

                 321                                                                           400
 human CB2 (306) IRSSAHHCLAHWKKCVRGLGSEAKEEAPRSSVTETEADGKITPWPDSRDLDLSDC------------------------
   rat CB2S (306) IRSAAQHCLTGWKKYLQGLGSEGKEEAPKSSVTETEAEVKTTTGPGSRTPGCSNC------------------------
   rat CB2L (306) IRSAAQHCLTGWKKYLQGLGSEGKEEAPKSSVTETEAETLVLKDKQELGGDCLLRTSSIHSPMLSLADSANRQDVRPHCP
 mouse CB2 (306) IRSAAQHCLIGWKKYLQGLGPEGKEEGPRSSVTETEADVKTT--------------------------------------
   cow CB2 (306) IRSSAHHRLARWKKCVRGLGPEGKGEIPRSSVTETEADVKTTPGLDSRELSWPDEL------------------------
Consensus  (321) IRSAAQHCL GWKKYLQGLGSEGKEEAPRSSVTETEADVKTT     SR

                 401                    425
 human CB2 (361) -------------------------
   rat CB2S (361) -------------------------
   rat CB2L (386) EELTWWCSVRRPISLPNKAGQSTLL
 mouse CB2 (348) -------------------------
   cow CB2 (362) -------------------------
Consensus  (401)
```

Fig. 2 (continued)

6 Structural Characteristics of the CB_2 Receptor

CB_2 receptors also belong to GPCR family 1a. Site-directed mutagenesis and receptor modeling studies suggested that, unlike other GPCRs, where the DRY motif and A244 in TM6 are important for receptor activation and where mutation of these residues lead to constitutive activity, mutagenesis of D130 in the DRY motif and A244 of the CB_2 receptor only abolishes ligand binding and no constitutive activity was observed (Feng and Song 2003). In contrast, C313 and C320 in the human CB_2 receptor are important for functional receptor coupling to adenylyl cyclase but not for ligand binding affinity. In addition, Y299 in the NPVIY motif of TM7 appears to be important for ligand binding and receptor function. It has also been demonstrated that the human CB_2 receptor undergoes agonist-induced phosphorylation of S352, which promotes its desensitization and internalization (Bouaboula et al. 1999b). Interestingly, this residue is lacking in mouse CB_2.

CB_2 receptor sequences are less conserved throughout evolution than those of CB_1 receptors, with the overall sequence homology between mammals including human, cattle, rat (short isoform) and mouse about 70% (Fig. 2b). The mouse and rat CB_2 receptors are 90% identical, but they are less homologous to the human CB_2 receptor, sharing 80% and 81% identity with the human receptor, respectively.

7 CB_1 and CB_2 Receptor Localization

The distribution and subcellular localization of CB_1 receptors are discussed at length in the chapter "Endocannabinoid Receptors: CNS Localization of the CB_1 Cannabinoid Receptor" by István Katona in this volume and so will not be further considered here. The CB_2 receptor was originally described as a "peripheral" cannabinoid receptor and was found at the highest levels in tissues of the immune system, such as spleen, tonsil, thymus and lymphoid tissues (Galiegue et al. 1995). Accurate assessment of CB_2 expression has been hampered by non-selective antibodies and by the fact that CB_2 expression is highly inducible, for example in cell culture. That is, the presence of CB_2 in a cultured cell does not necessarily imply that CB_2 receptors are found at signaling relevant levels in the native tissue. Thus studies purporting to show the presence of CB_2 by a single technique, particularly in the absence of appropriate controls, must be treated with skepticism. Preferable are studies that show (functional) expression by multiple approaches, for example by antibodies, rt-PCR, in situ hybridization, and/or pharmacological tools. With these caveats in mind, CB_2 mRNA is present in immune cells with a rank order of expression as follows: B cells > macrophage/monocytes > NK cells > T cells (Galiegue et al. 1995). Recently, CB_2 expression has been reported in keratinocytes (Ibrahim et al. 2005), gut neurons (Wright et al. 2008), and brainstem (Van Sickle et al. 2005). In addition, CB_2 receptors have been shown to be expressed or up-regulated under pathological states; examples include spinal cord and DRG tissues of animal pain models (Jhaveri et al. 2008; Wotherspoon et al. 2005; Zhang et al. 2003) and human multiple sclerosis CNS tissues (Benito et al. 2007). Evidence has been presented for both a neuronal and microglial localization of these induced CB_2 receptors.

8 Cellular Signaling of CB_1 and CB_2 Receptors

8.1 Inhibition of Adenylyl Cyclase – $G_{i/o}$ Coupling of CB_1 and CB_2 Receptors

CB_1 and CB_2 are both $G_{i/o}$-coupled GPCRs, and their activation leads to the inhibition of adenylyl cyclase and reduction in the production of cAMP (Howlett et al. 2002). If adenylyl cyclase activity is high prior to the activation of cannabinoid receptors, this will result in a decrease in cAMP levels. In practice for adenylyl cyclase assays measuring the activity of $G_{i/o}$-coupled GPCRs, the intracellular cAMP level is first raised by forskolin, an adenylyl cyclase activator, or a G_s-receptor agonist such as secretin or isoproterenol, allowing a sufficient assay window for measuring reduction of cAMP levels upon the activation of a $G_{i/o}$-coupled GPCRs.

8.2 *Cannabinoid Receptor Activation of MAP Kinases*

Activation of CB_1 and CB_2 receptors reliably leads to the activation of mitogen-activated protein kinases, particularly the extracellular signal-regulated kinases (ERK1/2) through a pertussis toxin-sensitive $G_{i/o}$ pathway (Howlett et al. 2002). In addition, Jnk and p38 MAP kinases are activated by these receptors (Howlett 2005).

8.3 *Crosstalk Between Cannabinoid and Other Receptors*

Crosstalk between the MAP kinase signaling pathways mediated by CB_2 receptor activation and MAP kinase activity evoked by other $G_{i/o}$-dependent receptors has been observed, as the CB_2 inverse agonist SR1,44,528 has been shown to inhibit the MAP kinase activity induced by other $G_{i/o}$-dependent receptors, such as a lysophosphatidic acid receptor (Bouaboula et al. 1999a). It is hypothesized that crosstalk between distinct signaling pathways that convergent to the activation of MAP kinase is possibly achieved by altering the stoichiometry of $G_{i/o}$ proteins that are available to other GPCRs when the $CB_2/G_{i/o}$ complex is promoted and stabilized by CB_2 receptor inverse agonists. Over-expression of CB_2 receptors can also alter modulation of ion channels by other $G_{i/o}$-linked GPCRs (Felder et al. 1995). Similar phenomena have been observed for CB_1 receptor attenuating modulation of calcium channels and MAP kinase by other $G_{i/o}$-linked receptors (Canals and Milligan 2008; Vasquez and Lewis 1999).

8.4 *Transactivation Between Cannabinoid Receptors and Tyrosine Kinase Receptors*

Transactivation of receptor tyrosine kinases (RTKs) is a frequent point of crosstalk between GPCR and RTK signaling and might be responsible for some of the growth-promoting effects of GPCR agonists. CB_1 receptors have been reported to transactivate TrkB (BDNF) receptors. CB_1/TrkB transactivation mediates endocannabinoid-induced chemotaxis in the absence of BDNF (Berghuis et al. 2005). Transactivation between CB_2 receptors and RTKs has not been reported, but likely occurs.

8.5 Cannabinoid Receptor-Mediated Modulation of Ion Channels

Most $G_{i/o}$-coupled receptors also inhibit a subset of voltage-gated calcium channels and activate inwardly rectifying potassium (GIRK) channels. CB_1 receptors follow this paradigm (Mackie et al. 1995). The marked presynaptic localization of CB_1 receptors in close proximity to voltage-gated calcium channels suggests that a major mode of action of CNS CB_1 receptors is the modulation of synaptic transmission (Nyiri et al. 2005). As discussed in the chapter "Endocannabinoid Signaling in Neural Plasticity" by Alger in this volume, this appears to be the case. CB_1 activation of GIRK channels is observed in heterologous expression systems (Mackie et al. 1995) and is likely in at least some neurons (Bacci et al. 2004; Kreitzer et al. 2002). The situation is more complicated with CB_2 receptors. One report examining transfected CB_2 receptor modulation of endogenous calcium and GIRK channels in AtT20 cells did not find effects of CB_2 agonists on these channels (but expression of CB_2 receptors did disrupt signaling of other GPCRs, the latter effect consistent with G protein sequestering (see above)) (Felder et al. 1995). However, another report examining over-expression of both CB_2 receptors and GIRKs in *Xenopus* oocytes did find CB_2-mediated activation of GIRK currents (Ho et al. 1999), suggesting that under some conditions CB_2 is capable of activating GIRK channels. With the likely presence of CB_2 in some neurons under some conditions, it will be important to determine if CB_2 can directly modulate ion channels.

9 Implications of Constitutive Receptor Activity, Protean Agonism, and Inverse Agonism

Receptor constitutive activity refers to the ability of a receptor to activate G proteins and downstream signaling pathways in the absence of agonist. It is generally believed that constitutive activity is due to receptors spontaneously assuming an active conformation in the absence of an agonist. However, one needs to keep in mind that endogenous ligands, if present in the tissues studied, will produce a similar effect in the absence of added ligand, an issue particularly relevant for lipid receptors where their ligands may be continuously produced in the course of membrane turnover or remodeling (Gbahou et al. 2003). Thus, constitutive activity means that a fraction of receptors are actively signaling in the absence of an agonist. The level of receptor constitutive activity is dependent upon the system, including factors such as receptor expression levels, cellular environment and the conditions of cell growth (Yao et al. 2006). The high levels of CB_1 expression in a variety of neurons means that constitutive activity of this receptor may be relevant in the clinical use of CB_1 inverse agonists. Receptor constitutive activity can be revealed by the use of inverse agonist ligands, as these ligands

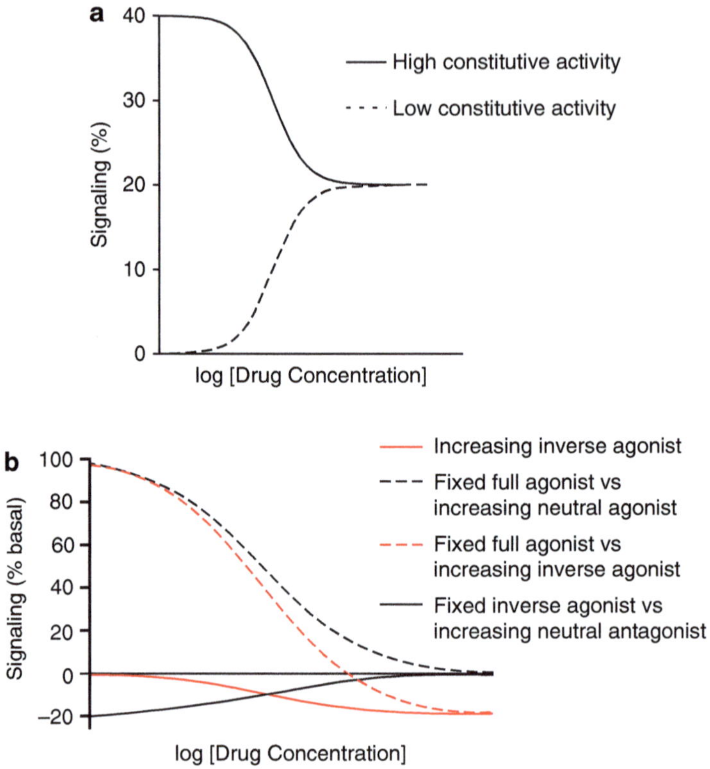

Fig. 3 *Protean agonism and inverse agonism.* (**a**) *Protean agonism.* In this example, the protean agonist is assumed to have an efficacy of 20%. In the case of high constitutive activity (40%), increasing concentrations of the protean agonist will decrease the observed signaling, appearing to be an inverse agonist. If the constitutive activity is low (0%) increasing concentrations of the protean agonist will increase the signaling, appearing to be an (partial) agonist. Note that if the baseline constitutive activity is 20%, the protean agonist will behave as a neutral antagonist. (**b**) *Interactions between a full agonist, neutral antagonist, and inverse agonist.* Increasing concentrations of a neutral antagonist will reverse the positive efficacy of an agonist (*dashed black line*) or the negative efficacy of an inverse agonist (*solid black line*), returning the system to its basal level. Increasing concentrations of an inverse agonist (*dashed red line*) will reverse the effect of a full agonist, eventually leading to negative efficacy. Increasing concentrations of an inverse agonist in the absence of other ligands (*solid red line*) will inhibit basal signaling activity, causing negative efficacy

stabilize a receptor conformation that promotes a lower activation state than the resting state, resulting in an apparent negative efficacy. It is worth noting that the apparent efficacy of a ligand is dependent upon the level of receptor constitutive activity of the receptor in the assay system used. Therefore, a partial agonist in one system can behave as an antagonist or an inverse agonist in others (that is, it can be a protean agonist – Fig. 3a). GTPγS and adenylyl cyclase assays are often used to evaluate receptor constitutive activity and for characterization of inverse agonists. A true neutral antagonist will block both agonist as well as inverse agonist activity, independent of the level of receptor constitutive activity (Fig. 3b). It has been speculated that there are very few true neutral antagonists for GPCRs (Kenakin 2004). Most apparent neutral antagonists are low affinity inverse agonists or their neutral antagonism is specific to the assay system in which they are characterized (Bond and Ijzerman 2006). Thus, it is important when characterizing a putative neutral antagonist that a variety of different conditions (that is, varying levels of receptor expression and second messenger systems) are evaluated.

Receptor constitutive activity is a physiologically and/or pathologically important phenomenon. The constitutive activity for the CB_2 receptor, although not extensively studied in tissues, has been demonstrated in recombinant cell lines expressing the CB_2 receptor (Yao et al. 2006) For example, SR1,44,528 has been shown to potentiate the gene expression induced by forskolin-induced cAMP responsive element (Portier et al. 1999), and in addition, AM630 produced a further increase in the forskolin-induced cAMP level (Ross et al. 1999), indicating constitutive activity of CB_2 in these recombinant systems that is readily reversed by the inverse agonists SR1,44,528 and AM630.

As mentioned above, protean agonists describe a group of ligands that behave as agonists in one system but as inverse agonists or neutral antagonists in another. For example, AM1,241 behaves as a partial agonist, neutral antagonist or inverse agonist at CB_2 receptors depending on the assay systems employed and assay conditions used (Yao et al. 2006).

10 Cannabinoid Receptor Ligands

10.1 Non-Selective CB_1/CB_2 Receptor Agonists

There are four well-developed classes of cannabinoid receptor agonist (Howlett et al. 2002): the classical cannabinoids, non-classical cannabinoids, aminoalkylindoles and eicosanoids. Classical cannabinoids are ABC-tricyclic benzopyrans. Classical cannabinoids may be found in nature, such as Δ^9-tetrahydrocannabinol (Δ^9THC) or may be synthetic, such as HU210 (11-hydroxy-Δ^8-tetrahydrocannabinol-dimethylheptyl) or DALN (desacetyl-levo-nantradol). Non-classical cannabinoids arose from extensive SAR work conducted at Pfizer thirty years ago. These compounds are characterized by the opening of the dihydropyran ring. Many of these compounds have high affinity for both CB_1 and CB_2 receptors. Of these, CP55,940 has played a major role in defining CB_1 receptor localization and function (Herkenham et al. 1991). Recently, compounds in the naphthalene class have reported to be potent agonists at CB_1 and CB_2 receptors with limited brain penetration (Dziadulewicz et al. 2007).

The classical and non-classical cannabinoids are structurally related to Δ^9THC, varying primarily in side chain modifications, some of which substantially increase receptor affinity (for example, the 3-dimethyl heptyl analogs). However, the next class of cannabinoid receptor ligands, the aminoalkylindoles, were developed as anti-inflammatory drugs and analgesics, and were only subsequently found to be cannabinoid receptor agonists (at both CB_1 and CB_2 receptors) (Compton et al. 1992). Of the aminoalkylindoles, WIN55,212-2 is the most frequently encountered. As discussed below, aminoalkylindoles have provided a route towards the synthesis of relatively selective CB_2 agonists. Not unexpectedly, given their structural differences from other cannabinoid receptor agonists, aminoalkylindoles bind to CB_1 receptors in a slightly different fashion (but still in a displaceable manner) than the other well-characterized CB_1 receptor agonists (Song and Bonner 1996). Consistent with this, WIN55,212-2 activation of CB_1 receptors promotes a different repertoire of cellular events (Compton et al. 1992), a fact that must be kept in mind when evaluating experiments performed with (high concentrations of) this agonist.

The final group of CB_1 receptor ligands are the eicosanoids. These eicosanoid derivatives collectively form the group of compounds known as endogenous cannabinoids (endocannabinoids) (Freund et al. 2003). The synthesis and degradation of the endocannabinoids is discussed in the chapter "The Life Cycle of the Endocannabinoids: Formation and Inactivation" by Alexander and Kendall, this volume. There are two major classes of endocannabinoids, the acylethanolamides and the acylesters. The prototypic member of the acylethanolamide family is *N*-arachidonoylethanolamine (anandamide, AEA) (Devane et al. 1992). However, a number of additional acylethanolamides, varying in chain length or extent of acyl chain saturation are found in vivo and have activity at CB_1 receptors (Felder et al. 1993). A hallmark of the acylethanolamides is that they have relatively low intrinsic efficacy at CB_1 receptors. An extensive literature exists on the SAR of acylethanolamides for CB_1 (Lin et al. 1998; Reggio 2002; Ryan et al. 1997). In general, a shorter acyl chain and decreasing degree of saturation leads to lower affinity. 2-Arachidonoyl glycerol (2AG) (and its 1/3 isomer) is the only acylester extensively studied (Stella et al. 1997; Sugiura et al. 1995). While 2AG's affinity for CB_1 receptors is similar to that of AEA, it is consistently found to have a higher intrinsic efficacy (Luk et al. 2004). In addition to the acylamides and esters, additional eicosanoid compounds have been reported to be endogenous CB_1 agonists. Two of these, virodhamine and noladin ether, were initially reported to be present in brain; however later studies have failed to consistently verify these initial reports (Richardson et al. 2007). In addition, there are a large number of acyl amino acid conjugates that have been reported to have varying efficacy at CB_1 receptors (Bradshaw and Walker 2005).

10.2 CB_1 Receptor Antagonists

The first and most extensively studied class of CB_1 antagonists is the 1,5-diarylpyrazoles, typified by rimonabant (SR141,716A) (Howlett et al. 2002). Other widely encountered members of this family include AM251 and AM281 (Howlett et al. 2002). These compounds generally show 100–1000-fold selectivity for CB_1 over CB_2 (depending on the assay system). They are also inverse agonists. Another early CB_1 antagonist is the substituted benzofuran, LY320,135. While much less studied than rimonabant, it has a lower affinity for CB_1 than rimonabant, but like rimonabant it shows strong selectivity for CB_1 and is an inverse agonist (Felder et al. 1998).

The ability of CB_1 antagonists to depress food consumption and promote weight loss has lead to robust efforts among pharmaceutical companies to develop additional CB_1 antagonists (Black 2004). Other than rimonabant, the compound furthest along in clinical development is Merck's substituted acyclic sulfonamide, taranabant or MK0364 (Addy et al. 2008). A Pfizer compound, CP945,598, has also been used in multiple clinical trials. Another antagonist that has been tested in man is the 3,4-diaryl-4,5-dihydropyrazole (SLV-319) (Foloppe et al. 2008).

The compounds discussed above all show inverse agonism under appropriate assay conditions and it has been hypothesized that some of the adverse effects of rimonabant and taranabant might be mediated by inverse agonism. In this regard it is interesting that a pyrazole analog, AM4113, which has high affinity for CB_1 receptors, does not show inverse agonism in the adenylyl cyclase assay but does suppress food intake and may have a lower incidence of pro-emetic behaviors (Bergman et al. 2008; Chambers et al. 2007; Sink et al. 2008). Whether neutral antagonists of CB_1 will have a therapeutic advantage over CB_1 inverse agonists is speculative and remains to be determined.

All of the CB_1 ligands described above are small, lipophilic molecules. However, a recent report identified the endogenous peptide, hemopressin, to be a novel CB_1 receptor inverse agonist (Heimann et al. 2007). The implications of this observation are profound and if these findings are confirmed they will force a re-thinking of the control CB_1 receptor function.

10.3 CB_2 Receptor Agonists

Significant efforts have focused on generating CB_2 receptor selective agonists as potential therapeutic agents, as it is believed that selective activation of CB_2 receptors will produce anti-inflammation, analgesia and other therapeutic benefits without the undesirable CNS side effects thought to be mainly mediated by the activation of CB_1 receptors. Many synthetic CB_2 receptor agonists have been developed with significant (but not absolute) selectivity over the CB_1 receptor. They can be divided into several classes according to their structures. Indoles represented by AM1,241 are thoroughly studied, and have been characterized in a variety of in vivo animal models to demonstrate CB_2-mediated efficacies. Although AM1,241-evoked analgesic efficacy has been reported to involve the μ-opioid receptor (Ibrahim et al. 2005), this phenomenon is not a consistent characteristic of CB_2-mediated analgesia, as other CB_2 receptor selective agonists in the class (A-796,260, A-836,339 (Yao et al. 2008; Yao et al. 2009) and L-768,242, as well as GW405,833) do not share this property (Whiteside et al. 2005). A class of synthetic Δ^9THC derivatives that is quite selective for CB_2 receptors emerged from SAR-based structural design. One of the well-characterized ligands in this class is JWH-133 (Marriott and Huffman 2008). JWH-133 has been shown to have anti-spasticity efficacy in animal models of multiple sclerosis (Baker et al. 2000). However, due to less than perfect selectivity, the effects are likely to be at least partly mediated by CB_1 receptors (Pryce and Baker 2007). Thiazolylidine compounds, such as the Taisho compounds (Ohta et al. 2008), and A-8,36,339 demonstrated excellent selectivity over the CB_1 receptor and have been shown to be efficacious in in vivo analgesic models (Yao et al. 2009).

10.4 CB_2 Receptor Antagonists

The most widely used CB_2 receptor selective antagonists are SR1,44,528, a pyrazole, and AM630, an indole. In in vitro pharmacological studies SR1,44,528 and AM630 have been shown to block CB_2 receptor activation by selective agonists (Rinaldi-Carmona et al. 1998; Shire et al. 1999). In addition, in in vivo studies these antagonists block CB_2 receptor-mediated actions (Ibrahim et al. 2005; Yao et al. 2008). JTE-907, a quinolinone-3-carboxamide, has been shown to be an inverse agonist at the CB_2 receptor (Ueda et al. 2005). The triaryl bis-sulfones (SCH2,26,336) are a new class of CB_2 antagonist (Lavey et al. 2005). Both JTE-907 and SCH2,26,336 have been shown to have anti-inflammatory effects (Lavey et al. 2005; Ueda et al. 2005). SCH2,26,336 has been radiolabeled and [^{35}S] SCH2,26,336 has been used in in vitro pharmacological characterization of the CB_2 receptor, as well as localization of CB_2 receptors by autoradiography in tissue sections (Gonsiorek et al. 2006).

10.5 Allosteric Modulators of Cannabinoid Receptors

The preceding discussion has focused on orthosteric ligands of the cannabinoid receptor. These are ligands that interact directly with the binding site whose occupancy activates the receptor. Another class of molecules that interact with receptors are allosteric modulators. These compounds bind to sites on the receptor distinct from the orthosteric binding site but induce conformational changes in the receptor that alter the properties of orthosteric ligands. A well-known example of an allosteric modulator would be a benzodiazepine acting on the $GABA_A$ receptor. Allosteric modulators of receptor function are potentially exciting therapeutic targets as they alter the function of endogenous ligands and may bypass some of disadvantages of orthostatic ligands (desensitization, up-regulation, etc.). Two families of allosteric modulators have been described for CB_1 receptors (Horswill et al. 2007; Price et al. 2005). No allosteric modulators of CB_2 receptors have been published. This is an active area of research and advances can be expected over the next few years.

11 Non-CB_1/Non-CB_2 Receptors

11.1 GPR55

The persistence of cannabinoid effects in CB_1 and/or CB_2 knockout mice suggests the existence of additional cannabinoid receptors (Begg et al. 2005). In addition, strong pharmacological evidence supports the presence of a vascular cannabinoid receptor distinct from CB_1 or CB_2 (Begg et al. 2005). Evidence has emerged over the past several years that GPR55 may be one such receptor. Although some controversy remains, this receptor can be formally considered a cannabinoid receptor based on its activation by anandamide and Δ^9THC at low micromolar concentrations (Lauckner et al. 2008; Ryberg et al. 2007; Waldeck-Weiermair et al. 2008). In addition, lysophosphatidylinositol (LPI), an endogenous lipid mediator, also activates this receptor (Lauckner et al. 2008; Oka et al. 2007; Waldeck-Weiermair et al. 2008). However, LPI is not a specific GPR55 agonist as it also activates TRPM8 at concentrations reported to activate GPR55 (Andersson et al. 2007). GPR55 stimulation releases calcium from intracellular stores via phospholipase C (Lauckner et al. 2008; Waldeck-Weiermair et al. 2008) and, in some cases, activates ERK1/2 MAP kinase (Oka et al. 2007; Waldeck-Weiermair et al. 2008). GPR55 mRNA is widely distributed at moderate to low levels in the CNS and is also found in the vasculature and other peripheral tissues (Ryberg et al. 2007). While GPR55 appears to fulfill the criteria of a cannabinoid receptor, its pharmacology is inconsistent with several of the non-CB_1/non-CB_2 effects mentioned above. Therefore, additional cannabinoid receptors clearly remain to be identified.

11.2 Interactions of Cannabinoids with Ion Channels

Numerous cannabinoids and cannabinoid receptor ligands have been found to interact with various ligand-gated and voltage-gated ion channels, typically in the low micromolar range (Akopian et al. 2008; Barann et al. 2002; Maingret et al. 2001; Oz et al. 2004; Poling et al. 1996; Ross 2003). While these interactions may have physiological relevance under some conditions, this topic is beyond the scope of the current review and the interested reader is referred to an excellent recent review (Oz 2006).

12 Conclusions

The most direct route to manipulate the endocannabinoid system is by engaging cannabinoid receptors with agonists or antagonists. However, in order to understand and interpret these interactions, a basic familiarity with the principles of receptor pharmacology, including selectivity, efficacy, functional selectivity, and allosteric modulation, is necessary. The past thirty years have seen a proliferation of CB_1 and CB_2 agonists and antagonists. A few of these, for example mixed CB_1/CB_2 agonists (Δ^9THC) and CB_1 antagonists, have therapeutic efficacy in man. Others, such as CB_2 agonists, have considerable therapeutic promise based on preclinical studies. Finally, non-orthosteric ligands, such as allosteric modulators, offer intriguing therapeutic possibilities. Certainly, the next few years will be a rich and exciting time for cannabinoid receptor pharmacology.

References

Addy C et al. (2008) The acyclic CB1R inverse agonist taranabant mediates weight loss by increasing energy expenditure and decreasing caloric intake. Cell Metab 7(1):68–78

Akopian AN et al. (2008) Cannabinoids desensitize capsaicin and mustard oil responses in sensory neurons via TRPA1 activation. J Neurosci 28(5):1064–1075

Anday JK, Mercier RW (2005) Gene ancestry of the cannabinoid receptor family. Pharmacol Res 52(6):463–466

Andersson H et al. (2003) Membrane assembly of the cannabinoid receptor 1: impact of a long N-terminal tail. Mol Pharmacol 64(3):570–577

Andersson DA, Nash M, Bevan S (2007) Modulation of the cold-activated channel TRPM8 by lysophospholipids and polyunsaturated fatty acids. J Neurosci 27(12):3347–3355

Bacci A, Huguenard JR, Prince DA (2004) Long-lasting self-inhibition of neocortical interneurons mediated by endocannabinoids. Nature 431(7006):312–316

Baker D et al. (2000) Cannabinoids control spasticity and tremor in a multiple sclerosis model. Nature 404(6773):84–87

Barann M et al. (2002) Direct inhibition by cannabinoids of human 5-HT3A receptors: probable involvement of an allosteric modulatory site. Br J Pharmacol 137(5):589–596

Begg M et al. (2005) Evidence for novel cannabinoid receptors. Pharmacol Ther 106(2):133–145

Benito C et al. (2007) Cannabinoid CB1 and CB2 receptors and fatty acid amide hydrolase are specific markers of plaque cell subtypes in human multiple sclerosis. J Neurosci 27(9):2396–2402

Berghuis P et al. (2005) Endocannabinoids regulate interneuron migration and morphogenesis by transactivating the TrkB receptor. Proc Natl Acad Sci USA 102(52):19115–19120

Bergman J et al. (2008) Some effects of CB1 antagonists with inverse agonist and neutral biochemical properties. Physiol Behav 93(4–5):666–670

Black SC (2004) Cannabinoid receptor antagonists and obesity. Curr Opin Investig Drugs 5(4):389–394

Bockaert J, Pin JP (1999) Molecular tinkering of G protein-coupled receptors: an evolutionary success. Embo J 18(7):1723–1729

Bond RA, Ijzerman AP (2006) Recent developments in constitutive receptor activity and inverse agonism, and their potential for GPCR drug discovery. Trends Pharmacol Sci 27(2):92–96

Bonhaus DW et al. (1998) Dual activation and inhibition of adenylyl cyclase by cannabinoid receptor agonists: evidence for agonist-specific trafficking of intracellular responses. J Pharmacol Exp Ther 287(3):884–888

Bouaboula M et al. (1999a) Gi protein modulation induced by a selective inverse agonist for the peripheral cannabinoid receptor CB2: implication for intracellular signalization cross-regulation. Mol Pharmacol 55(3):473–480
Bouaboula M, Dussossoy D, Casellas P (1999b) Regulation of peripheral cannabinoid receptor CB2 phosphorylation by the inverse agonist SR 144528. Implications for receptor biological responses. J Biol Chem 274(29):20397–20405
Bradshaw HB, Walker JM (2005) The expanding field of cannabimimetic and related lipid mediators. Br J Pharmacol 144(4):459–465
Breivogel CS et al. (2001) Evidence for a new G protein-coupled cannabinoid receptor in mouse brain. Mol Pharmacol 60(1):155–163
Brown SM, Wager-Miller J, Mackie K (2002) Cloning and molecular characterization of the rat CB2 cannabinoid receptor. Biochim Biophys Acta 1576(3):255–264
Canals M, Milligan G (2008) Constitutive activity of the cannabinoid CB1 receptor regulates the function of co-expressed Mu opioid receptors. J Biol Chem 283(17):11424–11434
Chambers AP et al. (2007) A neutral CB1 receptor antagonist reduces weight gain in rat. Am J Physiol Regul Integr Comp Physiol 293(6):R2185–R2193
Compton DR et al. (1992) Aminoalkylindole analogs: cannabimimetic activity of a class of compounds structurally distinct from delta 9-tetrahydrocannabinol. J Pharmacol Exp Ther 263(3):1118–1126
Devane WA et al. (1992) Isolation and structure of a brain constituent that binds to the cannabinoid receptor. Science 258(5090):1946–1949
Dziadulewicz EK et al. (2007) Naphthalen-1-yl-(4-pentyloxynaphthalen-1-yl) methanone: a potent, orally bioavailable human CB1/CB2 dual agonist with antihyperalgesic properties and restricted central nervous system penetration. J Med Chem 50(16):3851–3856
Ehlers CL et al. (2007) Association between single nucleotide polymorphisms in the cannabinoid receptor gene (CNR1) and impulsivity in southwest California Indians. Twin Res Hum Genet 10(6):805–811
Elphick MR, Egertova M (2005) The phylogenetic distribution and evolutionary origins of endocannabinoid signalling. Handb Exp Pharmacol 168:283–297
Fay JF, Dunham TD, Farrens DL (2005) Cysteine residues in the human cannabinoid receptor: only C257 and C264 are required for a functional receptor, and steric bulk at C386 impairs antagonist SR141716A binding. Biochemistry 44(24):8757–8769
Felder CC et al. (1993) Anandamide, an endogenous cannabimimetic eicosanoid, binds to the cloned human cannabinoid receptor and stimulates receptor-mediated signal transduction. Proc Natl Acad Sci USA 90(16):7656–7660
Felder CC et al. (1995) Comparison of the pharmacology and signal transduction of the human cannabinoid CB1 and CB2 receptors. Mol Pharmacol 48(3):443–450
Felder CC et al. (1998) LY320135, a novel cannabinoid CB1 receptor antagonist, unmasks coupling of the CB1 receptor to stimulation of cAMP accumulation. J Pharmacol Exp Ther 284(1):291–297
Feng W, Song ZH (2003) Effects of D3.49A, R3.50A, and A6.34E mutations on ligand binding and activation of the cannabinoid-2 (CB2) receptor. Biochem Pharmacol 65(7):1077–1085
Foloppe N et al. (2008) Discovery of a novel class of selective human CB1 inverse agonists. Bioorg Med Chem Lett 18(3):1199–1206
Freund TF, Katona I, Piomelli D (2003) Role of endogenous cannabinoids in synaptic signaling. Physiol Rev 83(3):1017–1066
Galandrin S, Oligny-Longpre G, Bouvier M (2007) The evasive nature of drug efficacy: implications for drug discovery. Trends Pharmacol Sci 28(8):423–430
Galiegue S et al. (1995) Expression of central and peripheral cannabinoid receptors in human immune tissues and leukocyte subpopulations. Eur J Biochem 232(1):54–61
Garcia DE et al. (1998) Protein kinase C disrupts cannabinoid actions by phosphorylation of the CB1 cannabinoid receptor. J Neurosci 18(8):2834–2841

Gbahou F et al. (2003) Protean agonism at histamine H3 receptors in vitro and in vivo. Proc Natl Acad Sci USA 100(19):11086–11091
Gonsiorek W et al. (2006) Characterization of peripheral human cannabinoid receptor (hCB2) expression and pharmacology using a novel radioligand, [^{35}S]Sch225336. J Biol Chem 281 (38):28143–28151
Govaerts SJ et al. (2004) Characterization of the pharmacology of imidazolidinedione derivatives at cannabinoid CB1 and CB2 receptors. Eur J Pharmacol 495(1):43–53
Griffin G, Tao Q, Abood ME (2000) Cloning and pharmacological characterization of the rat CB2 cannabinoid receptor. J Pharmacol Exp Ther 292(3):886–894
Hamdani N et al. (2008) The CNR1 gene as a pharmacogenetic factor for antipsychotics rather than a susceptibility gene for schizophrenia. Eur Neuropsychopharmacol 18(1):34–40
Heimann AS et al. (2007) Hemopressin is an inverse agonist of CB1 cannabinoid receptors. Proc Natl Acad Sci USA 104(51):20588–20593
Herkenham M et al. (1991) Characterization and localization of cannabinoid receptors in rat brain: a quantitative in vitro autoradiographic study. J Neurosci 11(2):563–583
Ho BY et al. (1999) Coupling of the expressed cannabinoid CB1 and CB2 receptors to phospholipase C and G protein-coupled inwardly rectifying K + channels. Receptor Channel 6(5):363–374
Hoenicka J et al. (2007) Association in alcoholic patients between psychopathic traits and the additive effect of allelic forms of the CNR1 and FAAH endocannabinoid genes, and the 3' region of the DRD2 gene. Neurotox Res 11(1):51–60
Horswill JG et al. (2007) PSNCBAM-1, a novel allosteric antagonist at cannabinoid CB1 receptors with hypophagic effects in rats. Br J Pharmacol 152(5):805–814
Howlett AC (2005) Cannabinoid receptor signaling. Handb Exp Pharmacol 168:53–79
Howlett AC et al. (2002) International Union of Pharmacology. XXVII. Classification of cannabinoid receptors. Pharmacol Rev 54(2):161–202
Hsieh C et al. (1999) Internalization and recycling of the CB1 cannabinoid receptor. J Neurochem 73(2):493–501
Ibrahim MM et al. (2005) CB2 cannabinoid receptor activation produces antinociception by stimulating peripheral release of endogenous opioids. Proc Natl Acad Sci USA 102 (8):3093–3098
Jhaveri MD et al. (2008) Evidence for a novel functional role of cannabinoid CB receptors in the thalamus of neuropathic rats. Eur J NeuroSci 27(9):2482
Jin W et al. (1999) Distinct domains of the CB1 cannabinoid receptor mediate desensitization and internalization. J Neurosci 19(10):3773–3780
Karsak M et al. (2005) Cannabinoid receptor type 2 gene is associated with human osteoporosis. Hum Mol Genet 14(22):3389–3396
Kenakin T (2004) Efficacy as a vector: the relative prevalence and paucity of inverse agonism. Mol Pharmacol 65(1):2–11
Kreitzer AC, Carter AG, Regehr WG (2002) Inhibition of interneuron firing extends the spread of endocannabinoid signaling in the cerebellum. Neuron 34(5):787–796
Lauckner JE, Hille B, Mackie K (2005) The cannabinoid agonist WIN55, 212-2 increases intracellular calcium via CB1 receptor coupling to Gq/11 G proteins. Proc Natl Acad Sci USA 102 (52):19144–19149
Lauckner JE et al. (2008) GPR55 is a cannabinoid receptor that increases intracellular calcium and inhibits M current. Proc Natl Acad Sci USA 105(7):2699–2704
Lavey BJ et al. (2005) Triaryl bis-sulfones as a new class of cannabinoid CB2 receptor inhibitors: identification of a lead and initial SAR studies. Bioorg Med Chem Lett 15(3):783–786
Lin S et al. (1998) Novel analogues of arachidonylethanolamide (anandamide): affinities for the CB1 and CB2 cannabinoid receptors and metabolic stability. J Med Chem 41 (27):5353–5361
Luk T et al. (2004) Identification of a potent and highly efficacious, yet slowly desensitizing CB1 cannabinoid receptor agonist. Br J Pharmacol 142(3):495–500

Mackie K et al. (1995) Cannabinoids activate an inwardly rectifying potassium conductance and inhibit Q-type calcium currents in AtT20 cells transfected with rat brain cannabinoid receptor. J Neurosci 15(10):6552–6561
Maingret F et al. (2001) The endocannabinoid anandamide is a direct and selective blocker of the background K(+) channel TASK-1. Embo J 20(1–2):47–54
Marriott KS, Huffman JW (2008) Recent advances in the development of selective ligands for the cannabinoid CB(2) receptor. Curr Top Med Chem 8(3):187–204
Martini L et al. (2007) Ligand-induced down-regulation of the cannabinoid 1 receptor is mediated by the G-protein-coupled receptor-associated sorting protein GASP1. Faseb J 21(3):802–811
Matsuda LA et al. (1990) Structure of a cannabinoid receptor and functional expression of the cloned cDNA. Nature 346(6284):561–564
Mauler F et al. (2002) Characterization of the diarylether sulfonylester (-)-(R)-3-(2-hydroxymethylindanyl-4-oxy) phenyl-4, 4, 4-trifluoro-1-sulfonate (BAY 38-7271) as a potent cannabinoid receptor agonist with neuroprotective properties. J Pharmacol Exp Ther 302(1):359–368
McAllister SD et al. (2002) A critical role for a tyrosine residue in the cannabinoid receptors for ligand recognition. Biochem Pharmacol 63(12):2121–2136
McAllister SD et al. (2003) An aromatic microdomain at the cannabinoid CB(1) receptor constitutes an agonist/inverse agonist binding region. J Med Chem 46(24):5139–5152
McPartland JM, Norris RW, Kilpatrick CW (2007) Coevolution between cannabinoid receptors and endocannabinoid ligands. Gene 397(1–2):126–135
Morello JP, Bouvier M (1996) Palmitoylation: a post-translational modification that regulates signalling from G-protein coupled receptors. Biochem Cell Biol 74(4):449–457
Munro S, Thomas KL, Abu-Shaar M (1993) Molecular characterization of a peripheral receptor for cannabinoid. Nature (London, United Kingdom) 6441:61–65
Niehaus JL et al. (2007) CB1 cannabinoid receptor activity is modulated by the cannabinoid receptor interacting protein CRIP 1a. Mol Pharmacol 72(6):1557–1566
Nyiri G et al. (2005) CB1 cannabinoid receptors are enriched in the perisynaptic annulus and on preterminal segments of hippocampal GABAergic axons. Neuroscience 136(3):811–822
Ohta H et al. (2008) Imine derivatives as new potent and selective CB2 cannabinoid receptor agonists with an analgesic action. Bioorg Med Chem 16(3):1111–1124
Oka S et al. (2007) Identification of GPR55 as a lysophosphatidylinositol receptor. Biochem Biophys Res Commun 362(4):928–934
Oz M (2006) Receptor-independent actions of cannabinoids on cell membranes: focus on endocannabinoids. Pharmacol Ther 111(1):114–144
Oz M et al. (2004) Differential effects of endogenous and synthetic cannabinoids on alpha7-nicotinic acetylcholine receptor-mediated responses in Xenopus Oocytes. J Pharmacol Exp Ther 310(3):1152–1160
Palczewski K et al. (2000) Crystal structure of rhodopsin: A G protein-coupled receptor. Science 289(5480):739–745
Poling JS et al. (1996) Anandamide, an endogenous cannabinoid, inhibits Shaker-related voltage-gated K+ channels. Neuropharmacology 35(7):983–991
Portier M et al. (1999) SR 144528, an antagonist for the peripheral cannabinoid receptor that behaves as an inverse agonist. J Pharmacol Exp Ther 288(2):582–589
Price MR et al. (2005) Allosteric modulation of the cannabinoid CB1 receptor. Mol Pharmacol 68(5):1484–1495
Pryce G, Baker D (2007) Control of spasticity in a multiple sclerosis model is mediated by CB1, not CB2, cannabinoid receptors. Br J Pharmacol 150(4):519–525
Reggio PH (2002) Endocannabinoid structure–activity relationships for interaction at the cannabinoid receptors. Prostaglandins Leukot Essent Fatty Acids 66(2–3):143–160
Richardson D et al. (2007) Quantitative profiling of endocannabinoids and related compounds in rat brain using liquid chromatography-tandem electrospray ionization mass spectrometry. Anal Biochem 360(2):216–226

Rinaldi-Carmona M et al. (1996) Characterization and distribution of binding sites for [^{3}H]-SR 141716A, a selective brain (CB1) cannabinoid receptor antagonist, in rodent brain. Life Sci 58 (15):1239–1247
Rinaldi-Carmona M et al. (1998) Modulation of CB1 cannabinoid receptor functions after a long-term exposure to agonist or inverse agonist in the Chinese hamster ovary cell expression system. J Pharmacol Exp Ther 287(3):1038–1047
Ross RA (2003) Anandamide and vanilloid TRPV1 receptors. Br J Pharmacol 140(5):790–801
Ross RA et al. (1999) Agonist-inverse agonist characterization at CB1 and CB2 cannabinoid receptors of L759633, L759656, and AM630. Br J Pharmacol 126(3):665–672
Ryan WJ et al. (1997) Potent anandamide analogs: the effect of changing the length and branching of the end pentyl chain. J Med Chem 40(22):3617–3625
Ryberg E et al. (2005) Identification and characterisation of a novel splice variant of the human CB1 receptor. FEBS Lett 579(1):259–264
Ryberg E et al. (2007) The orphan receptor GPR55 is a novel cannabinoid receptor. Br J Pharmacol 152(7):1092–1101
Schonbrunn A (2007) Selective agonism in somatostatin receptor signaling and regulation. Mol Cell Endocrinol 286(1–2):35–39
Seifert J et al. (2007) No association of CNR1 gene variations with susceptibility to schizophrenia. Neurosci Lett 426(1):29–33
Selinger Z (2008) Discovery of G protein signaling. Annu Rev Biochem 77:1–13
Shire D et al. (1995) An amino-terminal variant of the central cannabinoid receptor resulting from alternative splicing. J Biol Chem 270(8):3726–3731
Shire D et al. (1996) Structural features of the central cannabinoid CB1 receptor involved in the binding of the specific CB1 antagonist SR 141716A. J Biol Chem 271(12):6941–6946
Shire D et al. (1999) Cannabinoid receptor interactions with the antagonists SR 141716A and SR 144528. Life Sci 65(6–7):627–635
Shoemaker JL et al. (2005) Agonist-directed trafficking of response by endocannabinoids acting at CB2 receptors. J Pharmacol Exp Ther 315(2):828–838
Siegfried Z et al. (2004) Association study of cannabinoid receptor gene (CNR1) alleles and anorexia nervosa: differences between restricting and binging/purging subtypes. Am J Med Genet B Neuropsychiatr Genet 125(1):126–130
Sim LJ, Selley DE, Childers SR (1995) In vitro autoradiography of receptor-activated G proteins in rat brain by agonist-stimulated guanylyl 5'-[gamma-[^{35}S]thio]-triphosphate binding. Proc Natl Acad Sci USA 92(16):7242–7246
Sim LJ et al. (1996) Differences in G-protein activation by mu- and delta-opioid, and cannabinoid, receptors in rat striatum. Eur J Pharmacol 307(1):97–105
Sink KS et al. (2008) The novel cannabinoid CB1 receptor neutral antagonist AM4113 suppresses food intake and food-reinforced behavior but does not induce signs of nausea in rats. Neuropsychopharmacology 33(4):946–955
Sipe JC et al. (2005) Reduced endocannabinoid immune modulation by a common cannabinoid 2 (CB2) receptor gene polymorphism: possible risk for autoimmune disorders. J Leukoc Biol 78(1):231–238
Song ZH, Bonner TI (1996) A lysine residue of the cannabinoid receptor is critical for receptor recognition by several agonists but not WIN55212-2. Mol Pharmacol 49(5):891–896
Stella N, Schweitzer P, Piomelli D (1997) A second endogenous cannabinoid that modulates long-term potentiation. Nature 388(6644):773–778
Sugiura T et al. (1995) 2-Arachidonoylglycerol: a possible endogenous cannabinoid receptor ligand in brain. Biochem Biophys Res Commun 215(1):89–97
Sun Y, McGarrigle D, Huang XY (2007) When a G protein-coupled receptor does not couple to a G protein. Mol Biosyst 3(12):849–854
Tao Q, Abood ME (1998) Mutation of a highly conserved aspartate residue in the second transmembrane domain of the cannabinoid receptors, CB1 and CB2, disrupts G-protein coupling. J Pharmacol Exp Ther 285(2):651–658

Thomas BF et al. (1998) Comparative receptor binding analyses of cannabinoid agonists and antagonists. J Pharmacol Exp Ther 285(1):285–292
Tian X et al. (2005) The conformation, location, and dynamic properties of the endocannabinoid ligand anandamide in a membrane bilayer. J Biol Chem 280(33):29788–29795
Ueda Y et al. (2005) Involvement of cannabinoid CB(2) receptor-mediated response and efficacy of cannabinoid CB(2) receptor inverse agonist, JTE-907, in cutaneous inflammation in mice. Eur J Pharmacol 520(1–3):164–171
Ujike H et al. (2002) CNR1, central cannabinoid receptor gene, associated with susceptibility to hebephrenic schizophrenia. Mol Psychiatry 7(5):515–518
Urban JD et al. (2007) Functional selectivity and classical concepts of quantitative pharmacology. J Pharmacol Exp Ther 320(1):1–13
Valk PJ et al. (1997) The genes encoding the peripheral cannabinoid receptor and alpha-L-fucosidase are located near a newly identified common virus integration site, Evi11. J Virol 71(9):6796–6804
Van Sickle MD et al. (2005) Identification and functional characterization of brainstem cannabinoid CB2 receptors. Science 310(5746):329–332
Vasquez C, Lewis DL (1999) The CB1 cannabinoid receptor can sequester G-proteins, making them unavailable to couple to other receptors. J Neurosci 19(21):9271–9280
Waldeck-Weiermair M et al. (2008) Integrin clustering enables anandamide-induced Ca2+ signaling in endothelial cells via GPR55 by protection against CB1-receptor-triggered repression. J Cell Sci 121:1704–1717
Whiteside GT et al. (2005) A role for cannabinoid receptors, but not endogenous opioids, in the antinociceptive activity of the CB2-selective agonist, GW405833. Eur J Pharmacol 528(1–3):65–72
Wotherspoon G et al. (2005) Peripheral nerve injury induces cannabinoid receptor 2 protein expression in rat sensory neurons. Neuroscience 135(1):235–245
Wright KL, Duncan M, Sharkey KA (2008) Cannabinoid CB2 receptors in the gastrointestinal tract: a regulatory system in states of inflammation. Br J Pharmacol 153(2):263–270
Yao BB et al. (2006) In vitro pharmacological characterization of AM1241: a protean agonist at the cannabinoid CB2 receptor? Br J Pharmacol 149(2):145–154
Yao BB et al. (2008) In vitro and in vivo characterization of A-796260: a selective cannabinoid CB2 receptor agonist exhibiting analgesic activity in rodent pain models. Br J Pharmacol 153(2):390–401
Yao BB et al. (2009) Characterization of a cannabinoid CB2 receptor-selective agonist, A-836339 [2,2,3,3-tetramethyl-cyclopropanecarboxylic acid [3-(2-Methoxy-ethyl)-4,5-dimethyl-3H-thiazol-(2Z)-ylidene]-amide], using in vitro pharmacological assays, in vivo pain models, and pharmacological magnetic resonance imaging. J Pharmacol Exp Ther 328(1):141–151
Zhang J et al. (2003) Induction of CB2 receptor expression in the rat spinal cord of neuropathic but not inflammatory chronic pain models. Eur J Neurosci 17(12):2750–2754
Zuo L et al. (2007) CNR1 variation modulates risk for drug and alcohol dependence. Biol Psychiatry 62(6):616–626

Endocannabinoid Receptors: CNS Localization of the CB_1 Cannabinoid Receptor

István Katona

Contents

Abstract The evolution of plant metabolic pathways to invent compounds which distract predators, and the history of medicine to find treatments for diseases, often share a common logic. An attractive example to illustrate the rationale behind this is the *Cannabis sativa* plant, which was exploited for its widespread therapeutic effects for several thousand years, but historical "prescriptions" highlighted its distractive behavioral side-effects if abused. This chapter aims to explain the characteristically wide pharmacological and behavioral profile of the *Cannabis* plant by pointing to the ubiquitous anatomical distribution of CB_1 cannabinoid receptors, its predominant molecular target, throughout the nervous system. However, in contrast to their abundant regional and cellular localization, the subcellular arrangement of CB_1 receptors and the enzymes involved in the metabolism of its main endogenous ligand, 2-arachidonoylglycerol (2-AG), are strikingly polarized on the neuronal surface in the adult brain. Though there are still several unresolved issues,

I. Katona
Institute of Experimental Medicine, Hungarian Academy of Sciences, H-1083, Budapest, Szigony utca 43, Hungary
e-mail: katona@koki.hu

D. Kendall and S. Alexander (eds.), *Behavioral Neurobiology of the Endocannabinoid System*, Current Topics in Behavioral Neurosciences 1,
DOI: 10.1007/978-3-540-88955-7_3,

the known pieces of the puzzle outline a picture in which the biosynthetic machinery for 2-AG is accumulated in the somatodendritic compartment of neurons, whereas its receptor and degrading enzyme are both found on axon terminals. This molecular architecture suggests that a main physiological role of endocannabinoid signaling is the retrograde regulation of synaptic transmission, and the present chapter aims to summarize compelling evidence that it is an ancient and fundamental component of several distinct types of synapses throughout the nervous system.

Keywords Synapse • Retrograde • DAGL • DGL-alpha • 2-AG • MGL • CB1 cannabinoid receptor

Abbreviations

2-AG	2-Arachidonoylglycerol
DGL	Diacylglycerol lipase
FAAH	Fatty acid amide hydrolase
MGL	Monoacylglycerol lipase
NAE	*N*-Acylethanolamine
RER	Rough endoplasmic reticulum
VGCC	Voltage-gated calcium channels

1 Introduction

The utility of molecular neuroanatomy lies in its ability to aid in correctly interpreting pharmacological, physiological and behavioral experiments by identifying the precise localization of molecular components involved in certain signaling pathways. Undoubtedly, this task requires investigations at several organizational levels of the nervous system with the main goal of finding certain patterns in the distribution of given molecular elements either at the regional, cellular or subcellular level. Compartmentalized localization of a given signaling pathway may predict its cell physiological role or may forecast its behavioral and therapeutic importance.

Molecular neuroanatomy has supported developments in the understanding of behavioral neurobiology of the endocannabinoid system for nearly two decades. Interestingly enough, the first neuroanatomical study describing the position of a putative cannabinoid (CB) receptor by Herkenham and colleagues served as a key trigger for the molecular identification of the receptor itself (Herkenham et al. 1990). This landmark study highlighted the ubiquitous distribution of cannabinoid binding sites throughout the brain, providing an explanation for the wide spectrum of cannabinoid behavioral effects. Furthermore, a pattern in the density of cannabinoid binding clarified certain peculiar features of cannabinoid effects, e.g. the lack of lethal doses due to low density levels in lower brainstem areas that control cardiovascular and respiratory functions. Finally, the high abundance of cannabinoid binding sites was found to be comparable with the density of receptors for major neurotransmitters like glutamate or GABA, indicating well in advance that endogenous cannabinoid signaling would have a fundamental role in neuronal communication.

The subsequent molecular cloning of CB_1 receptors by Matsuda and colleagues paved the way for more detailed anatomical studies, which provided information, initially, on the cellular expression pattern at the mRNA level and – after the development of selective antibodies – on the subcellular localization pattern at the protein level (Matsuda et al. 1990; Tsou et al. 1998). These studies confirmed the widespread presence of CB_1 receptors throughout the central nervous system, but also highlighted characteristic cell-type-specific differences in its expression level. Further work on the subcellular distribution of CB_1 receptors uncovered the fact that the receptor protein is strikingly accumulated on the plasma membrane of axon terminals (Katona et al. 1999), which indicated that endocannabinoids may have a pivotal role in the regulation of synaptic neurotransmission. Indeed, this prediction has been confirmed by subsequent electrophysiological experiments (see the chapter "Endocannabinoid Signaling in Neural Plasticity" by Alger, this volume, for details). Finally, the molecular identification of metabolic enzymes involved in the synthesis or inactivation of endogenous cannabinoid molecules represented another milestone and opened the possibility of finding the start and finish of the endogenous signaling pathway mimicked by phytocannabinoids (Bisogno et al. 2003; Cravatt et al. 1996; Dinh et al. 2002; Okamoto et al. 2004; Simon and Cravatt 2008). Remarkably, follow-up molecular neuroanatomy studies soon confirmed that the biosynthetic and degrading enzymes for 2-arachidonoyl-glycerol (2-AG) are accumulated at synaptic junctions together with CB_1 receptors (Gulyas et al. 2004; Katona et al. 2006; Yoshida et al. 2006), in accordance with the emerging notion that 2-AG may be the predominant endogenous ligand of CB_1 receptors (Sugiura et al. 2006). While 2-AG is thought to be the key endocannabinoid molecule involved in the regulation of synaptic neurotransmission, the precise cell physiological role of anandamide, another endogenous cannabimimetic molecule, and its related bioactive congeners, the so-called *N*-acylethanolamines (NAEs), are still under intense investigation. These molecules have unique behavioral activity profiles and are suggested to activate various molecular targets; thus, molecular neuroanatomy studies in the future may facilitate understanding of the logic behind the existence of these distinct molecular signaling pathways. First attempts to localize some of the metabolic enzymes has already uncovered strikingly different subcellular localization (Egertova et al. 2008; Gulyas et al. 2004; Nyilas et al. 2008), indicating that the cell physiological function of NAEs is indeed distinct from the retrograde messenger role of 2-AG. Although the signaling pathways involving NAEs may certainly have a crucial role in behavioral neurobiology of the endocannabinoid system (see for example Maccarrone et al. 2008), current knowledge of the molecular architecture of these pathways is still very limited at the regional, cellular, and subcellular level. Therefore, the present chapter will focus on the compartmentalized localization of molecular elements involved in 2-AG signaling and will highlight the presence of this signaling pathway as a common feature of several distinct types of synapses throughout the nervous system.

2 Methodological Considerations for Localization Studies on the Endocannabinoid System

2.1 *Negative Controls for Positive Findings*

The exponential growth in the accessibility of easily (and often freely, see Austin et al. 2004) available knockout mouse lines has transformed the view on molecular neuroanatomy findings in the last decade. As validation became a basic requirement for molecular neuroanatomy studies (see for example Saper and Sawchenko 2003), localization data double-checked in knockout controls is among the most solid information available for neuroscience research. This is strikingly different from physiological or behavioral studies in which interpretation of phenotypic differences in a certain experimental paradigm requires consideration of potential perturbations in developmental processes, though recent advancements in conditional gene deletion studies will hopefully soon circumvent this problem.

Knockout animals for quite a few molecular components of the endocannabinoid system are available from several sources (Buckley et al. 2000; Cravatt et al. 2001; Hasegawa et al. 2004; Johns et al. 2007; Ledent et al. 1999; Leung et al. 2006; Marsicano et al. 2002; Zimmer et al. 1999). Because all three major labeling methods may potentially result in non-specific staining patterns on brain sections (pharmacological compounds used in radioligand binding studies, riboprobes used to visualize mRNA expression, antibodies used to label the position of proteins), these knockout animals provide a unique opportunity to confirm any positive experimental findings regardless of the nature of the visualizing agent used in a given study. If knockout controls are not available, an alternative solution is the application of at least two independent visualizing agents, e.g. two riboprobes corresponding to two non-overlapping sequences or two antibodies raised against distinct non-overlapping epitopes (see for example Katona et al. 2006). If the two independent tools visualize identical staining patterns, then the probability of false positive staining is significantly reduced (but not entirely excluded). Careful targeted analysis of potential background caused by the tag in the fusion protein used to generate the antibody may also reveal that certain staining patterns are irrelevant, as has been elegantly demonstrated in the case of astrocytic labeling for CB_1 receptors in the amygdala (McDonald and Mascagni 2001). Three widely used control methods to rule out false positive stainings are not definitive enough. Omission of the primary agents (radioligand, riboprobe or antibody) from the staining procedure reveals that the procedure itself is specific, but does not rule out the possibility that the agent binds to another molecular target in a non-specific manner. Competition experiments using excess "cold" (non-labeled) ligands and riboprobes confirm that the labeled and non-labeled agents recognize the same target or targets, but does not exclude that they compete for a wrong molecular identity. Finally, preabsorbtion tests with the immunizing protein for antibodies corroborate that a given antibody recognizes its epitope, but the disappearance of the staining pattern may simply be due to the fact that the antibody was titrated out by preabsorption and thereby has a reduced capability to bind to its non-specific target as well.

2.2 *Positive Controls for Negative Findings*

It is very difficult, if not impossible, to prove unequivocally the absence of a given molecule in certain brain regions, cell types, or subcellular compartments. Several factors may cause false negative findings. Most often, the quantity of a given molecule is not high enough at certain locations, so more sophisticated procedures or more sensitive agents are required to demonstrate its presence. This is especially true if the given molecule is also expressed at very high levels at another location. In this case, the quantity of the labeling agent may not be high enough to visualize smaller amounts of the molecule at another position as well. Research on the localization of CB_1 cannabinoid receptors followed the aforementioned trajectory, as first generation antibodies revealed its presence exclusively on GABAergic axon terminals (see for example Hajos et al. 2000; Katona et al. 1999); it took several years before a new generation of more sensitive antibodies and proper knockout controls could reveal unequivocally its presence also on glutamatergic axon terminals, where the number of CB_1 receptors is an order of magnitude lower compared with GABAergic axon terminals (Katona et al. 2006; Kawamura et al. 2006; Monory et al. 2006). Another explanation for false negative immunostainings may be the fact that a given molecule may have different binding partners at distinct subcellular compartments, and these binding partners may mask recognition by competing for an overlapping domain of the protein. Finally, expression levels of most genes are not constant in time and space; several factors including development stages, the internal state of the animal, or pathological processes during disease may result in lack of labeling. Certainly, in this case, absence of labeling may be physiologically relevant, but cannot be used to generalize a given distribution pattern. A noteworthy example is the developmental switch in the subcellular distribution of diacylglycerol lipase-α (DGL-α), a predominant biosynthetic enzyme of 2-AG, which is found in axons in the embryonic brain, but was shown to localize in dendrites after birth (Bisogno et al. 2003).

To circumvent the above difficulties, the best strategy to imply the absence of a given molecule is to study different experimental levels (e.g. absence of mRNA, protein, physiological effect) and paradigms; if all converge onto the same result, then the probability of reporting a false negative finding will be strongly reduced.

Nevertheless, the most appropriate way to discuss the absence of labeling is to use the term "the amount of the given molecule did not reach the detection threshold".

2.3 *Quantification of Positive Findings*

Quantitative molecular neuroanatomy is still a developing field because the same problems discussed above may also render precise quantification difficult. Nevertheless, quantification should always be a goal for neuroanatomists, because it may reveal or explain physiologically important differences, such as between brain regions, cell types or subcellular compartments; or during development and disease. Several approaches now help the investigator to quantify the amount of a given molecule, and these will be described in detail below along with the localization patterns of molecular players of endocannabinoid signaling.

3 Regional Distribution

The first two studies of Herkenham et al. (1990, 1991b) carried out radioligand binding on whole brain sections from several species using the tritiated CB_1 receptor agonist [^{3}H]-CP55940 (Fig. 1a). This compound is 45 times more potent on CB_1 receptors than Δ^9-THC, and, more importantly, it failed to stimulate [^{35}S]-GTPγS binding on brain membranes from CB_1 receptor knockout animals (Breivogel et al. 2001), suggesting that the regional localization pattern revealed by [^{3}H]-CP55940 binding must represent the presence of CB_1 receptors. Furthermore, immunostaining using a novel highly sensitive CB_1 antibody developed by the Watanabe group (Fukudome et al. 2004) reveals an almost entirely identical distribution pattern at the whole brain level (Fig. 1a, b), indicating that, indeed, the CB_1 protein may be responsible for the characteristic binding pattern of [^{3}H]-CP55940. The first studies on the regional distribution of CB_1 receptors established three conceptually important points, which are now – retrospectively – widely accepted as key features for explaining the physiological and pathophysiological importance of the endocannabinoid system. These features, summarized below, are the (a) high density (b) characteristic regional pattern, and (c) similarity across species (Herkenham et al. 1990, 1991b).

Although the brain utilizes a plethora of messenger molecules, the density of these molecules, their metabolic enzymes, and their receptors vary across a wide spectrum. Remarkably, the density of cannabinoid receptors in several brain areas was found to be in the range of whole-brain glutamate receptors, cortical GABA receptors, and striatal dopamine receptors (Herkenham et al. 1990), implying that endocannabinoid signaling requires the same amount of receptor proteins as classical neurotransmitters involved in basal synaptic neurotransmission, and highly exceeds the density of neuropeptide receptors, which are supposed to function

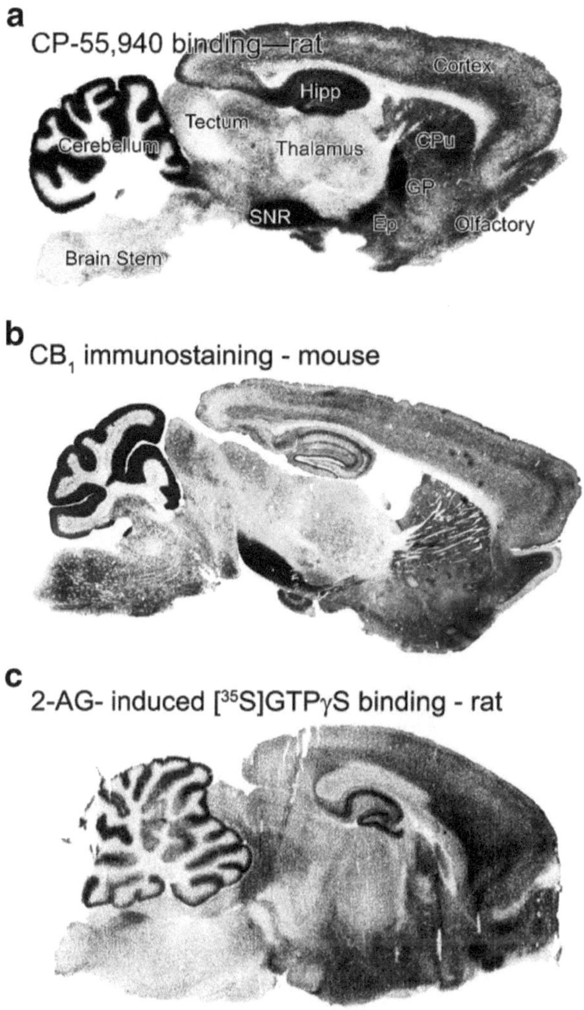

Fig. 1 Regional distribution of CB_1 cannabinoid receptors in the rodent brain. Three distinct experimental paradigms provide independent evidence that CB_1 cannabinoid receptors are widely distributed in the rodent brain. (**a**) Autoradiographic film image demonstrates that the CB_1 agonist CP55940 binds to CB_1 receptors in most brain areas, albeit with a different affinity. (**b**) A nearly identical pattern is visualized by immunostaining using the most sensitive antibody against CB_1 receptors. (**c**) Remarkably, the endogenous cannabinoid receptor ligand 2-AG also induces a similar pattern of CB_1 receptor activation, when its degrading enzyme is inhibited. Note that the three different experiments consistently show that high density of CB_1 is localized in the molecular layer of the cerebellum, in the basal ganglia (substantia nigra pars reticulata, SNR; caudate-putamen, CPU; globus pallidus, GP; hippocampus, Hipp) as well as in the neocortex. Interestingly, the thalamus and the brainstem contain only very few CB_1 receptors. Images in **a**, **b** and **c** were kindly provided by M. Herkenham (Herkenham et al. 1990); by M. Watanabe (Fukudome et al. 2004); and by J. Laitinen (Palomaki et al. 2007), respectively

instead as modulatory substances. In retrospect, this finding already suggested indirectly that endocannabinoid signaling may have a crucial role in the regulation of synaptic neurotransmission. On the other hand, its predominant endogenous ligand 2-AG was measured at a concentration of ~1–10 nmol g^{-1} in the brain (Stella et al. 1997), which is an order of magnitude lower in concentration than glutamate or GABA, two major synaptic neurotransmitters, and more closely resembles dopamine concentration. A potential explanation of this mismatch may be the peculiar synaptic physiological role of 2-AG. In contrast to glutamate or GABA, it is probably not involved in basal synaptic activity, but, rather, it is synthesized and released upon excess neuronal activity in an "on-demand" manner. Surprisingly, the brain concentration of anandamide is even lower (about ~10 pmol g^{-1}), which is in the range of neuromodulators (Cadas et al. 1996); thus, its physiological role may be even more specialized in time and space.

Brain areas are highly different in regard to the number of cells they contain, the number of synapses these cells receive, and the degree of plasticity their synapses express upon activity-dependent changes in neuronal function. Importantly, the general regional distribution of CB_1 receptors shows a peculiar pattern, which largely overlaps with other molecular elements implicated in synaptic neurotransmission and plasticity; furthermore, this pattern may explain several characteristic behavioral effects of phytocannabinoids. The highest density of CB_1 receptors was found in the cerebellum, especially in the molecular layer, where CB_1 was shown to be accumulated on the axon terminals of parallel fibers (Kawamura et al. 2006). In accordance with the striking cannabinoid effect on movement control, a very high density of CB_1 receptors was also found in the substantia nigra pars reticulata and in the globus pallidus; the underlying cellular localization of CB_1 was reported on the GABAergic axon terminals deriving from the striatum (Herkenham et al. 1991a). Finally, a very high density of CB_1 receptors was also found in the hippocampus, where the receptor protein is localized on both GABAergic and on glutamatergic axon terminals (Katona et al. 1999, 2006; Kawamura et al. 2006; Monory et al. 2006). Modest CB_1 receptor localization was reported in several other brain areas; for example, throughout the neocortex, in the amygdala, in the periaqueductal gray nucleus, in the medial hypothalamus, and in the superficial layers of the dorsal horn of the spinal cord (Herkenham et al. 1990). Remarkably, the brainstem contains only a very low density of CB_1 receptors, and several key medullary nuclei responsible for the organization of respiratory and cardiovascular functions seem to lack endocannabinoid signaling. This interesting paucity of a molecular pathway involved in synaptic plasticity may be explained by the well-known rigidity in the synaptic, neuronal, and network activity of these brain areas, which is indispensable for the rhythm generation in these key physiological processes. On a different note, this pattern may also explain why there is no history of fatal overdose due to cannabis consumption in contrast with other drugs of abuse (e.g. benzodiazepines or opiates).

Finally, a key observation of Herkenham et al. (1990) was the finding that CB_1 receptors show a largely similar regional distribution pattern across different mammalian species. This conservative localization pattern, which was later extended even to the subcellular level (namely CB_1 receptors are presynaptically localized at least in vertebrates), in accordance with the surprisingly high sequence identity at the protein level suggested that endocannabinoid signaling may be ancient and may also have a basic physiological role in the regulation of neuronal activity. Nevertheless, subtle differences were uncovered between species, but this may well explain the different contributions of a given brain area to the general lifestyle of the given species. The two most striking examples are the particularly high density of CB_1 receptors in the molecular layer of the cerebellum in dogs, as compared with a relatively moderate labeling in this layer in humans, whereas our emotionally sophisticated species shows a characteristically higher density of CB_1 receptors in the basolateral amygdaloid complex not found in other mammalian species.

Importantly, the regional distribution pattern of CB_1 receptors established in the early studies of Herkenham et al. (1990, 1991b) was nicely confirmed by functional autoradiography experiments mapping the precise anatomical loci of the intrinsic endocannabinoid pathway itself (Palomaki et al. 2007). In their elegant paradigm, Palomäki and colleagues used potent inhibitors of monoacylglycerol lipase (MGL), the degrading enzyme of 2-AG, to increase the lifetime of this endogenous cannabinoid in brain tissue prepared for functional autoradiography (Palomaki et al. 2007). Remarkably, elevated intrinsic 2-AG level resulted in an almost entirely similar brain distribution of CB_1 receptor activation as obtained by [^{3}H]-CP55940, the exogenous CB_1 receptor agonist (Fig. 1a–c). Further pharmacological experiments using inhibitors of DGL uncovered the fact that "on-demand" activation of 2-AG biosynthesis was responsible for the characteristic regional pattern of endogenous 2-AG-mediated activation of CB_1 receptors. In contrast to 2-AG, which is a full efficacy agonist of CB_1 receptors, experimental increase in the tissue level of anandamide, a partial agonist of CB_1 receptors, by blocking its degrading enzyme FAAH, did not induce CB_1 receptor activation in this experimental paradigm. Taken together, these findings suggest that the operation of an endogenous chemical messenger, 2-AG, and its signaling pathway, the DGL-CB_1-MGL route, may explain the pattern in the specific regional distribution of cannabinoid receptors and may underlie the characteristic behavioral effects of cannabinoids.

Although the regional distribution of cannabinoid signaling in the brain is well established by the above studies in the *postmortem* brain, exciting new technologies are opening the way for in vivo approaches investigating this signaling system. Promising novel radioligands were recently developed to assess CB_1 receptors in live animals and even in humans using positron emission tomography (PET) (Burns et al. 2007; Yasuno et al. 2008). Notably, one of these radioligands, [^{11}C]-MePPeP, was also confirmed to be specific for CB_1 receptors in knockout animals (Terry et al. 2008). These exciting new tools provide ample opportunity to examine CB_1 receptor functioning in several behavioral or disease paradigms in the live brain. Most importantly, the first preliminary findings largely reproduced the regional distribution of CB_1 receptors in live human and monkey brains as revealed by the earlier autoradiography studies on postmortem brain tissue (Burns et al. 2007); thus, future work using these radioligands will hopefully facilitate our understanding of the physiological and pathophysiological importance of endocannabinoid signaling in the near future.

4 Cellular Distribution

The cellular expression pattern of CB_1 receptors in the central nervous system shows two important features. First and foremost, it is ubiquitous, i.e. nearly all neuronal cell types express this receptor, with some notable exceptions (see below). Secondly, the expression level in selected cell types varies greatly even under normal conditions; certain physiological stimuli or diseases can robustly change the expression level. Before going into details, one has to call attention to the fact that the most reliable technique of determining whether a given cell type is expressing CB_1 is still in situ hybridization. Because the majority of the functional CB_1 protein population is localized on axon terminals, the detection of proteins in the cell body may be ambiguous and a lack of labeling, again, does not exclude expression by the given cell type.

The expression pattern of CB_1 receptors in the brain was reported in the first landmark paper of Matsuda et al. (1990), which also included information about the molecular identification of CB_1 receptors. The authors used ^{35}S-labeled radioactive oligonucleotides on brain sections and demonstrated that certain cells express enormous amounts of CB_1 mRNA, especially in cortical areas. Although they reported that some of these cells are granule cells in the dentate gyrus, further studies revealed that granule cells do not express CB_1 mRNA or protein; instead, the neurons with high expression levels belong to a select subpopulation of GABAergic interneurons (Katona et al. 1999; Marsicano and Lutz 1999; Tsou et al. 1999). Intriguingly, similarly strong CB_1 expression was also found in scattered neurons throughout the neocortex (Bodor et al. 2005; Marsicano and Lutz 1999), as well as in the basolateral amygdala (Katona et al. 2001; Marsicano and Lutz 1999; McDonald and Mascagni 2001), and detailed analysis of their neurochemical markers revealed the presence of the neuropeptide cholecystokinin, a characteristic neurochemical marker of a special GABAergic cell population responsible for perisomatic and dendritic inhibition (Cope et al. 2002; Freund and Buzsaki 1996; Hajos et al. 2000; Klausberger et al. 2003; Pawelzik et al. 2002). Taken together, this suggests that the cortical type of neuronal networks may need specialized GABAergic inhibitory cells to regulate the activity of principal neurons, and principal cells may utilize 2-AG to escape from inhibition under certain circumstances by CB_1-mediated suppression of GABA release (Kim and Alger 2004; Makara et al. 2005; Ohno-Shosaku et al. 2001; Wilson and Nicoll 2001).

Matsuda and colleagues (Matsuda et al. 1990) also reported high CB_1 expression in the ventromedial hypothalamus, which was confirmed by follow-up studies investigating the whole-brain distribution of CB_1 receptors in rats and humans (Mailleux and Vanderhaeghen 1992; Matsuda et al. 1993; Westlake et al. 1994). These latter studies provide an important, comprehensive summary of the cellular expression pattern of CB_1 in nearly all brain regions and should be used as key references similar to the radioligand binding studies of Herkenham et al. (1990, 1991b). Without listing all brain regions, the key message derived from these studies is that the CB_1 receptor is indeed numerous and ubiquitous in the brain.

Most notably, the cerebellar cortex expressed the highest amount of CB_1 receptors outside the forebrain, particularly in the granule cell layer, which corresponded very well with the high density of radioligand binding in the molecular layer, where the axons of granule cells, the so-called parallel fibers, terminate. In addition, widely distributed GABAergic interneurons were also shown to synthesize CB_1 mRNA in a manner similar to the forebrain. A high intensity in situ hybridization signal was also observed in the dorsolateral part of the striatum, corresponding to GABAergic medium spiny neurons projecting to the substantia nigra (Mailleux and Vanderhaeghen 1992; Matsuda et al. 1993). Further detailed studies on selected brain regions largely confirmed these papers and provided the precise details of cellular expression patterns both in quality and quantity by investigating co-localization with cell-type-specific markers (see for example very comprehensive works by Marsicano and Lutz (1999) for cortical areas; or by Hohmann and Herkenham (2000) for the striatum).

Besides the high CB_1-expressing cell types found only in selected brain regions, most principal cell types also express a much lower amount of the receptor mRNA throughout the central nervous system. This rendered the appreciation of the functional significance of CB_1 receptors on these cells difficult for a long period of time; however, recent elegant cell-type-specific knockout models have provided convincing evidence that their presence serves important functions; for example, in protecting neurons from excitotoxicity (Marsicano et al. 2003; Monory et al. 2006). It is important to emphasize that, although these neurons express CB_1 at a lower magnitude, they outnumber high CB_1-expressing cell types by an order of magnitude. Therefore, the total CB_1 mRNA amount in most brain regions may be summed from these two populations in roughly equal levels (50–50%). Because these distinct types of neurons fulfill entirely different physiological roles at the network level in all brain regions, studies simply measuring expressional changes in the CB_1 level from homogenized tissue sources may not have the resolution to reveal cell-type-specific changes: this requires more tedious experimental analysis using cell-type-specific knockout animals or high resolution anatomical analysis (Ludanyi et al. 2008; Monory et al. 2006).

Cellular expression patterns can also be studied at the protein level using selective antibodies. However, one must remain very cautious when drawing conclusions from these experiments, especially in the case of negative findings. The author had earlier experience with antibodies raised against CB_1 receptors that were confirmed to be specific in knockout animals, which were excellent for visualizing CB_1 receptors in cell bodies but not axon terminals, and vice versa. This suggests that the CB_1 receptor protein may have different conformations in these two cellular domains, and the probability of a given antibody to recognize its epitope may, therefore, vary widely. In addition, a recent, careful analysis revealed that four widely used commercial antibodies against CB_1 failed to recognize CB_1 proteins, in contrast to other antibodies raised in academic research labs (Grimsey et al. 2008).

Two detailed whole-brain studies using these "home-made" antibodies for immunocytochemistry revealed a highly similar cellular distribution pattern for the CB_1 receptor protein, as compared to its mRNA (Pettit et al. 1998; Tsou et al.

1998). Intensely stained neurons were found in cortical structures, whereas weakly CB_1-positive cells were found throughout the brain, with a tendency for lower expression level cells to be located in the brainstem in comparison to midbrain and forebrain structures. Two years later, Egertova and Elphick (2000) extended these observations using another antibody to selectively label the axonal distribution pattern of CB_1 receptors but not the cell bodies producing the receptors. Briefly, CB_1-immunoreactive fibers cover several specific brain areas, and selected layers in these areas, in dense meshworks in agreement with the cellular expression pattern; whereas lower brain areas exhibited far fewer CB_1 receptors. Notable differences between mRNA and protein distributions revealed by the above studies can be explained by the presence of CB_1 receptors on the axons of projection neurons, e.g. high expression of CB_1 mRNA in neurons in the anterior olfactory nuclei may explain the high density of CB_1 protein in the internal granular layer of the olfactory bulb where otherwise CB_1 expression is under the detection threshold.

Finally, though with caution, one also has to emphasize that while CB_1 cannabinoid receptors appear to be a general feature of most cell types, we should note that there are a few neurons that are exceptions, in which neither the CB_1 mRNA levels nor the CB_1 protein levels in the cell bodies reach the detection threshold, and physiological studies also indicate a lack of cannabinoid receptors. These cell types include the midbrain dopaminergic neurons (Herkenham et al. 1991a), Purkinje cells of the cerebellum (Egertova et al. 1998), granule cells of the dentate gyrus (Katona et al. 2006; Marsicano and Lutz 1999), parvalbumin-positive interneurons of cortical areas (Bodor et al. 2005; Katona et al. 2000, 2001; Katona et al. 1999; Marsicano and Lutz 1999; McDonald and Mascagni 2001; Tsou et al. 1999) (but not in the basal ganglia, where parvalbumin-positive cells express CB_1 (Hohmann and Herkenham 2000; Uchigashima et al. 2007)), and cholinergic neurons of the striatum (Hohmann and Herkenham 2000; Uchigashima et al. 2007). Unfortunately, the reason for the absence of CB_1 on the boutons of these neurons is unknown. In addition, the presence of the related CB_2 receptors has also been demonstrated in the CNS in recent studies (Van Sickle et al. 2005); however, neither anatomical nor physiological evidence for the synaptic localization of CB_2 receptors is yet available.

5 Subcellular Distribution

The first indirect anatomical evidence that CB_1 receptors may be targeted to axonal processes is derived from the early work of Herkenham et al. (1991a). Selective ibotenic acid lesions of the CB_1-expressing striatal neurons resulted in a massive loss of cannabinoid binding sites in the substantia nigra and in the pallidum, indicating that striato-pallidal and striato-nigral projection neurons carry CB_1 receptors on their axonal processes (see also Matyas et al. 2006 for a more direct demonstration). Further immunostaining studies visualized dense meshworks of CB_1-positive fibers throughout the central nervous system (Egertova and Elphick

2000; Tsou et al. 1998). Higher resolution studies using electron microscopy confirmed in cortical areas that these immunostained profiles are indeed axons; more precisely, the vast majority of labeling was found on the plasma membrane of axon terminals, suggesting that CB_1 receptors may have a pivotal role in the regulation of synaptic transmission (Fig. 2e–h) (Katona et al. 1999, 2000, 2001;

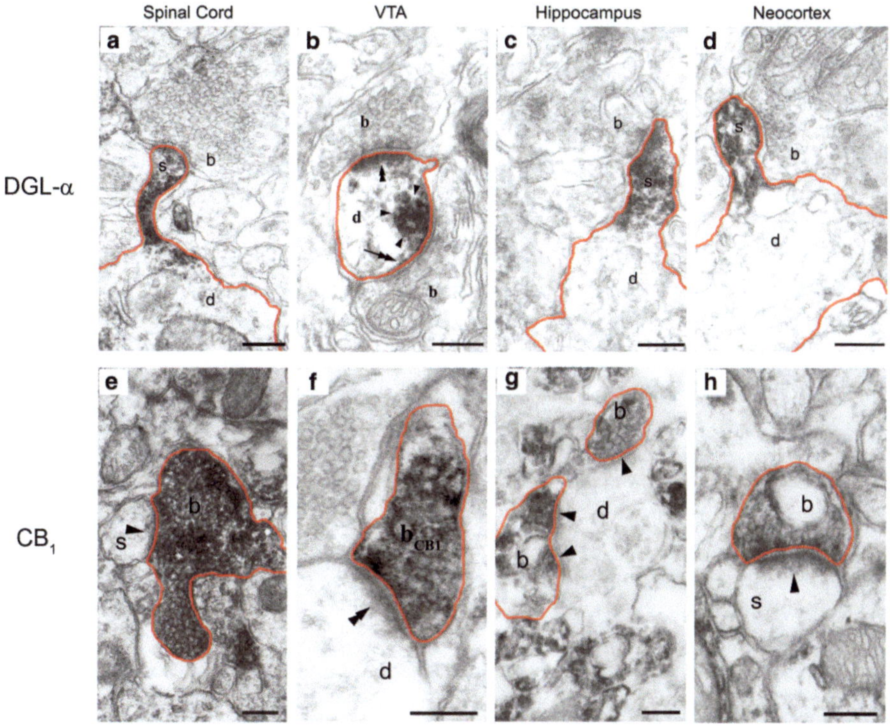

Fig. 2 Postsynaptic synthetic enzyme and presynaptic receptor for 2-AG are conserved features of excitatory synapses throughout the CNS. (**a–d**) Electron microscopic images demonstrate that DGL-α, the predominant biosynthetic enzyme of the endocannabinoid 2-AG, is accumulated at the postsynaptic side of excitatory synapses. The dense black end product of the immunoperoxidase reaction represents the position of DGL-α, and it shows a highly compartmentalized distribution limited to the spine head (s) in case of the spinal cord (**a**), hippocampus (**c**) or neocortex (**d**). Even in aspiny neurons like the dopaminergic cells of the ventral tegmental area (VTA), DGL-α is concentrated perisynaptically (**b**). Note that all of these DGL-α-containing postsynaptic profiles receive asymmetrical synapses from DGL-α-negative boutons (**b**). (**e–h**) High-resolution electron micrographs show that CB_1 cannabinoid receptors, the molecular targets of 2-AG, are located presynaptically on glutamatergic axon terminals (**b**) throughout the CNS. The dense end product of CB_1-immunostaining is restricted to boutons, whereas postsynaptic profiles, spines (s) or dendritic shafts (**d**) are always CB_1-immunonegative. The strongly asymmetrical postsynaptic density (*arrowheads*) demonstrates that these are excitatory, glutamatergic synapses. Red outlines assist the identification of the postsynaptic or presynaptic profiles (spines vs. axon terminals), which contain immunostaining for either DGL-α (**a–d**) or CB_1 (**e–h**), respectively. Scale bars: **a–h**, 0.2μm. Images were modified from Katona et al. (2006) and Mátyás et al. (2008) or kindly provided by Rita Nyilas (**a**,**e**) and Barna Dudok (**d**,**h**)

Hajos et al. 2000). Further anatomical studies extended this finding to other brain areas or even to peripheral tissues (see for example Calignano et al. 2000; Vizi et al. 2001), and now it is very well established that most types of axon terminals bear presynaptic CB_1 receptors (but see the few exceptions above). While demonstration of co-localization of neurochemical markers with CB_1 receptors at the boutonal level was a very difficult task a few years ago, recent advances in confocal microscopy now provide ample opportunities to test various markers and to establish which axon terminals carry presynaptic CB_1 receptors (see for example two exemplary studies by Kawamura et al. 2006; Uchigashima et al. 2007). This widespread distribution of presynaptic CB_1 receptors was totally unexpected a few years ago; however, in light of the crucial physiological role of 2-AG in the negative feedback regulation of neurotransmitter release (see the chapter "Endocannabinoid Signaling in Neural Plasticity" by Alger, this volume, for details), it is conceivable retrospectively. For details regarding given types of axon terminals, a number of reviews provide a comprehensive summary of the anatomical, pharmacological and physiological evidence for presynaptic cannabinoid receptors and for their role in the regulation of synaptic plasticity (Chevaleyre et al. 2006; Freund et al. 2003; Schlicker and Kathmann 2001).

Axon terminals carry a conserved elaborate machinery mediating basal synaptic transmission, but are also equipped with various molecular signaling pathways, which help to adjust the efficacy of synaptic communication at a scale over an order of magnitude. Presynaptic CB_1 receptors seem to be involved in at least two pathways, both of which result in the attenuation of neurotransmitter release, albeit at a different time scale. Activation of presynaptic CB_1 receptors may result in the inhibition of voltage-gated calcium channels (VGCCs), of which N-type VGCCs seem to be particularly important (Wilson et al. 2001); engagement of this signaling pathway causes short-term synaptic depression. In addition, presynaptic CB_1 receptors may also down-regulate the activity of the adenylyl-cyclase-protein kinase A signaling pathway, which results in the long-term depression of synaptic neurotransmitter release (Chevaleyre et al. 2007). Whether these two molecular cascades evoking synaptic depression at a different time scale can be initiated by the same population of CB_1 receptors is unknown. On the other hand, high-resolution neuroanatomical studies suggest that there are two peaks of distribution of these receptors on given axon terminals, opening the possibility that there are two macromolecular signaling complexes involving CB_1 receptors. Interestingly, CB_1 receptors are positioned perisynaptically on hippocampal GABAergic axon terminals, but were not found intrasynaptically close to the vesicle docking sites (Nyiri et al. 2005). This distribution pattern fits well with the predicted subaxonal distribution of N-type VGCCs on the same type of axon terminals (Hefft and Jonas 2005) and is also in agreement with the notion that VGCC-mediated short-term depression is a homosynaptic phenomenon, i.e. the endocannabinoid ligand may arrive from the neighboring postsynaptic domains. Besides perisynaptic CB_1 receptors, a second population is also accumulated further away from the perisynaptic zone (Nyiri et al. 2005), which may be an ideal position in which to detect ligands coming from heterosynaptic sources, in harmony with the findings that endocannabinoid-mediated long-term depression on GABAergic axon terminals is a heterosynaptic phenomenon (Chevaleyre and Castillo 2003). Importantly, identical dual peaks in the subaxonal distribution of CB_1 receptors were also reported on juvenile cerebellar parallel fibers; however the receptor localization becomes more concentrated perisynaptically in the adult cerebellum (Kawamura et al. 2006).

Besides presynaptically located CB_1 receptors, there is growing evidence that CB_1 receptors may also be found in the somatodendritic domain. While there is no unequivocal evidence that CB_1 is present on distal dendritic segments or close to postsynaptic specializations, complementary experimental approaches revealed its presence within the cell body or in thick proximal dendritic shafts of cortical GABAergic interneurons. Immunogold staining uncovered the fact that CB_1 receptor is present on the rough endoplasmic reticulum (RER), Golgi complex, multivesicular bodies and the endosome–lysosome system inside the cell bodies and proximal dendrites (Katona et al. 1999, 2001; Bodor et al. 2005). This suggests that the freshly synthesized CB_1 receptors are recognized by the antibodies, which may explain the labeling on the RER or Golgi complex. In addition, Leterrier and colleagues provided interesting data recently showing that EGFP-tagged CB_1 receptor proteins undergo a continuous recycling process between subcellular structures and somatic plasma membrane, which plays an important role in the subsequent axonal transport of CB_1 proteins (Leterrier et al. 2006) and may also explain the localization of CB_1 receptors on multivesicular bodies (Bodor et al. 2005; Katona et al. 2001). Finally, an exciting physiological paradigm discovered by Bacci and co-workers suggests that certain cortical interneuron types may undergo lasting self-inhibition via endocannabinoid signaling and intracellular CB_1 receptor activation (Bacci et al. 2004), though the anatomical substrates for this phenomenon require further investigations.

6 Distribution of Other Molecular Components Involved in 2-AG Signaling

Pharmacological and physiological studies converge on the notion that 2-AG may be the primary endogenous ligand of presynaptic CB_1 receptors (Kim and Alger 2004; Makara et al. 2005; Melis et al. 2004; Straiker and Mackie 2007). Recent breakthrough discoveries identified the enzymes responsible for the synthesis – two diacylglycerol lipases (DGL-α and DGL-β) (Bisogno et al. 2003) – and inactivation – MGL (Dinh et al. 2002) – of synaptic 2-AG (for details see the chapter “The Life Cycle of the Endocannabinoids: Formation and Inactivation” by Alexander and Kendall, this volume). In accordance with the retrograde transmitter function of 2-AG, subsequent high-resolution anatomical studies provided direct evidence that the synthetic enzyme of 2-AG is postsynaptic (Katona et al. 2006; Yoshida et al. 2006), whereas the degrading enzyme is presynaptic (Gulyas et al. 2004), implying that 2-AG must follow a retrograde route through the synaptic cleft. Interestingly, in the case of excitatory synapses, DGL-α is not only found in postsynaptic dendritic spines, but it is also accumulated in a perisynaptic annulus around the postsynaptic density (Katona et al. 2006; Yoshida et al. 2006), where it is inserted into a perisynaptic signaling machinery together with $mGlu_5$ receptors and Homers (Jung et al. 2007). Remarkably, this specialized molecular architecture for retrograde regulation of neurotransmitter release is becoming widely appreciated as a common principle of excitatory synapses throughout the nervous system (Fig. 2). In addition to the various pharmacological and physiological evidence, anatomical studies revealed that the molecular components of this negative feedback pathway is conserved at glutamatergic synapses in the spinal cord (Fig. 2a, e) (Nyilas et al. 2009), midbrain (Fig. 2b, f) (Matyas et al. 2008), cerebellum (Yoshida et al. 2006), nucleus accumbens (Matyas et al. 2007), striatum (Uchigashima et al. 2007), hippocampus (Fig. 2c, g) (Katona et al. 2006; Yoshida et al. 2006) and in the prefrontal (Lafourcade et al. 2007) and somatosensory cortex (Fig. 2d, h) (Dudok et al. 2007). Regarding GABAergic synapses, there is experimental evidence for postsynaptic DGL-α in the ventral tegmental area (Matyas et al. 2008) and in the striatum (Uchigashima et al. 2007). Whether this biosynthetic enzyme of 2-AG is also present at other synapses is unknown, but the lack of labeling at cortical GABAergic synapses with the first generation of DGL-α antibodies indicate that its expression level is much lower compared to glutamatergic synapses. Nevertheless, this issue must be revisited in the future if more sensitive antibodies are available, because pharmacological evidence suggests that diacylglycerol lipases are involved in retrograde 2-AG signaling at cortical GABAergic synapses as well (Hashimotodani et al. 2008). In accordance with this, the presence of MGL in GABAergic axon terminals has already been demonstrated at the electron microscopic level (Gulyas et al. 2004).

While the evidence presented above demonstrates that the molecular architecture for retrograde 2-AG signaling is a conserved feature of synapses throughout the adult brain, the situation may be different in the embryonic brain. Bisogno and colleagues suggested that both DGLs are found in growing axonal tracts in the embryonic spinal cord and optic nerve (Bisogno et al. 2003), and this was confirmed in telencephalic axons at E14.5, although the expression of DGLs was shifted to the dendritic shafts already at E18.5 in the telencephalon (Berghuis et al. 2007). Further immunostaining revealed a high density of CB_1 receptors in the growth cones, where it has been shown to regulate maturation of both GABAergic and glutamatergic axons (Berghuis et al. 2007; Mulder et al. 2008).

7 Conclusions

Anatomical studies in the last decade have revealed the precise position of several molecular components of the endocannabinoid system at the regional, cellular and subcellular level. These studies contributed significantly to the notion that 2-AG is a retrograde messenger at most synapses of the nervous system. The underlying molecular architecture involves (a) postsynaptic synthesis of 2-AG by DGL-α, (b) a presynaptic effector, the CB_1 cannabinoid receptor, and (c) presynaptic inactivation of 2-AG by MGL (Fig. 3). Ample anatomical evidence demonstrates that this molecular architecture is a general feature of synapses, and together with physiological and pharmacological data it is now clear that 2-AG signaling may be as common a principle of synaptic transmission as it is of glutamate or GABA signaling.

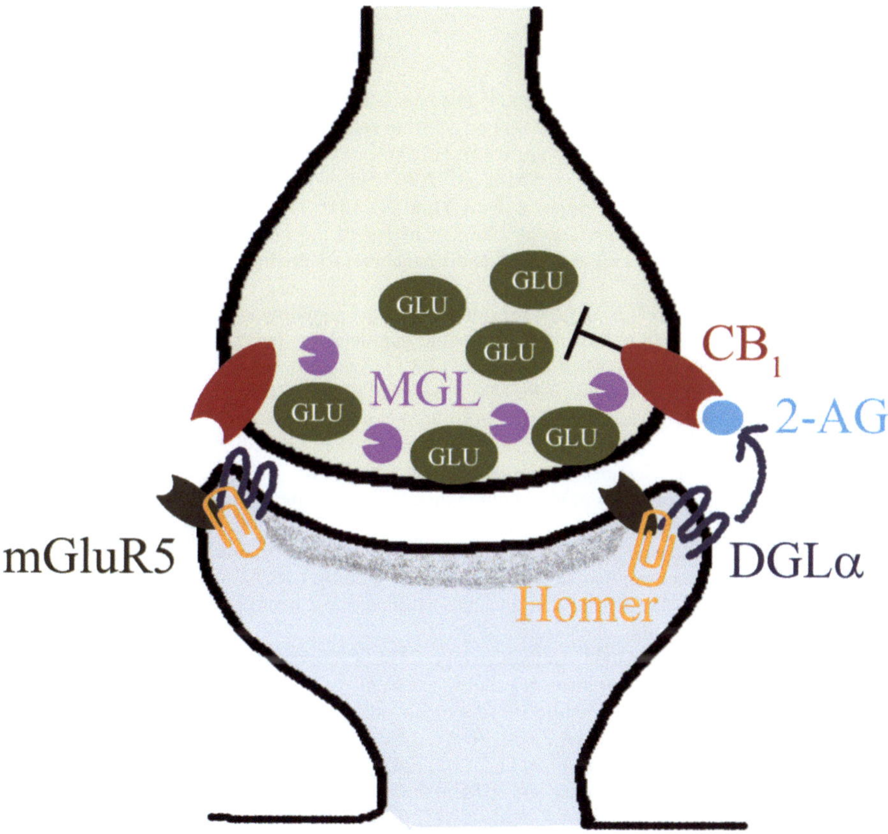

Fig. 3 Molecular architecture of synaptic 2-AG signaling pathway. The schematic diagram illustrates the synaptic position of molecular elements involved in retrograde 2-AG signaling. Spillover of glutamate from the synaptic cleft will activate perisynaptically located metabotropic glutamate receptors (predominantly $mGlu_5$ receptors), which are anchored together with 2-AG's synthetic enzyme diacylglycerol lipase (DGL-α) via the scaffolding protein Homer (note that other Homer-binding partners are not indicated for reasons of clarity). DGL-α produces 2-AG, which travels retrogradely through the synaptic cleft to activate presynaptic CB_1 cannabinoid receptors, which then suppress neurotransmitter release from the axon terminal. Finally, 2-AG is inactivated by the degrading enzyme monoacylglycerol (MGL), which is also localized in the axon terminal. This striking spatial organization of 2-AG signaling provides direct anatomical support for the view that 2-AG is a retrograde transmitter at glutamatergic synapses

Acknowledgements This work was supported by ETT 561/2006 and by the János Bolyai scholarship. The author is very grateful to Drs. Tamás F. Freund, Norbert Hájos, Ken Mackie, Daniele Piomelli and Masahiko Watanabe for their long-term collaborative support of his work on the endocannabinoid system. The author is also indebted to Ms. Anikó Ludányi, Rita Nyilas, Gabriella Urbán and Mr. Barna Dudok for their help with the preparation of figures and to Drs. Miles Herkenham, Jarmo Laitinen and Masahiko Watanabe for their kind contribution of figures to this book chapter.

References

Austin CP, Battey JF, Bradley A et al. (2004) The knockout mouse project. Nat Genet 36:921–924

Bacci A, Huguenard JR, Prince DA (2004) Long-lasting self-inhibition of neocortical interneurons mediated by endocannabinoids. Nature 431:312–316

Berghuis P, Rajnicek AM, Morozov YM et al. (2007) Hardwiring the brain: endocannabinoids shape neuronal connectivity. Science (New York, NY) 316:1212–1216

Bisogno T, Howell F, Williams G et al. (2003) Cloning of the first sn1-DAG lipases points to the spatial and temporal regulation of endocannabinoid signaling in the brain. J Cell Biol 163:463–468

Bodor AL, Katona I, Nyiri G et al. (2005) Endocannabinoid signaling in rat somatosensory cortex: laminar differences and involvement of specific interneuron types. J Neurosci 25:6845–6856

Breivogel CS, Griffin G, Di Marzo V et al. (2001) Evidence for a new G protein-coupled cannabinoid receptor in mouse brain. Mol Pharmacol 60:155–163

Buckley NE, McCoy KL, Mezey E et al. (2000) Immunomodulation by cannabinoids is absent in mice deficient for the cannabinoid CB(2) receptor. Eur J Pharmacol 396:141–149

Burns HD, Van Laere K, Sanabria-Bohorquez S et al. (2007) [18F]MK-9470, a positron emission tomography (PET) tracer for in vivo human PET brain imaging of the cannabinoid-1 receptor. Proc Natl Acad Sci USA 104:9800–9805

Cadas H, Gaillet S, Beltramo M et al. (1996) Biosynthesis of an endogenous cannabinoid precursor in neurons and its control by calcium and cAMP. J Neurosci 16:3934–3942

Calignano A, Katona I, Desarnaud F et al. (2000) Bidirectional control of airway responsiveness by endogenous cannabinoids. Nature 408:96–101

Chevaleyre V, Castillo PE (2003) Heterosynaptic LTD of hippocampal GABAergic synapses: a novel role of endocannabinoids in regulating excitability. Neuron 38:461–472

Chevaleyre V, Takahashi KA, Castillo PE (2006) Endocannabinoid-Mediated Synaptic Plasticity in the CNS. Annu Rev Neurosci 29:37–76

Chevaleyre V, Heifets BD, Kaeser PS et al. (2007) Endocannabinoid-mediated long-term plasticity requires cAMP/PKA signaling and RIM1alpha. Neuron 54:801–812

Cope DW, Maccaferri G, Marton LF et al. (2002) Cholecystokinin-immunopositive basket and Schaffer collateral-associated interneurones target different domains of pyramidal cells in the CA1 area of the rat hippocampus. Neuroscience 109:63–80

Cravatt BF, Giang DK, Mayfield SP et al. (1996) Molecular characterization of an enzyme that degrades neuromodulatory fatty-acid amides. Nature 384:83–87

Cravatt BF, Demarest K, Patricelli MP et al. (2001) Supersensitivity to anandamide and enhanced endogenous cannabinoid signaling in mice lacking fatty acid amide hydrolase. Proc Natl Acad Sci USA 98:9371–9376

Dinh TP, Carpenter D, Leslie FM et al. (2002) Brain monoglyceride lipase participating in endocannabinoid inactivation. Proc Natl Acad Sci USA 99:10819–10824

Dudok B, Urban GM, Mackie K et al. (2007) Molecular basis of retrograde endocannabinoid signaling at neocortical glutamatergic synapses. Society for Neuroscience. Abstract 37: 244.7/G19

Egertova M, Elphick MR (2000) Localisation of cannabinoid receptors in the rat brain using antibodies to the intracellular C-terminal tail of CB. J Comp Neurol 422:159–171
Egertova M, Giang DK, Cravatt BF et al. (1998) A new perspective on cannabinoid signalling: complementary localization of fatty acid amide hydrolase and the CB_1 receptor in rat brain. Proceedings 265:2081–2085
Egertova M, Simon GM, Cravatt BF et al.(2008) Localization of N-acyl phosphatidylethanolamine phospholipase D (NAPE-PLD) expression in mouse brain: A new perspective on N-acylethanolamines as neural signaling molecules. J Comp Neurol 506:604–615
Freund TF, Buzsaki G (1996) Interneurons of the hippocampus. Hippocampus 6:347–470
Freund TF, Katona I, Piomelli D (2003) Role of endogenous cannabinoids in synaptic signaling. Physiol Rev 83:1017–1066
Fukudome Y, Ohno-Shosaku T, Matsui M et al. (2004) Two distinct classes of muscarinic action on hippocampal inhibitory synapses: M2-mediated direct suppression and M1/ M3-mediated indirect suppression through endocannabinoid signalling. Eur J Neurosci 19:2682–2692
Grimsey NL, Goodfellow CE, Scotter EL et al. (2008) Specific detection of CB_1 receptors; cannabinoid CB_1 receptor antibodies are not all created equal!. J Neurosci Methods 171:78–86
Gulyas AI, Cravatt BF, Bracey MH et al. (2004) Segregation of two endocannabinoid-hydrolyzing enzymes into pre- and postsynaptic compartments in the rat hippocampus, cerebellum and amygdala. Eur J NeuroSci 20:441–458
Hajos N, Katona I, Naiem SS et al. (2000) Cannabinoids inhibit hippocampal GABAergic transmission and network oscillations. Eur J Neurosci 12:3239–3249
Hasegawa K, Martin F, Huang G et al. (2004) PEST domain-enriched tyrosine phosphatase (PEP) regulation of effector/memory T cells. Science (New York, NY) 303:685–689
Hashimotodani Y, Ohno-Shosaku T, Maejima T et al. (2008) Pharmacological evidence for the involvement of diacylglycerol lipase in depolarization-induced endocanabinoid release. Neuropharmacology 54:58–67
Hefft S, Jonas P (2005) Asynchronous GABA release generates long-lasting inhibition at a hippocampal interneuron-principal neuron synapse. Nat Neurosci 8:1319–1328
Herkenham M, Lynn AB, Little MD et al. (1990) Cannabinoid receptor localization in brain. Proc Natl Acad Sci USA 87:1932–1936
Herkenham M, Lynn AB, de Costa BR et al. (1991a) Neuronal localization of cannabinoid receptors in the basal ganglia of the rat. Brain Res 547:267–274
Herkenham M, Lynn AB, Johnson MR et al. (1991b) Characterization and localization of cannabinoid receptors in rat brain: a quantitative in vitro autoradiographic study. J Neurosci 11:563–583
Hohmann AG, Herkenham M (2000) Localization of cannabinoid CB(1) receptor mRNA in neuronal subpopulations of rat striatum: a double-label in situ hybridization study. Synapse (New York, NY) 37:71–80
Johns DG, Behm DJ, Walker DJ et al. (2007) The novel endocannabinoid receptor GPR55 is activated by atypical cannabinoids but does not mediate their vasodilator effects. Br J Pharmacol 152:825–831
Jung KM, Astarita G, Zhu C et al. (2007) A key role for diacylglycerol lipase-alpha in metabotropic glutamate receptor-dependent endocannabinoid mobilization. Mol Pharmacol 72:612–621
Katona I, Sperlagh B, Sik A et al. (1999) Presynaptically located CB_1 cannabinoid receptors regulate GABA release from axon terminals of specific hippocampal interneurons. J Neurosci 19:4544–4558
Katona I, Sperlagh B, Magloczky Z et al. (2000) GABAergic interneurons are the targets of cannabinoid actions in the human hippocampus. Neuroscience 100:797–804
Katona I, Rancz EA, Acsady L et al. (2001) Distribution of CB_1 cannabinoid receptors in the amygdala and their role in the control of GABAergic transmission. J Neurosci 21:9506–9518
Katona I, Urban GM, Wallace M et al. (2006) Molecular composition of the endocannabinoid system at glutamatergic synapses. J Neurosci 26:5628–5637

Kawamura Y, Fukaya M, Maejima T et al. (2006) The CB_1 cannabinoid receptor is the major cannabinoid receptor at excitatory presynaptic sites in the hippocampus and cerebellum. J Neurosci 26:2991–3001

Kim J, Alger BE (2004) Inhibition of cyclooxygenase-2 potentiates retrograde endocannabinoid effects in hippocampus. Nat Neurosci 7:697–698

Klausberger T, Magill PJ, Marton LF et al. (2003) Brain-state- and cell-type-specific firing of hippocampal interneurons in vivo. Nature 421:844–848

Lafourcade M, Elezgarai I, Mato S et al. (2007) Molecular components and functions of the endocannabinoid system in mouse prefrontal cortex. PLoS ONE 2:e709

Ledent C, Valverde O, Cossu G et al. (1999) Unresponsiveness to cannabinoids and reduced addictive effects of opiates in CB_1 receptor knockout mice. Science (New York, NY) 283:401–404

Leterrier C, Laine J, Darmon M et al. (2006) Constitutive activation drives compartment-selective endocytosis and axonal targeting of type 1 cannabinoid receptors. J Neurosci 26:3141–3153

Leung D, Saghatelian A, Simon GM et al. (2006) Inactivation of N-acyl phosphatidylethanolamine phospholipase D reveals multiple mechanisms for the biosynthesis of endocannabinoids. Biochemistry 45:4720–4726

Ludanyi A, Eross L, Czirjak S et al. (2008) Downregulation of the CB_1 cannabinoid receptor and related molecular elements of the endocannabinoid system in epileptic human hippocampus. J Neurosci 28:2976–2990

Maccarrone M, Rossi S, Bari M et al. (2008) Anandamide inhibits metabolism and physiological actions of 2-arachidonoylglycerol in the striatum. Nat Neurosci 11:152–159

Mailleux P, Vanderhaeghen JJ (1992) Distribution of neuronal cannabinoid receptor in the adult rat brain: a comparative receptor binding radioautography and in situ hybridization histochemistry. Neuroscience 48:655–668

Makara JK, Mor M, Fegley D et al. (2005) Selective inhibition of 2-AG hydrolysis enhances endocannabinoid signaling in hippocampus. Nat Neurosci 8:1139–1141

Marsicano G, Lutz B (1999) Expression of the cannabinoid receptor CB_1 in distinct neuronal subpopulations in the adult mouse forebrain. Eur J Neurosci 11:4213–4225

Marsicano G, Wotjak CT, Azad SC et al. (2002) The endogenous cannabinoid system controls extinction of aversive memories. Nature 418:530–534

Marsicano G, Goodenough S, Monory K et al. (2003) CB_1 cannabinoid receptors and on-demand defense against excitotoxicity. Science (New York, NY) 302:84–88

Matsuda LA, Lolait SJ, Brownstein MJ et al. (1990) Structure of a cannabinoid receptor and functional expression of the cloned cDNA. Nature 346:561–564

Matsuda LA, Bonner TI, Lolait SJ (1993) Localization of cannabinoid receptor mRNA in rat brain. J Comp Neurol 327:535–550

Matyas F, Yanovsky Y, Mackie K et al. (2006) Subcellular localization of type 1 cannabinoid receptors in the rat basal ganglia. Neuroscience 137:337–361

Matyas F, Watanabe M, Mackie K et al. (2007) Molecular architecture of the cannabinoid signaling system in the core of the nucleus accumbens. Ideggyogyaszati szemle 60:187–191

Matyas F, Urban GM, Watanabe M et al. (2008) Identification of the sites of 2-arachidonoylglycerol synthesis and action imply retrograde endocannabinoid signaling at both GABAergic and glutamatergic synapses in the ventral tegmental area. Neuropharmacology 54:95–107

McDonald AJ, Mascagni F (2001) Localization of the CB_1 type cannabinoid receptor in the rat basolateral amygdala: high concentrations in a subpopulation of cholecystokinin-containing interneurons. Neuroscience 107:641–652

Melis M, Perra S, Muntoni AL et al. (2004) Prefrontal cortex stimulation induces 2-arachidonoyl-glycerol-mediated suppression of excitation in dopamine neurons. J Neurosci 24:10707–10715

Monory K, Massa F, Egertova M et al. (2006) The endocannabinoid system controls key epileptogenic circuits in the hippocampus. Neuron 51:455–466

Mulder J, Aguado T, Keimpema E et al. (2008) Endocannabinoid signaling controls pyramidal cell specification and long-range axon patterning. Proceedings of the National Academy of Sciences of the United States of America

Nyilas R, Dudok B, Urban GM et al. (2008) Enzymatic machinery for endocannabinoid biosynthesis associated with calcium stores in glutamatergic axon terminals. J Neurosci 28:1058–1063

Nyilas R, Gregg LC, Mackie K, Watanabe M, Zimmer A, Hohmann AG and Katona I (2009) Molecular architecture of endocannabinoid signaling at nociceptive synapses mediating analgesia. European Journal of Neuroscience, 29:1964–1978.

Nyiri G, Cserep C, Szabadits E et al. (2005) CB_1 cannabinoid receptors are enriched in the perisynaptic annulus and on preterminal segments of hippocampal GABAergic axons. Neuroscience 136:811–822

Ohno-Shosaku T, Maejima T, Kano M (2001) Endogenous cannabinoids mediate retrograde signals from depolarized postsynaptic neurons to presynaptic terminals. Neuron 29:729–738

Okamoto Y, Morishita J, Tsuboi K et al. (2004) Molecular characterization of a phospholipase D generating anandamide and its congeners. J Biol Chem 279:5298–5305

Palomaki VA, Lehtonen M, Savinainen JR et al. (2007) Visualization of 2-arachidonoylglycerol accumulation and cannabinoid CB_1 receptor activity in rat brain cryosections by functional autoradiography. J Neurochem 101:972–981

Pawelzik H, Hughes DI, Thomson AM (2002) Physiological and morphological diversity of immunocytochemically defined parvalbumin- and cholecystokinin-positive interneurones in CA1 of the adult rat hippocampus. J Comp Neurol 443:346–367

Pettit DA, Harrison MP, Olson JM et al. (1998) Immunohistochemical localization of the neural cannabinoid receptor in rat brain. J Neurosci Res 51:391–402

Saper CB, Sawchenko PE (2003) Magic peptides, magic antibodies: guidelines for appropriate controls for immunohistochemistry. J Comp Neurol 465:161–163

Schlicker E, Kathmann M (2001) Modulation of transmitter release via presynaptic cannabinoid receptors. Trends Pharmacol Sci 22:565–572

Simon GM, Cravatt BF (2008) Anandamide biosynthesis catalyzed by the phosphodiesterase GDE1 and detection of glycerophospho-N-acyl ethanolamine precursors in mouse brain. J Biol Chem 283:9341–9349

Stella N, Schweitzer P, Piomelli D (1997) A second endogenous cannabinoid that modulates long-term potentiation. Nature 388:773–778

Straiker A, Mackie K (2007) Metabotropic suppression of excitation in murine autaptic hippocampal neurons. J Physiol 578:773–785

Sugiura T, Kishimoto S, Oka S et al. (2006) Biochemistry, pharmacology and physiology of 2-arachidonoylglycerol, an endogenous cannabinoid receptor ligand. Prog Lipid Res 45:405–446

Terry G, Liow JS, Chernet E et al. (2008) Positron emission tomography imaging using an inverse agonist radioligand to assess cannabinoid CB(1) receptors in rodents. NeuroImage 41:690–698

Tsou K, Brown S, Sanudo-Pena MC et al. (1998) Immunohistochemical distribution of cannabinoid CB_1 receptors in the rat central nervous system. Neuroscience 83:393–411

Tsou K, Mackie K, Sanudo-Pena MC et al. (1999) Cannabinoid CB_1 receptors are localized primarily on cholecystokinin-containing GABAergic interneurons in the rat hippocampal formation. Neuroscience 93:969–975

Uchigashima M, Narushima M, Fukaya M et al. (2007) Subcellular arrangement of molecules for 2-arachidonoyl-glycerol-mediated retrograde signaling and its physiological contribution to synaptic modulation in the striatum. J Neurosci 27:3663–3676

Van Sickle MD, Duncan M, Kingsley PJ et al. (2005) Identification and functional characterization of brainstem cannabinoid CB2 receptors. Science (New York, NY) 310:329–332

Vizi ES, Katona I, Freund TF (2001) Evidence for presynaptic cannabinoid CB(1) receptor-mediated inhibition of noradrenaline release in the guinea pig lung. Eur J Pharmacol 431:237–244

Westlake TM, Howlett AC, Bonner TI et al. (1994) Cannabinoid receptor binding and messenger RNA expression in human brain: an in vitro receptor autoradiography and in situ hybridization histochemistry study of normal aged and Alzheimer's brains. Neuroscience 63:637–652

Wilson RI, Nicoll RA (2001) Endogenous cannabinoids mediate retrograde signalling at hippocampal synapses. Nature 410:588–592

Wilson RI, Kunos G, Nicoll RA (2001) Presynaptic specificity of endocannabinoid signaling in the hippocampus. Neuron 31:453–462

Yasuno F, Brown AK, Zoghbi SS et al. (2008) The PET radioligand [11C]MePPEP binds reversibly and with high specific signal to cannabinoid CB_1 receptors in nonhuman primate brain. Neuropsychopharmacology 33:259–269

Yoshida T, Fukaya M, Uchigashima M et al. (2006) Localization of diacylglycerol lipase-alpha around postsynaptic spine suggests close proximity between production site of an endocannabinoid, 2-arachidonoyl-glycerol, and presynaptic cannabinoid CB_1 receptor. J Neurosci 26:4740–4751

Zimmer A, Zimmer AM, Hohmann AG et al. (1999) Increased mortality, hypoactivity, and hypoalgesia in cannabinoid CB_1 receptor knockout mice. Proc Natl Acad Sci USA 96:5780–5785

Pharmacological Tools in Endocannabinoid Neurobiology

Marco Mor and Alessio Lodola

Contents

Abstract Pharmacological and biochemical investigations on the endocannabinoid system are facilitated by the availability of compounds which interact with its constituents in specific and understandable ways. This chapter describes the main representatives of several classes of chemicals employed as pharmacological tools in this field, focusing on small organic compounds having, where possible, a drug-like structure. Many compounds having different intrinsic activity and selectivity towards the G-protein coupled receptors (GPCR) CB_1 and CB_2 are now available and are currently employed in research protocols. Recently, allosteric ligands for

M. Mor (✉) and A. Lodola
Dipartimento Farmaceutico, Università degli Studi di Parma, viale G. P. Usberti 27/A Campus Universitario, I-43100 Parma, Italy
e-mail: marco.mor@unipr.it

D. Kendall and S. Alexander (eds.), *Behavioral Neurobiology of the Endocannabinoid System*, Current Topics in Behavioral Neurosciences 1,
DOI: 10.1007/978-3-540-88955-7_4, © Springer-Verlag Berlin Heidelberg 2009

CB_1 receptor and selective ligands for GPR55, a newly characterised GPCR, have also been described in the literature. As for compounds affecting endocannabinoid levels in living tissues, many classes of selective and, in some cases, drug-like inhibitors of FAAH are available, while only compounds with poor selectivity or in vivo activity are known to inhibit other enzymes involved in endocannabinoid catabolism, such as NAAA or MGL, and in endocannabinoid biosynthesis.

Keywords CB1 ligands • CB2 ligands • Anandamide Transport • FAAH inhibitors • MGL inhibitors • NAAA inhibitors

Abbreviations

2-AG	2-Arachidonoylglycerol
AEA	Anandamide
CCP	*N*-(cyclohexylcarbonyl)pentadecylamine
DAG	Diacylglycerol
ECB	Endocannabinoid
FAAH	Fatty acid amide hydrolase
GPCR	G protein-coupled receptor
MGL	Monoacylglycerol lipase
NAAA	*N*-Acylethanolamine acid amidase
NAE	*N*-Acylethanolamine
NAPE-PLD	*N*-Acylphosphatidylethanolamine phospholipase D
OEA	*N*-Oleoylethanolamine
PA	Phosphatidic acid
PEA	*N*-Palmitoylethanolamine
PLC	Phospholipase C
TRPV1	Transient receptor potential vanilloid type 1
Δ^9-THC	Δ^9-Tetrahydrocannabinol

1 Cannabinoid Receptor Ligands

The identification of cannabinoid receptors and the characterization of their endogenous ligands has triggered an exponential growth of studies exploring the endocannabinoid (ECB) system and its regulatory functions in physiological and pathological processes. Such studies have been significantly facilitated, for both in vitro experiments and animal models, by the introduction of selective cannabinoid receptor ligands and inhibitors of ECB metabolism and transport. In the last decade, the ECB system was found to be involved in several physiological functions, both in the central nervous system (CNS) and in the peripheral one as well as in other organs. Modulation of the activity of ECBs revealed therapeutic promises in a wide range of diseases and pathological conditions (Elsohly 2002), possibly overcoming the traditional limit to the development of cannabinoid-based medicines, i. e. the socially intolerable psychoactive properties of plant-derived or synthetic agonists, mainly mediated by CB_1 receptor activation. However, this problem does not emerge when the therapeutic scope is accomplished by treatment with CB_1 receptor antagonists, as in obesity, or with agents enhancing the action of ECBs through the blockade of their metabolism or transport. Moreover, the use of selective CB_2 receptor agonists, lacking psychoactive properties, may represent another promising opportunity for certain pathological conditions. Full exploitation of the therapeutic promises for drugs acting on the ECB system, however, still needs experimental work with in vitro and in vivo models, which could elucidate the benefits and the risks associated with selective perturbations in the activity of cannabinoid receptors and of ECB distribution and metabolism.

In this chapter, compounds are described which are commonly employed as pharmacological tools in the study of the ECB system, or that may have potential usefulness in this sense. In particular, attention is focused on drug-like chemicals, also providing brief outlines of the possible therapeutic benefits of some compounds. The chapter is organised in two main sections, the former dedicated to cannabinoid receptor ligands, the latter describing the compounds affecting ECB metabolism.

2 Natural Products and Non-Selective Ligands

Among the more than 60 cannabinoids contained in extracts from the plant *Cannabis sativa*, the best known is an isomer of tetrahydrocannabinol, Δ^9-THC (Fig. 1) (Gaoni and Mechoulam 1964), which is also its main psychoactive component. It is an agonist at both CB_1 and CB_2 receptors and it partly mimics the actions of the endogenous cannabinoids (Howlett et al. 2002). The therapeutical potential of several different components still remains to be fully explored (Mechoulam 2005; Thomas et al. 2005). Among these, cannabidiol has raised remarkable interest in spite of its lack of affinity for CB_1 or CB_2 receptors, mainly for its anti-proliferative activity and its potentiation of cannabinoid actions (Mechoulam et al. 2007). Furthermore, the finding that species different from *C. sativa* have components able to bind to CB receptors opens new and interesting possibilities for pharmacological research (Gertsch et al. 2006).

The identification of Δ^9-THC as the main active component of *C. sativa* stimulated the preparation of a range of synthetic compounds having similar biological activity and different grades of chemical diversity. These efforts led to the identification of novel chemical classes able to bind CB_1 and CB_2 receptors with different profiles of selectivity and intrinsic activity, as well as facilitating the identification of the natural modulators of CB receptors, the so-called ECBs (Lambert and Fowler 2005). Cannabinoid receptor ligands can be grouped in four classes – classical,

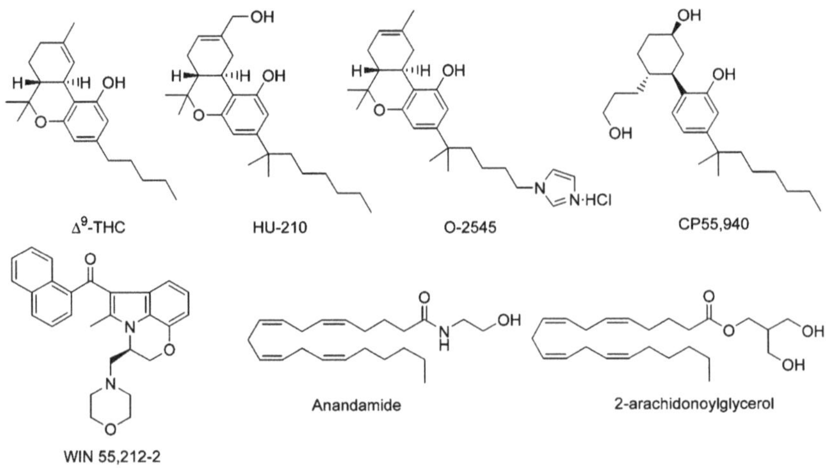

Fig. 1 Some plant-derived, synthetic or endogenous cannabinoids

non-classical, aminoalkylindole and eicosanoids – which are briefly summarised here (Pacher et al. 2006).

2.1 *Classical Cannabinoids*

These include plant-derived compounds or synthetic analogues having a dibenzopyran scaffold, such as Δ^9-THC and HU210 (Howlett et al. 2002). Δ^9-THC (Fig. 1) binds with similar affinity to CB_1 and CB_2 receptors and behaves as a partial agonist at both of these receptor subtypes (Felder et al. 1995), with lower efficacy at CB_2 than at CB_1 receptors (Bayewitch et al. 1996). HU210 shows affinity values for CB_1 and CB_2 receptors exceeding those of many other cannabinoids. It is a potent cannabinoid receptor agonist and its pharmacological effects in vivo are exceptionally long-lasting. The enhanced affinity and efficacy shown by HU210 at cannabinoid receptors can be largely attributed to the replacement of the pentyl side chain of Δ^8-THC (an isomer of Δ^9-THC resembling it both in its affinities for CB_1 and CB_2 receptors and in CB_1 receptor efficacy) with the dimethylheptyl group. The poor water solubility of classical cannabinoids requires the use of solubilizing agents for many pharmacological uses. Addition of an imidazole at the side chain of Δ^8- THC allows the use of a water-soluble hydrochloride salt (O-2545, Fig. 1), maintaining high affinity and efficacy at CB_1 and CB_2 receptors (Martin et al. 2006).

2.2 *Non-Classical Cannabinoids*

These present limited modifications in their scaffold, generally being bicyclic analogues of Δ^9-THC without the central pyran ring. The most commonly used non-classical cannabinoid is CP55940 (Fig. 1), which has CB_1 and CB_2 affinities in the low nanomolar range and exhibits relatively high efficacy at both receptor subtypes. In its [^{3}H]-labelled form, CP55940 was employed to demonstrate definitively the existence of high-affinity, saturable, stereospecific binding sites for synthetic cannabinoid agonists in rat brain and it is currently employed in competitive binding experiments (Devane et al. 1988).

2.3 *Aminoalkylindoles*

The prototype of this group is WIN55212-2 (Fig. 1), which does not show clear structural similarity with classical, non-classical or eicosanoid cannabinoids. It has been reported that WIN55212-2 binds to the CB_1 receptor in a different way from-classical and non-classical cannabinoids (Song and Bonner 1996), even if mutual displacement is observed between WIN55212-2 and non-aminoalkylindole cannabinoids at CB_1 binding site (Kuster et al. 1993). Like CP55940, WIN55212-2 exhibits relatively high efficacy at CB_1 and CB_2 receptors and possesses CB_1 and CB_2 affinities in the low nanomolar range. However, it has slightly greater affinity for CB_2 than for CB_1 receptors.

2.4 *Eicosanoids*

The prototypes and most investigated members of this group are the ECBs *N*-arachidonoylethanolamine (anandamide or AEA) and 2-arachidonoylglycerol (2-AG), whose chemical structures are reported in Fig. 1 (Piomelli 2003). AEA acts as an endogenous ligand for CB_1 and CB_2 receptors, and for other targets, such as the transient receptor potential vanilloid type 1 (TRPV1) channel. AEA affinity for the CB_1 receptor is higher than for the CB_2 receptor. When protected from enzymatic hydrolysis, it shows a CB_1 affinity comparable to that of Δ^9-THC. Depending on the tissue and biological response being measured, it can behave as a partial or full agonist at CB_1 receptors. It has low efficacy at CB_2 receptors, where it may even act as an antagonist, depending on the interacting G proteins (Gonsorek et al. 2000). 2-AG is a full agonist at both CB_1 and CB_2 receptors. Some authors consider 2-AG the true ligand for CB_2 receptors, as AEA binds poorly to these receptors. The basal levels of 2-AG in the brain are much higher (about two orders of magnitude) than those of AEA, suggesting that only a fraction of the total is involved in ECB signalling (Sugiura et al. 2006).

Other putative ECBs are noladin ether (2-arachidonylglyceryl ether), *N*-arachidonoyldopamine and virodhamine (*O*-arachidonoylethanolamine); their biological significance and biochemical characteristics are still to be fully investigated (Piomelli 2003).

3 CB_1-Selective Ligands

Δ^9-THC and most of the cannabinoid ligands described so far show similar affinities for CB_1 and CB_2 receptors. Only recently, synthetic ligands able to discriminate between the two receptor isoforms emerged, including agonists and antagonists. The development of potent and highly selective CB_1 and CB_2 receptor antagonists is particularly noteworthy as it provided critically important tools to explore the physiological functions of these receptor subtypes (Rinaldi-Carmona et al. 1994; Rinaldi-Carmona et al. 1998).

3.1 Agonists

AEA shows some CB_1 selectivity and represents a template for CB_1-selective agonists, but its use for in vivo experiments is limited by reduced metabolic stability. The replacement of a hydrogen atom with a methyl group at the 1' position of AEA led to (*R*)-(+)-methanandamide, a rather CB_1-selective agonist endowed with improved metabolic stability. Other modifications of the ethanolamine moiety led to the potent CB_1-selective agonists arachidonoyl-2'-chloroethylamide (ACEA) and arachidonoylcyclopropylamide (ACPA), both exhibiting good CB_1 efficacy (Fig. 2) (Hillard et al. 1999).

Pharmacological investigations showed that ACEA and ACPA produce hypothermia in mice, an effect that can be inhibited by co-administration of the selective CB_1 receptor antagonist rimonabant (Fig. 3). However, unlike methanandamide, neither ACEA nor ACPA show resistance to enzymatic hydrolysis (Di Marzo et al. 2001), even if it has been shown that the addition of a methyl group to the 1' carbon of ACEA markedly decreases the susceptibility of this molecule to FAAH-mediated hydrolysis (Jarrahian et al. 2000). ACPA and ACEA also showed poor affinity for CB_2 receptors, confirming that selectivity for CB_1 receptor over CB_2 receptors can be achieved by designing structural analogues of AEA (Di Marzo et al. 2001).

3.2 Antagonists and Inverse Agonists

The gold standard among CB_1-selective ligands that counteract the effects of cannabinoid agonists is the diarylpyrazole rimonabant (SR141716A, Fig. 3)

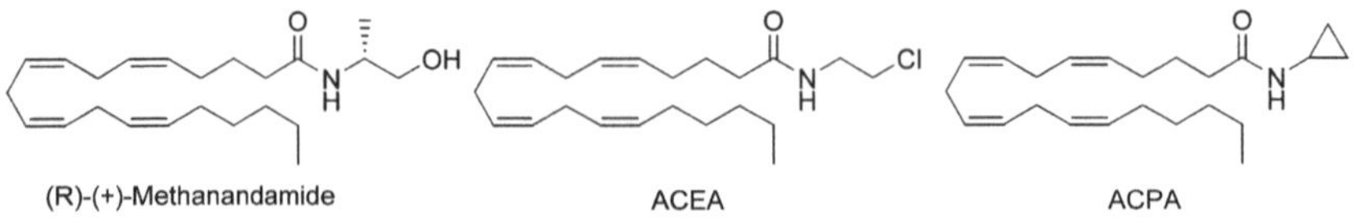

Fig. 2 CB_1-selective agonists based on the anandamide template

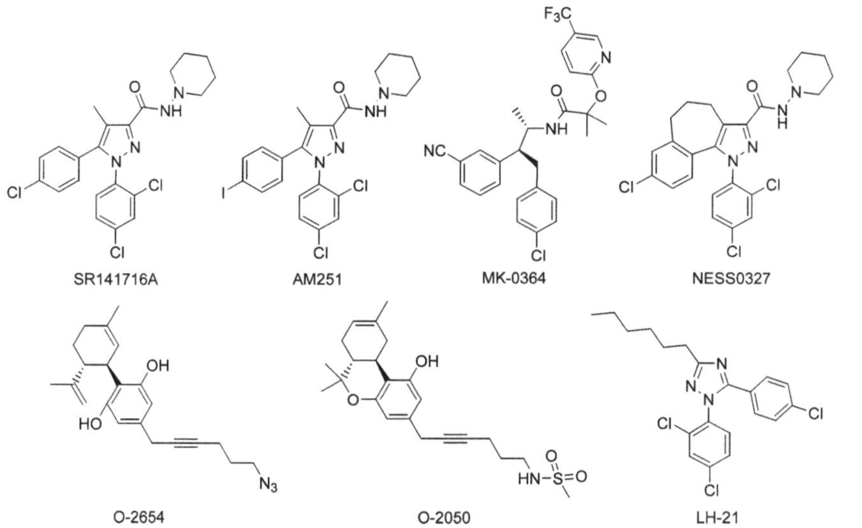

Fig. 3 CB_1-selective antagonists/inverse agonists

(Rinaldi-Carmona et al. 1994), which was also the first compound of this class to reach the market as an anti-obesity drug. It is a highly potent and selective CB_1 receptor ligand that readily prevents or reverses CB_1-mediated effects both in vitro and in vivo (Pertwee 2005a) and is now regarded as the prototypical CB_1 antagonist/inverse agonist. Other notable CB_1-selective antagonists/inverse agonists are AM251 and AM281, both belonging to the same diarylpyrazole class as rimonabant, but differing in the presence of an iodine, instead of a chlorine, on the 5-phenyl ring (AM251, Fig. 3) or in the presence of a morpholine ring instead of a piperidine one (AM281). Several pharmacological studies employed AM251, which has similar affinity and intrinsic activity at CB_1 receptors to rimonabant, while it has been claimed to be more selective, as rimonabant retains some effects in CB_1 knockout mice (Muccioli 2007). Taken together, the two pyrazole derivatives rimonabant and AM251 allowed for a better understanding of the ECB system, and represented optimal tools for exploration of the therapeutic potential of CB_1 antagonists/inverse agonists, eventually leading to the introduction of rimonabant in the clinical market. Despite their initial classification as CB_1 antagonists, there is significant evidence that rimonabant, AM251 and AM281 are inverse agonists (Pertwee 2005b). In fact, these CB_1 receptor ligands can not only attenuate the effects of CB_1 receptor agonists, but also elicit responses by themselves in some CB_1 receptor-containing tissues that are opposite in direction from those elicited by CB_1 receptor agonists, reducing the constitutive activity of CB_1 receptors.

Rimonabant was used in numerous animal models to elucidate both the role of the ECB system and the therapeutic promise of CB_1 receptor "antagonists". Much of the pioneering research in this field was performed using this compound, taking advantage of its selectivity and oral bioavailability. Registered as Acomplia, it has been approved for use in several countries for the treatment of obesity, though the Food and Drug Administration has expressed concern about the potential for adverse neurological and psychiatric effects, considering the widespread distribution of CB_1 receptors in the brain (Woods 2007). Efforts to generate new CB_1 antagonists/inverse agonists led to several interesting compounds, such as the highly potent and selective ligand MK-0364 (taranabant, Fig. 3) (Fong et al. 2007), a new clinical candidate for the treatment of obesity whose therapeutical potential is under investigation.

The clinical relevance of inverse agonism exhibited by rimonabant and its congeners is a matter of debate, and the availability of neutral antagonists is an urgent need in cannabinoid research, to differentiate the effects on ECB tone from those on receptor constitutive activity in animal tissues. Even if compounds with a purely antagonist profile and an established use as pharmacological tools are still lacking, neutral antagonism in biochemical assays and interesting in vivo effects have been reported for some compounds (Fig. 3), such as O-2654 (Thomas et al. 2004), O-2050 (Gardner and Mallet 2006), LH-21 (Pavon et al. 2006), and AM4113 (Sink et al. 2008).

Recently, some series of conformationally constrained derivatives have also been synthesised to achieve potency improvements by blocking their structure in the putative active conformation (Thomas et al. 2006). This strategy led to NESS0327 (Fig. 3), which has been reported to be a neutral antagonist with exceptional affinity for the CB_1 receptor in the femtomolar range. Furthermore, CB_1/CB_2 affinity ratio showed that NESS0327 is more than 60,000 times more selective for the CB_1 receptor, whereas rimonabant only showed a 285-fold selectivity (Ruiu et al. 2003). More extensive studies are needed to determine whether these new compounds will prove to be superior to rimonabant when used in vivo.

Wider dissertations regarding CB_1 receptor blockage and its therapeutical potential are available throughout this book and in dedicated reviews (Muccioli 2007; Jagerovic et al. 2008).

3.3 Allosteric Modulators

Besides CB_1 antagonists and inverse agonists, there are other compounds that can modulate the effects of CB_1 receptor agonists through different mechanisms. Experiments with the synthetic indole derivative Org 27569 (Fig. 4) and its close

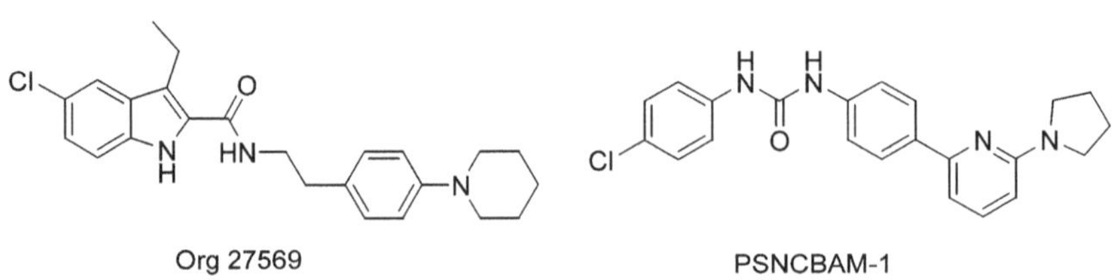

Fig. 4 Allosteric modulators of CB_1 receptors

analogues Org 29647 and Org 27759 revealed that these ligands bind allosterically to the CB_1 receptor and elicit a conformational change that increases agonist affinity for the orthosteric binding site (Price et al. 2006). These findings indicated the presence of a new way of modulating CB_1 receptor activation by endogenously released ECBs. Indeed, a new allosteric modulator of the CB_1 receptor, PSNCBAM-1 (Fig. 4), is able to increase CB_1 affinity of orthosteric agonists, antagonising their functional activity. Its effects on food intake and body weight in rats provided a first report of in vivo activity for an allosteric CB_1 receptor antagonist (Horswill et al. 2007). Future research will certainly address the question of whether allosteric modulators of the CB_1 receptor are advantageous in some pathological scenarios by comparison with the currently available orthosteric ligands (Ross 2007).

4 CB_2-Selective Ligands

4.1 *Agonists*

The well-known differences in receptor distribution and signal transduction mechanisms between CB_1 and CB_2 receptors are likely to account for the relative absence of CNS side effects induced by CB_2 agonists. These considerations suggest that novel therapies targeting CB_2 receptors may achieve their therapeutic potential without the risks related to the socially intolerable psychoactive properties of CB_1 agonists. Significant drug discovery efforts have thus been directed towards the development and characterization of CB_2-selective agonists, both in vitro and in vivo, mainly with the aim of evaluating and validating the CB_2 receptor as a target for the treatment of inflammatory and neuropathic pain (Guindon and Hohmann 2008). HU308 (Fig. 5) was one of the first CB_2-selective agonists employed, exhibiting low affinity for CB_1 receptors (Hanus et al. 1999). HU308 exhibits anti-inflammatory and peripheral anti-hyperalgesic properties, which are reversed by the CB_2 antagonist SR144528 but not by the CB_1 antagonist rimonabant. HU308 does not show CNS activity in the classical tetrad of behavioural tests

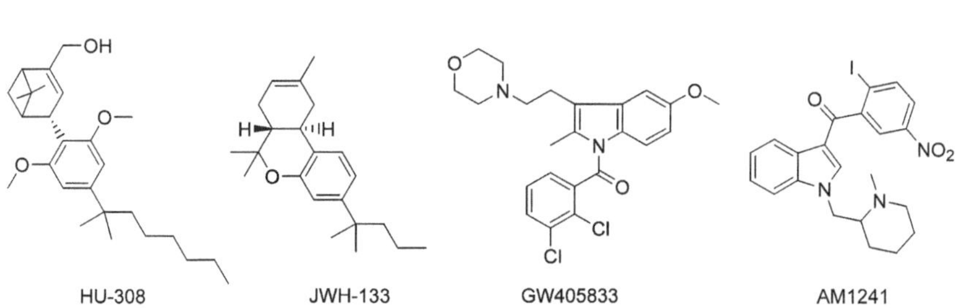

Fig. 5 CB_2-selective agonists

assessing signs of CB_1 receptor activation associated with Δ^9-THC (Gaoni and Mechoulam 1964).

Other CB_2-selective agonists (Fig. 5) widely used as pharmacological tools are the classical cannabinoid, JWH133, and the less selective aminoalkylindole, JWH015 (Howlett et al. 2002; Pertwee 2005b). Both behave as potent CB_2 agonists in functional assays, inhibiting both inflammatory and neuropathic hyperalgesia through a CB_2-dependent mechanism (Huffman et al. 1999; Jonsson et al. 2006). The indole derivative GW405833 (also indicated as L768242) behaves as a potent partial agonist at the CB_2 receptor and produces anti-hyperalgesic effects in several rodent models of pain (Valenzano et al. 2005). Another indole derivative, AM1241, is reported to give CB_2-mediated anti-hyperalgesic effects in multiple models of persistent nociception, including those induced by tissue and nerve injury, while avoiding the classical signs of CB_1 activation. While the racemate of this compound behaves either as an agonist or an inverse agonist, depending on receptor species, its (*S*) enantiomer showed slightly lower receptor affinity, but full agonist behaviour on cAMP accumulation in cell lines expressing recombinant human, rat, and mouse CB_2 receptors (Bingham et al. 2007).

The structural requirements for ligands binding at CB_2 receptors and the therapeutic potential of CB_2 receptor agonists have been thoroughly discussed in the recent literature (Whiteside et al. 2007; Poso and Huffman 2008), and several novel structures of potent and selective compounds which may be employed as pharmacological tools are being reported with an increasing rate (Giblin et al. 2007; Manera et al. 2006; Murineddu et al. 2006; Khanolkar et al. 2007; Gonsiorek et al. 2007; Ermann et al. 2008; Ohta et al. 2008; Yao et al. 2008; Kikuchi et al. 2008).

4.2 *Antagonists*

Fewer classes of selective CB_2 receptor antagonists have been reported so far (Muccioli 2007). Two prototypical selective antagonists are SR144528, developed from a pyrazole scaffold (Rinaldi-Carmona et al. 1998), and AM630, having an indole nucleus (Fig. 6). Both are extensively used as pharmacological tools to block CB_2 activation, but they are not pure antagonists, showing inverse intrinsic activity in some cannabinoid receptor-containing bioassay systems (Howlett et al. 2002). Potent CB_2 receptor ligands based on a quinoline carboxamide moiety have also been reported, such as JTE907 (Fig. 6) which is endowed with high selectivity for the CB_2 receptor and inverse-agonistic properties. JTE907 has been employed for in vivo investigations, showing anti-pruritic activity in a model of dermatitis when orally administered to mice (Iwamura et al. 2001; Maekawa et al. 2006). More recently, a new class of selective CB_2 receptor ligands was developed, based on a triaryl bis-sulfone backbone exemplified by Sch.336 (Fig. 6). These compounds exhibit inverse-agonistic properties at CB_2 receptors, with Sch.336 inhibiting leukocyte trafficking in rodents and blocking lung eosinophilia in mice, a model for allergic asthma (Lunn et al. 2006). A novel radioligand with the same structure,

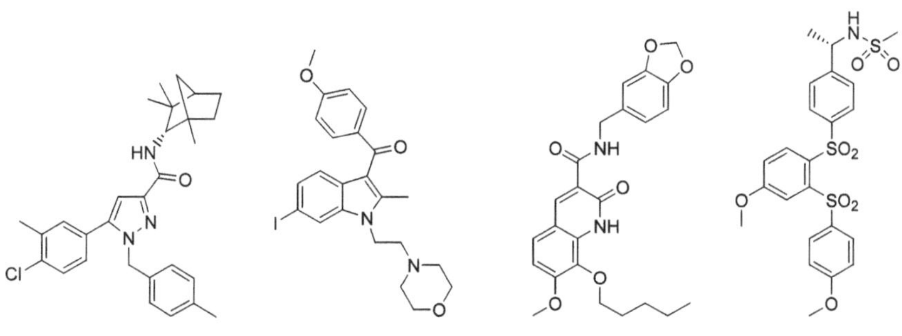

SR144528 AM630 JTE-907 Sch.336

Fig. 6 CB_2-selective antagonists

named [^{35}S]-Sch225336, has also been recently proposed as a useful tool for binding assays on CB_2 receptors (Gonsiorek et al. 2006).

5 Ligands of Other ECB Receptors

Besides CB_1 and CB_2 receptors, ECBs and exogenous cannabinoids target other binding sites, which may explain why the pharmacology of cannabinoids cannot be fully explained by their activity at those two receptors only. The field is further complicated by the heterogeneity of endogenous compounds chemically related to the ECBs 2-AG and AEA, having their own spectra of targets. These compounds, thoroughly described in the previous chapters of this book, include members of the *N*-acylethanolamine (NAE) family such as *N*-palmitoylethanolamine (PEA) and *N*-oleoylethanolamine (OEA) shown in Fig. 7, which also activate the nuclear receptor PPARα (Fu et al. 2003; Lo Verme et al. 2005). Additional cannabinoid targets include the ion channel TRPV1 and non-CB_1/CB_2 G-protein coupled receptors, such as GPR119 and GPR55. While the pharmacology of many of these targets has been recently reviewed elsewhere (Brown 2007), in this context it is worth mentioning, as a promising pharmacological tool, the high selectivity for GPR55 shown by compound O-1602 (Fig. 7) This is an agonist with nanomolar potency at the GPR55 receptor, giving no sign of CB_1 or CB_2 activation at micromolar concentrations (Ryberg et al. 2007).

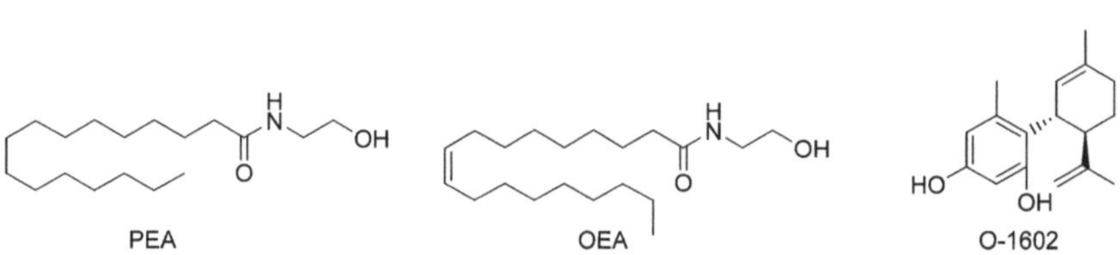

PEA OEA O-1602

Fig. 7 PEA, OEA and the selective GPR55 agonist, O-1602

6 Modulators of ECB Metabolism

ECBs are produced on demand, as a response to elevated intracellular calcium levels. AEA and 2-AG are synthesised from membrane phospholipids through at least two distinct enzymatic reactions (Di Marzo et al. 1996, 1999). Once synthesised, AEA and 2-AG have to reach their membrane receptors to evoke cannabimimetic effects. To control ECB levels, effective mechanisms for their synthesis, trafficking and removal are present inside the cells (Piomelli 2003). Many enzymes involved in the degradation of ECBs have been identified and characterised, and their roles have been in some cases elucidated, exploiting the availability of potent and selective inhibitors. On the other hand, the details of ECB biosynthetic pathways are still more elusive, and a smaller number of well-characterised compounds able to selectively act at this level is available. A thorough description of ECB biosynthesis and inactivation is reported in Chapter 1 of this book. Here, we will focus on the pharmacological tools which may be employed to assess the role of these enzymatic components of the ECB system. The first part of this section looks at the modulators of ECB catabolism, whereas the second part reports some findings on agents blocking ECB synthesis.

6.1 ECB Catabolism

To terminate ECB signalling, these endogenous compounds must be chemically inactivated. If we focus on the two main ECBs, the amide AEA and the ester 2-AG, this is eventually accomplished by enzymatic hydrolysis. While these lipidic agents can undergo different metabolic transformations, their inactivation appears to be mainly regulated by a two-step process: ECBs must be transported into cells by a facilitated transport system (Beltramo et al. 1997) and then hydrolyzed by the action of different enzymes (Bari et al. 2006). AEA is mainly hydrolyzed, to arachidonic acid and ethanolamine, by the action of a hydrolase selective for fatty acid amides, named fatty acid amide hydrolase (FAAH) (McKinney and Cravat 2005), whereas 2-AG is degraded to arachidonic acid and glycerol, mainly by the action of a monoglyceride lipase (MGL) (Dinh et al. 2002a). Other enzymes causing the hydrolysis of AEA and 2-AG have been identified, such as FAAH-2, a homologue of FAAH expressed in periphery and only present in mammals (Wei et al. 2006), and of *N*-acylethanolamine-hydrolyzing acid amidase (NAAA) (Tsuboi et al. 2005), another enzyme catalysing the hydrolysis of linear fatty acid amides in humans and rodents. While the catabolism of AEA, 2-AG and other lipid-deriving signalling molecules will disclose unexpected complexity and interconnection, the mostly established pharmacological tools are those affecting AEA transport or inhibition of FAAH, MGL and NAAA.

6.1.1 Inhibition of Anandamide Transport

AEA transport in neurons is structurally specific, displays classical saturation kinetics, and is inhibited by AEA analogues (Beltramo et al. 1997). The first inhibitor of the cellular uptake of AEA to be developed was *N*-(4-hydroxyphenyl) arachidonoylamide (AM404, Fig. 8). AM404 produces a variety of in vivo actions, including potentiation of the effects on hypothermia in FAAH knockout mice of low doses of AEA, which are blocked by CB_1 antagonists (Fowler et al. 2005). However, AM404 is not particularly selective, as it also inhibits FAAH, binds to CB_1 receptors and activates the vanilloid receptor TRPV1 at concentrations similar or below those at which it inhibits AEA uptake (Lambert and Fowler 2005). Many acyl-based compounds have been investigated for their ability to inhibit AEA uptake, leading to the identification of the arachidonoyl derivatives UCM707 and VDM11 and the oleoyl derivative OMDM-2 (Fig. 8) (Lopez-Rodrguez et al. 2003; Ortar et al. 2003).

The existence of a specific transporter had been questioned, and in fact hydrolytic activity of FAAH may be responsible for the maintenance of a concentration gradient of AEA. In this regard, FAAH inhibition by compounds also acting as reuptake blockers can be functionally relevant and, to some extent, limits their usefulness as pharmacological tools (Glaser et al. 2003). However, detailed investigations demonstrated that AM1172 (Fig. 8), a hydrolysis-resistant analogue of AM404, inhibits AEA internalisation without interacting with FAAH in brain neurons and astrocytoma cells (Fegley et al. 2004), confirming that membrane transport and intracellular hydrolysis are two discrete steps in AEA deactivation. It has been claimed, however, that in different conditions AM1172 may be a weak

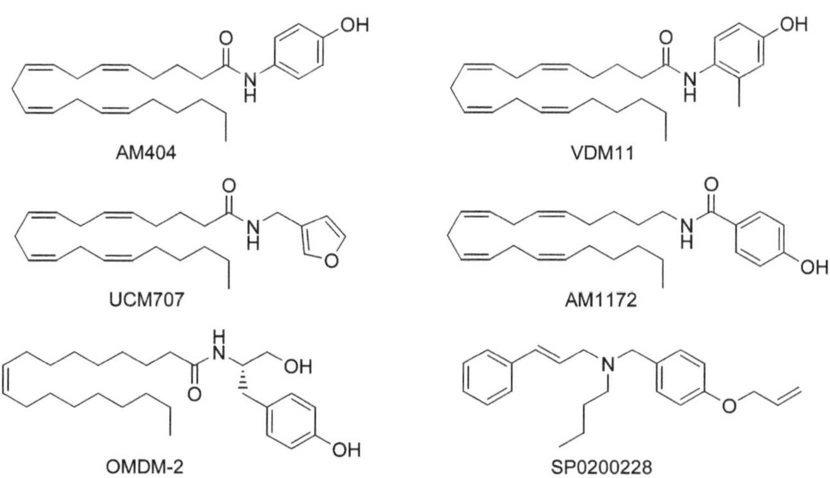

Fig. 8 Inhibitors of anandamide transport

inhibitor of FAAH (Hillard et al. 2007; Fowler et al. 2004; Vandevoorde and Fowler 2005).

A new generation of ECB transport inhibitors is strongly needed, particularly with improved drug-likeness and selectivity compared to the cited fatty acid amides. A different chemotype is represented by a class of trialkylamines (e.g. SP0200228, Fig. 8) that inhibit AEA accumulation and amplify the analgesic effects of AEA (Piomelli 2005). However, research in this area is delayed by the elusive nature of the ECB transport system, whose molecular identity is still undefined.

6.1.2 FAAH Inhibitors

The characterization of FAAH and MGL in many tissues boosted the search of drug-like selective inhibitors, to be used for a characterization of the physiological and pathological roles of ECBs and as potential new drugs. In fact, partly as a result of experiments with FAAH and MGL inhibitors, there is now evidence that ECB levels are unbalanced in certain pathological conditions, supporting the hypothesis that such inhibitors can have therapeutic applications (Piomelli 2005).

Immediately after the discovery of AEA, a widely used inhibitor of its enzymatic hydrolysis was the non-selective, irreversible serine protease inhibitor, phenylmethylsulphonyl fluoride (PMSF, Fig. 9) (Deutsch and Chin 1993), which also inhibits MGL, but only at higher concentrations (Ho and Hillard 2005). More potent covalent inhibitors of FAAH, such as methyl arachidonyl fluorophosphonate

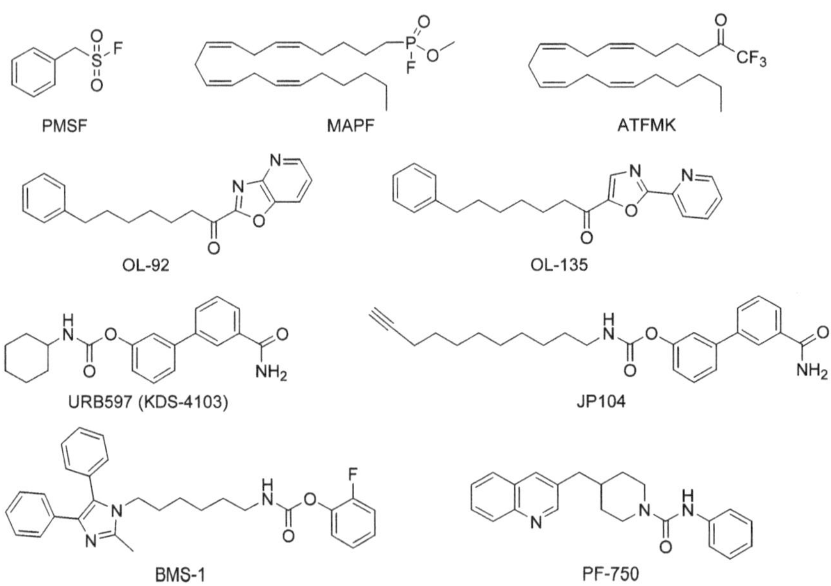

Fig. 9 Representatives of main FAAH inhibitor classes

(MAFP) (De Petrocellis et al. 1997) and palmitylsulfonyl fluoride (AM374) (Deutsch et al. 1997), were also used, but all these compounds were poorly selective. Looking for more selective inhibitors, less reactive compounds were taken into consideration. An electron-withdrawing trifluoromethyl group close to a carbonyl makes arachidonoyl trifluoromethyl ketone (ATFMK) a reversible FAAH inhibitor, considered as a transition state analogue (Koutek et al. 1994). A similar strategy, also exploiting selective recognition at the FAAH catalytic site, led to the class of α-ketoheterocycles, which includes very potent and selective inhibitors such as OL-92 and OL-135 (Boger et al. 2005) (Fig. 9). These are competitive inhibitors presumably acting via reversible hemiketal formation with the active site Ser241 (Boger et al 2000). OL-92 and OL-135 proved to be exceptionally potent and selective FAAH inhibitors in vitro, enhancing ECB signalling and producing analgesia in vivo (Lichtman et al. 2004).

Amino acid conjugates of arachidonic acid also provided FAAH inhibitors, possibly acting as competing substrates (Lambert and Fowler 2005). This is the case for *N*-arachidonoylserotonin, which displays dual activity as a FAAH inhibitor and a TRPV1 antagonist, being highly effective against both acute and chronic peripheral pain (Maione et al. 2007).

Among the FAAH inhibitors developed so far, the *O*-aryl carbamates represent the most promising class of clinical candidates for the treatment of CNS and peripheral disorders. The *O*-aryl carbamate class includes potent and selective inhibitors based on the *N*-cyclohexylcarbamic acid *O*-aryl ester scaffold (Kathuria et al. 2003; Mor et al. 2004), such as URB597 (KDS-4103, Fig. 9), which have been intensively investigated to reveal a pharmacological profile characterised by a unique combination of analgesic, anxiolytic-like, and anti-depressant-like properties (Piomelli et al. 2006; Jayamanne et al. 2006; Gobbi et al. 2005; Bortolato et al. 2007). URB597 inhibits FAAH by carbamoylation of the active nucleophile Ser241, giving irreversible inactivation as revealed by mass spectrometry (Alexander and Cravatt 2005) and computational investigations (Lodola et al. 2008). Furthermore, URB597 inhibits FAAH in vitro at nanomolar concentrations, showing no displacement of radiolabelled ligands from CB_1 and CB_2 receptors. Consistently, URB597 does not evoke catalepsy or hypothermia in vivo, supporting the hypothesis that selective enhancement of ECB transmission by chemical inactivation of FAAH does not produce the typical signs of intoxication by exogenous cannabinoids (Russo et al. 2007; Kathuria et al. 2003). Several classes of carbamate derivatives and related compounds, exemplified here by JP104 and BMS-1 (Fig. 9), have been developed by academic and industrial groups. For more detailed information, the reader is referred to review articles dedicated to FAAH inhibitors (Vandevoorde 2008; Labar and Michaux 2007).

Recently, the issue of in vivo selectivity has been raised for URB597 and other FAAH inhibitors. It has been shown that these derivatives display multiple off-targets, such as rat liver carboxylesterases (Zhang et al. 2007), due to the tendency of the carbamic fragment to react with different serine hydrolases. Although the impact of these observations for the clinical development of this class of compounds has not been assessed, efforts were made to reduce the intrinsic reactivity of

FAAH-carbamoylating agents, exploiting the strong nucleophilicity of its active serine, which is part of an unusual serine–serine–lysine catalytic triad (Lodola et al. 2005). A novel class of inhibitors based on the chemically stable urea group has been recently disclosed. As an example from this class, PF-750 was found to inhibit FAAH in a time-dependent manner by covalently modifying the nucleophile Ser241, avoiding any interactions with all the other mammalian serine hydrolases tested so far (Ahn et al. 2007).

6.1.3 MGL Inhibitors

MGL is a traditional serine hydrolase endowed with a serine–histidine–aspartate catalytic triad (Karlsson et al. 1997) and it is responsible for 2-AG hydrolysis in rat cerebellar membranes (Saario et al. 2004). Several compounds able to inhibit MGL have been reported so far, but their therapeutic potential remains almost unexplored because of their poor selectivity and/or limited in vivo potency (Saario and Laitinen 2007b). The situation is also complicated by the presence of more than one molecular entity endowed with MGL activity (Muccioli et al. 2007). The first compounds reported to inhibit MGL hydrolytic activity were identified among non-specific serine hydrolase inhibitors, such as MAFP, PMSF, AM374 (Saario and Laitinen 2007a; Dinh et al. 2002b). Commonly used lipase inhibitors, such as RHC-80267 or tetrahydrolipstatin (THL or orlistat, Fig. 10), are also known to inhibit 2-AG hydrolysis in rat cerebellar membranes, but their inhibitory effect on DAGLα, an enzyme responsible for 2-AG synthesis, hampers their use for in vivo investigations (Bisogno et al. 2006).

Looking for selective inhibitors of MGL, various analogues of 2-AG have been synthesised, leading to the identification of α-methyl-1-arachidonoylglycerol (α-Me-1-AG), which inhibits both cytoplasmic and membrane MGL at micromolar concentrations, with a weak effect on FAAH activity (Ghafouri et al. 2004).

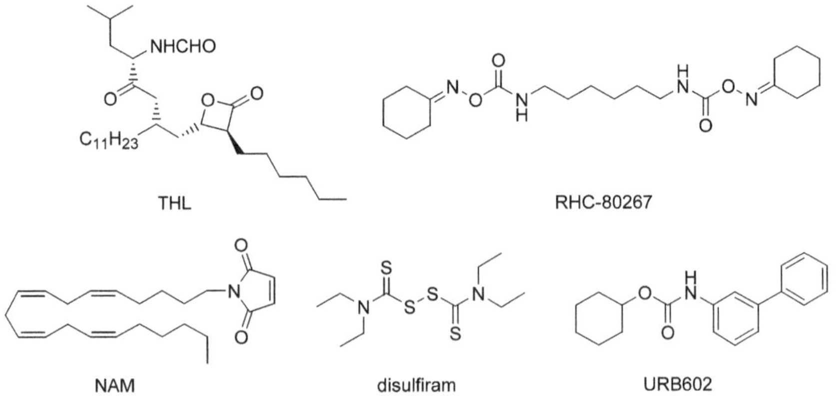

Fig. 10 MGL inhibitors

α-Me-1-AG and other 1-AG derivatives were also synthesised, but potency improvements were not associated with selectivity towards FAAH inhibition (Vandevoorde 2008).

Sulfhydryl-specific compounds, such as *p*-chloromercuribenzoic acid (*p*-CMB) or mercury chloride, inhibit MGL, indicating that cysteine residues may have a key role in regulating MGL hydrolytic activity. Thus, *N*-arachidonylmaleimide (NAM, Fig. 10) inhibits 2-AG hydrolysis in rat cerebellar membranes with nanomolar potency (Saario et al. 2005). It has been proposed that the maleimide scaffold can covalently bind a cysteine sulfhydryl group by a Michael addition mechanism, thus explaining the irreversible inhibition of the enzyme (Saario et al. 2005). These findings prompted researchers to screen disulfide-containing agents, leading to the observations that disulfiram (a well-known inhibitor of aldehyde dehydrogenase, used for decades to treat alcoholism) inhibits human purified MGL, and that its inhibition is reversed by the reducing agent dithiothreitol (DTT) (Labar et al. 2007). The efficacy of sulfhydryl-specific compounds is underlined by the erroneous identification of a putative inhibitor (URB754, structure not shown) as an MGL inhibitor, while the observed activity was actually due to a bis(dimethylthio) mercury impurity present in a commercial sample of this compound (Makara et al. 2005, 2007; Tonidandel et al. 2006).

At present, the most strongly characterised inhibitor of MGL is the carbamate derivative URB602 (Fig. 10), giving MGL inhibition at micromolar concentrations and no detectable FAAH inhibition, nor [^{3}H]-WIN55212-2 displacement from CB_1 or CB_2 receptors at 100 μM (Hohmann et al. 2005). Biochemical investigations indicated that URB602 inhibits recombinant MGL through a rapid non-competitive and reversible mechanism, excluding the formation of a stable carbamoylated adduct at the MGL active site (King et al. 2007). Despite its low potency, URB602 was utilised in vivo to provide evidence for the involvement of 2-AG in pain suppression. URB602 enhanced non-opioid stress-induced analgesia in the tail-flick test, when administered via microinjection into the periaqueductal grey matter (Hohmann et al. 2005), or intrathecally into the lumbar spinal cord (Suplita et al. 2006). Moreover, URB602 decreased pain behaviour in a formalin-induced inflammatory pain model (Guindon et al. 2007). URB602 was also recently reported to inhibit MGL activity in microglial cells with micromolar potency (Muccioli et al. 2007).

6.1.4 NAAA Inhibitors

Beside the central role played by FAAH in the degradation of AEA and other *N*-acylethanolamines such as PEA and OEA, another important enzyme shown to be involved in the degradation of NAEs in macrophages and peripheral tissues is NAAA (Tsuboi et al. 2007). Potent and specific NAAA inhibitors are needed as tools to clarify the biological role of this enzyme, but at present no drug-like potent and selective NAAA inhibitor has been reported. NAAA is less sensitive to PMSF and MAFP than FAAH or MGL (Ueda et al. 1999, 2001), but it is inhibited by the

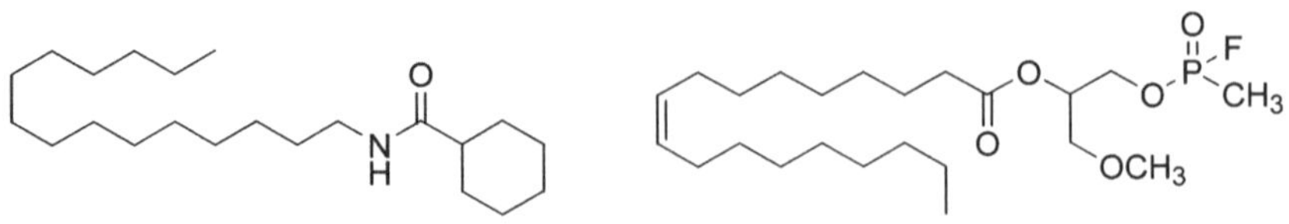

CCP O-3841

Fig. 11 NAAA and DAG lipase α inhibitors

sulfhydryl-specific agent *p*-CMB (Ueda et al. 2001), consistently with the presence of a catalytic cysteine in the active site (Tsuboi et al. 2005). Several NAE derivatives were shown to inhibit NAAA with some selectivity over FAAH. The most potent inhibitor identified within this class is *N*-(cyclohexylcarbonyl)pentadecylamine (CCP, Fig. 11), which inhibits NAAA at micromolar concentrations, and does not inhibit FAAH at concentrations up to 100 μM (Tsuboi et al. 2004). New compounds endowed with higher potency, good selectivity and a pharmacokinetic profile allowing in vivo administration are still required to assess the physiological role of this interesting acid amidase.

6.2 ECB Synthesis

Since AEA and 2-AG are synthesised on demand, and increased production and release of these ECBs is responsible for unwanted signs and symptoms of certain disorders, selective inhibitors of their biosynthesis would not only constitute important experimental tools, but may also have therapeutic perspectives.

NAEs, including AEA, are principally produced by two successive enzymatic reactions: (a) *N*-acylation of phosphatidylethanolamine to generate *N*-acylphosphatidylethanolamine (NAPE) by Ca^{2+}-dependent *N*-acyltransferase (NAT); (b) hydrolysis of NAPE to yield NAE, by a specific phospholipase D (NAPE-PLD) (Okamoto et al. 2007).

Hydrolysis, by a DAG lipase, of the ester bond at *sn*-1 position of a diacylglycerol (DAG) containing a *sn*-2 arachidonoyl chain produces 2-AG. In turn, DAG can be generated either from phosphoinositides by a specific phospholipase C (PLC) or from phosphatidic acid (PA) by PA phosphohydrolases. At present, two different isozymes of DAG lipase (α and β) have been identified (Bisogno et al. 2003).

Potent and selective inhibitors of the enzyme NAT have yet to be discovered, whereas it is known that NAPE-PLD activity can be blocked by *p*-CMB, which suggests the presence of catalytically important cysteine residues in its active site (Wang et al. 2006). Recently, it was reported that the lipase inhibitor THL shows detectable inhibition of NAPE-PLD at micromolar concentrations (Bisogno et al. 2006).

Some inhibitors are available for enzymes involved in the synthesis of 2-AG. THL and MAFP inhibit DAG lipase α with nanomolar potencies, but they are not selective and thus of limited utility as pharmacological tools. A noteworthy compound is the phosphonate ester O-3841 (Fig. 11), which inhibits DAG lipase α in the nanomolar range, lacks any detectable inhibitory effect on NAPE-PLD, FAAH, MGL, and does not interact with CB_1 and CB_2 receptors up to the concentration of 25 μM (Bisogno et al. 2006). These findings suggest that O-3841 may be used as a pharmacological tool to investigate the role of 2-AG biosynthesis on physiological processes and pathological conditions.

References

Ahn K, Johnson DS, Fitzgerald LR et al. (2007) Novel mechanistic class of fatty acid amide hydrolase inhibitors with remarkable selectivity. Biochemistry 46:13019–13030

Alexander JP, Cravatt BF (2005) Mechanism of carbamate inactivation of FAAH: implications for the design of covalent inhibitors and *in vivo* functional probes for enzymes. Chem Biol 12:1179–1187

Bari M, Battista N, Fezza F et al. (2006) New insights into endocannabinoid degradation and its therapeutic potential. Mini Rev Med Chem 6:257–268

Bayewitch M, Rhee M-H, Avidor-Reiss T et al. (1996) (-)-Δ^9-Tetrahydrocannabinol antagonizes the peripheral cannabinoid receptor-mediated inhibition of adenylyl cyclase. J Biol Chem 27:19902–19905

Beltramo M, Stella N, Calignano A et al. (1997) Functional role of high-affinity anandamide transport, as revealed by selective inhibition. Science 277:1094–1097

Bingham B, Jones PG, Uveges AJ et al. (2007) Species-specific *in vitro* pharmacological effects of the cannabinoid receptor 2 (CB_2) selective ligand AM1241 and its resolved enantiomers. Br J Pharmacol 151:1061–1070

Bisogno T, Howell F, Williams G et al. (2003) Cloning of the first sn1-DAG lipases points to the spatial and temporal regulation of endocannabinoid signaling in the brain. J Cell Biol 163:463–468

Bisogno T, Cascio MG, Saha B et al. (2006) Development of the first potent and specific inhibitors of endocannabinoid biosynthesis. Biochim Biophys Acta 1761:205–212

Boger DL, Sato H, Lerner AE et al. (2000) Exceptionally potent inhibitors of fatty acid amide hydrolase: the enzyme responsible for degradation of endogenous oleamide and anandamide. Proc Natl Acad Sci USA 97:5044–5049

Boger DL, Miyauchi H, Du W et al. (2005) Discovery of a potent, selective, and efficacious class of reversible alpha-ketoheterocycle inhibitors of fatty acid amide hydrolase effective as analgesics. J Med Chem 48:1849–1856

Bortolato M, Mangieri RA, Fu J et al. (2007) Antidepressant-like activity of the fatty acid amide hydrolase inhibitor URB597 in a rat model of chronic mild stress. Biol Psychiatry 62:1103–1110

Brown AJ (2007) Novel cannabinoid receptors. Br J Pharmacol 152:567–575

De Petrocellis L, Melck D, Ueda N et al. (1997) Novel inhibitors of brain neuronal and basophilic anandamide amidohydrolase. Biochem Biophys Res Commun 231:82–88

Deutsch DG, Chin SA (1993) Enzymatic synthesis and degradation of anandamide, a cannabinoid receptor agonist. Biochem Pharmacol 46:791–796

Deutsch DG, Lin S, Hill WA et al. (1997) Fatty acid sulfonyl fluorides inhibit anandamide metabolism and bind to the cannabinoid receptor. Biochem Biophys Res Commun 3:217–221

Devane WA, Dysarz FA III, Johnson MR et al. (1988) Determination and characterization of a cannabinoid receptor in rat brain. Mol Pharmacol 34:605–613

Di Marzo V, De Petrocellis L, Sepe N et al. (1996) Biosynthesis of anandamide and related acylethanolamides in mouse J774 macrophages and N18 neuroblastoma cells. Biochem J 316:977–984

Di Marzo V, Bisogno T, De Petrocellis L et al. (1999) Biosynthesis and inactivation of the endocannabinoid 2-arachidonoylglycerol in circulating and tumoral macrophages. Eur J Biochem 264:258–267

Di Marzo V, Bisogno T, De Petrocellis L et al. (2001) Highly selective CB(1) cannabinoid receptor ligands and novel CB(1)/VR(1) vanilloid receptor "hybrid" ligands. Biochem Biophys Res Commun 281:444–451

Dinh TP, Carpenter D, Leslie FM et al. (2002a) Brain monoglyceride lipase participating in endocannabinoid inactivation. Proc Natl Acad Sci USA 99:10819–10824

Dinh TP, Freund TF, Piomelli D (2002b) A role for monoglyceride lipase in 2-arachidonoylglycerol inactivation. Chem Phys Lipids 31:149–158

Elsohly MA (2002) Chemical constituents of cannabis. In: Grotenherman F, Russo E (eds) Cannabis and cannabinoids. Pharmacology, toxicology and therapeutic potential. Haworth Press, Binghamton, NY

Ermann M, Riether D, Walker ER et al. (2008) Arylsulfonamide CB_2 receptor agonists: SAR and optimization of CB_2 selectivity. Bioorg Med Chem Lett 18:1725–1729

Fegley D, Kathuria S, Mercier R et al. (2004) Anandamide transport is independent of fatty-acid amide hydrolase activity and is blocked by the hydrolysis-resistant inhibitor AM1172. Proc Natl Acad Sci USA 101:8756–8761

Felder CC, Joyce KE, Briley EM et al. (1995) Comparison of the pharmacology and signal transduction of the human cannabinoid CB_1 and CB_2 receptors. Mol Pharmacol 48:443–450

Fong TM, Guan XM, Marsh DJ et al. (2007) Antiobesity efficacy of a novel cannabinoid-1 receptor inverse agonist, N-[(1S, 2S)-3-(4-chlorophenyl)-2-(3-cyanophenyl)-1-methylpropyl]-2-methyl-2-[[5-(trifluoromethyl) pyridin-2-yl]oxy]propanamide (MK-0364), in rodents. J Pharmacol Exp Ther 321:1013–1022

Fowler CJ, Tiger G, Ligresti A et al. (2004) Selective inhibition of anandamide cellular uptake versus enzymatic hydrolysis: a difficult issue to handle. Eur J Pharmacol 492:1–11

Fowler CJ, Holt S, Nilsson O et al. (2005) The endocannabinoid signaling system: pharmacological and therapeutic aspects. Pharmacol Biochem Behav 81:248–262

Fu J, Gaetani S, Oveisi F et al. (2003) Oleoylethanolamide regulates feeding and body weight through activation of the nuclear receptor PPAR-alpha. Nature 425:90–93

Gaoni Y, Mechoulam R (1964) Isolation, structure and partial synthesis of an active constituent of hashish. J Am Chem Soc 86:1646–1647

Gardner A, Mallet PE (2006) Suppression of feeding, drinking, and locomotion by a putative cannabinoid receptor 'silent antagonist'. Eur J Pharmacol 530:103–106

Gertsch J, Raduner S, Altmann KH (2006) New natural noncannabinoid ligands for cannabinoid type-2 (CB_2) receptors. J Recept Signal Transduct Res 26:709–730

Ghafouri N, Tiger G, Razdan RK et al. (2004) Inhibition of monoacylglycerol lipase and fatty acid amide hydrolase by analogues of 2-arachidonoylglycerol. Br J Pharmacol 143:774–784

Giblin GM, O'Shaughnessy CT, Naylor A et al. (2007) Discovery of 2-[(2, 4-dichlorophenyl) amino]-N-[(tetrahydro-2H-pyran-4-yl) methyl]-4-trifluoromethyl)-5-pyrimidinecarboxamide, a selective CB2 receptor agonist for the treatment of inflammatory pain. J Med Chem 50:2597–2600

Glaser S, Abumrad N, Fatade F et al. (2003) Evidence against the presence of an anandamide transporter. Proc Natl Acad Sci USA 100:4269–4274

Gobbi G, Bambico FR, Mangieri et al. (2005) Antidepressant-like activity and modulation of brain monoaminergic transmission by blockade of anandamide hydrolysis. Proc Natl Acad Sci USA 102:18620–18625

Gonsiorek W, Hesk D, Chen SC et al. (2006) Characterization of peripheral human cannabinoid receptor (hCB_2) expression and pharmacology using a novel radioligand, $[^{35}S]$Sch225336. J Biol Chem 281:28143–28151

Gonsiorek W, Lunn CA, Fan X et al. (2007) Sch35966 is a potent, selective agonist at the peripheral cannabinoid receptor (CB_2) in rodents and primates. Br J Pharmacol 151:1262–1271

Gonsorek W, Lunn C, Fan X et al. (2000) Endocannabinoid 2-AG is a full agonist through human type 2 cannabinoid receptor: antagonism by anandamide. Mol Pharmacol 57:1045–1050

Guindon J, Hohmann AG (2008) Cannabinoid CB_2 receptors: a therapeutic target for the treatment of inflammatory and neuropathic pain. Br J Pharmacol 153:319–334

Guindon J, Desroches J, Beaulieu P (2007) The antinociceptive effects of intraplantar injections of 2-arachidonoyl glycerol are mediated by cannabinoid CB_2 receptors. Br J Pharmacol 150:693–701

Hanus L, Breuer A, Tchilibon S et al. (1999) HU-308: a specific agonist for CB_2, a peripheral cannabinoid receptor. Proc Natl Acad Sci USA 96:14228–14233

Hillard CJ, Manna S, Greenberg MJ et al. (1999) Synthesis and characterization of potent and selective agonists of the neuronal cannabinoid receptor (CB_1). J Pharmacol Exp Ther 289:1427–1433

Hillard CJ, Shi L, Tuniki VR et al. (2007) Studies of anandamide accumulation inhibitors in cerebellar granule neurons: comparison to inhibition of fatty acid amide hydrolase. J Mol Neurosci 33:18–24

Ho W-SV, Hillard CJ (2005) Modulators of endocannabinoid enzymic hydrolysis and membrane transport. In: Pertwee RG (ed) Cannabinoids. Handbook of experimental pharmacology. Springer, Heidelberg

Hohmann AG, Suplita RL, Bolton NM et al. (2005) An endocannabinoid mechanism for stress-induced analgesia. Nature 435:1108–1112

Horswill JG, Bali U, Shaaban S et al. (2007) PSNCBAM-1, a novel allosteric antagonist at cannabinoid CB_1 receptors with hypophagic effects in rats. Br J Pharmacol 152:805–814

Howlett AC, Barth F, Bonner TI et al. (2002) International union of pharmacology. XXVII. Classification of cannabinoid receptors. Pharmacol Rev 54:161–202

Huffman JW, Liddle J, Yu S et al. (1999) 3-(10, 10-Dimethylbutyl)-1-deoxy-Δ8-THC and related compounds: synthesis of selective ligands for the CB2 receptor. Bioorg Med Chem 7:2905–2914

Iwamura H, Suzuki H, Ueda Y et al. (2001) *In vitro* and *in vivo* pharmacological characterization of JTE-907, a novel selective ligand for cannabinoid CB_2 receptor. J Pharmacol Exp Ther 296:420–425

Jagerovic N, Fernandez-Fernandez C, Goya P (2008) CB_1 cannabinoid antagonists: structure-activity relationships and potential therapeutic applications. Curr Top Med Chem 8:205–230

Jarrahian A, Manna S, Edgemond WS et al. (2000) Structure–activity relationships among N-arachidonylethanolamine (Anandamide) head group analogues for the anandamide transporter. J Neurochem 74:2597–2606

Jayamanne A, Greenwood R, Mitchell VA et al. (2006) Actions of the FAAH inhibitor URB597 in neuropathic and inflammatory chronic pain models. Br J Pharmacol 147:281–288

Jonsson KO, Persson E, Fowler CJ (2006) The cannabinoid CB_2 receptor selective agonist JWH133 reduces mast cell oedema in response to compound 48/80 *in vivo* but not the release of β-hexosaminidase from skin slices *in vitro*. Life Sci 78:598–606

Karlsson M, Contreras JA, Hellman U et al. (1997) cDNA cloning, tissue distribution, and identification of the catalytic triad of monoglyceride lipase. Evolutionary relationship to esterases, lysophospholipases, and haloperoxidases. J Biol Chem 272:27218–27223

Kathuria S, Gaetani S, Fegley et al. (2003) Modulation of anxiety through blockade of anandamide hydrolysis. Nat Med 9:76–81

Khanolkar AD, Lu D, Ibrahim M et al. (2007) Cannabilactones: a novel class of CB2 selective agonists with peripheral analgesic activity. J Med Chem 50:6493–6500

Kikuchi A, Ohashi K, Sugie Y et al. (2008) Pharmacological evaluation of a novel cannabinoid 2 (CB_2) ligand, PF-03550096, *in vitro* and *in vivo* by using a rat model of visceral hypersensitivity. J Pharmacol Sci 106:219–224

King AR, Duranti A, Tontini A et al. (2007) URB602 inhibits monoacylglycerol lipase and selectively blocks 2-arachidonoylglycerol degradation in intact brain slices. Chem Biol 14:1357–1365

Koutek B, Prestwich GD, Howlett AC et al. (1994) Inhibitors of arachidonoyl ethanolamide hydrolysis. J Biol Chem 269:22937–22940

Kuster JE, Stevenson JI, Ward SJ et al. (1993) Aminoalkylindole binding in rat cerebellum: selective displacement by natural and synthetic cannabinoids. J Pharmacol Exp Ther 264:1352–1363
Labar G, Michaux C (2007) Fatty acid amide hydrolase: from characterization to therapeutics. Chem Biodivers 4:1882–1902
Labar G, Bauvois C, Muccioli GG et al. (2007) Disulfiram is an inhibitor of human purified monoacylglycerol lipase, the enzyme regulating 2-arachidonoylglycerol signaling. Chembiochem 23:1293–1297
Lambert D, Fowler CJ (2005) The endocannabinoid system: drug targets, lead compounds, and potential therapeutic applications. J Med Chem 48:1–29
Lichtman AH, Leung D, Shelton C et al. (2004) Reversible inhibitors of fatty acid amide hydrolase that promote analgesia: evidence for an unprecedented combination of potency and selectivity. J Pharmacol Exp Ther 311:441–448
Lo Verme J, Fu J, Astarita G et al. (2005) The nuclear receptor peroxisome proliferator-activated receptor-alpha mediates the anti-inflammatory actions of palmitoylethanolamide. Mol Pharmacol 67:15–19
Lodola A, Mor M, Hermann JC et al. (2005) QM/MM modelling of oleamide hydrolysis in fatty acid amide hydrolase (FAAH) reveals a new mechanism of nucleophile activation. Chem Commun (Camb) 35:4399–4401
Lodola A, Mor M, Rivara S et al. (2008) Identification of productive inhibitor binding orientation in fatty acid amide hydrolase (FAAH) by QM/MM mechanistic modelling. Chem Commun (Camb) 2:214–216
Lopez-Rodrguez ML, Viso A, Ortega-Gutierrez S et al. (2003) Design, synthesis, and biological evaluation of new inhibitors of the endocannabinoid uptake: comparison with effects on fatty acid amidohydrolase. J Med Chem 46:1512–1522
Lunn CA, Fine JS, Rojas-Triana A et al. (2006) A novel cannabinoid peripheral cannabinoid receptor-selective inverse agonist blocks leukocyte recruitment *in vivo*. J Pharmacol Exp Ther 316:780–788
Maekawa T, Nojima H, Kuraishi Y et al. (2006) The cannabinoid CB_2 receptor inverse agonist JTE-907 suppresses spontaneous itch-associated responses of NC mice, a model of atopic dermatitis. Eur J Pharmacol 542:179–183
Maione S, De Petrocellis L, de Novellis V et al. (2007) Analgesic actions of *N*-arachidonoyl-serotonin, a fatty acid amide hydrolase inhibitor with antagonistic activity at vanilloid TRPV1 receptors. Br J Pharmacol 150:766–781
Makara JK, Mor M, Fegley D et al. (2005) Selective inhibition of 2-AG hydrolysis enhances endocannabinoid signaling in hippocampus. Nat Neurosci 8:1139–1141
Makara JK, Mor M, Fegley D et al. (2007) Selective inhibition of 2-AG hydrolysis enhances endocannabinoid signaling in hippocampus. Nat Neurosci 10:134
Manera C, Benetti V, Castelli MP et al. (2006) Design, synthesis, and biological evaluation of new 1, 8-naphthyridin-4(1H)-on-3-carboxamide and quinolin-4(1H)-on-3-carboxamide derivatives as CB2 selective agonists. J Med Chem 49:5947–5957
Martin BR, Wiley JL, Beletskaya I et al. (2006) Pharmacological characterization of novel water-soluble cannabinoids. J Pharmacol Exp Ther 318:1230–1239
McKinney MK, Cravat BF (2005) Structure and function of fatty acid amide hydrolase. Annu Rev Biochem 74:411–432
Mechoulam R (2005) Plant cannabinoids: a neglected pharmacological treasure trove. Br J Pharmacol 146:913–915
Mechoulam R, Peters M, Murillo-Rodriguez E et al. (2007) Cannabidiol-recent advances. Chem Biodivers 4:1678–1692
Mor M, Rivara S, Lodola A et al. (2004) Cyclohexylcarbamic acid 3'- or 4'-substituted biphenyl-3-yl esters as fatty acid amide hydrolase inhibitors: synthesis, quantitative structure-activity relationships, and molecular modelling studies. J Med Chem 47:4998–5008
Muccioli GG (2007) Blocking the cannabinoid receptors: drug candidates and therapeutic promises. Chem Biodivers 4:1805–1827
Muccioli GG, Xu C, Odah E et al. (2007) Identification of a novel endocannabinoid-hydrolysing enzyme expressed by microglial cells. J Neurosci 27:2883–2889

Murineddu G, Lazzari P, Ruiu S et al. (2006) Tricyclic pyrazoles. 4. Synthesis and biological evaluation of analogues of the robust and selective CB2 cannabinoid ligand 1-(2',4'-dichlorophenyl)-6-methyl-N-piperidin-1-yl-1,4-dihydroindeno[1,2-c]pyrazole-3 carboxamide. J Med Chem 49:7502–7512

Ohta H, Ishizaka T, Tatsuzuki M et al. (2008) Imine derivatives as new potent and selective CB_2 cannabinoid receptor agonists with an analgesic action. Bioorg Med Chem 16:1111–1124

Okamoto Y, Wang J, Morishita J et al. (2007) Biosynthetic pathways of the endocannabinoid anandamide. Chem Biodivers 4:1842–1857

Ortar G, Ligresti A, De Petrocellis L et al. (2003) Novel selective and metabolically stable inhibitors of anandamide cellular uptake. Biochem Pharmacol 65:1473–1481

Pacher P, Batkay S, Kunos G (2006) The endocannabinoid system as an emerging target of pharmacotherapy. Pharmacol Rev 58:390–462

Pavon FJ, Bilbao A, Hernández-Folgado L et al. (2006) Antiobesity effects of the novel in vivo neutral cannabinoid receptor antagonist 5-(4-chlorophenyl)-1-(2,4-dichlorophenyl)-3-hexyl-1H-1,2,4-triazole-LH 21. Neuropharmacology 51:358–366

Pertwee RG (2005a) Inverse agonism and neutral antagonism at cannabinoid CB_1 receptors. Life Sci 76:1307–1324

Pertwee RG (2005b) Pharmacological action of cannabinoid. In: Pertwee RG (ed) Cannabinoids. Handbook of experimental pharmacology. Springer, Heidelberg

Piomelli D (2003) The molecular logic of endocannabinoid signalling. Nat Rev Neurosci 4:873–884

Piomelli D (2005) The endocannabinoid system: a drug discovery perspective. Curr Opin Investig Drugs 6:672–679

Piomelli D, Tarzia G, Duranti A et al. (2006) Pharmacological profile of the selective FAAH inhibitor KDS-4103 (URB597). CNS Drug Rev 12:21–38

Poso A, Huffman JW (2008) Targeting the cannabinoid CB_2 receptor: modeling and structural determinants of CB_2 selective ligands. Br J Pharmacol 153:335–346

Price MR, Baillie GL, Thomas A et al. (2006) Allosteric modulation of the cannabinoid CB_1 receptor. Mol Pharmacol 68:1484–1495

Rinaldi-Carmona M, Barth F, Heaulme M et al. (1994) SR141716A, a potent and selective antagonist of the brain cannabinoid receptor. FEBS Lett 350:240–244

Rinaldi-Carmona M, Barth F, Millan J et al. (1998) SR144528, the first potent and selective antagonist of the CB_2 cannabinoid receptor. J Pharmacol Exp Ther 284:644–650

Ross RA (2007) Tuning the endocannabinoid system: allosteric modulators of the CB_1 receptor. Br J Pharmacol 152:565–566

Ruiu S, Pinna AG, Marchese G et al. (2003) Synthesis and characterization of NESS 0327: a novel putative antagonist of the CB_1 cannabinoid receptor. J Pharmacol Exp Ther 306:363–370

Russo R, Lo Verme J, La Rana G et al. (2007) The fatty-acid amide hydrolase inhibitor URB597 (cyclohexylcarbamic acid 3'-carbamoylbiphenyl-3-yl ester) reduces neurophatic pain after oral administration. J Pharmacol Exp Ther 322:236–242

Ryberg E, Larsson N, Sjögren S et al. (2007) The orphan receptor GPR55 is a novel cannabinoid receptor. Br J Pharmacol 152:1092–1101

Saario SM, Laitinen JT (2007a) Monoglyceride lipase as an enzyme hydrolyzing 2-arachidonoyl-glycerol. Chem Biodivers 4:1903–1913

Saario SM, Laitinen JT (2007b) Therapeutic potential of endocannabinoid-hydrolysing enzyme inhibitors. Basic Clin Pharmacol Toxicol 101:287–293

Saario SM, Savinainen JR, Laitinen JT et al. (2004) Monoglyceride lipase-like enzymatic activity is responsible for hydrolysis of 2-arachidonoylglycerol in rat cerebellar membranes. Biochem Pharmacol 67:1381–1387

Saario SM, Salo OM, Nevalainen T et al. (2005) Characterization of the sulfhydryl-sensitive site in the enzyme responsible for hydrolysis of 2-arachidonoyl-glycerol in rat cerebellar membranes. Chem Biol 12:649–656

Sink KS, McLaughlin PJ, Wood JA et al. (2008) The novel cannabinoid CB(1) receptor neutral antagonist AM4113 suppresses food intake and food-reinforced behavior but does not induce signs of nausea in rats. Neuropsychopharmacology 33:946–955

Song ZH, Bonner TI (1996) A lysine residue of the cannabinoid receptor is critical for receptor recognition by several agonists but not WIN55212–2. Mol Pharmacol 49:891–896
Sugiura T, Kishimoto S, Oka S et al. (2006) Biochemistry, pharmacology and physiology of 2-arachidonoylglycerol, an endogenous cannabinoid receptor ligand. Prog Lipid Res 45:405–446
Suplita RL 2nd, Gutierrez T, Fegley D et al. (2006) Endocannabinoids at the spinal level regulate, but do not mediate, nonopioid stress-induced analgesia. Neuropharmacology 50:372–379
Thomas A, Ross RA, Saha B et al. (2004) 6"-Azidohex-2"-yne-cannabidiol: a potential neutral, competitive cannabinoid CB_1 receptor antagonist. Eur J Pharmacol 487:213–221
Thomas A, Stevenson LA, Wease KN et al. (2005) Evidence that the plant cannabinoid Δ^9-tetrahydrocannabivarin is a cannabinoid CB_1 and CB_2 receptor antagonist. Br J Pharmacol 146:917–926
Thomas B, Zhang Y, Brackeen M et al. (2006) Conformational characteristics of the interaction of SR141716A with the CB_1 cannabinoid receptor as determined through the use of conformationally constrained analogs. AAPS J 8:E665–E671
Tonidandel L, Tarzia G, Antonietti F et al. (2006) On the formation of [H3C-S-S-CH3]+* ions from the bis(dimethylthio) mercury molecular ion. Rapid Commun Mass Spectrom 20:3154–3158
Tsuboi K, Hilligsmann C, Vandevoorde S et al. (2004) *N*-cyclohexanecarbonylpentadecylamine: a selective inhibitor of the acid amidase hydrolysing N-acylethanolamines, as a tool to distinguish acid amidase from fatty acid amide hydrolase. Biochem J 379(Pt 1):99–106
Tsuboi K, Sun YX, Okamoto Y et al. (2005) Molecular characterization of N-acylethanolamine-hydrolyzing acid amidase, a novel member of the choloylglycine hydrolase family with structural and functional similarity to acid ceramidase. J Biol Chem 280:11082–11092
Tsuboi K, Takezaki N, Ueda N (2007) The N-acylethanolamine-hydrolyzing acid amidase (NAAA). Chem Biodivers 4:1914–1925
Ueda N, Yamanaka K, Terasawa Y et al. (1999) An acid amidase hydrolyzing anandamide as an endogenous ligand for cannabinoid receptors. FEBS Lett 454:267–270
Ueda N, Yamanaka K, Yamamoto S (2001) Purification and characterization of an acid amidase selective for *N*-palmitoylethanolamine, a putative endogenous anti-inflammatory substance. J Biol Chem 276:35552–35557
Valenzano KJ, Tafesse L, Lee G et al. (2005) Pharmacological and pharmacokinetic characterization of the cannabinoid receptor 2 agonist, GW405833, utilizing rodent models of acute and chronic pain, anxiety, ataxia and catalepsy. Neuropharmacology 48:658–672
Vandevoorde S (2008) Overview of the chemical families of fatty acid amide hydrolase and monoacylglycerol lipase inhibitors. Curr Top Med Chem 8:247–267
Vandevoorde S, Fowler CJ (2005) Inhibition of fatty acid amide hydrolase and monoacylglycerol lipase by the anandamide uptake inhibitor VDM11: evidence that VDM11 acts as an FAAH substrate. Br J Pharmacol 145:885–893
Wang J, Okamoto Y, Morishita J et al. (2006) Functional analysis of the purified anandamide-generating phospholipase D as a member of the metallo-beta-lactamase family. J Biol Chem 281:12325–12335
Wei BQ, Mikkelsen TS, McKinney MK et al. (2006) A second fatty acid amide hydrolase with variable distribution among placental mammals. J Biol Chem 281:36569–36578
Whiteside GT, Lee GP, Valenzano KJ (2007) The role of the cannabinoid CB_2 receptor in pain transmission and therapeutic potential of small molecule CB_2 receptor agonists. Curr Med Chem 14:917–936
Woods SC (2007) The endocannabinoid system: novel pathway for cardiometabolic Risk-factor reduction. JAAPA 20(Suppl Endocannabinoid):7–10
Yao BB, Hsieh GC, Frost JM et al. (2008) *In vitro* and *in vivo* characterization of A-796260: a selective cannabinoid CB_2 receptor agonist exhibiting analgesic activity in rodent pain models. Br J Pharmacol 153:390–401
Zhang D, Saraf A, Kolasa T et al. (2007) Fatty acid amide hydrolase inhibitors display broad selectivity and inhibit multiple carboxylesterases as off-targets. Neuropharmacology 52:1095–1105

Genetic Models of the Endocannabinoid System

Krisztina Monory and Beat Lutz

Contents

Abstract The endocannabinoid (ECB) system comprises cannabinoid receptors, ECBs and the whole machinery for the synthesis and degradation of ECBs. It has emerged as an important signalling system in the nervous system, controlling numerous physiological processes, including synaptic transmission, learning and memory, reward, feeding, neuroprotection, neuroinflammation, and neural development. This system is also implicated in various diseases of the nervous system, and thus has become a promising therapeutic target. The use of genetically modified mice has contributed crucially to our rapidly expanding knowledge of the ECB system. In this chapter, the existing mouse mutants targeting the ECB system will be discussed in detail. The use of conditional mutants has given an additional dimension to the analysis of the system, and, it is hoped, will finally enable us to understand this widespread and complex system in the context of intricate networks where different brain regions and neurotransmitter systems interact tightly with each other.

K. Monory and B. Lutz (✉)
Department of Physiological Chemistry, Johannes Gutenberg University Mainz, Duesbergweg 6, D-55099, Mainz, Germany
e-mail: blutz@uni-mainz.de

D. Kendall and S. Alexander (eds.), *Behavioral Neurobiology of the Endocannabinoid System*, Current Topics in Behavioral Neurosciences 1,
DOI: 10.1007/978-3-540-88955-7_5,

Keywords Gene targeting • Knockout technology • Conditional knockout • Cre recombinase • Cre/loxP system • Behaviour • Experimental design • Compensatory mechanism

Abbreviations

2-AG	2-Arachidonoyl-glycerol
AAV	Adeno-associated virus
AEA	Arachidonoyl-ethanolamine (anandamide)
CaMK	Calcium/calmoduline-dependent kinase
DAGL	Diacylglycerol lipase
ES	Embryionic stem (cell)
FAAH	Fatty acid amide hydrolase
GABA	Gamma aminobutyric acid
MGL	Monoacyl-glycerol lipase

1 Introduction

Δ^9-Tetrahydrocannabinol (THC), the major pharmacologically active compound of marijuana (Gaoni and Mechoulam 1964), has an impressively wide range of pharmacological effects in mammals including humans. For some time, this multitude of effects was believed to be the consequence of the lipophilic nature of THC molecules, which would change membrane fluidity and thus non-specifically affect neuronal communication (Hillard et al. 1985). However, the availability of radiolabelled synthetic cannabinoid ligands in the 1980s enabled the demonstration of specific cannabinoid binding sites in the brain (Howlett et al. 1988). A few years later, two different cannabinoid receptors were identified (CB_1 and CB_2 receptors); the CB_1 receptor is highly enriched in the central nervous system (CNS) (Matsuda et al. 1990), while the CB_2 receptor is prominently expressed in immune cells (Munro et al. 1993). Radioligand binding (Herkenham et al. 1990), immunolabelling (Tsou et al. 1998; Katona et al. 1999) and in situ hybridisation (Marsicano and Lutz 1999) showed a widespread expression of CB_1 receptors in the adult rodent brain, especially in the cerebellum, basal ganglia, cerebral cortex, hippocampus and amygdala. Electron microscopy showed that the CB_1 receptor is mainly localised on axon terminals (Katona et al. 1999), and electrophysiology demonstrated that the CB_1 receptor is able to modify neurotransmitter release by presynaptic mechanisms (Wilson and Nicoll 2001).

The CB_1 receptor is present at highest density on presynaptic terminals of cholecystokinin-positive GABAergic inhibitory interneurons (Tsou et al. 1998; Katona et al. 1999). Electrophysiological studies and coexpression analysis, however, have revealed the complexity of the ECB system. CB_1 receptors are also present in many other types of neurons, including glutamatergic (Sullivan 1999; Monory et al. 2006); cholinergic (Degroot et al. 2006), serotonergic (Häring et al. 2007), noradrenergic (Oropeza et al. 2007) and possibly on dopaminergic (Degroot et al. 2006) neurons. This may explain the many diverse effects of *Cannabis* and also suggests an intrinsic complexity of the ECB system in the regulation of neurotransmission. However, several other different factors increase further the complexity of physiology and pharmacology of the ECB system.

First, after the initial belief of mutually exclusive expression of CB_1 receptors in the CNS and CB_2 receptors in the peripheral nervous system, it has become evident that the expression pattern of both receptors is more complex. Namely, the CB_1 receptor was found in several peripheral, non-neuronal tissues (Cota et al. 2003; Bensaid et al. 2003), while the CB_2 receptor was also identified in the brain (Van Sickle et al. 2005; Onaivi et al. 2008). Moreover, several lines of evidence have suggested the existence of yet-to-be identified cannabinoid receptors (Brown 2007). Indeed, the orphan G protein-coupled receptor GPR55 seems to be one of the suspected new cannabinoid receptors (Baker et al. 2006; Lauckner et al. 2008). Furthermore, variants of CB_1 receptor protein were shown to derive from different splice forms in humans (Shire et al. 1995; Ryberg et al. 2005). However, the physiological relevance of these variants has still to be elucidated.

Furthermore, there are different endogenous ligands that can activate cannabinoid receptors. Five ECBs have been identified to date: anandamide (AEA), 2-arachidonoylglycerol (2AG), noladin ether, virodhamine and *N*-arachidonoyl dopamine (NADA) (Di Marzo 2008). They are polyunsaturated fatty acids derived from arachidonic acid. Undoubtedly, AEA and 2AG are the two best characterised ECBs, but it is expected that novel lipids may be identified as ligands of cannabinoid receptors in the near future.

For the different ECBs, different synthesising and degrading enzymes are engaged (Di Marzo 2008). As ECBs are lipids, primarily generated from membrane phospholipid precursors, they are not stored nor released like "classical" neurotransmitters, but are synthesised on-demand after increase of intracellular Ca^{2+} concentration or after stimulation of metabotropic glutamate and acetylcholine receptors. These enzymes are the *N*-acylphosphatidylethanolamine-specific phospholipase D (NAPE-PLD) for AEA, and diacylglycerol lipase (DGL) α and β for 2AG. Both DGL isoforms are postsynaptically localised, but expressed in a different temporal manner, the α isoform being the main isoform postnatally present, while the β isoform is the main isoform prenatally. These synthesising enzymes are presumably anchored to the plasma membrane (DGL) or to internal membrane compartments (NAPE-PLD). DGL and NAPE-PLD are the main but not the only synthetic enzymes of AEA and 2AG. For example, 2AG can also be synthesised by lyso-PLC (Sugiura et al. 1995). For AEA, two alternative pathways have been proposed (Liu et al. 2008).

After receptor binding, ECBs are presumably taken up quickly by the cells and are inactivated. The uptake is possibly mediated by a membrane transporter protein, though it has not yet been identified. Similarly to their synthesis, degradation of

ECBs also follows different pathways (Di Marzo 2008). AEA is degraded by FAAH-1 and FAAH-2, the latter found only in humans (Wei et al. 2006). 2AG is degraded by monoacylglycerol lipase (MGL). In both cases, the end product of the degradation is arachidonic acid. Supporting the concept of retrograde transmission of ECBs, where 2AG appears to be the main player as a retrograde transmitter, MGL is presynaptically localised, while FAAH is postsynaptically associated with membranes of cytoplasmic organelles.

The dynamics of the ECB system activity is determined by a multitude of factors and poses numerous questions: Which biosynthetic pathway is initiated under which circumstances? Which receptor is activated by which ligand? How fast are ECBs taken up by the cell and how fast is the degradation? Such questions have also to be brought into the context of the anatomical complexity of the various components of the ECB system in the nervous system.

To tackle the multifaceted process of ECB signalling, state-of-the-art genetics offers appropriate and powerful tools. Gene inactivation in all cells of the body or only in certain cell types or brain regions should help to decipher physiological and pathophysiological processes in which ECBs participate. Indeed, in the last few years, a growing number of mutant mouse lines have become available in the field of ECB research. In the following paragraphs, we will first outline the background of the transgenic techniques and the available mutants. Then, we discuss the advantages as well as possible pitfalls of genetic approaches; we will address several considerations in the context of behavioural experiments with mutant mice and talk further about possible caveats of pharmacology. Finally, we would also like to point out possible aspects of the use of other advanced genetic models in analysing the ECB system.

2 Genetic Models

2.1 *Available Tools*

During the last two decades, targeted mutagenesis in mice has become one of the most powerful tools for studying gene function in mammals, as illustrated by the 2007 Nobel Prize in Physiology or Medicine awarded to Mario Capecchi, Martin Evans and Oliver Smithies (Mak 2007). Their discoveries and technological advancements in introducing highly specific modifications in the mouse genome by the use of homologous recombination in embryonic stem cells revolutionised medical research and have led to fundamental discoveries in all fields of mammalian biology, ranging from embryonic development to the generation of animal models for human diseases.

The first milestone in the development of the gene targeting technology was the establishment of totipotent cell lines (i.e. embryonic stem cells) derived from mouse blastocysts. If kept in appropriate culture conditions, these cells can

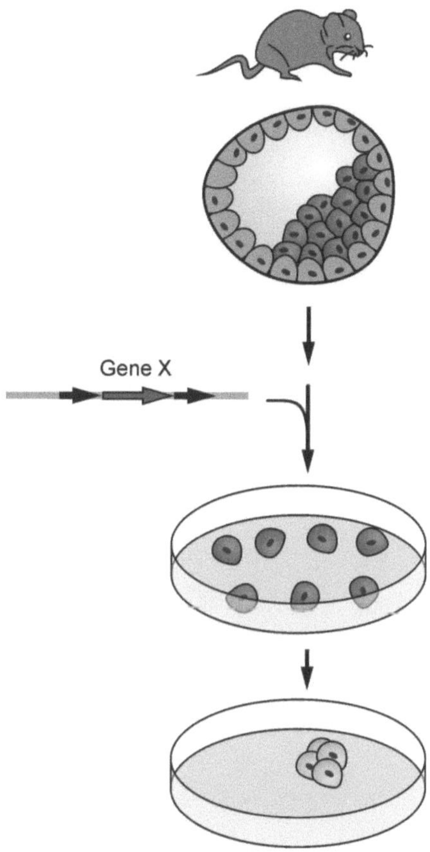

Fig. 1 Genetic manipulation of embryonic stem cells. Embryonic stem (ES) cells are derived from mouse blastocysts and are kept under conditions that allow the maintenance of their totipotency. A gene targeting construct containing selection markers is electroporated into ES cells, plated on culture dish in the presence of a selection medium. Only cells that received the targeting construct by homologous recombination survive selection, forming a colony of cells which contains the genetic modification in a heterozygous state. Here, the targeted Gene X (*grey arrow*) is flanked by two loxP sites (*black arrows*)

contribute to all cell types of the body, including germ cells. The second seminal discovery was the observation by Capecchi and colleagues that mammalian somatic cells possess the enzymatic machinery for mediating homologous recombination between exogenously introduced DNA sequences and homologous sequences in the genome. Both discoveries led to the first gene disruption experiments in mice (Thomas and Capecchi 1987; Doetschman et al. 1987), now currently referred to as gene "knockout" technology (Figs. 1 and 2).

The number of knockout mouse lines has grown almost exponentially after the establishment of this technique, which helped scientists to understand protein functions in the context of the entire animal, avoiding many of the limitations of in vitro models or pharmacological tools. Gene knockout technology has enabled the study of specific gene products in a highly complex context such as the nervous system, leading to seminal discoveries in neurosciences, including learning and memory, cognition and synaptic plasticity. As an example, the study of learning and memory in gene targeted mutant mice has led to insights into key mechanisms

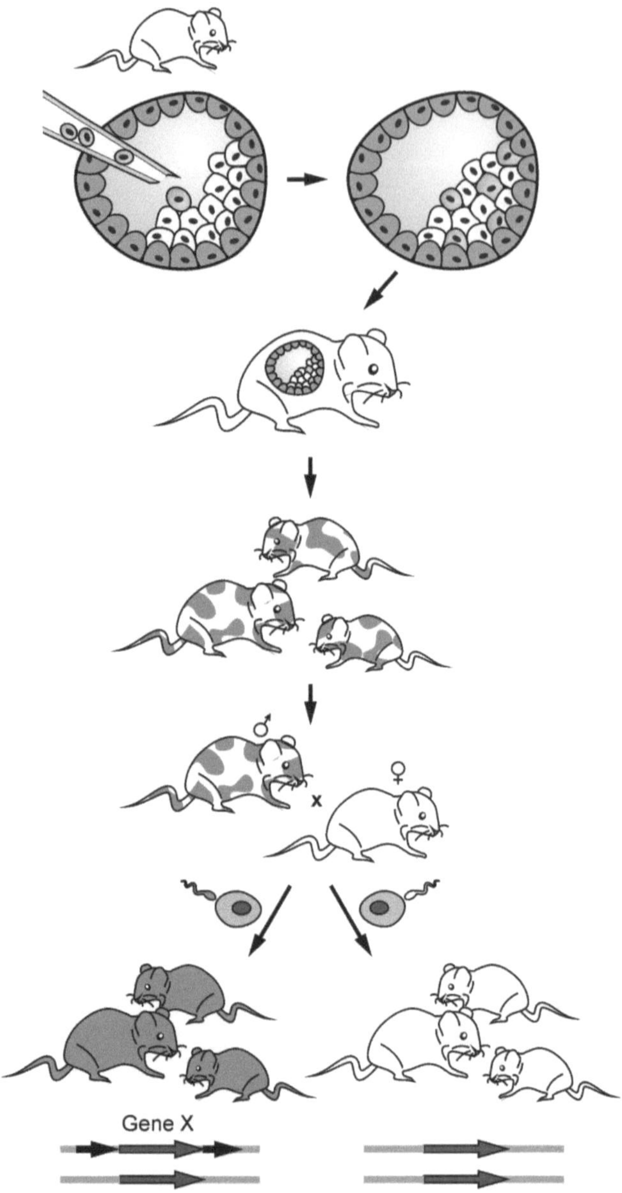

Fig. 2 Generation of mutant mice by gene targeting of embryonic stem cells. Genetically modified ES cells (as male karyotype) are injected into blastocysts, which are then transferred into the uterus of foster mothers. As the blastocysts are obtained from a mouse strain different from that used for the establishment of ES cell lines, the offspring are chimaeras (as mixture of ES cell-derived cells and of inner cell mass from the injected blastocyst). This is later visible by the coat colour. The crossing of male chimaeras with wild-type mice of the appropriate strain will lead to offspring with a distinct coat colour, showing that the injected ES cells contributed fully to the animal via the gametes of the chimaera. Genotyping of these mice will show whether the genetically modified Gene X is present in a heterozygous state. Gene X (*grey*) is flanked by two loxP sites (*black arrows*)

underlying the brain's ability to select, to store and to recall information (Silva et al. 1992; Grant et al. 1992).

Despite the progress enabled by this new technology, some of its intrinsic limitations have become evident. First, complete knockout mice lack the targeted gene product during their entire lifespan, i.e. from embryogenesis to adult life. It has been recognised that many genes with distinct expression and functions during embryogenesis also have specific functions in the adult. If the role during development is essential, the mutation might cause embryonic lethality. Developmental deficits might also induce particular phenotypes in the adult. Thus, such phenotypes may not be easily interpreted at the mechanistic level, as early developmental disturbances may distort or strengthen the phenotypes observed in the adult. Observable defects in the embryo, however, may also be ameliorated due to compensatory mechanisms. Biological processes are highly regulated to maintain cellular homeostasis. The disruption of a gene in the entire organism might trigger the up- or down-regulation of related or unrelated gene products. This, again, might mask or distort the observed phenotype in these mice and hamper the interpretation of the phenotype observed.

In addition, in these mutants, the targeted gene is lost in all cells of the body. Frequently, a single gene product is responsible for distinct functions in different organs or cell types. But even if the biochemical activity of a protein *per se* is similar or is the same in different cell types, the distinct physiological role of the gene product might depend on the context, i.e. the tissue or cell type. In particular in the brain, containing a myriad of different cell types that communicate extensively with each other, it makes an important difference whether a certain protein is present, e.g. in an inhibitory or an excitatory neuron, or in glial cells. Moreover, complex brain functions are a result of co-ordinated interactions of many regions and cell types. Thus, a single gene product might differentially participate in an activated neuronal circuit, depending on the physiological or pathophysiological context; e.g. emotional behaviour will recruit other neuronal circuits and brain areas than reward-related behaviour. However, these distinct functions are unlikely to be uncovered with "conventional" knockout mice.

For the functional dissection of the nervous system, it is therefore essential to relate gene function to particular anatomical brain regions and cell types. This is made possible by the technique of conditional mutagenesis. The most widespread technique for generating conditional knockout mice is the Cre/loxP system (Sauer and Henderson 1988) (Fig. 3); however, other systems are also emerging (Feil 2007). Cre recombinase is a P1 bacteriophage-derived enzyme which is able to mediate sequence-specific recombination between a 34-bp-long sequence referred to as loxP (locus of crossover in P1) (Sternberg and Hamilton 1981).

In the conditional mutagenesis approach, two transgenic mouse lines have to be established and crossed (Fig. 4). First, the so-called "floxed" mouse has to be generated, where the gene or region of interest is flanked by two loxP sites. Introducing two loxP sites upstream and downstream in the same orientation induces the deletion of the loxP-enclosed sequence, when the Cre recombinase

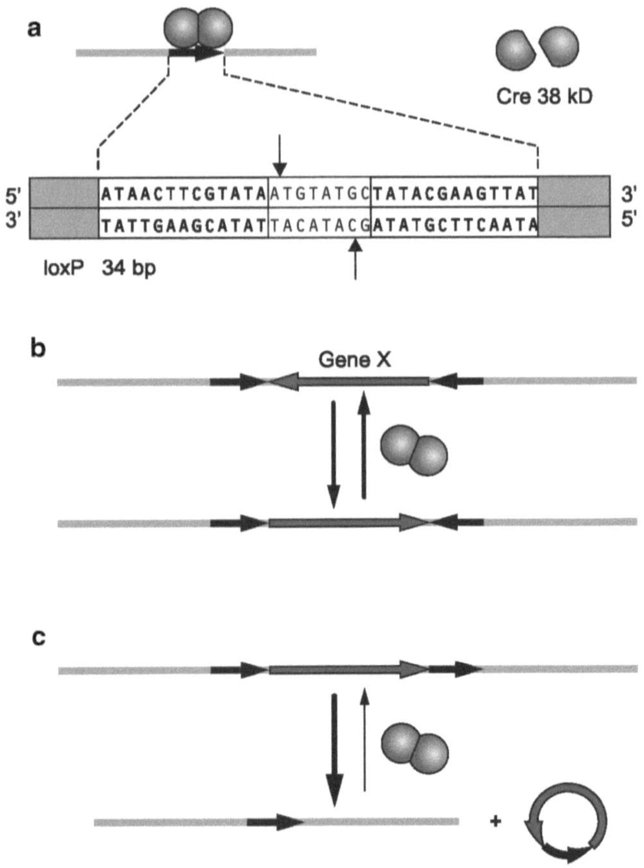

Fig. 3 Principle of the Cre/loxP system. (**a**) One loxP site consists of 34 bp, composed of two 13-bp-long repeats forming a palindrome, which is interrupted by an 8-bp-long non-palindromic sequence. This 8-bp sequence determines the orientation of the loxP site. Cre recombinase binds as a dimer to the loxP site. After binding of one Cre recombinase monomer to each inverted repeat, the DNA strands are cleaved in the spacer region (*vertical arrows*) and exchanged between two loxP sites. (**b**) When loxP sites are oriented "head-to-head", Cre recombinase catalyses an inversion of the sequence "Gene X" between the two loxP sites. (**c**) When the two loxP sites are oriented "head-to-tail", Cre recombinase catalyses the excision of the intervening sequence, leading to a loss of Gene X

protein is present. If appropriately performed, the introduction of these two short loxP sequences should not cause any phenotype. Thus, the "floxed" mice behave like wild-type mice. Second, a mouse line is needed with a cell-type specific expression of Cre recombinase. To this end, specific regulatory sequences should drive the expression of Cre recombinase in the second transgenic mouse line (Fig. 4). In the offspring from crossing of these two mouse lines, the target gene will be deleted in all cells where Cre recombinase is expressed, while the gene

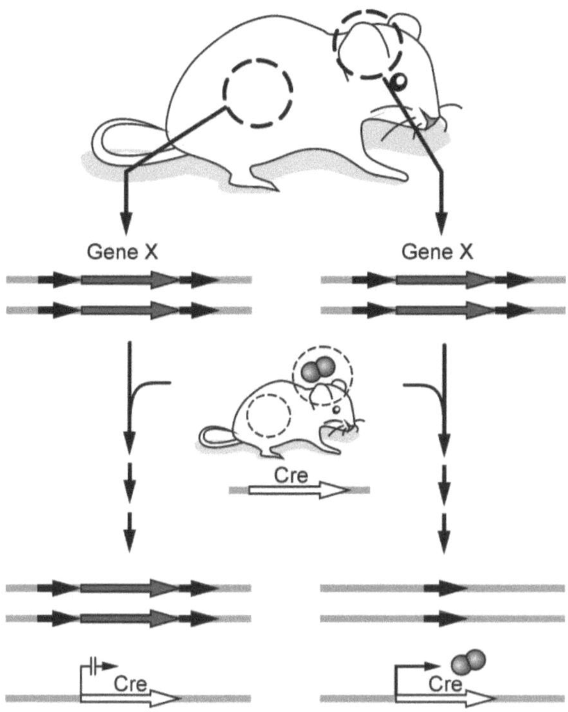

Fig. 4 Principle of the generation of cell- and tissue-specific gene inactivation. Essential sequences of Gene X are flanked by two loxP sites. Mice with the so-called "floxed" allele will be crossed with a transgenic mouse line expressing Cre recombinase in a cell- and/or tissue-specific manner. After an appropriate breeding schedule, mice will be obtained which either lack the expression of Cre recominase not shown or which express Cre recombinase in specific cells, leading to the excision of the loxP-flanked sequences, i.e. Gene X

will keep the wild-type expression pattern in all cells missing Cre recombinase expression. Currently, the "zoo" of Cre recombinase-expressing transgenic mouse lines is continuously growing. A recent review gave an exhaustive list of lines that are useful in the analysis of the nervous system (Gaveriaux-Ruff and Kieffer 2007). In Table 1, transgenic lines are listed that are particularly interesting in the analysis of the ECB system or that have been established only very recently.

In a more advanced approach, not only spatial, but temporal regulation of gene excision is possible. Generating a fusion protein of Cre recombinase with a modified ligand-binding domain of the oestrogen receptor renders the system inducible by tamoxifen (Feil et al. 1996). Cytoplasmic binding of the heat-shock protein Hsp90 to the ligand-binding domain traps the protein to the cytoplasm and inhibits the enzymatic activity of Cre recombinase by conformational inhibition (Fig. 5). When tamoxifen binds to the modified ligand-binding domain of the oestrogen receptor, Hsp90 dissociates, and the fusion protein translocates to the nucleus, where Cre recombinase can excise the "floxed" target sequence. Tamoxifen can be injected into mice intraperitoneally at any chosen time in postnatal (Erdmann et al. 2007), but also in prenatal life (Erdmann et al. 2008). Thus, gene function can be studied after developmental processes are finished. If this fusion protein is expressed under the control of specific regulatory sequences, the gene deletion will be spatially as well as temporally regulated (Feil 2007).

Table 1 Available mouse mutants targeting components of the endocannabinoid system

Targeted gene	Genetic modification	Reference
CB_1	Null mutant	Ledent et al. (1999)
CB_1	Null mutant	Zimmer et al. (1999)
CB_1	Null mutant	Marsicano et al. (2002)
CB_1	Null mutant	Robbe et al. (2002)
CB_1	Conditional mutant *CB_1R deleted from principal forebrain neurons*	Marsicano et al. (2003)
CB_1	Conditional mutant *CB_1R deleted from forebrain GABAergic neurons*	Monory et al. (2006)
CB_1	Conditional mutant *CB_1R deleted from cortical glutamatergic neurons*	Monory et al. (2006)
CB_1	Conditional mutant *CB_1R deleted from dopamine receptor D1 expressing neurons*	Monory et al. (2007)
CB_1	Conditional mutant *CB_1R deleted from nociceptive dorsal root neurons*	Agarwal et al. (2007)
CB_1	Conditional mutant *CB_1R deleted from all neural precursors*	Maresz et al. (2007)
CB_1	Conditional mutant *CB_1R deleted from T cells*	Maresz et al. (2007)
CB_1	Conditional mutant *CB_1R deleted in hepatocytes*	Jeong et al. (2008)
CB_2	Null mutant	Buckley et al. (2000)
CB_2	Null mutant	Wotherspoon et al. (2005)
GPR55	Null mutant	Staton et al. (2008)
FAAH	Null mutant	Cravatt et al. (2001)
FAAH	NES-FAAH *FAAH expression under the control of the neural-specific enolase promoter*	Cravatt et al. (2004)
NAPE-PLD	Null mutant	Leung et al. (2006)
TRPV1	Null mutant	Caterina et al. (2000)
TRPV1	Null mutant	Davis et al. (2000)

Abbreviation: CB_1R. CB_1 receptor

2.2 *Null Mutant Mouse Lines*

The first gene targeting addressing the function of the ECB system was performed for CB_1 receptors. There are four different knockout mouse lines available (Ledent et al. 1999; Zimmer et al. 1999; Marsicano et al. 2002; Robbe et al. 2002). In three of these lines, distinct parts of the CB_1 receptor gene were replaced by the 3-phosphoglycerate kinase-neomycin (PGK-Neo) resistance cassette. The following parts were abolished in the CB_1 receptor gene: amino acids 32–448 (Zimmer et al. 1999), the first 233 amino acids of the receptor (Ledent et al. 1999), and part of the 5' intronic region plus the protein encoding region up to the sixth transmembrane domain (Robbe et al. 2002), respectively. Thus, in any of these lines, small parts of the CB_1 receptor protein-encoding sequences remained in the mutated gene locus. Whether or not some small peptide parts of CB_1 receptor are still translated is

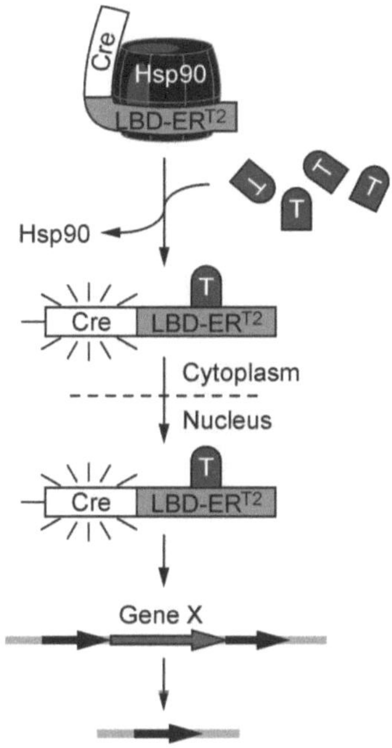

Fig. 5 Principle of ligand-inducible Cre recombinase. Cre recombinase is fused to the modified ligand binding domain of the oestrogen receptor (LBD-ER^{T2}), the latter being unable to bind any endogenous steroid hormones. This fusion protein binds the heat-shock protein 90 (Hsp90), leading to a conformation of the Cre recombinase that prevents enzymatic activity. Application of tamoxifen leads to binding to the LBD-ER^{T2}, and Hsp90 is released from the fusion protein. Next, Cre recombinase reconstitutes its enzymatic activity, translocates into the nucleus, recognises the two loxP sites, and excises Gene X

difficult to evaluate, but functional studies, e.g. ligand binding (Zimmer et al. 1999), electrophysiological experiments (Robbe et al. 2002) and behavioural experiments (Zimmer et al. 1999; Ledent et al. 1999), strongly indicated complete loss of CB_1 receptor function. In the fourth line (Marsicano et al. 2002), two loxP sites were introduced, one into the intron preceding the exon encoding CB_1 receptor protein and the other one into the 3'UTR immediately after the open reading frame. Afterwards, these loxP flanked sequences were removed by Cre recombinase-mediated excision, leading to the complete loss of CB_1 receptor protein-encoding sequences. The PGK-Neo selection cassette, which was flanked by two FRT recombination sites, was also introduced by homologous recombination, but was not removed, and thus it is still present in the 3'UTR of this null mutant line. In addition to the targeting strategies, differences exist regarding the genetic background. Two lines are in the inbred strain C57BL/6J (Zimmer et al. 1999; Robbe et al. 2002), one in C57BL/6N (Marsicano et al. 2002), and one in the outbred strain CD1 (Ledent et al. 1999).

The analysis of these mutant CB_1 receptor mice provided a myriad of extraordinary insights into the numerous functions of the ECB system as mediated by CB_1 receptors in the nervous system. These functions include the roles of CB_1 receptors in drug addiction and reward (Ledent et al. 1999; Racz et al. 2003; Wang et al. 2003), pharmacological effects of THC (Zimmer et al. 1999), cognitive processes (Reibaud et al. 1999; Bilkei-Gorzo et al. 2005), emotional behaviours (Marsicano et al. 2002; Haller et al. 2002), neuroprotection (Marsicano et al. 2003), pain (Ledent et al. 1999; Zimmer et al. 1999), and feeding behaviour (Di Marzo et al. 2001; Cota et al. 2003; Ravinet et al. 2004). Furthermore, CB_1

receptor-deficient mice were important to prove the involvement of ECBs in the regulation of synaptic transmission, such as short-term and long-term suppression of GABA transmission (Wilson et al. 2001; Marsicano et al. 2002). In most of the studies, the phenotypic changes in CB_1 receptor knock-out mice were able to be mimicked by using specific CB_1 receptor antagonists, such as rimonabant and AM251. In recent years, CB_1 receptor mutant mouse lines have become a crucial tool for proving CB_1 receptor functions in peripheral, non-neuronal tissues, such as liver (Osei-Hyiaman et al. 2005), adipocytes (Cota et al. 2003) and skin (Karsak et al. 2007).

The inactivation of the CB_2 receptor also used the strategy of a PGK-Neo replacement vector (Buckley et al. 2000). By homologous recombination, it is reported that a fragment of 341 bp of the exon was deleted. Accordingly, this led to a loss of the C-terminal amino acids from positions 217 to 347, but leaving the splice acceptor of the protein-encoding exon unaffected as well as the sequences coding for the first 216 amino acids of the CB_2 receptor protein. These amino acids potentially code for the N-terminal CB_2 receptor protein, containing the first five transmembrane domains. This CB_2 receptor-deficient mouse line showed no binding activity in the spleen for the radiolabeled non-selective CB_1 receptor/CB_2 receptor agonist CP55940, suggesting that ligand binding to CB_2 receptor is indeed completely lost in this line, but the existence of remaining CB_2 receptor protein-encoding sequences requires further attention.

This mutant line has emerged as an important tool for studying CB_2 receptor functions in non-immune cells. In particular, the functional presence of CB_2 receptor in brain neurons (Van Sickle et al. 2005; Onaivi et al. 2008), liver (Julien et al. 2005) and bone (Ofek et al. 2006) gave new insights into the roles of this receptor.

A second CB_2 receptor-deficient mouse line was also established (Wotherspoon et al. 2005) and is distributed by The Jackson Laboratory (Bar Harbor, USA; for further information see http://www.informatics.jax.org/external/ko/deltagen/614_MolBio.html). Based on the available data sheet, this gene targeting also used a PGK-Neo replacement vector, leading to the deletion of sequences encoding amino acids 26–140, thus deleting the first 3 transmembrane domains. The first 25 amino acids might still be translated. After the inserted PGK-Neo cassette, the rest of the CB_2 receptor coding region remained in the genome. Theoretically, there might be a splicing over the PGK-Neo cassette, but this is rather unlikely. Thus, only minor remaining parts of the CB_2 receptor protein are present in this mouse line, likely not to interfere with cellular processes.

The orphan G protein-coupled receptor GPR55 (Sawzdargo et al. 1999) has recently been proposed to be a third cannabinoid receptor (Baker et al. 2006; Ryberg et al. 2007), and a null mutant mouse line was recently generated (Staton et al. 2008). A combined LacZ PGK-Neo cassette was used to replace a major part of the GPR55 open reading frame. The deletion removed the sequences coding for amino acids 39–281 of GPR55, representing the second to sixth transmembrane domains. Thus, the mutation retains the first 118 bp of the GPR55 open reading frame encoding 39 amino acids, containing 20 of the 22 amino acids of the first transmembrane domain. In recent experiments, it was shown that GPR55-

deficient mice lack inflammatory mechanical hyperalgesia and neuropathic hypersensitivity (Staton et al. 2008), suggesting GPR55 as a potential target in pain treatments.

Unlike CB_1 and CB_2 receptors (Matsuda et al. 1990; Munro et al. 1993), which are encoded by one exon, the protein-coding sequence of FAAH spans across 15 exons of the gene (Wan et al. 1998). For the FAAH null mutant (Cravatt et al. 2001), standard targeted gene disruption procedures were used and replaced the first exon of the FAAH gene (encoding amino acids 1–65) and 2 kb of upstream sequence with the PGK-Neo cassette. Loss of FAAH protein was evidenced by immunostaining, and by enzymatic activity measurements. The mutants basically lacked the degradation of AEA and oleamide, leading to 15-fold increased levels of AEA in the brain (Cravatt et al. 2001). The second FAAH gene (FAAH-2) is not relevant in this context, as it exists only in humans (Wei et al. 2006). FAAH-deficient mice served as an excellent model in order to understand the physiological consequences of increased AEA levels. They were analysed in several behavioural paradigms, including pain (Cravatt et al. 2001), seizure susceptibility (Clement et al. 2003) spatial memory (Varvel et al. 2007), anxiety (Moreira et al. 2008), emotionality (Naidu et al. 2007), nicotine reward (Merritt et al. 2008), ethanol drinking (Basavarajappa et al. 2006), and in experimental autoimmune encephalitis (Maresz et al. 2007, Webb et al. 2008). In addition, these mice were very useful in elucidating the roles of ECBs in embryonic proliferation of cortical neurons and in radial migration (Mulder et al. 2008), and in proliferation and differentiation of neural progenitors in the adult brain (Aguado et al. 2006). Finally, the analysis of FAAH-deficient mice led to the description of novel AEA degradation pathways (Mulder and Cravatt 2006).

In order to dissociate the peripheral and the central nervous system functions of FAAH, a transgenic mouse line was established in which FAAH is expressed under the control of the neural-specific enolase (NSE) promoter (Cravatt et al. 2004). This transgenic line was crossed with the FAAH knock-out line to in order to obtain $FAAH^{-/-}$ mice that contain the NSE-FAAH transgene. Thus, this approach is a conditional rescue experiment of FAAH, re-expressing FAAH in the central nervous system, but lacking it in all peripheral organs. The analysis of these mice revealed that the anti-inflammatory effects of FAAH deficiency were mediated by peripheral FAAH.

To date, NAPE-PLD is the only ECB-synthesising enzyme which was inactivated by gene targeting (Leung et al. 2006). Using a PGK-Neo replacement strategy, exon 4, which codes for amino acids 98–313, was completely deleted. As NAPE-PLD consists of 396 amino acids, most of the coding region was lost, in fact leading to the loss of the enzyme as detected by Western blot. However, this study revealed that NAPE-PLD is not the only enzyme in the biosynthesis of ECBs. Apparently, a Ca^{2+}-independent PLD activity must exist in NAPE-PLD-deficient mice, accepting substrates including the AEA precursor C20:4 NAPE (Leung et al. 2006). Clearly, this investigation points to serious difficulties in the genetic targeting of the synthesising enzymes, caused by redundant pathways.

The transient receptor potential vanilloid receptor 1 (TRPV1, formerly called VR1) (Caterina et al. 1997) is a heat- and proton-sensitive cation channel implicated mostly in noxious heat sensation. In 1999, however, it was shown that the ECB AEA is also able to specifically activate TRPV1 channels (Zygmunt et al. 1999). Since then, there has been an ever-increasing amount of data showing the interplay between the ECB and the endovanilloid systems (Starowicz et al. 2007). TRPV1 knockout mice became available from two different laboratories (Caterina et al. 2000; Davis et al. 2000). These mice were generated by homologous recombination, replacing part of the TRPV1 sequence with a PGK-Neo selection cassette. The 839 amino acid long TRPV1 protein is encoded by 15 exons. Caterina et al. (2000) replaced the receptor's fifth and sixth transmembrane domains and its pore loop, while Davis et al. (2000) replaced transmembrane domains 2–4. Following the original reports on deficits of pain perception in these mice, not unexpectedly recent investigations discovered also roles of TRPV1 in the CNS, including an involvement in emotional memory and anxiety (Marsch et al. 2007) and synaptic transmission (Gibson et al. 2008).

2.3 Conditional CB_1 Receptor Mutants

Gene targeting using homologous recombination in embryonic stem cells offers unprecedented precision in manipulating single genes and in investigating the in vivo roles of gene products in mice. As discussed above, this has proved to be true in the case of the ECB system, too. However, it has become clear that conventional gene targeting has several limitations. The loss of the gene product throughout development and the lack of spatial and temporal specificity of the generated mutation might give rise to complex, secondary phenotypical alterations. This is clearly a disadvantage in the functional analysis of genes associated with complex brain functions. One current solution of this problem is the generation of conditional mutants where the mutation will be present only in certain cell populations and/or only after a certain time point.

Unfortunately, at present, ECB system-related conditional mutants are only available for the CB_1 receptor gene. Several conditional CB_1 receptor mutant lines have been established in the last few years and have been instrumental in deciphering detailed functions of CB_1 receptors in numerous expression sites in neuroprotection (Marsicano et al. 2003; Monory et al. 2006), synaptic plasticity (Domenici et al. 2006; Azad et al. 2008), stress responses (Steiner et al. 2008b), pain perception (Agarwal et al. 2007), a model of multiple sclerosis (Maresz et al. 2007), in the response to THC (Monory et al. 2007), responses to cocaine (Corbille et al. 2007), in neural development (Berghuis et al. 2007; Mulder et al. 2008), and in hepatic functions of CB_1 receptor (Jeong et al. 2008).

Neuroprotective effects of the ECB system were shown in different models. In particular, in the kainic acid-induced acute seizure model, CB_1 receptor null mutants showed greatly increased seizures compared to their wild-type littermates

(Marsicano et al. 2003). To understand which CB_1 receptor-containing neuronal population is responsible for this function, a series of experiments was carried out with three different CB_1 receptor conditional mutant lines (Monory et al. 2006) (Table 2). Mice lacking CB_1 receptor on GABAergic neurons ($CB_1^{f/f;Dlx5/6\text{-}Cre}$, also

Table 2 Useful Cre recombinase-expressing transgenic lines for the analysis of the nervous system

Regulatory elements	Sites of recombination	Reference
AgPR-Cre (Tg)	Neurons expressing agouti-related peptide (hypothalamus)	Gropp et al. (2005)
CaMKIIα-iCre (BAC)	Principal projecting forebrain neurons (cerebral cortex, hippocampus, thalamus, striatum)	Casanova et al. (2001)
CaMKIIα-iCreER^{T2} (BAC)	Principal projecting forebrain neurons (cerebral cortex, hippocampus)	Erdmann et al. (2007)
DAT-iCre (BAC)	Dopaminergic neurons (in areas A8, A9, A10, A11, A12 and A16)	Turiault et al. (2007)
DBH-Cre (PAC)	Noradrenergic neurons (locus coeruleus, sypmpathetic ganglia)	Parlato et al. (2007)
Dlx5/6-Cre (Tg)	Forebrain GABAergic neurons (interneurons in cerebral cortex and hippocampus, and in striatum)	Monory et al. (2006)
Dlx5/6-Cre-IRES-GFP (Tg)	Forebrain GABAergic neurons (interneurons in cerebral cortex and hippocampus, and in striatum)	Stenman et al. (2003)
D1-Cre (YAC)	Dopamine receptor D1-expressing neurons	Lemberger et al. (2007)
GFAP-CreER^{T2} (Tg)	Astrocytes	Hirrlinger et al. (2006)
GLAST-CreER^{T2} (KI)	Astroglia and radial glia	Mori et al. (2006)
Emx-Cre (PAC)	Dosal telencephalon	Iwasato et al. (2004)
K_v3.2-Cre (BAC)	Thalamic projections neurons	Anderson et al. (2005)
Na_v1.8-Cre (BAC)	Nociceptive neurons in dorsal root ganglia	Agarwal et al. (2004)
Nestin-Cre (Tg)	Neuronal and glial cell precursor, neural stem cells	Tronche et al. (1999)
Nestin-CreER^{T2} (Tg)	Neuronal and glial cell precursor, neural stem cells	Imayoshi et al. (2006)
NEX-Cre (KI)	Glutamatergic neurons in cerebral cortex and hippocampus	Goebbels et al. (2006)
Pcp2-Cre (BAC)	Purkinje cells and retinal rod bipolar neurons	Zhang et al. (2004)
Peripherin-Cre (Tg)	Peripheral nervous system	Zhou et al. (2002)
PLP-CreER^{T2} (Tg)	Oligodendrocytes	Leone et al. (2003)
POMC-Cre (BAC)	Arcuate nucleus in hypothalmus, nucleus of the solitary tract	Balthasar et al. (2004)
P0 (Tg)	Myelinating Schwann cells	Akagi et al. (1997)
SF1-Cre (BAC)	Ventromedial hypothalamus	Dhillon et al. (2006)
Sim1-Cre (BAC)	Posterior hypothalamus (line 2)	Balthasar et al. (2005)
Six3-Cre#69 (Tg)	Layer 4 sensory cortex, hypothalamus	Liao and Xu (2008)

Abbreviations: BAC, bacterial artificial chromosome; GFP, green fluorescence protein; IRES, internal ribosomal entry site; KI, knock-in; PAC, phage artificial chromosome; Tg, promoter transgene; YAC, yeast artificial chromosome named as GABA-$CB_1^{-/-}$) did not differ from wild-types, demonstrating that CB_1 receptor on these neurons does not participate in the protection from kainic acid-induced seizures. However, the mutants that lack CB_1 receptor on principal forebrain neurons and in cortical glutamatergic neurons ($CB_1^{f/f;CaMKII\alpha\text{-}Cre}$, also named as CaMK-$CB_1^{-/-}$; and $CB_1^{f/f;Nex\text{-}Cre}$, also named as Glu-$CB_1^{-/-}$, respectively) had the same phenotype as the complete knockout animals, suggesting that the protective function of the CB_1 receptor is mediated by the cortical glutamatergic neurons.

The neuroprotective function of CB_1 receptors is partly caused by the decreased glutamate release from the glutamatergic axon terminals. This notion was explored by Domenici et al. (2006) using CB_1 receptor conditional mutants. They found that in slices of the basolateral amygdala, the CA1 region of the hippocampus, and the primary somatosensory cortex of wild-type mice, application of a CB_1 receptor agonist reduced evoked excitatory postsynaptic responses. This effect was not seen in mice lacking CB_1 receptors in all principal forebrain neurons (CaMK-$CB_1^{-/-}$). However, CB_1 receptor agonist reduced glutamatergic responses in slices obtained from mice lacking CB_1 receptors exclusively in GABAergic neurons (GABA-$CB_1^{-/-}$), thus excluding the involvement of CB_1 receptor expressed on GABAergic neurons in this effect of the drug. On the other hand, CB_1 receptors on GABAergic neurons play an important role in the modulation of long-term depression (LTD) in the amygdala (Azad et al. 2008). Exogenous CB_1 receptor agonist treatment blocked LTD in the amygdala; an effect that was not present in CB_1 receptor null mutants and in GABA-$CB_1^{-/-}$ mice. These results showed that the CB_1 receptor expressed on either glutamatergic or GABAergic neurons plays a differential role in the control of synaptic transmission and plasticity.

Likewise, in brain development, both aforementioned populations of CB_1 receptor play important roles. Berghuis et al. (2007) revealed that the CB_1 receptor is enriched in axonal growth cones of GABAergic interneurons in the rodent cortex during late gestation and that cannabinoids induce chemorepulsion and collapse of axonal growth cones of these GABAergic interneurons. To strengthen these findings, GABA-$CB_1^{-/-}$ mice were investigated later in development. Indeed, impaired target selection of cortical GABAergic interneurons onto glutamatergic principal neurons in the hippocampus was found. Very recently, the role of ECB signalling in cortical pyramidal cell development was investigated with the help of Glu-$CB_1^{-/-}$ mice (Mulder et al. 2008). In this study, CB_1 receptor null mutants and Glu-$CB_1^{-/-}$ mice were shown to develop axon fasciculation deficits. Accordingly, in wild-type mice, ECB signalling was shown to be operational in subcortical proliferative zones from embryonic day 12 in the mouse telencephalon, controlling the proliferation of pyramidal cell progenitors, and radial migration of postmitotic pyramidal cells. These experiments identified ECBs as axon guidance cues and demonstrate that ECB signalling regulates synaptogenesis and target selection in vivo. Apart from clarifying the ECBs' roles in brain development, these studies may help understanding the damaging effects of marijuana smoking during pregnancy and early postnatal development until puberty.

Marijuana and its main pharmacologically active component THC have a number of pharmacological effects in the adult. This multitude of effects is the result of complex interactions between different neuronal populations and circuits. To tackle this question, conditional mutants are excellent tools. Testing a series of conditional mutants, each carrying a deletion of the CB_1 receptor in different neuronal populations, for their response to high dose (10 mg/kg) of THC (Monory et al. 2007) has brought cannabinoid research closer to understanding the mechanisms of the pharmacological effect of this drug. Mice lacking CB_1 receptors in GABAergic neurons (GABA-$CB_1^{-/-}$) responded to THC in a similar way to wild-type littermates, whereas deletion of the receptor in all principal forebrain neurons (CaMK-$CB_1^{-/-}$) abolished or strongly reduced the behavioural and autonomic responses to the drug. Deleting CB_1 receptors only from glutamatergic cortical neurons (Glu-$CB_1^{-/-}$) strongly affected locomotor and hypothermic effects of THC, but left THC-induced nociception and catalepsy unaffected, suggesting that these effects are probably mediated by non-cortical projecting neurons. Deletion of CB_1 receptors from the majority of striatal neurons and a subpopulation of cortical glutamatergic neurons (D1-$CB_1^{-/-}$), on the other hand, blocked the cataleptic effect of the drug. However, there are still open questions regarding the in vivo THC pharmacology in mice. For example, cataleptic effects of THC are abolished in D1-$CB_1^{-/-}$, where the CB_1 receptor is missing in dopamine receptor D1-expressing striatal medium spiny neurons and in a small group of layer VI cortical pyramidal cells (Monory et al. 2007). However, cataleptic effects of THC are present in GABA-$CB_1^{-/-}$ where all GABAergic neurons (including medium spiny neurons) lack CB_1 receptors. Similarly, this effect is present in the Glu-$CB_1^{-/-}$, where cortical pyramidal cells (including those in layer VI) lack the CB_1 receptor. However, in CaMK-$CB_1^{-/-}$, where both striatal GABAergic and cortical glutamatergic neurons lack CB_1 receptor expression, THC was not able to induce catalepsy. These apparently puzzling results point to the importance of neuronal circuits, rather than specific neuronal populations, in mediating complex brain functions.

In the above study, THC-induced analgesia was not affected by deleting CB_1 receptors from either GABAergic, cortical glutamatergic or dopaminoceptive neurons. However, THC had no nociceptive effects in CaMK-$CB_1^{-/-}$. This suggests that non-cortical, non-striatal projecting neurons might play an important role in CB_1 receptor-mediated nociception. However, as the recombination pattern of CB_1 receptors in the spinal cord of CaMK-$CB_1^{-/-}$is not known, it is unclear how much of these effects are mediated by spinal versus supraspinal CB_1 receptors. The participation of CB_1 receptors in pain was recently studied (Agarwal et al. 2007). These authors generated a conditional mutant line in which the CB_1 receptor was specifically deleted in nociceptive neurons localised in dorsal root ganglia, while preserving its expression in the CNS. The nociceptor-specific loss of CB_1 receptors substantially reduced the analgesia produced by local and systemic, but not intrathecal, delivery of cannabinoids, suggesting that the peripheral rather than the central terminals of nociceptors are the important site of cannabinergic modulation. However, a smaller albeit important part of nociception is mediated by centrally expressed CB_1 receptors.

The participation of specific CB_1 receptor populations was also addressed in experimental autoimmune encephalomyelitis (EAE) (Maresz et al. 2007). To this end, two new CB_1 receptor conditional mouse lines were generated; one lacks CB_1 receptor expression in the entire CNS and another lacks CB_1 receptor in T cells. Additionally, in this study, CB_2 receptor-deficient mice were also included. Results showed that CB_1 receptors in the CNS, but not on T cells, were able to ameliorate disease symptoms, while the presence of CB_2 receptors on T cells was critical in suppressing autoimmune reaction in EAE.

The analysis of these conditional CB_1 receptor mutants also provided new insights into the role of the ECB system in behavioural and endocrine stress responses (Steiner et al. 2008b). Glu-$CB_1^{-/-}$ showed decreased passive stress coping (i.e. decreased immobility) in the forced swim test (FST), while GABA-$CB_1^{-/-}$ and CaMK-$CB_1^{-/-}$ behaved as wild-type littermate controls. Interestingly, FST-induced corticosterone secretion was only increased in CaMK-$CB_1^{-/-}$, but not in Glu-$CB_1^{-/-}$ and GABA-$CB_1^{-/-}$, indicating that behavioural and neuroendocrine acute stress coping in response to FST engage different neuronal subpopulations containing CB_1 receptors. While the CB_1 receptors on GABAergic terminals are not crucially involved in these responses, CB_1 receptors on glutamatergic terminals appear to have differential functions: CB_1 receptors on cortical neurons are accountable for the behavioural responses and on subcortical neurons for the endocrine responses. In addition, these conditional mutants may help in understanding the puzzling effects of the CB_1 receptor antagonist rimonabant in the FST, where antidepressant-like effects were reported (Steiner et al. 2008a).

Recent investigations using conditional CB_1 receptor mutants were also able to show the functional importance of CB_1 receptor in hepatocytes (Jeong et al. 2008, Osei-Hyiaman et al. 2008). Diet- and ethanol-induced steatosis did not occur in the conditional mutants. Indirect calorimetry furthermore showed that CB_1 receptor deletion in hepatocytes leads to increased fat burning, which was further enhanced by systemic rimonabant treatment (Osei-Hyiaman et al. 2008), indicating that organs other than the liver are involved in metabolic processes controlled by CB_1 receptors.

2.4 Complications with Cre Recombinase-Expressing Lines

Conditional mutants are valuable tools in deciphering the exact function of CB_1 receptors and other components of the ECB system at their numerous expression sites. However, even these sophisticated tools are not without limitations.

One such limitation is the unwanted ectopic Cre recombinase expression, causing target gene deletion in cell populations that were not intended to be included. Such recombination may lead to serious problems in the interpretation of the resulting phenotype. It is therefore very important to analyse thoroughly the expression pattern of the targeted gene in the conditional mutants, before a detailed phenotype analysis takes place.

Another possible problem is that the Cre recombinase-expressing transgene might cause some unspecific effects apart from excising the desired target gene. For this reason, it is very important to perform control experiments with mice that express only Cre recombinase, but do not contain the targeted "floxed" alleles. The phenotype of these mice should be compared to wild-type littermates in order to ensure that Cre recombinase expression alone has no effect on a specific behaviour studied.

Transgenes that engage a cell-type specific expression of Cre recombinase in the adult may also be active in germ cells to some extent under certain circumstances. Subsequently, this causes germ line deletion of the target gene. Offspring resulting from such a gamete will have a complete knockout on the allele that is derived from the affected gamete. Should this mutation not be recognised in time, the mutation could spread and soon the breeding colony would contain increasing numbers of complete knockout mice. Certainly, these mice would still express Cre recombinase in the normal, Mendelian manner. Consequently, genotyping for the presence of Cre recombinase itself would not reveal this problem. It is, therefore, highly important to genotype conditional mutants regularly for the existence of the "floxed" target gene, too.

3 Methodological Considerations on Behavioural Experiments with Mutant Mice

There are several important considerations when planning to carry out behavioural experiments with transgenic mice. These include sample size, genetic background, age and sex of the animals, breeding scheme, and choice of proper controls. The following paragraphs are not exhaustive, but do reflect numerous issues which are often encountered by researchers performing behavioural experiments with transgenic animals. Several recent overview articles have also covered this theme (Crawley 2008; Sousa et al. 2006).

The individual variations in behaviour are considerable, even though congenic lines are used in most of the cases. Therefore, the number of animals per experimental group has to be large enough for the standard deviation not to exceed the level that would conceal between-group differences. In practical terms, this mostly means a sample size of 10–12 (Crawley and Paylor 1997). Very few tests give such uniform data that less than 10 mice are enough for the experiments.

To reduce within-group variability, it is important that the genetic background of the animals is as homogenous as possible. Scores of new mutant mouse lines are generated every year using embryonic stem cell techniques. The genomic composition of these animals depends on the origin of genomic DNA used to generate the targeting construct, the origin of the ES cell line, and the strain of mice for mating the chimaeras and for conducting the backcrossing (Bockamp et al. 2002). The very robust ES cell lines that contribute most effectively to the germ line following blastocyst injections are derived from the 129/Sv mouse strain, while this strain is only seldom used by behavioural scientists. Therefore, before any behavioural experiment can be done, the new transgenic mice have to be backcrossed through a minimum of 5–6 generations into the desired wild-type strain.

The choice of the wild-type strain for backcrossing depends on the type of experiments that are to be carried out, as genetic differences between mouse strains can have a great impact on the observed phenotype. To begin with, certainly, as the goal is genetic uniformity, an inbred strain is called for. However, there are many available inbred strains (and sub-strains), and their characteristics have to be kept in mind for the right choice (Brooks et al. 2004, 2005; Nguyen and Gerlai 2002; Taft et al. 2006). A mouse strain that has difficulties in learning *per se* will not be suitable for studying a gene function in learning and memory. Similarly, a mouse strain that will not drink ethanol cannot be used for alcohol addiction studies.

Another consideration is the fertility and productiveness of the strain in question: if there are only 2–3 pups in one litter, there might be only one or less animal per litter with the desired experimental genotype. This would require the establishment of about 30–35 breeding pairs to secure ten mutant and ten wild-type control mice (considering that normally only males are used for behavioural experiments). Bearing in mind the costs, the animal facility space requirements and the manpower needed, together with the risk of not observing a phenotype in the experiment, this is by and large not a viable plan.

Wild-type strains are available from several companies that have their own colonies. These colonies, though originating from the same founder, might develop some gene mutations over time. Some of these mutations are simply not known, but others are identified. For example, the C57BL/6J line from Harlan carries a null mutation for α-synuclein (Specht and Schoepfer 2001; Wotjak 2003), a gene known to be involved in learning and memory as well as in neurodegenerative diseases such as Parkinson and Alzheimer disease. Therefore, if the mutation could potentially influence the outcome of the experiments (e.g. studies on Parkinson disease in this case) choosing another strain or another supplier is advisable.

Choosing the proper controls for a behavioural experiment is vital. In this regard, there is often a conflict between good scientific practise and available resources. However, the importance of littermate controls cannot be emphasised enough. Why is it so important? As mentioned above, the genetic background of a transgenic mouse line is never 100% identical to a chosen wild-type strain. Therefore, by comparing your mutants to a certain wild-type strain, the effects of several variables are assessed – most of whose identity one does not even know. Importantly, complex behaviours (especially anxiety traits) often depend on early life experiences. Animals that grew up in different environments (e.g. a supplier company's vs. a laboratory's animal facility) will certainly behave differently in a test. Yet this observed phenotype might not have anything to do with the mutation the transgenic animal carries. Importantly, early experiences also include the effects of maternal behaviour. A genetic mutation might influence maternal care in many different ways – starting from extent of fertility through proper milk composition to enough nursing provided. Thus, even breeding mutants and wild-types separately could introduce variables that are not the direct consequences of the mutated gene in the observed individual but in the mother. The best way to exclude these confounding variables is to set up a breeding scheme where the generated mutant and control mice will be littermates. When breeding knockouts, this means heterozygous mating. The disadvantage of the method is that only 25% of the offspring is knockout, 25% is wild-type littermate and the rest, 50%, is "unwanted" heterozygous. This setup also requires more storage space in the animal facility, more manpower and higher costs for genotyping. Still, in the long run, this method pays back, as it gives clearer results with less room for misinterpretation. Whatever the decision, however, one rule has to be kept in mind: when it comes to publication, the mating scheme and the origin of the control group must be unambiguously traceable for the reader.

On the other hand, with the advent of more sophisticated genetic tools, double transgenic mice, such as conditional mutants, are becoming more widespread. However, setting up a breeding scheme with double heterozygous parents would result in 1 in 16 double transgenic offspring. This is clearly unrealistic in most laboratories. In this case, scientists have to weigh up which genotypes are the most important experimentally and design a breeding scheme where the relevant mutant and its control counterpart are littermates. Taking conditional knockouts as an example, a good solution is breeding mice that are transgenic or wild-type for the Cre recombinase allele, respectively, and all have the "floxed" allele in a homozygous manner.

The other considerations such as gender and age of the experimental animals do not differ in experiments with transgenic or wild-type mice. Often, relatively young adult males are used. These choices have mostly practical reasons, and though they are justifiable, it is important to keep in mind that they will to some extent limit the possible interpretations of the results. Regarding the analysis of the ECB system, it has to be kept in mind that changes in the importance of the system may occur depending on age (Wang et al. 2003; Bilkei-Gorzo et al. 2005).

Behavioural experiments are addressing complex issues, and thus the phenotypes observed are the results of complex – and frequently redundant – mechanisms. When studying knockout animals, redundancy (i.e. compensatory mechanisms) is especially an aspect to keep in mind. Therefore, the magnitude of difference between the mutant group and the controls is critical. Mutations that cause strong phenotypes in basal level conditions are mostly seriously debilitating, and thus are not easily used to tackle the subtle regulatory processes of complex behaviours. On the other hand, subtle and hidden phenotypes are obviously more difficult to deal with. The trick is to challenge the system in a way that will reveal differences between the experimental groups - a requirement easy to demand but often difficult to fulfil.

4 Caveats in Genetics and Pharmacology

To study a biological system in vivo, pharmacology and genetics are two obvious approaches to choose from. However, it is important to regard them as complementary rather than competing methods, as both approaches have some weaknesses and advantages. Therefore, using these two approaches in parallel and confirming the data obtained with the other method strengthens conclusions greatly.

What are then the main limitations of pharmacology? Pharmacology is always "dirty". There is not one drug that has 100% specificity to a given binding site.

Therefore, there will always be some other site influenced as well as that studied, potentially confounding the obtained observations. This site can be another subtype of the studied protein, but it can even be a completely different class of molecule. As an example from the ECB system, AEA is very promiscuous, as it binds, e.g., to CB_1 receptors, CB_2 receptors, and TRPV1. This may also be relevant in cases where a genetic deletion causes altered AEA levels, e.g. in CB_1 receptor-deficient mice (Di Marzo et al. 2000) or in FAAH-deficient mice (Cravatt et al. 2001). Another example is the specificity of the CB_1 receptor antagonist/inverse agonist rimonabant, for which non-CB_1 receptor/non-CB_2 receptor sites were proposed (Jin et al. 2004; Haller et al. 2004).

Another point is whether the receptor in question is constitutively active in vivo. If so, instead of simply blocking the receptor with an antagonist resulting in an effect opposite to agonist stimulation, agonist treatment will prevent in this case the action of tonically released endogenous ligands on the receptor. A similar effect is seen when inverse agonists are used for pharmacological intervention. These molecules stabilise the receptor in a third conformational state different from the agonist-activated or the antagonist-inactivated ones. Importantly, several cannabinoid antagonists, most prominently rimonabant, are considered as inverse agonists. Even though site-directed mutagenesis studies could show which amino acids might be responsible for this effect in the amino acid chain of CB_1 receptor (McAllister et al. 2003; Hurst et al. 2002), in an in vivo situation there is still an open question as to how much the effects of rimonabant are due to its inverse agonist effect and how much to its blocking the effects of tonically released ECBs (Hentges et al. 2005).

This kind of problem is less prevalent in a genetically modified model. If a certain protein is genetically inactivated, the observed phenotype should be a specific consequence of its absence. Shouldn't it? However, there are some important aspects to consider here, too. First, there might be some compensatory mechanisms that try to counterbalance the loss of an important protein – this mostly manifests in the up-regulation of related gene(s) in the same family. For example, this is nicely shown for the leucin-zipper transcription factors CREB/CREM (Mantamadiotis et al. 2002). For cannabinoid receptors, such an up-regulation has not yet been reported. Another important aspect is the developmental effects of genes. Inactivation of genes that participate in developmental processes might leave the animal with malformed organs, aberrant wiring, changed hormonal secretion or altered motor ability, to name but a few. If these effects are not known or hidden, the resulting phenotype might be interpreted as the effect of the lack of acute activation of the investigated gene product. This can be tested by pharmacological treatment of wild-type animals with an antagonist. If the phenotype is the result of an acute lack of the gene product, it will be reproducible by pharmacological blockade of the protein. For example, this is illustrated by the involvement of CB_1 receptors in extinction of aversive memories (Marsicano et al. 2002) and protection against kainic acid-induced seizures (Marsicano et al. 2003).

On the other hand, possible compensatory processes or developmental effects in knockout mice may explain discrepant results between genetic and pharmacological invalidation. It was reported that CB_1 receptor-deficient mice have different phenotypes as compared to pharmacologically treated wild-type mice, as illustrated by the fact that CB_1 receptor-deficient mice show a depression-like behaviour (Steiner et al. 2008c), while the CB_1 receptor antagonist rimonabant induces antidepressant-like behaviour as monitored in the FST (Steiner et al. 2008a). Another example is that genetic inactivation of CB_1 receptors and pharmacological treatment with rimonabant resulted in differential effects on "non-associative" memory and forebrain monoamine concentrations in mice (Thiemann et al. 2007).

A possible solution to circumvent developmental effects in genetic approaches is gene inactivation with a spatiotemporal specificity. Gene inactivation in the adult stage should ensure that phenotypic alterations are not caused by developmental deficits, although compensatory processes in the adult brain cannot be excluded. In this respect, even conditional mutants might not be fully devoid of developmental effects (Berghuis et al. 2007; Mulder et al. 2008).

5 Perspectives

The combination of genetic and pharmacological experiments has given an impressive number of novel insights into the physiological and pathophysiological roles of the ECB system. This regulatory system, however, contains a high intrinsic complexity, which requires special attention in the design of genetic experiments. (a) As ECB signalling is active throughout neural development, genetic inactivation should occur in the adult stage in case of investigations on adult processes. (b) Particularly in the nervous system, ECB signalling acts in a sophisticated spatiotemporal manner. Thus, spatial and temporal specificity in inactivation experiments are highly important in order to allow firm conclusions from these genetic experiments. (c) As the synthesising and degradation machinery of ECBs appears to be redundant, the genetic analysis of these components will be very difficult and remains a special challenge. (d) The application of adeno-associated virus and other viral systems appears to be very useful in order to inactivate "floxed" genes in specific brain regions (Monory et al. 2006), to introduce RNA silencing constructs (Xia et al. 2004), or to overexpress genes in a spatiotemporal manner (Klugmann et al. 2005). It is hoped that the combination of the techniques discussed will give novel insights into this fascinating neuroregulatory system.

References

Agarwal N, Offermanns S, Kuner R (2004) Conditional gene deletion in primary nociceptive neurons of trigeminal ganglia and dorsal root ganglia. Genesis 38:122–129

Agarwal N, Pacher P, Tegeder I et al. (2007) Cannabinoids mediate analgesia largely via peripheral type 1 cannabinoid receptors in nociceptors. Nat Neurosci 10:870–879

Aguado T, Palazuelos J, Monory K et al. (2006) The endocannabinoid system promotes astroglial differentiation by acting on neural progenitor cells. J Neurosci 26:1551–1561

Akagi K, Sandig V, Vooijs M et al. (1997) Cre-mediated somatic site-specific recombination in mice. Nucleic Acids Res 25:1766–1773

Anderson MP, Mochizuki T, Xie J et al. (2005) Thalamic Cav3.1 T-type Ca2+ channel plays a crucial role in stabilizing sleep. Proc Natl Acad Sci USA 102:1743–1748

Azad SC, Kurz J, Marsicano G et al. (2008) Activation of CB_1 specifically located on GABAergic interneurons inhibits LTD in the lateral amygdala. Learn Mem 15:143–152

Baker D, Pryce G, Davies WL et al. (2006) In silico patent searching reveals a new cannabinoid receptor. Trends Pharmacol Sci 27:1–4

Balthasar N, Coppari R, McMinn J et al. (2004) Leptin receptor signaling in POMC neurons is required for normal body weight homeostasis. Neuron 42:983–991

Balthasar N, Dalgaard LT, Lee CE et al. (2005) Divergence of melanocortin pathways in the control of food intake and energy expenditure. Cell 123:493–505

Basavarajappa BS, Yalamanchili R, Cravatt BF et al. (2006) Increased ethanol consumption and preference and decreased ethanol sensitivity in female FAAH knockout mice. Neuropharmacology 50:834–844

Bensaid M, Gary-Bobo M, Esclangon A et al. (2003) The cannabinoid CB(1) receptor antagonist SR141716 increases Acrp30 mRNA expression in adipose tissue of obese fa/fa rats and in cultured adipocyte cells. Mol Pharmacol 63:908–914

Berghuis P, Rajnicek AM, Morozov YM et al. (2007) Hardwiring the brain: endocannabinoids shape neuronal connectivity. Science 316:1212–1216

Bilkei-Gorzo A, Racz I, Valverde O et al. (2005) Early age-related cognitive impairment in mice lacking cannabinoid CB_1 receptors. Proc Natl Acad Sci USA 102:15670–15675

Bockamp E, Maringer M, Spangenberg C et al. (2002) Of mice and models: improved animal models for biomedical research. Physiol Genomics 11:115–132

Brooks SP, Pask T, Jones L et al. (2004) Behavioural profiles of inbred mouse strains used as transgenic backgrounds. I: motor tests. Genes Brain Behav 3:206–215

Brooks SP, Pask T, Jones L et al. (2005) Behavioural profiles of inbred mouse strains used as transgenic backgrounds. II: cognitive tests. Genes Brain Behav 4:307–317

Brown AJ (2007) Novel cannabinoid receptors. Br J Pharmacol 152:567–575

Buckley NE, McCoy KL, Mezey E et al. (2000) Immunomodulation by cannabinoids is absent in mice deficient for the cannabinoid CB2 receptor. Eur J Pharmacol 396:141–149

Casanova E, Fehsenfeld S, Mantamadiotis T et al. (2001) A CamKIIalpha iCre BAC allows brain-specific gene inactivation. Genesis 31:37–42

Caterina MJ, Schumacher MA, Tominaga M et al. (1997) The capsaicin receptor: a heat-activated ion channel in the pain pathway. Nature 398:816–824

Caterina MJ, Leffler A, Malmberg AB et al. (2000) Impaired nociception and pain sensation in mice lacking the capsaicin receptor. Science 288:306–313

Clement AB, Hawkins EG, Lichtman AH et al. (2003) Increased seizure susceptibility and proconvulsant activity of anandamide in mice lacking fatty acid amide hydrolase. J Neurosci 23: 3916–3923

Corbille AG, Valjent E, Marsicano G et al. (2007) Role of cannabinoid type 1 receptors in locomotor activity and striatal signaling in response to psychostimulants. J Neurosci 27:6937–6947

Cota D, Marsicano G, Tschöp M et al. (2003) The endogenous cannabinoid system affects energy balance via central orexigenic drive and peripheral lipogenesis. J Clin Invest 112:423–431

Cravatt BF, Demarest K, Patricelli MP et al. (2001) Supersensitivity to anandamide and enhanced endogenous cannabinoid signaling in mice lacking fatty acid amide hydrolase. Proc Natl Acad Sci USA 98:9371–9376

Cravatt BF, Saghatelian A, Hawkins EG et al. (2004) Functional disassociation of the central and peripheral fatty acid amide signaling systems. Proc Natl Acad Sci USA 101:10821–10826

Crawley JN (2008) Behavioral phenotyping strategies for mutant mice. Neuron 57:809–818

Crawley JN, Paylor R (1997) A proposed test battery and constellations of specific behavioral paradigms to investigate the behavioral phenotypes of transgenic and knockout mice. Horm Behav 31:197–211

Davis JB, Gray J, Gunthorpe MJ et al. (2000) Vanilloid receptor-1 is essential for inflammatory thermal hyperalgesia. Nature 405:183–187

Degroot A, Kofalvi A, Wade MR et al. (2006) CB_1 receptor antagonism increases hippocampal acetylcholine release: site and mechanism of action. Mol Pharmacol 70:1236–1245

Dhillon H, Zigman JM, Ye C et al. (2006) Leptin directly activates SF1 neurons in the VMH, and this action by leptin is required for normal body-weight homeostasis. Neuron 49:191–203

Di Marzo V (2008) Endocannabinoids: synthesis and degradation. Rev Physiol Biochem Pharmacol 160:1–24

Di Marzo V, Breivogel CS, Tao Q et al. (2000) Levels, metabolism, and pharmacological activity of anandamide in CB(1) cannabinoid receptor knockout mice: evidence for non-CB(1), non-CB(2) receptor-mediated actions of anandamide in mouse brain. J Neurochem 75:2434–2444

Di Marzo V, Goparaju SK, Wang L et al. (2001) Leptin-regulated endocannabinoids are involved in maintaining food intake. Nature 410:822–825

Doetschman T, Gregg RG, Maeda N et al. (1987) Targetted correction of a mutant HPRT gene in mouse embryonic stem cells. Nature 330:576–578

Domenici MR, Azad SC, Marsicano G et al. (2006) Cannabinoid receptor type 1 located on presynaptic terminals of principal neurons in the forebrain controls glutamatergic synaptic transmission. J Neurosci 26:5794–5799

Erdmann G, Schutz G, Berger S (2007) Inducible gene inactivation in neurons of the adult mouse forebrain. BMC Neurosci 8:63

Erdmann G, Schutz G, Berger S (2008) Loss of glucocorticoid receptor function in the pituitary results in early postnatal lethality. Endocrinology 149:3446–3451

Feil R (2007) Conditional somatic mutagenesis in the mouse using site-specific recombinases. Handb Exp Pharmacol 178:3–28

Feil R, Brocard J, Mascrez B et al. (1996) Ligand-activated site-specific recombination in mice. Proc Natl Acad Sci USA 93:10887–10890

Gaoni Y, Mechoulam R (1964) Isolation, structure and partial synthesis of an active constituent of hashish. J Am Chem Soc 86:1646–1647

Gaveriaux-Ruff C, Kieffer BL (2007) Conditional gene targeting in the mouse nervous system: Insights into brain function and diseases. Pharmacol Ther 113:619–634

Gibson HE, Edwards JG, Page RS et al. (2008) TRPV1 channels mediate long-term depression at synapses on hippocampal interneurons. Neuron 57:746–759

Goebbels S, Bormuth I, Bode U et al. (2006) Genetic targeting of principal neurons in neocortex and hippocampus of NEX-Cre mice. Genesis 44:611–621

Grant SG, O'Dell TJ, Karl KA et al. (1992) Impaired long-term potentiation, spatial learning, and hippocampal development in fyn mutant mice. Science 258:1903–1910

Gropp E, Shanabrough M, Borok E et al. (2005) Agouti-related peptide-expressing neurons are mandatory for feeding. Nat Neurosci 8:1289–1291

Haller J, Bakos N, Szirmay M et al. (2002) The effects of genetic and pharmacological blockade of the CB_1 cannabinoid receptor on anxiety. Eur J NeuroSci 16:1395–1398

Haller J, Varga B, Ledent C et al. (2004) CB_1 cannabinoid receptors mediate anxiolytic effects: convergent genetic and pharmacological evidence with CB_1-specific agents. Behav Pharmacol 15:299–304

Häring M, Marsicano G, Lutz B et al. (2007) Identification of the cannabinoid receptor type 1 in serotonergic cells of raphe nuclei in mice. Neuroscience 146:1212–1219

Hentges ST, Low MJ, Williams JT (2005) Differential regulation of synaptic inputs by constitutively released endocannabinoids and exogenous cannabinoids. J Neurosci 25:9746–9751

Herkenham M, Lynn AB, Little MD et al. (1990) Cannabinoid receptor localization in brain. Proc Natl Acad Sci USA 87:1932–1936

Hillard CJ, Harris RA, Bloom AS (1985) Effects of the cannabinoids on physical properties of brain membranes and phospholipid vesicles: fluorescence studies. J Pharmacol Exp Ther 232:579–588

Hirrlinger PG, Scheller A, Braun C et al. (2006) Temporal control of gene recombination in astrocytes by transgenic expression of the tamoxifen-inducible DNA recombinase variant CreERT2. Glia 54:11–20

Howlett AC, Johnson MR, Melvin LS et al. (1988) Nonclassical cannabinoid analgetics inhibit adenylate cyclase: development of a cannabinoid receptor model. Mol Pharmacol 33:297–302

Hurst DP, Lynch DL, Barnett-Norris J et al. (2002) N-(piperidin-1-yl)-5-(4-chlorophenyl)-1-(2,4-dichlorophenyl)-4-methyl-1H-pyrazole-3-carboxamide (SR141716A) interaction with LYS 3.28(192) is crucial for its inverse agonism at the cannabinoid CB_1 receptor. Mol Pharmacol 62:1274–1287

Imayoshi I, Ohtsuka T, Metzger D et al. (2006) Temporal regulation of Cre recombinase activity in neural stem cells. Genesis 44:233–238

Iwasato T, Nomura R, Ando R et al. (2004) Dorsal telencephalon-specific expression of Cre recombinase in PAC transgenic mice. Genesis 38:130–138

Jeong WI, Osei-Hyiaman D, Park O et al. (2008) Paracrine activation of hepatic CB_1 receptors by stellate cell-derived endocannabinoids mediates alcoholic fatty liver. Cell Metab 7:227–235

Jin K, Xie L, Kim SH et al. (2004) Defective adult neurogenesis in CB_1 cannabinoid receptor knockout mice. Mol Pharmacol 66:204–208

Julien B, Grenard P, Teixeira-Clerc F et al. (2005) Antifibrogenic role of the cannabinoid receptor CB2 in the liver. Gastroenterology 128:742–755

Karsak M, Gaffal E, Date R et al. (2007) Attenuation of allergic contact dermatitis through the endocannabinoid system. Science 316:1494–1497

Katona I, Sperlágh B, Sík A et al. (1999) Presynaptically located CB_1 cannabinoid receptors regulate GABA release from axon terminals of specific hippocampal interneurons. J Neurosci 19:4544–4558

Klugmann M, Symes CW, Leichtlein CB et al. (2005) AAV-mediated hippocampal expression of short and long Homer 1 proteins differentially affect cognition and seizure activity in adult rats. Mol Cell Neurosci 28:347–360

Lauckner JE, Jensen JB, Chen HY et al. (2008) GPR55 is a cannabinoid receptor that increases intracellular calcium and inhibits M current. Proc Natl Acad Sci USA 105:2699–2704

Ledent C, Valverde O, Cossu G et al. (1999) Unresponsiveness to cannabinoids and reduced addictive effects of opiates in CB_1 receptor knockout mice. Science 283:401–404

Lemberger T, Parlato R, Dassesse D et al. (2007) Expression of Cre recombinase in dopaminoceptive neurons. BMC Neurosci 8:4

Leone DP, Genoud S, Atanasoski S et al. (2003) Tamoxifen-inducible glia-specific Cre mice for somatic mutagenesis in oligodendrocytes and Schwann cells. Mol Cell Neurosci 22:430–440

Leung D, Saghatelian A, Simon GM et al. (2006) Inactivation of N-acyl phosphatidylethanolamine phospholipase D reveals multiple mechanisms for the biosynthesis of endocannabinoids. Biochemistry 45:4720–4726

Liao GY, Xu B (2008) Cre recombinase-mediated gene deletion in layer 4 of murine sensory cortical areas. Genesis 46:289–293

Liu J, Wang L, Harvey-White J et al. (2008) Multiple pathways involved in the biosynthesis of anandamide. Neuropharmacology 54:1–7

Mak TW (2007) Gene targeting in embryonic stem cells scores a knockout in Stockholm. Cell 131:1027–1031

Mantamadiotis T, Lemberger T, Bleckmann SC et al. (2002) Disruption of CREB function in brain leads to neurodegeneration. Nat Genet 31:47–54

Maresz K, Pryce G, Ponomarev ED et al. (2007) Direct suppression of CNS autoimmune inflammation via the cannabinoid receptor CB_1 on neurons and CB2 on autoreactive T cells. Nat Med 13:492–497

Marsch R, Foeller E, Rammes G et al. (2007) Reduced anxiety, conditioned fear, and hippocampal long-term potentiation in transient receptor potential vanilloid type 1 receptor-deficient mice. J Neurosci 27:832–839

Marsicano G, Lutz B (1999) Expression of the cannabinoid receptor CB_1 in distinct neuronal subpopulations in the adult mouse forebrain. Eur J NeuroSci 11:4213–4225

Marsicano G, Wotjak CT, Azad SC et al. (2002) The endogenous cannabinoid system controls extinction of aversive memories. Nature 418:530–534

Marsicano G, Goodenough S, Monory K et al. (2003) CB_1 cannabinoid receptors and on-demand defense against excitotoxicity. Science 302:84–88

Matsuda LA, Lolait SJ, Brownstein MJ et al. (1990) Structure of a cannabinoid receptor and functional expression of the cloned cDNA. Nature 346:561–564

McAllister SD, Rizvi G, Anavi-Goffer S et al. (2003) An aromatic microdomain at the cannabinoid CB(1) receptor constitutes an agonist/inverse agonist binding region. J Med Chem 46:5139–5152

Merritt LL, Martin BR, Walters C et al. (2008) The endogenous cannabinoid system modulates nicotine reward and dependence. J Pharmacol Exp Ther 326(2):483–492

Monory K, Massa F, Egertova M et al. (2006) The endocannabinoid system controls key epileptogenic circuits in the hippocampus. Neuron 51:455–466

Monory K, Blaudzun H, Massa F et al. (2007) Genetic dissection of behavioural and autonomic effects of Delta(9)-tetrahydrocannabinol in mice. PLoS Biol 5:e269

Moreira FA, Kaiser N, Monory K et al. (2008) Reduced anxiety-like behaviour induced by genetic and pharmacological inhibition of the endocannabinoid-degrading enzyme fatty acid amide hydrolase (FAAH) is mediated by CB_1 receptors. Neuropharmacology 54:141–150

Mori T, Tanaka K, Buffo A et al. (2006) Inducible gene deletion in astroglia and radial glia–a valuable tool for functional and lineage analysis. Glia 54:21–34

Mulder AM, Cravatt BF (2006) Endocannabinoid metabolism in the absence of fatty acid amide hydrolase (FAAH): discovery of phosphorylcholine derivatives of N-acyl ethanolamines. Biochemistry 45:11267–11277

Mulder J, Aguado T, Keimpema E et al. (2008) Endocannabinoid signaling controls pyramidal cell specification and long-range axon patterning. Proc Natl Acad Sci USA 105:8760–8765

Munro S, Thomas KL, Abu-Shaar M (1993) Molecular characterization of a peripheral receptor for cannabinoids. Nature 365:61–65

Naidu PS, Varvel SA, Ahn K et al. (2007) Evaluation of fatty acid amide hydrolase inhibition in murine models of emotionality. Psychopharmacology (Berl) 192:61–70

Nguyen PV, Gerlai R (2002) Behavioural and physiological characterization of inbred mouse strains: prospects for elucidating the molecular mechanisms of mammalian learning and memory. Genes Brain Behav 1:72–81

Ofek O, Karsak M, Leclerc N et al. (2006) Peripheral cannabinoid receptor, CB2, regulates bone mass. Proc Natl Acad Sci USA 103:696–701

Onaivi ES, Ishiguro H, Gong JP et al. (2008) Brain neuronal CB2 cannabinoid receptors in drug abuse and depression: from mice to human subjects. PLoS ONE 3:e1640

Oropeza VC, Mackie K, Van Bockstaele EJ (2007) Cannabinoid receptors are localized to noradrenergic axon terminals in the rat frontal cortex. Brain Res 1127:36–44

Osei-Hyiaman D, Depetrillo M, Pacher P et al. (2005) Endocannabinoid activation at hepatic CB_1 receptors stimulates fatty acid synthesis and contributes to diet-induced obesity. J Clin Invest 115:1298–1305

Osei-Hyiaman D, Liu J, Zhou L et al. (2008) Hepatic CB(1) receptor is required for development of diet-induced steatosis, dyslipidemia, and insulin and leptin resistance in mice. J Clin Invest 118:3160–3169

Parlato R, Otto C, Begus Y et al. (2007) Specific ablation of the transcription factor CREB in sympathetic neurons surprisingly protects against developmentally regulated apoptosis. Development 134:1663–1670

Racz I, Bilkei-Gorzo A, Toth ZE et al. (2003) A critical role for the cannabinoid CB_1 receptors in alcohol dependence and stress-stimulated ethanol drinking. J Neurosci 23:2453–2458

Ravinet TC, Delgorge C, Menet C et al. (2004) CB_1 cannabinoid receptor knockout in mice leads to leanness, resistance to diet-induced obesity and enhanced leptin sensitivity. Int J Obes Relat Metab Disord 28:640–648

Reibaud M, Obinu MC, Ledent C et al. (1999) Enhancement of memory in cannabinoid CB_1 receptor knock-out mice. Eur J Pharmacol 379:R1–R2

Robbe D, Kopf M, Remaury A et al. (2002) Endogenous cannabinoids mediate long-term synaptic depression in the nucleus accumbens. Proc Natl Acad Sci USA 99:8384–8388

Ryberg E, Vu HK, Larsson N et al. (2005) Identification and characterisation of a novel splice variant of the human CB_1 receptor. FEBS Lett 579:259–264

Ryberg E, Larsson N, Sjogren S et al. (2007) The orphan receptor GPR55 is a novel cannabinoid receptor. Br J Pharmacol 152:1092–1101

Sauer B, Henderson N (1988) Site-specific DNA recombination in mammalian cells by the Cre recombinase of bacteriophage P1. Proc Natl Acad Sci USA 85:5166–5170

Sawzdargo M, Nguyen T, Lee DK et al. (1999) Identification and cloning of three novel human G protein-coupled receptor genes GPR52, PsiGPR53 and GPR55: GPR55 is extensively expressed in human brain. Brain Res Mol Brain Res 64:193–198

Shire D, Carillon C, Kaghad M et al. (1995) An amino-terminal variant of the central cannabinoid receptor resulting from alternative splicing. J Biol Chem 270:3726–3731

Silva AJ, Paylor R, Wehner JM et al. (1992) Impaired spatial learning in alpha-calcium-calmodulin kinase II mutant mice. Science 257:206–211

Sousa N, Almeida OF, Wotjak CT (2006) A hitchhiker's guide to behavioral analysis in laboratory rodents. Genes Brain Behav 5(Suppl 2):5–24

Specht CG, Schoepfer R (2001) Deletion of the alpha-synuclein locus in a subpopulation of C57BL/6 J inbred mice. BMC Neurosci 2:11

Starowicz K, Nigam S, Di Marzo V (2007) Biochemistry and pharmacology of endovanilloids. Pharmacol Ther 114:13–33

Staton PC, Hatcher JP, Walker DJ et al. (2008) The putative cannabinoid receptor GPR55 plays a role in mechanical hyperalgesia associated with inflammatory and neuropathic pain. Pain 139 (1):225–236

Steiner MA, Marsicano G, Nestler EJ et al. (2008a) Antidepressant-like behavioral effects of impaired cannabinoid receptor type 1 signaling coincide with exaggerated corticosterone secretion in mice. Psychoneuroendocrinology 33:54–67

Steiner MA, Marsicano G, Wotjak CT et al. (2008b) Conditional cannabinoid receptor type 1 mutants reveal neuron subpopulation-specific effects on behavioral and neuroendocrine stress responses. Psychoneuroendocrinology 33(8):1165–1170

Steiner MA, Wanisch K, Monory K et al. (2008c) Impaired cannabinoid receptor type 1 signaling interferes with stress-coping behavior in mice. Pharmacogenomics J 8:196–208

Stenman J, Toresson H, Campbell K (2003) Identification of two distinct progenitor populations in the lateral ganglionic eminence: implications for striatal and olfactory bulb neurogenesis. J Neurosci 23:167–174

Sternberg N, Hamilton D (1981) Bacteriophage P1 site-specific recombination. I. Recombination between loxP sites. J Mol Biol 150:467–486

Sugiura T, Kondo S, Sukagawa A et al. (1995) 2-Arachidonoylglycerol: a possible endogenous cannabinoid receptor ligand in brain. Biochem Biophys Res Commun 215:89–97

Sullivan JM (1999) Mechanisms of cannabinoid-receptor-mediated inhibition of synaptic transmission in cultured hippocampal pyramidal neurons. J Neurophysiol 82:1286–1294

Taft RA, Davisson M, Wiles MV (2006) Know thy mouse. Trends Genet 22:649–653

Thiemann G, Fletcher BC, Ledent C et al. (2007) The genetic versus pharmacological invalidation of the cannabinoid CB(1) receptor results in differential effects on 'non-associative' memory and forebrain monoamine concentrations in mice. Neurobiol Learn Mem 88:416–423

Thomas KR, Capecchi MR (1987) Site-directed mutagenesis by gene targeting in mouse embryo-derived stem cells. Cell 51:503–512

Tronche F, Kellendonk C, Kretz O et al. (1999) Disruption of the glucocorticoid receptor gene in the nervous system results in reduced anxiety. Nat Genet 23:99–103

Tsou K, Brown S, Sanudo-Pena MC et al. (1998) Immunohistochemical distribution of cannabinoid CB1 receptors in the rat central nervous system. Neuroscience 83:393–411

Turiault M, Parnaudeau S, Milet A et al. (2007) Analysis of dopamine transporter gene expression pattern–generation of DAT-iCre transgenic mice. FEBS J 274:3568–3577

Van Sickle MD, Duncan M, Kingsley PJ et al. (2005) Identification and functional characterization of brainstem cannabinoid CB2 receptors. Science 310:329–332

Varvel SA, Wise LE, Niyuhire F et al. (2007) Inhibition of fatty-acid amide hydrolase accelerates acquisition and extinction rates in a spatial memory task. Neuropsychopharmacology 32:1032–1041

Wan M, Cravatt BF, Ring HZ et al. (1998) Conserved chromosomal location and genomic structure of human and mouse fatty-acid amide hydrolase genes and evaluation of clasper as a candidate neurological mutation. Genomics 54:408–414

Wang L, Liu J, Harvey-White J et al. (2003) Endocannabinoid signaling via cannabinoid receptor 1 is involved in ethanol preference and its age-dependent decline in mice. Proc Natl Acad Sci USA 100:1393–1398

Webb M, Luo L, Ma JY et al. (2008) Genetic deletion of Fatty Acid Amide Hydrolase results in improved long-term outcome in chronic autoimmune encephalitis. Neurosci Lett 439:106–110

Wei BQ, Mikkelsen TS, McKinney MK et al. (2006) A second fatty acid amide hydrolase with variable distribution among placental mammals. J Biol Chem 281:36569–36578

Wilson RI, Nicoll RA (2001) Endogenous cannabinoids mediate retrograde signalling at hippocampal synapses. Nature 410:588–592

Wilson RI, Kunos G, Nicoll RA (2001) Presynaptic specificity of endocannabinoid signaling in the hippocampus. Neuron 31:453–462

Wotherspoon G, Fox A, McIntyre P et al. (2005) Peripheral nerve injury induces cannabinoid receptor 2 protein expression in rat sensory neurons. Neuroscience 135:235–245

Wotjak CT (2003) C57BLack/BOX? The importance of exact mouse strain nomenclature. Trends Genet 19:183–184

Xia H, Mao Q, Eliason SL et al. (2004) RNAi suppresses polyglutamine-induced neurodegeneration in a model of spinocerebellar ataxia. Nat Med 10:816–820

Zhang XM, Ng AH, Tanner JA et al. (2004) Highly restricted expression of Cre recombinase in cerebellar Purkinje cells. Genesis 40:45–51

Zhou L, Nepote V, Rowley DL et al. (2002) Murine peripherin gene sequences direct Cre recombinase expression to peripheral neurons in transgenic mice. FEBS Lett 523:68–72

Zimmer A, Zimmer AM, Hohmann AG et al. (1999) Increased mortality, hypoactivity, and hypoalgesia in cannabinoid CB_1 receptor knockout mice. Proc Natl Acad Sci USA 96:5780–5785

Zygmunt PM, Petersson J, Andersson DA et al. (1999) Vanilloid receptors on sensory nerves mediate the vasodilator action of anandamide. Nature 400:452–457

Endocannabinoid Signaling in Neural Plasticity

Bradley E. Alger

Contents

Abstract Plasticity refers to a physiologically measured change that may last for short or long periods of time. Endocannabinoids (ECBs) are prevalent throughout most of the brain, and modulate synaptic transmission in many ways. This chapter will focus on the roles of ECBs in neural plasticity in the mammalian brain. The topics covered can be divided loosely into two themes: how ECBs regulate synaptic

B.E. Alger
Departments of Physiology and Psychiatry, University of Maryland School of Medicine, 655 W. Baltimore St./BRB 5-025, Baltimore, MD, 21201, USA
e-mail: balgerlab@gmail.com

D. Kendall and S. Alexander (eds.), *Behavioral Neurobiology of the Endocannabinoid System*, Current Topics in Behavioral Neurosciences 1,
DOI: 10.1007/978-3-540-88955-7_6, © Springer-Verlag Berlin Heidelberg 2009

plasticity, and how ECBs' actions themselves are regulated by neuronal activity. Because ECBs regulate synaptic plasticity, the modifiability of ECB mobilization constitutes a form of "metaplasticity" (as reported by Abraham and Bear (Trends Neurosci 19:126–130, 1996)), i.e., an upstream process that determines the nature and extent of synaptic plasticity. Many of their basic functions are still being discovered, and while there is consensus on large issues, many points of divergence exist as well. This chapter concentrates on developments in the roles of ECBs in synaptic plasticity that have come to light since the major review by Chevaleyre et al. (Annu Rev Neurosci 29:37–76, 2006).

Keywords DSI • DSE • LTD • iLTD • Inhibition • Seizure

1 Introduction

1.1 Definitions, Scope, Limitations, and Caveats

The ECB system consists of the principal brain cannabinoid receptor, CB_1R, its endogenous ligands – 2-arachidonylglycerol (2-AG) and anandamide (AEA) – as well as a transporter process, metabolic and catabolic enzymes. Neither the mainly peripheral CB_2R (Pertwee 2005) nor the newly discovered, putative cannabinoid receptor, GPR55 (Ryberg et al. 2007), have received much attention from CNS physiologists thus far, and will not be dealt with here. Present physiological techniques cannot unambiguously distinguish among ECB synthesis, release, and transport processes, and I use 'mobilization' to encompass all steps between initial stimulation of the ECB system and activation of CB_1R. Cannabinoids, including ECBs, can act via non-CB_1R mechanisms, but these will not be covered.

The terminology ECB-STD and ECB-LTD (or iLTD) (Chevaleyre et al. 2006) to identify short-term and long-term plasticities initiated by ECBs has been retained. ECBs can be mobilized by a rise in $[Ca^{2+}]_i$, activation of numerous G-protein coupled receptors (GPCRs), or the combined actions of the two. GPCRs that trigger ECBs mobilization include metabotropic receptors for glutamate (Varma et al. 2001; Maejima et al. 2001), acetylcholine (Kim et al. 2002), dopamine (Yin and Lovinger 2006; Kreitzer and Malenka 2005), cholecystokinin (Foldy et al. 2007), oxytocin (Oliet et al. 2007), and glucocorticoids (Di et al. 2005), to name a few. Various stimuli use different biochemical pathways for ECB mobilization, and when necessary they are distinguished by superscripts: ECB_{mGluR}, ECB_{mAChR}, ECB_{Ca}, etc.

The work reviewed here was done on in vitro preparations from the rat or mouse brain, mainly with whole-cell electrophysiological recording methods. Generally, acute slices (300–400 μm thick) were used, but in a few cases dissociated tissue culture or organotypic slices were studied. Lovinger and colleagues have developed a "isolated neuron/bouton" preparation that has provided novel insights (Zhu and Lovinger 2005). Slice preparation and maintenance techniques are fairly similar across the various laboratories, yet substantial points of divergence can be found. Experimental temperatures range from 22 to 34°C; developmental ages range from neonatal to fully adult; intrapipette contents differ, sometimes widely; storage and recording chambers, flow rates, and drug application methods often change from laboratory to laboratory for sometimes unexplained reasons. Understandably, but unfortunately, there have been few systematic studies on whether or how such experimental variables affect results. On the one hand, the diversity of methods fosters confidence in the robustness of replicated observations. On the other hand, the occasional disagreements and assertions that some seemingly minor experimental factor is critically important emphasize the need for cautious interpretation. Despite the explosion of interest in the cellular physiology of the endocannabinoid system, these are still relatively "early days" and consensus is a work in progress.

1.2 ECBs: Basic Principles

The biochemistry and pharmacology of the ECB system are covered elsewhere in this volume (in the chapter "The life cycle of the endocannabinoids: formation and inactivation" by Alexander & Kendall; in the chapter "Endocannabinoid Receptor Pharmacology" by Mackie & Yao), and the reader is referred to those chapters for details. CB_1R is the principal brain ECB receptor, and is a heterotrimeric G-protein coupled receptor. Most CB_1Rs are located on presynaptic terminals (see the chapter "Endocannabinoid Receptors: CNS Localization of the CB_1 Cannabinoid Receptor" by Katona, this volume), and activation of CB_1R always inhibits transmitter release. In the mammalian brain, release of glutamate or GABA has received the most attention thus far, although glycine release is inhibited in the brain stem (Mukhtarov et al. 2005). A major mechanism by which CB_1R activation inhibits transmitter release is inhibition of presynaptic voltage-gated Ca channels (VGCCs), primarily N-type. Increases in presynaptic K channel activity occur at some synapses (Kreitzer et al. 2002). Cerebellar depolarization-induced suppression of inhibition (DSI) involves a suppression of tetrodotoxin (TTX)-insensitive miniature inhibitory post-synaptic currents (mIPSCs) (Llano et al. 1991; Diana and Marty 2003) and activation of CB_1Rs reduces TTX-insensitive mIPSC frequency (Takahashi and Linden 2000). Hence, in addition to presynaptic Ca channels, CB_1R activation could inhibit release by inhibiting a vesicle release step downstream of Ca influx. Suppression of mIPSCs is sensitive to the $[Ca^{2+}]_i$ in the terminals (Yamasaki et al. 2006), so this may itself be a regulated step. Diana and Marty (2003) estimated that, at Purkinje cell–interneuron synapses, suppression of the release machinery accounted for 13.4%, depression of interneuron firing, 23.2%, and depression of the probability of release given an action potential, 63.4% of the total synaptic depression caused by CB_1R activation. Long-term suppression of release (LTD or inhibitory LTD, iLTD) involves a variety of effectors (see Sect. 2.6).

Biochemical investigations have generated an enormous amount of well-validated information about synthesis of ECBs (in the chapter "The life cycle of the endocannabinoids: formation and inactivation" by Alexander & Kendall, this volume). Stimuli for ECBs often trigger phospholipase C (PLC) activity, generating diacylglycerol for diacylglycerol lipase (DGL) to cleave, yielding 2-AG. Yet studies of ECB-mediated neural response plasticity imply that new levels of experimental resolution may be necessary to understand the ECB system at the cellular physiological level. For example, while much evidence implicates 2-AG as the ECB in many systems, it is now accepted that PLC is not required for ECB_{Ca} production. Neither PLC inhibitors, nor deletion of PLC in mutant mice, affect ECB_{Ca} despite abolishing ECB_{GPCR} (Hashimotodani et al. 2005). Whether DGL itself is required for ECB_{Ca} remains controversial, with positive and negative effects of DGL inhibition having been reported. Some inhibitors of ECB metabolism that are effective when applied extracellularly are ineffective when applied intracellularly (Edwards et al. 2006), throwing basic assumptions about how the system functions at the moment into question. Undoubtedly, data from multiple techniques will be required before a complete picture is available.

Cannabinoids cannot be collected or assayed at the single cell level, therefore key components in the toolkit of endocannabinoid researchers are the CB_1R antagonists, AM-251 and rimonabant, which are inverse agonists that can produce effects on their own and not true receptor antagonists (Pertwee 2005; see Sect. 2.9). Recent work highlights two additional caveats to using them: AM251 is a putative agonist at GPR55 (Ryberg et al. 2007), and rimonabant is an effective antagonist at the vanilloid receptor, TRPV1 (e.g. Gibson et al. 2008). Since their non-specific actions do not overlap, both antagonists should routinely be used to confirm results.

2 ECBs Regulate Synaptic Plasticity

2.1 Short-Term Plasticity

The first example of ECB-mediated short term plasticity was depolarization-induced suppression of inhibition (DSI) (Pitler and Alger 1992; Llano et al. 1991) and its major properties had been delineated (Alger and Pitler 1995) before Wilson and Nicoll (2001) in hippocampal slices, and Ohno-Shosaku et al. (2001) in dissociated hippocampal culture, found that DSI is mediated by ECBs. At the same time Kreitzer and Regehr (2001) reported the discovery of depolarization-induced suppression of excitation (DSE) in the cerebellum and showed that it was also mediated by ECBs. These phenomena involve a transient suppression of synaptic transmission that follows a substantial increase in $[Ca^{2+}]_i$ in the receiving neuron. The hallmarks of DSI and DSE are that they are retrograde signal processes, with ECBs originating in a postsynaptic target cell, crossing the synapse in the reverse direction from conventional neurotransmitter travel, and suppressing the release of neurotransmitters (Alger 2002; Freund et al. 2003 for reviews). Inhibition of presynaptic Ca influx by ECBs has been measured in cerebellar parallel fibers (Kreitzer and Regehr 2001; Brown et al. 2003, 2004), and is the likely cellular mechanism for the short-term ECB phenomena. Presynaptic N-type Ca channels are affected by CB_1R activation in many instances (Wilson et al. 2001), but other Ca channels are inhibited by ECBs in cerebellum (Brown et al. 2004) as well. The magnitude and duration of DSI and DSE is dependent on temperature (Kreitzer et al. 2002) and rise in postsynaptic $[Ca^{2+}]_i$ (Pitler and Alger 1992; Wang and Zucker 2001; Brenowitz and Regehr 2003). A prolonged $[Ca^{2+}]_i$ rise can lower the peak $[Ca^{2+}]_i$ pulse required for ECB_{Ca} mobilization (Brenowitz et al. 2006), so the system is modifiable. DSI and DSE have now been reported to occur in numerous brain regions, and appear to have similar properties everywhere. Although widespread, the ability of cells to mobilize ECBs under conditions of transient, high $[Ca^{2+}]_i$ rises is not universal, even if the cells can mobilize ECBs with other stimuli. For instance, the medium spiny neurons of the dorsal lateral striatum do not produce DSE, although they can readily undergo ECB-LTD (Yin and Lovinger 2006; Kreitzer and Malenka 2005).

Typically, the relevant postsynaptic Ca for ECB mobilization comes through high voltage-activated VGCCs, probably postsynaptic N-type channels (Lenz et al. 1998), however calcium from intracellular stores may contribute in some cells (Robbe et al. 2002; Melis et al. 2004b), particularly in young or immature tissue (Isokawa and Alger 2006).

Activation of G protein-coupled and ionotropic glutamate receptors can trigger ECB mobilization (Brown et al. 2003; Ohno-Shosaku et al. 2007). Brief bursts of afferent stimulation induce ECB-dependent transient suppression of parallel fiber inputs onto Purkinje cells, a phenomenon that is similar to DSE and has been called SSE (Brown et al. 2003). If the concentration of agonist is high, then ECBs are mobilized in a relatively $[Ca^{2+}]_i$-independent way (Maejima et al. 2001; Kim et al. 2002). If the concentration of agonist is low, then direct mobilization may not occur, but the products of the G-protein receptor activation can synergize with Ca_i^{2+} to produce a very marked increase in ECB mobilization over what the rise in $[Ca^{2+}]_i$ alone could accomplish (Varma et al. 2001; Kim et al. 2002; Ohno-Shosaku et al. 2003). A molecular model that can account for this synergistic interaction proposes that PLCβ isoforms (β1 in hippocampus (Hashimotodani et al. 2005), β4 in cerebellum (Maejima et al. 2005)) act as coincidence detectors, i.e., they are activated by both $[Ca^{2+}]_i$ and G-protein products. They can therefore integrate the two kinds of signals, and the summed stimuli produce larger responses than either could alone. This appealing model may not explain all of the interactions between Ca and GPCR activation that lead to ECB mobilization, however: a Ca-dependent priming step is required to enable mGluRs to mobilize ECBs primary step that cannot be accounted for by coincidence detection is required (see Sect. 3.3). In many cases, the ECBs released by GPCRs affect synapses that were not responsible for triggering ECB mobilization, e.g., suppression of hippocampal GABAergic synapses by activation of mGluRs (Varma et al. 2002) or mAChRs (Kim et al. 2002). These are examples of "heterosynaptic" ECB actions (Chevaleyre et al. 2006). It has been suggested that normal intracellular action potential activity does not generate a large enough $[Ca^{2+}]_i$ rise to trigger ECB mobilization without the concurrence of glutamatergic synaptic activity (Hampson et al. 2003), however postsynaptic action potential bursts are effective stimuli for DSI of carbachol-induced IPSP/Cs (Pitler and Alger 1992; Reich et al. 2005). Whatever the details of the molecular model, it seems very likely that coordination between increases in postsynaptic $[Ca^{2+}]_i$ and neurotransmitters, often GPCR activators, may be the most prevalent stimulus for ECB mobilization in the brain.

Brief presynaptic trains of stimuli induced very localized $[Ca^{2+}]_i$ signals and ECB release from cerebellar Purkinje cells (Brown et al. 2003). CB_1Rs are present on parallel fibers, and by directly monitoring presynaptic Ca influx into parallel fiber terminals as an assay of the ECB effect, it was found that ionotropic glutamate receptors as well as mGluR1 contributed to ECB mobilization (Brown et al. 2003). ECBs affected only activated synapses, implying this was largely a "homosynaptic" effect (Chevaleyre et al. 2006), and providing an example of the extremely localized nature of ECB signaling.

2.2 Short-Term Target-Dependent Plasticity

Target-dependent plasticity refers to cases in which the postsynaptic target cell influences the type or degree of plasticity expressed by the incoming presynaptic contacts it receives. This is especially clear when a given afferent fiber system contacts more than one target cell in a given region, and the synaptic plasticity differs at each target. ECBs participate in target-dependent plasticity in the cerebellum, where the parallel fibers contact Purkinje cells and golgi cells (Beierlein et al. 2007). At these synapses either post-tetanic potentiation or depression was produced only at parallel fiber-to-Purkinje cell synapses; the golgi cell synaptic input was essentially unchanged by the same stimuli. Whether the Purkinje cell inputs were enhanced or depressed depended on the locus of the stimulation and the resulting degree of mGluR activation produced. Parallel fiber stimulation in the molecular layer activated numerous proximate synapses, which enabled glutamate spill-over to summate, activate mGluRs and mobilize ECBs from the Purkinje cells (Marcaggi and Attwell 2005). This resulted in ECB-dependent, stimulus-induced suppression of inhibition (Beierlein et al. 2007), which accounted for post-tetanic depression. Blocking CB_1R with AM251 uncovered post-tetanic potentiation, showing that the potentiation is an intrinsic property of these synapses that is masked when ECBs prevent glutamate release from the terminals. The parallel fibers onto golgi cells did express CB_1R that could be activated by exogenous cannabinoids, but the golgi cells seemed to be incapable of generating ECBs. In summary, the target-dependent plasticity was largely attributable to differential mobilization of ECBs.

2.3 Long-Term Plasticity

Although DSI itself is short-lasting, it can markedly affect postsynaptic excitability (Wagner and Alger 1996) and enhance the inducibility of long-term plasticity, such as LTP (Carlson et al. 2002). If a stimulus train that was too weak to initiate LTP was delivered during DSI, LTP was induced. The ECB-induced disinhibition enabled normally subthreshold excitation to become suprathreshold for LTP induction. The possibility that ECBs could themselves induce long-term plasticity was first established in the dorsal striatum by Gerdeman et al. (2002) and in nucleus accumbens (NAc) by Robbe et al. (2002). Inhibitory LTD of rat basolateral amygdala cells in vitro seemed to correlate with the resistance to extinction of fear conditioning in the behaving animal (Marsicano et al. 2002).

2.3.1 Striatum

Induction of LTD is caused by mGluR activation at excitatory glutamatergic synapses onto the striatal medium spiny neurons, and a retrograde messenger was known to be involved (Gubellini et al. 2004). LTD was initiated with brief-high frequency stimulus trains paired with postsynaptic depolarizations, and was induced in a postsynaptic, Ca-dependent way. LTD was expressed presynaptically as a decrease in the probability of glutamate release, and dopamine D2 receptor activation was mandatory. Exogenous CB_1R agonists inhibited glutamate release (Gerdeman and Lovinger 2001) and stimulation of dopaminergic afferents generated AEA in the striatum (Giuffrida et al. 1999), so ECBs were a good candidate messenger. Gerdeman et al. (2002) found that several brief high frequency stimulus trains induced LTD that was expressed presynaptically, absent in $CB_1R^{-/-}$ mice, and blocked by the CB_1R antagonist, rimonabant. Strongly buffering $[Ca^{2+}]_i$ prevented ECB-LTD induction, which besides showing that postsynaptic $[Ca^{2+}]_i$ was essential to it, argued that ECBs generated by other nearby cells could not travel enough to affect the EGTA-loaded cell. Bath application of the putative CB transporter blocker, AM404, rescued LTD induction, supporting the proposal that ECBs were key players in LTD induction, and revealing that when ECB removal was prevented, ECBs from other cells could affect multiple cells. The transporter is a major factor in defining the extremely local sphere of ECB actions, which as argued previously (Alger 2002) is a key feature of the ECB system. Interestingly, *intracellular* application of ECB transporter blockers did not facilitate, but suppressed ECB-iLTD (Ronesi et al. 2004) (see Sect. 3.4). Restrictions in the spread of ECBs permit the single cells originating them to undergo major long-term plasticities while neighboring cells remain unaffected. In this way information coding may be selectively addressed to cells that happen to mobilize ECBs at the same time.

ECB-LTD in the dorsal lateral striatum depends on activation of L-type VGCCs, and D2 receptors as well as group I mGluRs (Yin and Lovinger 2006; Kreitzer and Malenka 2005) (perhaps specifically mGluR1), internal stores of calcium, and postsynaptic PLC activation (Yin and Lovinger 2006). The bistable resting potential of the medium spiny cells may critically regulate ECB-LTD (Kreitzer and Malenka 2005). In the "up" state, a resting potential near −50 mV, L-type VGCCs are activated, and mGluR-induced ECB-LTD was easily induced; in the "down" state, near −70 mV, mGluR-LTD ECB-LTD was not readily induced, probably because the L channels are not activated at the negative membrane potential. Moreover, D2 activation markedly enhanced the state-dependent-LTD induction. While several of these features have been replicated, there are some controversial aspects of ECB-LTD in the dorsal striatum. A disagreement about the adequacy of CB_1R activation to induce LTD is covered in detail below.

2.3.2 Nucleus Accumbens (NAc)

Initiation of ECB-LTD of glutamatergic synapses in the NAc requires considerably greater synaptic stimulation (13 Hz for 10 min) than is necessary in other brain regions (Robbe et al. 2002). Pharmacological and genetic tests confirmed the involvement of CB_1R in the process, and showed that induction required postsynaptic $[Ca^{2+}]_i$ increases and maintenance involved presynaptic suppression of release. In Nac, mGluR5 is the predominant mGluR receptor, and ECB-LTD was abolished by the specific mGluR5 antagonist, MPEP, as well as a broad spectrum mGluR antagonist. The group I mGluR agonist DHPG mimicked and occluded ECB-LTD; DHPG effects were prevented by rimonabant. Unlike other systems in which Ca entry via VGCCs is required (see e.g. Gerdeman et al. 2002), in NAc, calcium from ryanodine-sensitive calcium stores mediates ECB-LTD initiation. And, unlike the dorsal striatum, D2 receptors play no role in the NAc. Hence, ECB-LTD in NAc has a number of distinctive features.

2.3.3 Hippocampus

Chevaleyre and Castillo (2003) showed that ECB-iLTD could be induced at inhibitory synapses in the hippocampus with brief high frequency stimulus trains, although longer-lasting theta-burst trains (Chevaleyre and Castillo 2004), or a 5-min bout of low-frequency (1 Hz) stimulation (Zhu and Lovinger 2005), will also induce ECB-iLTD. A 10-min application of the mGluR agonist DHPG induces ECB-iLTD, and mGluR antagonists block both synaptic and DHPG-induced iLTD (Chevaleyre and Castillo 2003; Edwards et al. 2006). ECB-iLTD could be inhibited by AM251, or extracellular pretreatment of slices with PLC or DGL inhibitors, although intracellular application was ineffective (Edwards et al. 2006). Intracellular application of DGL inhibitors do block other ECB_{GPCR} actions in hippocampus (Edwards et al. 2006) or cerebellum (Safo and Regehr 2005), so inefficiencies of intracellular delivery are not obviously at work. By applying AM251 at various times after brief field stimulation, it was found that ECB-iLTD induction required many minutes of CB_1R activation (Chevaleyre and Castillo 2003). Once established, ECB-iLTD cannot be blocked by either CB_1R or mGluR antagonists, showing that it is independent of continued receptor activation.

A key physiological feature of ECB-iLTD is that it can account for the EPSP-spike (E-S) potentiation that had been noted by Bliss and Lomo (1973) as a distinct dimension of LTP. E-S potentiation means that a given field EPSP is capable of triggering a larger population spike after LTP induction than before. By definition this was a different mechanism from the increase in the EPSP itself, since the hallmark of E-S coupling is that the EPSP size is held constant when making the comparison. Decreases in inhibition can account for E-S coupling (Abraham et al. 1987), but the mechanism of the persistent decrease in inhibition remained elusive. ECB-iLTD was recognized as being ideally suited for this role, and AM251 prevented induction of E-S potentiation (Chevaleyre and Castillo 2003). The facilitatory effects of ECB-iLTD could be localized to small regions of the dendrites. Fine focal theta-burst stimulation of glutamatergic fibers enabled Chevaleyre and Castillo (2004) to define affected dendritic areas as limited to 10 μm in length in which mGluR-dependent ECB release would induce iLTD, and concomitantly, LTP of excitatory synapses. A two-pathway experiment revealed that even though LTP was directly induced in only a small region very close to the theta-burst stimulating electrode, iLTD affected a broader dendritic penumbra. In this region, the decrease in inhibition lowered the threshold for LTP induction.

There is a relatively low level of CB_1R expression on hippocampal glutamatergic terminals (Kawamura et al. 2006), which accounts for the relatively small degree of ECB-mediated DSE in the hippocampus (Ohno-Shosaku et al. 2002). However, early in development from PN2-10, a heterosynaptic, glutamatergic ECB-LTD is associated with a homosynaptic cyclic AMP-dependent protein kinase (PKA) -dependent LTP (Yasuda et al. 2008). This developmentally transient form of plasticity declines with age until it is absent in mature hippocampus. It is unusual in having a very slow onset (many tens of minutes), and being associated with a decrease in the fiber volley – a measure of action potentials in presynaptic axons – that is prevented by the K channel blockers Ba, 4-AP and dendrotoxin. ECB-LTD was also prevented by these blockers.

2.3.4 Cerebellum

At parallel fiber–Purkinje cell synapses, ECBs released from the Purkinje cells regulate transmitter release presynaptically by DSE (Kreitzer and Regehr 2001). Pairing of brief bursts of parallel fiber stimuli and climbing fiber stimuli for 30 trials leads to ECB-LTD of the parallel fiber synapses (Safo and Regehr 2005). Bath application or intracellular infusion of DGL inhibitors abolished both short-term and long-term ECB effects induced by synaptic stimulation without, however, altering DSE. A rise in Purkinje cell $[Ca^{2+}]_i$ was required for ECB mobilization. The similarities with ECB-(i)LTD in striatum or hippocampus end at this point, because ultimate expression of cerebellar ECB-LTD is expressed postsynaptically, whereas the other expression mechanisms are presynaptic (see Sect. 2.6).

Besides participating in parallel fiber LTD induction, ECBs regulate parallel fiber plasticity in another, quite different way, by preventing parallel fiber LTP expression (van Beugen et al. 2006). When stimulated by themselves, parallel fibers undergo a presynaptic, PKA -dependent, form of LTP. Coactivation of climbing fibers during parallel fiber tetanization prevented LTP induction. CB_1R inhibition rescued LTP, and WIN55212-2 mimicked climbing fiber stimulation in blocking LTP. The dual roles of climbing fiber ECB effects, promoting LTD of the parallel fiber synapse while at the same time suppressing the presynaptic induction of LTP at parallel fiber terminals, are seen as complementary actions: by preventing LTP at a synapse destined for LTD, the climbing fiber LTP inhibition acts as a "safety lock" that ensures that synaptic weakening occurs.

2.3.5 Amygdala

In the basolateral amygdala, ECBs mobilized by low-frequency stimulus trains that activate mGluR1s induce ECB-iLTD (Azad et al. 2004). Neither PLC nor DGL inhibition affected iLTD, but it was enhanced in fatty acid amide hydrolase (FAAH) knock-out mice, implicating AEA, and not 2-AG, in the process. Postsynaptic inhibitors of adenylyl cylase or PKA inhibited iLTD induction, suggesting that the triggering of this cascade by mGluR1 mobilized AEA. Interestingly, in the lateral amygdala exogenous cannabinoids activating CB_1Rs on inhibitory interneurons can abolish LTD of excitatory synapses (Azad et al. 2008). The mechanism of this effect, and a possible role for ECBs, are not worked out.

2.3.6 Ventral Tegmental Area (VTA)

ECBs mobilized from dopamine cells in the VTA can induce DSE of glutamatergic synapses (Melis et al. 2004b). In addition, brief train stimulation of afferents from prefrontal cortex to the VTA lead to an ECB- and CB_1R-dependent decrease in the excitatory post-synaptic currents (EPSCs) onto dopaminergic neurons (Melis et al. 2004a). ECB mobilization was triggered by mGluR1 activation, and was blocked by postsynaptic infusion of an ECB transporter blocker. A rise in postsynaptic $[Ca^{2+}]_i$ was essential; Ca from ryanodine-sensitive internal stores was involved. The ECB appeared to be 2-AG, as EPSC depression was prevented when the recording electrode contained a DGL inhibitor, but PLC inhibition had no effect. Activation of D2 receptors enhanced ECB mobilization, but was not required for it. Initial reports of ECB actions in VTA did not include long-term effects. However, repeated cocaine treatment facilitates LTP induction in the VTA brain slice by persistently suppressing GABAergic synapses. Recent evidence (Pan et al. 2008) shows that cocaine stimulates mGluR1- and D2-dependent ECB-iLTD of VTA GABAergic eIPSCs.

2.3.7 Cortex

DSI (Trettel and Levine 2002; Fortin et al. 2004) and DSE (Fortin and Levine 2007) affect GABAergic and glutamatergic synapses in neocortex, with evidence of regional selectivity, and ECB-spike-timing dependent plasticity (STDP) occurs in somatosensory cortex (Nevian and Sakmann 2006). ECB-LTD of glutamatergic synapses was induced by low-frequency stimulation in layer 2/3 of visual cortex (Crozier et al. 2007). This LTD was initiated by activation of NMDA receptors (NMDARs), since AP5 or intracellular dizolcipine blocked it. Prior monocular deprivation occluded layer 2/3 ECB-LTD, and blocking CB_1Rs prevents the ocular dominance shift (Liu et al. 2008) suggesting that ECB-LTD contributes to the loss of visual responsiveness in the monocular deprivation model.

2.4 *Mechanisms of ECB-Dependent Long-Term Plasticity*

Although ECB-LTD occurs at numerous excitatory and inhibitory synapses, relatively little is known about its induction or maintenance mechanisms. ECB-LTD induction requires persistent activation of CB_1R; in experimental studies an mGluR agonist must be applied for between 5 and 10 min (10 is typical) to guarantee that LTD will occur. Field stimulation of glutamatergic afferents, which can induce LTD after bursts of stimuli lasting only 1 or 2 s (Chevaleyre and Castillo 2003), seems to be much more efficient. To some extent the very large time difference between exogenous application of mGluR agonist and synaptic stimulation is misleading, however. Since the *maintenance* of ECB-LTD does not depend on persistent activation of CB_1R, the time interval during which AM251 is effective in blocking ECB-LTD represents the duration of the *induction* phase; i.e., CB_1R must remain activated for at least that long for ECB-LTD induction to occur. It is possible to measure the duration of CB_1R activation by applying AM251 at various time intervals after field stimulation. This revealed that even a few seconds of field stimulation caused activation of CB_1R lasting for many minutes (Chevaleyre and Castillo 2003; Ronesi et al. 2004). In other words, there was no big temporal disparity between the duration of time that direct mGluR agonist application, or field stimulation, actually activates CB_1R.

2.5 *Sufficiency of CB_1R Activation for ECB-LTD (or -iLTD) Induction*

Given that long-duration CB_1R activation is a requirement for ECB-LTD induction, one can inquire why this is necessary, and whether long-duration CB_1R binding per se is sufficient for induction. A straightforward approach is to apply a CB_1R agonist for many minutes and ask if ECB-LTD is induced. The experiment has been done in several ways, and the results have been inconsistent. In cerebellum, parallel fiber LTD is ECB-dependent, but CB_1R activation alone by the synthetic agonist, WIN55212-2, is insufficient for induction (Safo and Regehr 2005); the response returns to baseline once WIN55212-2 is removed. Similarly, loading hippocampal CA1 pyramidal cells with the G-protein activator, GTPγS, caused persistent ECB mobilization and activation of CB_1R, yet application of AM251 returned the eIPSCs to expected control amplitudes (Kim et al. 2002). Additionally, eliciting persistent, ECB-dependent eIPSC suppression by repetition of overlapping DSI trials for 10 min had no lasting effects (Edwards et al. 2006). Soon after the last DSI trial, the eIPSCs returned to control amplitudes with no evidence of ECB-iLTD. Finally, prolonged stimulation of ECB_{mAChR} with carbachol suppressed eIPSCs continuously for up to 20 min in CA1. Again, full recovery of eIPSCs to control levels occurred shortly after the mAChR agonist was removed and atropine applied. Stimulation with an mGluR agonist reliably produced ECB-iLTD. It appeared that persistent CB_1R activation was insufficient for the long-term effects, and that some other consequence of mGluR activation led to ECB-iLTD initiation (Edwards et al. 2006). In contrast, in NAc (Robbe et al. 2002) and hippocampus (Chevaleyre and Castillo 2003; Chevaleyre et al. 2007), minutes-long application of WIN55212-2 reportedly can cause a significant, and apparently irreversible suppression of IPSCs in CA1. However, lipophilic compounds such as WIN55212-2 are difficult to remove completely from slices, and the possibility of lingering WIN55212-2 was difficult to eliminate.

Within the striatum, contradictory results have also been obtained. Continuous eEPSC suppression caused by 20 min of WIN55212-2 application or by loading postsynaptic medium spiny cells with AEA (Ronesi et al. 2004) (which escapes and persistently suppresses incoming glutamate release) reportedly could be fully reversed by addition of AM251. However, in the same preparation, Kreitzer and Malenka (2005) found that WIN55212-2 alone did induce LTD.

Recent reports in cerebellum (Safo and Regehr 2005), striatum (Singla et al. 2007) and hippocampus (Yin et al. 2006) offer a possible resolution to some of the conflicting findings: it turns out that establishment of ECB-LTD requires CB_1R activation plus concomitant presynaptic activity. WIN55212-2 application in the absence of presynaptic stimulation caused reversible eEPSC depression, whereas synaptic stimulation delivered throughout the WIN55212-2 application caused long-term, AM251-resistant depression after the WIN55212-2 was removed. Somewhat surprisingly, a very low frequency of stimulation (0.05 Hz) was sufficient for this form of ECB-LTD induction (Singla et al. 2007). The long intervals between stimuli would seem to preclude build-up of an intracellular chemical factor, and the explanation for this efficacy is unknown. In any case, some of the negative results reported by earlier work may have reflected the absence of adequate presynaptic co-stimulation to provide whatever condition is needed in concert with CB_1R activation to induce ECB-LTD. In the experiments of Singla et al. (2007) the unknown element seemed to involve an increase in presynaptic $[Ca^{2+}]_i$, because if WIN55212-2 were applied when extracellular calcium ion concentration ($[Ca^{2+}]_e$) was reduced, concurrent stimulation did not cause LTD (assessed after restoration of normal $[Ca^{2+}]_e$). Filling the postsynaptic cell with the calcium chelator, BAPTA, did not affect LTD induction via ECB and stimulation, supporting the inference that LTD induction took place exclusively via presynaptic $[Ca^{2+}]_i$-dependent mechanisms.

Indeed, combined activation of L-type Ca channels, modest postsynaptic activation, and synaptic transmission is sufficient to induce striatal ECB-LTD (Adermark and Lovinger 2007a). An L-channel activator, FPL64176, could induce persistent, CB_1R-dependent LTD if the cells were depolarized to −50 mV and simultaneously stimulated at a low frequency. In general, the requirement for simultaneous stimulation confers synapse specificity of LTD targeting: only those synapses releasing neurotransmitter during CB_1R would be susceptible to LTD (Singla et al. 2007). Still unresolved are what exactly the co-stimulation does, and why some neurotransmitters that release ECBs do not induce LTD even when co-stimulation is given (Edwards et al. 2006).

2.6 *Molecular Mechanisms of ECB-LTD (or iLTD) Maintenance*

Maintenance of striatal ECB-LTD has been investigated in a reduced striatal slice preparation in which the cell bodies of the cortical afferent fibers had been removed (Yin et al. 2006). This ruled out the participation of gene transcription in cortical cell somata. ECB-LTD maintenance was prevented by bath application of protein translation inhibitors, but loading them into the postsynaptic cell had no effect. Neither cycloheximide nor anisomycin affected basal transmission, the activation of CB_1R, or of mGluR. Similarly, postsynaptic loading of transcription inhibitors also failed to affect ECB-LTD. The data suggested that local axonal protein translation was essential for ECB-LTD. The target(s) of these proteins were not clear.

Chevaleyre et al. (2007) investigated the presynaptic mechanisms of ECB-iLTD induction in the hippocampus. Pharmacological interference with cAMP-PKA system prevented ECB-iLTD induction. CB_1R activation often inhibits adenylyl cyclase, and indeed forskolin opposed the effect of WIN55212-2, and PKA inhibitors occluded it. Neither manipulation affected DSI, suggesting that postsynaptic ECB mobilization was not affected and that the mechanism by which CB_1R activation induces ECB-iLTD was different from the presynaptic DSI mechanism, i.e., primarily blockade of N-type Ca channels (Wilson and Nicoll 2001). Internal postsynaptic application of a PKA inhibitor failed to affect ECB-LTD, lending support to the conclusion that the PKA effects were presynaptic. To investigate the ability of CB_1Rs to target the GABA release machinery, TTX- and Cd-insensitive mIPSCs were studied in elevated $[Ca^{2+}]_e$. Both WIN55212-2 and PKA inhibitors reduced mIPSC frequency but not amplitude, and forskolin prevented WIN55212-2's actions. RIM1α is an active zone protein that is required for presynaptic LTP induction (Castillo et al. 2002) and is a substrate for PKA. In RIM1$\alpha^{-/-}$ mice, ECB-iLTD could not be induced (Chevaleyre et al. 2007), but DSI and short-term ECB_{mGluR} were normal, showing that ECB mobilization and CB_1Rs were unaffected by the knockout. Apparently, decreased phosphorylation of RIM1α by PKA is at least partly responsible for ECB-iLTD induction. The RIM1$\alpha^{-/-}$ mice also had attenuated sensitivity to PKA inhibitors, and WIN55212-2, as if the iLTD process were fully saturated. After extending their conclusions to the basolateral amygdala, Chevaleyre et al. proposed that the PKA-RIM1α mechanism of ECB-iLTD may be a very general one.

Nevertheless other possibilities exist. In cerebellum, parallel fiber LTD is induced by a process involving nitric oxide (NO) and is expressed by a postsynaptic downregulation of AMPA receptors (AMPARs) (Ito 2001). This LTD is dependent on the activation of presynaptic CB_1R, and postsynaptic DGL (Safo and Regehr 2005), implying that 2-AG generated in the Purkinje cell is a key element. Blocking NO synthase with L-NAME confirmed that ECB-LTD is also NO-dependent. NO acted at a step downstream from the ECBs, since bypassing ECB synthesis and activating CB_1R directly with WIN55212-2 did not allow for LTD induction if NO synthesis was blocked. The apparent connection between CB_1R and NO is unknown. In fact, this issue has been further complicated by the finding that the NO synthase cascade is not localized to parallel fiber terminals, but rather to interneurons (Shin and Linden 2005). The study by van Beugen et al. (2006) (Sect. 2.3) offers a resolution: presynaptic CB1R activation simply suppresses presynaptic parallel fiber LTP induction, and thereby enhances the appearance of postsynaptic, NO-dependent LTD by preventing its occlusion by LTP. Inasmuch as parallel fiber LTP is dependent on presynaptic cAMP, PKA, and RIM1α (Castillo et al. 2002), the simplifying hypothesis would be that the major biochemical target of presynaptic CB_1R is the cAMP system, rather than NO. This is an appealing idea that can unify our understanding of the long-term ECB-dependent regulation of transmission at inhibitory and excitatory synapses. One gap remains to be filled: inhibition of PKA simply prevents LTP at excitatory synapses without causing LTD (van Beugen et al. 2006), whereas it causes LTD at inhibitory synapses (Chevaleyre et al. 2007).

In any event, the proposed downstream involvement of NO in cerebellar ECB-LTD (Safo and Regehr 2005) would be very different from the case of DSI in the hippocampus, where NO has been proposed to be upstream of ECB synthesis under conditions when mAChRs are also activated (Makara et al. 2007). In these experiments, blocking NO blocked DSI, and activating the NO pathway mimicked and occluded DSI. Neuronal NO synthase was found localized immediately postsynaptic to presynaptic terminals expressing NO-sensitive guanylate cyclase (Szabadits et al. 2007). NO released from the pyramidal cell would activate presynaptic guanylate cyclase. Indeed, guanylate cyclase activation caused cGMP accumulation in these terminals (Makara et al. 2007). Elements of the ECB system must be downstream of this step, but at present this connection remains mysterious. Unanswered questions include how and why mAChR activation could trigger the switch between an NO-independent DSI mechanism to an NO-dependent one. There is apparently no evidence that NO is involved in establishment of hippocampal ECB-iLTD, although strong, persistent activation of mAChRs, which copiously generates ECBs, does not cause iLTD (Edwards et al. 2006), perhaps arguing against this possibility.

Thus CB_1R activation is potentially coupled to at least four very different effectors, involving multiple biochemical pathways in the presynaptic cells: (1) direct G-protein-dependent suppression of Ca channels that mediates short-term DSI or DSE; (2) inhibition of cAMP production, leading to decreases in PKA, and an effect on the vesicle release machinery mediated by RIM1α; (3) activation of some component(s) of the NO signaling cascade; (4) opening of voltage-gated, presynaptic K channels through unknown biochemical pathways (exogenous cannabinoids do increase K currents via a PKA-dependent pathway in tissue culture (Mu et al. 2000)). A $[Ca^{2+}]_i$-dependent K conductance is turned on by CB1R autoreceptors in neocortical interneurons suggesting that CB_1R could be coupled to more than one K conductance (Bacci et al. 2004) (see Sect. 2.7).

2.7 *Spike-Timing Dependent Plasticity (STDP)*

Induction of Hebbian forms of plasticity depends on the occurrence of both pre- and postsynaptic activity. For the class of STDPs, the timing between pre- and postsynaptic events is critical (Dan and Poo 2004). The concept of STDP not only suggests that events must occur within a particular temporal window, but also that the order of events, whether the pre- or the postsynaptic cell activity occurs first, determines the type of plasticity (potentiation or depression) that is induced.

Timing-dependent LTD (tLTD) takes place in neocortical layer 5 principal neurons when the postsynaptic cell firing precedes presynaptic cell firing by 20–200 ms. Sjostrom et al. (2003) discovered that ECBs set the width of this temporal window. NMDAR activation was also required for tLTD, however tLTD only occurred if the ECBs, released by activity in the postsynaptic cell, were bound to CB_1Rs during the presynaptic activity. Interestingly, the released glutamate activated presynaptic NMDA autoreceptors, and hence the coincidence of presynaptic CB_1R and NMDAR activation was critical for tLTD. In fact, actual postsynaptic action potential firing is not required for this form of STDP – correctly timed subthreshold depolarizations are sufficient (Sjostrom et al. 2004) – making “spike-timing dependent plasticity” a misnomer.

In layer 2/3 pyramidal cells of somatosensory cortex, ECBs are also involved in STDP, although the mechanism is quite different from that in the layer 5 cells (Nevian and Sakmann 2006). In a study of the relationship between synaptic spine $[Ca^{2+}]_i$ and the long-term STDP produced at the synapse, Nevian and Sakmann found there was no simple correlation between them. STDP was produced by 60 pairings (at 0.1 Hz) in which a burst of three dc-triggered action potentials would precede or follow an eEPSP. When the burst occurred 50 ms before the eEPSP, LTD was induced; when it followed the eEPSP by 10 ms, LTP was induced. Two-photon excitation fluorescence microscopy revealed that spine $[Ca^{2+}]_i$ changes associated with the same stimuli that induced synaptic plasticity (though measured in different experiments) produced different effects. Calcium influx via NMDARs was essential for LTP induction, and not LTD. Conversely, calcium influx via VGCCs mediated LTD not LTP. In neither case did the peak $[Ca^{2+}]_i$ amplitudes predict LTP or LTD; either change could be associated with a given $[Ca^{2+}]_i$, so the timing dependence was not conferred by differences in $[Ca^{2+}]_i$. mGluRs were necessary for LTD, but not LTP induction, and the mGluR antagonist did not affect $[Ca^{2+}]_i$. LTD was blocked by AM251 and by the PLC inhibitor, U73122, also without changes in spine $[Ca^{2+}]_i$. The conclusion was that STDP induction of LTD was caused by ECBs mobilized by VGCC-mediated Ca influx in combination with mGluR activation. The combined effects summated during the long induction protocol. IP_3-sensitive Ca stores contribute (Bender et al. 2008) to ECB-LTD in somatosensory cortex, and presynaptic NMDARs are probably also involved.

ECB-dependent STDP also occurs in the dorsal cochlear nucleus (Tzounopoulos et al. 2007), but only in the cartwheel interneurons, not in the principal cells for which they provide feedforward inhibition. The difference was not in the general source of the synaptic inputs, which are the same for both target cells, i.e., this is another example of target-dependent plasticity (see Sect. 2.2). The same excitatory afferent parallel fiber system induced opposing kinds of plasticity in the two cell types: a Hebbian LTP in the principal (fusiform) cells, and an "anti-Hebbian" LTD in the interneurons. The Hebbian LTP was a conventional, NMDAR and calmodulin-dependent protein kinase (CaMKII)-dependent phenomenon, requiring glutamate from the presynaptic cell and sufficient postsynaptic depolarization to permit postsynaptic NMDAR activation. Anti-Hebbian LTD at the parallel fiber–cartwheel interneuron synapse was caused when a postsynaptic spike reliably followed the EPSP. A crucial aspect of anti-Hebbian LTD induction was the very precise nature of the timing, which demands occurrence of the postsynaptic spike within a 10 ms window centered on the EPSP. This narrow window is set by the co-occurrence and mutual cancelation of LTP and ECB-dependent LTD, except at the shortest intervals where LTD dominates. Blocking CB_1R led to the production of only LTP; whereas preventing LTP increased the timing window for LTD induction. A puzzle was why the ECBs only affected parallel fiber inputs to the interneurons. Physiological release of ECBs (DSE), or application of a synthetic CB_1R agonist affected the parallel fiber inputs to both cartwheel and principal cells, albeit to a significantly larger extent at the synapses onto the interneurons. Electron microscopic analysis of labeled CB_1R revealed many fewer receptors on the parallel fiber terminals onto principal cells, thus accounting for the difference. An important open question is why the timing window is so narrow. The kinetics of ECB mobilization, even if faster than sometimes thought (cf. Wilson and Nicoll 2001; Heinbockel et al. 2005), are an order of magnitude too long. It will be interesting to learn what aspect of ECB mobilization conveys the temporal sensitivity to this STDP.

Exogenous cannabinoids disrupt the temporal coordination of cell assemblies, assessed as changes in the local EEG in the hippocampus without markedly altering the absolute firing rates of either pyramidal cells or interneurons (Robbe et al. 2006). This effect was explained by the applied agonists' ability to access the CB_1Rs on glutamatergic terminals that do not seem to be the main target of endocannabinoids in hippocampus. Somewhat surprisingly, application of the

CB_1R antagonist, rimonabant, applied on its own did not alter the rhythms, although it did prevent the actions of the exogenous cannabinoids. The implication that ECBs normally play no role in rhythm generation would have profound significance for the understanding of this system, and will no doubt be followed up in future work.

2.8 *Interneurons Mobilize ECBs*

The interneuron is a key player in many forms of neuronal plasticity and CB_1R-expressing interneurons are regulated by ECBs mobilized by other cells. Whether interneurons can exercise the same sort of autoregulation has been uncertain until relatively recently. An investigation of hippocampal CA1 stratum radiatum interneurons (Hoffman et al. 2003) concluded against the idea. Because the interneuron inputs were sensitive to exogenous CB_1R agonists, it appeared that these cells could not generate ECBs. Bacci et al. (2004), recording from low threshold-spiking, cholecystokinin (CCK)-expressing, neocortical interneurons, obtained a strikingly different result. Those interneurons do mobilize ECBs following stimuli that induced a large increase in $[Ca^{2+}]_i$, but the CB_1Rs that were activated were present on the interneurons themselves, and functioned as autoreceptors. By activating Ca-dependent K channels apparently on or near the interneuronal somata, the ECBs hyperpolarized and inhibited the interneurons. A notable feature of this self-inhibition was its very long duration (>10 min at 32°C), which seemed to be partly maintained by persistent action of ECBs as it remained sensitive to AM251 for many minutes.

While the results just described confirmed that interneurons were capable of mobilizing ECBs, it was not clear if the ECBs were also used for regulating interneuron synaptic inputs. Golgi cells in the cerebellum cannot mobilize ECBs (Beierlein et al. 2007), but two other local cerebellar interneurons, the basket and stellate cells, can regulate the strength of their excitatory parallel fiber inputs by mobilizing ECBs (Beierlein and Regehr 2006). Direct depolarization of the interneurons or brief synaptic stimulation induced DSE or SSE. Prevention of the effects either by CB_1R antagonists, or inhibition of DGL, implicated ECBs and particularly 2-AG in mediating the presynaptic inhibition. Interestingly, the synaptic release not only of mGluRs but also of NMDARs were fully capable of mobilizing ECBs, and it was necessary to block both receptors to prevent synaptic stimulation from initiating the ECB effects. Inasmuch as both interneurons are activated by the parallel fibers, this system provides for a feedforward inhibition of Purkinje cells, and therefore ECB actions would decrease this feedforward inhibition.

Ali (2007) carried out a paired-recording study of Schaffer-collateral-associated, CCK-expressing, interneurons in hippocampal CA1. She found the cells were interconnected with facilitating inhibitory synapses: brief stimulation of one cell led to increasingly large IPSCs in the target cell. Since the output of these interneurons is also directed towards the pyramidal cell dendrites, the facilitating IPSCs would, by depressing the interneuron, disinhibit the pyramidal cells. If however the receiving interneuron were strongly stimulated independently, it would release ECBs, thus depressing the incoming facilitating IPSCs, and maintaining or heightening the pyramidal cell inhibition. Hence the combination of facilitating output, innervation by CB_1R-expressing GABAergic nerve terminals, and the capability of mobilizing ECBs constitutes a rich repertoire of tools whereby this network of CCK cells can modulate pyramidal cell firing.

A major twist on the idea that interneurons could directly mobilize ECBs and regulate their incoming synaptic input is that CCK cells in the hippocampus can regulate their own synaptic output, by triggering ECB mobilization from their target pyramidal cells (Foldy et al. 2006). In paired cell recordings, direct application of CCK reduced GABA release from CCK cells, but not from the parvalbumin (PV)-expressing basket cell interneurons. Most importantly, AM251 abolished the CCK-induced IPSC suppression. The ECBs appeared to originate from the pyramidal cells, because suppressive effect of CCK could be blocked by including the G-protein inhibitor, GDPβS, in the pyramidal cell recording pipette.

It is interesting that, although these other studies confirmed that CB_1R-expressing interneurons in other parts of the brain are competent to mobilize ECBs, thus far only the neocortical cells appear to respond with CB1R-mediated self-inhibition (Bacci et al. 2004) and the finding has not yet been replicated in the neocortex. This is surprising because the cortical interneurons are CCK-expressing and, like the CCK interneurons in other parts of the brain, express CB_1R, and as noted, some of them can release ECBs (Ali 2007). The ability of interneurons to mobilize ECBs allows them to regulate their involvement in circuit behaviors, particularly in the oscillations in which they play prominent roles.

3 Plasticity of the ECB System

3.1 *Use-Dependence of CB_1R Efficacy*

Even before ECBs were shown to be the retrograde messengers for DSI, it was clear that DSI could be regulated by presynaptic mechanisms. At a low concentration, 100 μM, 4-aminopyridine (4-AP) blocks only a few K channel subunits, yet at this concentration 4-AP abolished DSI (Alger et al. 1996). Other K channel blockers at much higher concentrations, e.g. 10 mM tetraethylammonium (TEA), were unable to do the same, implying that 4-AP's effect was fairly specific on an A- or perhaps D-type K current (Varma et al. 2002). One interpretation of this result was that the DSI messenger reduced synaptic transmitter release by activating presynaptic 4-AP sensitive channels. Another interpretation of 4-AP's effect was that it enhanced Ca influx into the terminal by blocking the K channels. In this scenario, the large influx of Ca would offset the suppression of Ca influx caused by DSI (cf., Klapstein and Colmers 1992). This hypothesis predicted that lowering extracellular $[Ca^{2+}]_e$ in the presence of 4-AP would largely restore DSI, which was in fact observed (Varma et al. 2002), although full reversal was not obtained. The DSI-opposing effects of low concentrations of Na conductance inhibitors may have a similar explanation (Alger et al. 1996; Varma et al. 2002). Despite some residual uncertainty about the underlying mechanism of the K channel blockade on DSI, these data revealed a potential for use-dependence of DSI. The discovery that DSI is mediated by ECBs implied that CB_1R-mediated actions generally would similarly be use-dependent. Indeed, the CB_1R ligand THC increased A-type K currents (Deadwyler et al. 1995, some of which are 4-AP-sensitive, supporting this possibility. Moreover, Ba and 4-AP opposed the IPSC suppression caused by WIN55212-2 (Hoffman and Lupica 2000).

Confirmation of the use-dependence of DSI came during paired recordings from identified CCK-interneurons and pyramidal cells (Foldy et al. 2006). The experiments showed directly that unitary (u) IPSC suppression caused either by WIN55212-2 or DSI could largely be lifted by increasing the firing frequency of the interneuron. Brief (200 ms long) trains of directly induced action potentials at frequencies $\geq$20 Hz were necessary. With this protocol, the uIPSC amplitudes and failure rates approached normal rates even in the presence of WIN55212-2. The uIPSC suppression caused by the N-type Ca channel blocker, ω-conotoxin-GVIA, could not be overcome, consistent with the data (Wilson et al. 2001) that transmitter release by CB_1R-expressing (i.e., CCK) interneurons (Freund et al. 2003) takes place exclusively via the N-type Ca channel. DSI could be abbreviated by 40 Hz stimulation, although not fully abolished. When the protocol was changed to 15 pulse trains of 100 Hz stimulation, even complete (100% uIPSC suppression) DSI could be largely, though not completely, erased (Foldy et al. 2006). The data clearly demonstrated that DSI could be modulated by activity in the presynaptic interneuron. A question remains concerning the mechanism by which high frequency stimulation restores transmission in the face of CB_1R activation. While increasing the preterminal $[Ca^{2+}]_i$ seems certain to play a role, it might not be the only factor. G-protein-dependent blockade of N-type Ca channels is voltage-dependent, and can be relieved by strong depolarizations (Bean 1989; Ikeda 1991). This may also be important for use-dependence of CB_1R suppression. In principle, these factors could be distinguished by manipulations of preterminal $[Ca^{2+}]_i$, which would not affect the voltage-dependent relief while dramatically altering the CB_1R effects on $[Ca^{2+}]_i$.

3.2 *Tonic CB_1R Activation and ECB Regulation*

Participation of ECBs in neuronal plasticity is shaped by the mechanisms of ECB mobilization. Thus far the emphasis has been on instances of ECB mobilization triggered by a sudden, strong increase in $[Ca^{2+}]_i$, by certain GPCRs, or by the coincidence of both stimuli. All of these mechanisms have a definable point of onset. However, ECBs may also be tonically mobilized by an as yet incompletely characterized, but persistent, Ca-dependent process with no clear point of onset. Interruption of this ongoing mechanism, in effect a negative regulation of ECB actions, represents another form of ECB-dependent neuronal plasticity.

Tonic actions of neurotransmitters or neuromodulators are inferred when a receptor antagonist alone produces effects that are opposite to the effect caused by the receptor agonist. Initial reports of ECB actions in the brain seemed to discount the possibility of tonic ECB actions: application of CB_1R antagonists did little or nothing on their own (e.g. Wilson and Nicoll 2001; Kim et al. 2002). The question of the tonic actions of ECBs is potentially tricky, because the CB_1R antagonists are inverse agonists (Pertwee 2005). In principle, CB_1R inverse agonists could induce effects that are the inverse of the agonists even in the absence of the agonist. They would do this by locking CB_1R, a G-protein binding receptor, into the GDP-bound state (Bouaboula et al. 1997; Vasquez and Lewis 1999). If CB_1R is constantly shuttling between GTP (active) and GDP (inactive) binding even in the agonist-unbound state, there could be a CB_1R "tone" that would be removed by the inverse agonist as it gradually trapped the receptor in the inactive state. Inhibition of CB_1R activated by tonically released ECB would have the same physiological effects as inverse agonism of intrinsically activated CB_1R. Nevertheless, the two mechanisms would be subject to strikingly different kinds of regulation. Despite this theoretical possibility, there seems to be no physiological evidence that CB_1R tone is set by intrinsic receptor activity, rather it is set by tonic mobilization of ECBs. In hippocampal CA1 (Wilson and Nicoll 2001) and NAc (Robbe et al. 2002), tonic ECB actions were only detected after the ECB transporter was blocked, implying that while ECBs may be tonically released, the transporter normally prevents CB_1R from being activated. In hippocampus, constitutive activation of the presynaptic, 2-AG degradative enzyme, monoacylglycerol lipase (Dinh et al. 2002), also plays a major role in preventing tonic activation of CB_1R, which is revealed when the enzyme is inhibited (Hashimotodani et al. 2007). Tonic release might be discovered in other areas when ECB transport or degradation are blocked.

In studying transmission between synaptically coupled GABA interneuron and CA3 pyramidal cell pairs, Losonczy et al. (2004) noted instances in which induction of a single action potential, or even a modest train of interneuronal action potentials, failed to initiate any synaptic response in the pyramidal cell. At first it appeared that the cells were simply not synaptically connected. However, a high frequency stimulus train lasting hundreds of milliseconds produced a gradually intensifying, erratic barrage of IPSCs in the pyramidal cell. Clearly the cells were synaptically coupled, but at low stimulus frequencies this was undetectable. Application of a CB_1R antagonist revealed strong coupling even at low stimulus frequencies, implying that the connection was actively "muted" by ECBs. The ability of high stimulus frequencies to reveal the synaptic connections by overcoming the CB_1R-induced suppression of GABA release demonstrated the use-dependence of ECB actions (see Sect. 3.1). In addition, since no overt stimulus of ECB mobilization had been applied, the experiments showed that the ECBs were tonically released.

These conclusions were confirmed and extended in the CA1 region (Neu et al. 2007). Paired CCK interneuron–pyramidal cell recordings revealed substantial variability in the probability of release from these interneurons. In many cases the connection was effectively silent until a CB_1R antagonist was applied. Although, in principle, tonically released ECBs might have come from any cell in the neighborhood of the recorded interneuron, they seemed to come only from the coupled pyramidal cell, because loading the postsynaptic cell with BAPTA abolished the tonic effects. Had ECBs spilled over from nearby cells to the target interneuron, then high BAPTA in a single pyramidal cell should have been ineffective. Tonic ECB release was not secondary to tonic activation of mAChR, mGluR, or NMDAR, all of which can induce ECB release in the hippocampus. Thus these experiments confirmed not only the reality of the tonic release phenomenon, but also supported the concept that ECB signaling is a local phenomenon; ECBs do not spill over from one cell to another under normal circumstances. Previous failures to observe tonic ECB release can probably be explained by the heterogeneity and relatively small numbers of interneurons susceptible to tonic release.

In identified pro-opiomelanocortin (POMC)-expressing neurons in the arcuate nucleus of the hypothalamus, Hentges et al. (2005) discovered that AM251 increased the baseline amplitudes of eIPSCs but not of eEPSCs. AM251 had no effect if the POMC cell had been loaded with 10 mM BAPTA, thus ruling out constitutive intrinsic activation of CB_1R as a mechanism, and showing that $[Ca^{2+}]_i$-dependent ECB mobilization was responsible, and that ECBs did not spill over from other cells. Another emerging theme highlighted in the POMC experiments was that, although eEPSCs were not suppressed by tonic ECBs, they were suppressed by WIN55212-2. When the ECB transport blocker VDM-11 was present, ECB-mediated suppression of eEPSCs did occur and this was suppressed by intracellular BAPTA in the recorded cell. Apparently CB_1Rs on excitatory terminals are located far enough from other sources of ECBs that, even with uptake blocked, ECBs cannot travel to them. As in neocortex and hippocampus (Marsicano et al. 2003; Monory et al. 2006), it is possible that hypothalamic CB_1Rs on glutamatergic terminals serve mainly as a back-up neuroprotective system that limits further glutamate release under conditions, such as seizures, when massive release of ECB-stimulating factors occurs.

Tonic release of ECBs has also been detected in the hypothamic paraventricular and supraoptic nuclei, specifically on the oxytocin (OT)-producing magnocellular neurosecretory cells (Oliet et al. 2007). GABAergic inputs onto these cells (but not onto the vasopressin-producing cells) typically have a low probability of release. Previous work had suggested that oxytocin action in these regions suppressed the inputs via a retrograde signaling process (Kombian et al. 1997). This seemed paradoxical, since the oxytocin receptors are exclusively localized on the postsynaptic OT cells, and not on incoming nerve terminals. A CB_1R agonist mimicked and occluded the ability of OT to suppress glutamate release in the supraoptic nucleus, while a CB_1R antagonist blocked them (Hirasawa et al. 2004), demonstrating that OT is another endogenous agent that causes retrograde effects indirectly by releasing ECBs. In the magnocellular cells of both the supraoptic and paraventricular nuclei, ECBs tonically suppress GABAergic transmission (Oliet et al. 2007), but tonic ECB mobilization was secondary to tonic release of OT. Either an OTR or a CB_1R antagonist significantly increased the probability of GABA release from the interneurons, whereas agonists decreased it. Effects of other CB_1R antagonists were mutually occlusive. Interestingly, the OT-dependent ECB mobilization was also $[Ca^{2+}]_i$-dependent, and could be blocked by chelating postsynaptic calcium. From the perspective of synaptic plasticity, the ECBs caused the OT cells to act as low pass filters: GABAergic synapses, initially having a low neurotransmitter release probability, demonstrated marked facilitation when stimulated at high frequencies (> 20 Hz). Thus, because of tonic ECB release, GABAergic inhibition would normally be blocked, and the OT cells would readily fire. However, high frequency stimulation of the interneuron would break through the ECB-suppression of GABA release and inhibit OT cell firing. By tonically releasing ECBs, the OT cells participated in a feedback loop that regulated their own firing pattern.

Having been observed by at least five different laboratories in five different brain regions, it must be accepted that tonic ECB release is a genuine experimental phenomenon. Whether or not it is a physiological phenomenon, that is, to what extent it occurs under physiological circumstances when principal cells do not have electrodes stuck in them, is not yet clear. In view of the repeated demonstrations that the tonic ECB mobilization originates in the recorded principal cell, and is sensitive to its state of $[Ca^{2+}]_i$ buffering, G-protein activation, etc., this must be a concern. Assuming it is physiologically relevant, the concept of persistent ECB suppression of certain synapses, and with it the capabilities for use-dependent frequency filtering of inputs, make it possible for ECBs to play a wider variety of regulatory roles than previously realized.

3.3 Plasticity of ECB Mobilization

Recognition that ECBs were not stored in membrane-bound vesicles and yet could be increased by various forms of stimuli in biochemical experiments led to the idea that they are produced "on demand" to meet immediate physiological needs. The on-demand hypothesis makes some predictions that have not always been met when tested in cellular physiological systems: that application of an appropriate stimulus should directly lead to ECB synthesis and that synthesis and release are essentially coupled, with ECBs being released as soon as they are produced. The on-demand hypothesis does not obviously predict plasticity of ECB mobilization.

Repetitive synaptic stimulation (Zhu and Lovinger 2007) or transient application of a group I mGluR agonist DHPG (Edwards et al. 2008) persistently enhanced submaximal hippocampal DSI in CA1. Zhu and Lovinger also showed that low frequency (1 Hz) synaptic stimulation in stratum radiatum for 5 min also induced iLTD at these GABAergic synapses. Both iLTD and enhancement of DSI were prevented by pretreatment with mGluR antagonists. Short-term DHPG application had a similar effect, and if mGluR antagonists were applied as DHPG was washed from the chamber, the DSI increase persisted (Edwards et al. 2008). Strong activation of mAChRs for several minutes increased DSI but did not have a lasting effect, implying that activation of ECB_{GPCR} alone was not sufficient to upregulate the ECB_{Ca} system. Since DSI can facilitate LTP induction (Carlson et al. 2002), upregulation of DSI can have lasting consequences. DSI and other forms of ECB_{Ca} are transient in nature, and may therefore offer greater flexibility in certain forms of neuronal network modification than long-lasting plasticities.

In many cells, application of an mGluR agonist may not mobilize ECBs even at high concentrations and even though the cells are otherwise capable of mobilizing them. Edwards et al. (2008) found that when the cells are first "primed" with a brief intense influx of Ca, then the same mGluR stimulus leads to robust ECB mobilization. At first glance the priming process has a lot in common with the molecular coincidence detector mechanism (Hashimotodani et al. 2005). Some critical features distinguish coincidence detection from priming, however. First, the coincidence detector model demands a strict temporal overlap in the elevation of $[Ca^{2+}]_i$ and the activation of the G-protein coupled receptor. Priming does not require such overlap and the two stimuli can be temporally separated by many minutes and facilitation of ECB mobilization will still occur. Indeed, if a cell is stimulated to produce a large Ca influx, and then allowed to fill with a high concentration of Ca chelator for tens of minutes, subsequent application of an mGluR agonist will evoke a robust ECB response. The induction of ECB-iLTD could be primed as well. Priming was not induced by an mAChR agonist, and it was suggested that the Ca-dependent step was closely linked to intracellular pathways accessed by group I mGluRs, although these pathways have not been identified. In summary, priming represents an upstream regulatory process that adjusts the responsiveness of the ECB_{mGluR} system, i.e., it is a form of "metaplasticity" (Abraham and Bear, 1996).

3.4 ECB Transport as a Synaptically Modifiable Process

The mechanism by which ECBs traffic between cells is not certain. The first direct evidence that ECBs could actually travel from a postsynaptic cell to presynaptic terminals seems to have been provided by Gerdeman et al. (2002), who directly loaded AEA into striatal cells, and observed CB_1R-dependent depression of glutamatergic synaptic input. However, AEA or 2-AG loaded into the postsynaptic cell could not escape and activate presynaptic CB_1R receptors if a ECB transporter blocker was also loaded (Ronesi et al. 2004). Since inhibiting the transporter from within prevents ECBs from reaching the CB_1R on the adjacent synaptic terminals, it appears that ECB extrusion from postsynaptic cells may depend on transporter-mediated facilitated diffusion. Extracellular and intracellular application of the ECB transporter blocker have diametrically opposite effects on ECB-LTD initiation; the former potentiates (Gerdeman et al. 2002) and the latter inhibits induction (Ronesi et al. 2004). The difference is attributable to differences in the direction of

ECB movements in the two cases: when applied extracellularly, transporter blockers will inhibit the uptake of ECBs into surrounding cells, thus retarding their clearance and prolonging their activation of CB_1R. The ECB transporter could be a substrate for regulation of ECB actions.

This has received support in a study of medium spiny neurons loaded with either AEA or 2-AG via the whole-cell pipette (Adermark and Lovinger 2007b) to investigate ECB release. Inhibition of afferent glutamatergic or $GABA_A$ergic inputs provided the bioassay for ECBs. Neither ECB diffused out of the cell and inhibited synaptic input significantly if the synapses were stimulated only infrequently with single pulses. Remarkably, synaptic responses evoked with double pulses delivered at the same rate were quickly depressed in a CB_1R-sensitive way. Direct activation of presynaptic CB_1R by WIN55212-2 was not stimulus-dependent, and conversely manipulations of postsynaptic cell properties – membrane potential, or $[Ca^{2+}]_i$ – did not alter ECB-mediated synaptic inhibition. The effects of loaded ECBs were prevented by co-loading the cell with ECB transport inhibitors, VDM-11 or AM404. Evidently, the rate of transporter-dependent postsynaptic release of ECBs was a function of afferent stimulation rate. The situation was similar at both excitatory and inhibitory synapses, except that inhibition of GABA release was much more sensitive to ECB release and, while stimulation did facilitate it, the release was not sensitive to the pattern of afferent activation. It is not clear if release or endogenous generation of cannabinoids is regulated by the same process, and the connection between the frequency of afferent stimulation and transport is mysterious.

3.5 *The ECB System and Seizures*

Seizures represent hyperexcitable brain states in which massive neuronal activity releases large quantities of neurotransmitters and neuromodulators into the extracellular space. Glutamate in particular can have numerous deleterious effects that lead to neurotoxicity. By acting on several receptor subtypes, glutamate and other neurotransmitters can also release ECBs, as can cellular depolarization, rise in intracellular $[Ca^{2+}]_i$, and other concomitants of seizures. As reviewed elsewhere in this volume "Genetic Models of the Endocannabinoid System" (Lutz), seizure-induced ECB release, by acting principally on CB_1Rs on glutamatergic terminals, can blunt the release of glutamate, and thereby retard and restrict the extent of neurotoxic damage. This work was based on the use of novel genetic mouse models involving targeted CB_1R deletions in various cell populations.

Seizures have other effects on the ECB system, and some have been assessed in physiological studies of the ECB system. In a developmental model, a single febrile seizure persistently upregulated ECB-mediated DSI in the hippocampal CA1 region (Chen et al. 2003). The strength of DSE was not affected, suggesting that the increase in DSI might not represent an increase in ECB mobilization. Responses to the exogenous cannabinoid, WIN55212-2, were also enhanced, implying that increased DSI might reflect an increase in CB_1R number, and Western blot analysis revealed that in fact CB_1R number was increased by the seizure. As the numbers of CB_1R-expressing nerve terminals were not altered, the conclusion was that the density of CB_1Rs per terminal must have increased. Although the data were clear, the result was somewhat puzzling in view of the extremely high density of CB_1Rs that are normally found on the interneuron terminals (Freund et al. 2003). The relationship between CB_1R number and its response is not worked out, so a twofold increase in receptor number could perhaps explain the data. It will be interesting to learn if other aspects of CB_1R functioning may also be affected by seizure activity. A follow-up study, using tetanic stimulation of in vitro slices to model the seizure, confirmed the increase in CB_1Rs, although quantitatively the effect was smaller than the febrile seizure model (Chen et al. 2007). A prior in vivo seizure prevented subsequent in vitro tetanus-induced enhancement of DSI. The in vitro study added the novel information that CB_1R activation itself was necessary for the elevation of CB_1R number. Treatment with AM251 during the in vivo seizure stimulation prevented the in vitro increase in DSI.

These studies revealed complex regulation of the ECB system by seizures. The proposal was that seizure-induced upregulation of CB_1Rs on inhibitory terminals would, by suppressing inhibition, contribute to the development of the postseizure, hyperexcitable state. Prevention of CB_1R activation with a CB_1R antagonist during the seizure might increase excitability at that time, but be helpful in the long run by preventing the development of persistent hyperexcitability that is a deleterious sequel to febrile seizures.

Clearly the potential scenarios emerging from the genetic and physiological studies outlined above present very different pictures of the roles of CB_1Rs in epilepsy. Recently a comparative study of hippocampal tissue from human epileptic and control brains has reported results that appear to be in general agreement with the studies of genetically engineered mice (Ludanyi et al. 2008). Quantitative PCR and electron microscopy revealed a significant down-regulation of CB_1R, DGLα, and a CB_1R-interacting protein (CRIP-1a). There were no changes in other relevant enzymes. In a cellular comparison, it was found in the hippocampal dentate gyrus that there was a robust reduction in CB_1R levels associated with glutamatergic terminals with no changes in the receptors on GABAergic terminals. In accordance with the studies on modified mice, these data point to the loss of an ECB-mediated neuroprotective function resulting from repeated seizures. The physiological and therapeutic implications are complex. Administering CB_1R agonists might not be very effective if the CB_1Rs most important for excitability control are simply missing.

3.6 Interactions Between AEA and 2-AG

Although consensus is developing that 2-AG is the major ECB in the brain, AEA is an agonist of CB_1R and is produced by various stimuli, including $[Ca^{2+}]_i$ and neurotransmitters. It could be that in different brain regions one or the other predominates. Yet the relationship between AEA and 2-AG has never been clarified, and it is possible that the two interact in some way. Maccarrone et al. (2008) have reported that mGluR5 (not mGluR1) activation in the striatum mobilizes ECBs and also increases CB_1R binding, perhaps by affecting receptor recycling. Contrary to a previous report (Giuffrida et al. 1999), stimulation increased 2-AG, but not AEA levels. The mGluR agonist also increased the activity of DGL and reduced MGL activity, both effects being associated with an increase in 2-AG levels. Most interestingly, increases in AEA, induced either by exogenous AEA or by downregulation of FAAH, decreased 2-AG and 2-AG-mediated actions. The common understanding that both AEA and 2-AG are primarily endocannabinoids might suggest competitive or other interactions among ECB-regulatory pathways. On the contrary, Maccarrone et al. (2008), showed that all of AEA's effects on 2-AG were mediated by TRPV1 channels, being abolished by pharmacological antagonism or genetic deletion of TRPV1 channels. AEA acting in its capacity as an endovanilloid (Starowicz et al. 2007) was responsible for the downregulation of 2-AG. It appears that TRPV1 inhibits glutathione-stimulated DGL, and hence the increases in 2-AG stimulated by DHPG. This critically important study demands follow-up, which, if confirmed, will certainly galvanize a major reevaluation of previous results pertaining to AEA and FAAH throughout the brain. In the context of ECB-mediated plasticity, it is easy to imagine how various mechanisms of AEA up- or downregulation would modify 2-AG-mediated synaptic plasticity.

AEA may be involved in a different form of regulation in the striatum. Ade and Lovinger (2007) showed that high-frequency stimulation (HFS) of glutamatergic afferents in dorsolateral induced LTP in young animals (PN12-14), but LTD in PN16-34 animals. This developmental shift in plasticity was correlated with changes in AEA levels. In young animals, stimulation increased only AEA levels without affecting 2-AG, and applied AEA permitted LTD induction. In the older animals, blocking CB_1R during HFS-induced LTD, and inhibiting the synthetic enzyme, DGL, had no effect on LTD. The authors suggest that a developmental increase in AEA may be the key factor in the shift from LTP to LTD induction with age.

4 Conclusion

This chapter has attempted an overview of the rapidly expanding field of ECBs and synaptic plasticity. The area is growing in both depth and breadth: firmly established phenomena, such as ECB-LTD and -iLTD are being investigated in greater cellular and molecular detail, and new phenomena, such as tonic ECB actions, the role of glia, and the interactions between AEA and 2-AG are coming to light. It appears that the field is in the phase of exuberant growth that characterizes developing systems; an eventual pruning back of some of the more extravagant claims, recognition of hidden connections between apparently disparate data, and a further opening up of new vistas of regulation, both of and by ECBs, are all likely to occur.

References

Abraham WC, Bear MF (1996) Metaplasticity: the plasticity of synaptic plasticity. Trends Neurosci 19:126–130
Abraham WC, Gustafsson B, Wigstrom H (1987) Long-term potentiation involves enhanced synaptic excitation relative to synaptic inhibition in guinea-pig hippocampus. J Physiol (Lond) 394:367–380
Adermark L, Lovinger DM (2007a) Combined activation of L-type Ca2+ channels and synaptic transmission is sufficient to induce striatal long-term depression. J Neurosci 27:6781–6787
Ade KK, Lovinger DM (2007) Anandamide regulates postnatal development of long-term synaptic plasticity in the rat dorsolateral striatum. J Neurosci 27(9):2403–2409.
Adermark L, Lovinger DM (2007b) Retrograde endocannabinoid signaling at striatal synapses requires a regulated postsynaptic release step. Proc Natl Acad Sci USA 104:20564–20569
Alger BE (2002) Retrograde signaling in the regulation of synaptic transmission: focus on endocannabinoids. Prog Neurobiol 68:247–286
Alger BE, Pitler TA (1995) Retrograde signaling at GABAA-receptor synapses in the mammalian CNS. Trends Neurosci 18:333–340
Alger BE, Pitler TA, Wagner JJ et al. (1996) Retrograde signalling in depolarization-induced suppression of inhibition in rat hippocampal CA1 cells. J Physiol (Lond) 496:197–209
Ali AB (2007) Presynaptic inhibition of GABAA receptor mediated unitary IPSPs by cannabinoid receptors at synapses between CCK-positive interneurons in rat hippocampus. J Neurophysiol 98:861–869
Azad SC, Monory K, Marsicano G et al. (2004) Circuitry for associative plasticity in the amygdala involves endocannabinoid signaling. J Neurosci 24:9953–9961
Azad SC, Kurz J, Marsicano G et al. (2008) Activation of CB1 specifically located on GABAergic interneurons inhibits LTD in the lateral amygdala. Learn Memory 15:143–152
Bacci A, Huguenard JR, Prince DA (2004) Long-lasting self-inhibition of neocortical interneurons mediated by endocannabinoids. Nature 431:312–316
Bean BP (1989) Neurotransmitter inhibition of neuronal calcium currents by changes in channel voltage dependence. Nature 340:153–156
Beierlein M, Regehr WG (2006) Local interneurons regulate synaptic strength by retrograde release of endocannabinoids. J Neurosci 26:9935–9943
Beierlein M, Fioravante D, Regehr WG (2007) Differential expression of posttetanic potentiation and retrograde signaling mediate target-dependent short-term synaptic plasticity. Neuron 54:949–959
Bender VA, Bender KJ, Brasier DJ et al. (2008) Two coincidence detectors for spike timing-dependent plasticity in somatosensory cortex. J Neurosci 26:4166–4177
Bliss TV, Lomo T (1973) Long-lasting potentiation of synaptic transmission in the dentate area of the anaesthetized rabbit following stimulation of the perforant path. J Physiol (Lond) 232:331–356
Bouaboula M, Perrachon S, Milligan L et al. (1997) A selective inverse agonist for central cannabinoid receptor inhibits mitogen-activated protein kinase activation stimulated by insulin or insulin-like growth factor 1. Evidence for a new model of receptor/ligand interactions. J Biol Chem 272:22330–22339
Brenowitz SD, Regehr WG (2003) Calcium dependence of retrograde inhibition by endocannabinoids at synapses onto Purkinje cells. J Neurosci 23:6373–6384

Brenowitz SD, Best AR, Regehr WG (2006) Sustained elevation of dendritic calcium evokes widespread endocannabinoid release and suppression of synapses onto cerebellar Purkinje cells. J Neurosci 26:6841–6850
Brown SP, Brenowitz SD, Regehr WG (2003) Brief presynaptic bursts evoke synapse-specific retrograde inhibition mediated by endogenous cannabinoids. Nat Neurosci 6:1048–1057
Brown SP, Safo PK, Regehr WG (2004) Endocannabinoids inhibit transmission at granule cell to Purkinje cell synapses by modulating three types of presynaptic calcium channels. J Neurosci 24:5623–5631
Brown SP, Brenowitz SD, Regehr W (2003) Brief presynaptic bursts evoke synapse-specific retrograde inhibition mediated by endogenous cannabinoids. Nat Neurosci 6:1048–1057
Carlson G, Wang Y, Alger BE (2002) Endocannabinoids facilitate the induction of LTP in the hippocampus. Nat Neurosci 5:723–724
Castillo PE, Schoch S, Schmitz F et al. (2002) RIM1α is required for presynaptic long-term potentiation. Nature 415:327–330
Chen K, Ratzliff A, Hilgenberg L et al. (2003) Long-term plasticity of endocannabinoid signaling induced by developmental febrile seizures. Neuron 39:599–611
Chen K, Neu A, Howard AL et al. (2007) Prevention of plasticity of endocannabinoid signalling inhibits persistent limbic hyperexcitability caused by developmental seizures. J Neurosci 27:46–58
Chevaleyre V, Castillo PE (2003) Heterosynaptic LTD of hippocampal GABAergic synapses. A novel role of endocannabinoids in regulating excitability. Neuron 38:461–472
Chevaleyre V, Castillo PE (2004) Endocannabinoid-mediated metaplasticity in the hippocampus. Neuron 43:871–881
Chevaleyre V, Takahashi KA, Castillo PE (2006) Endocannabinoid-mediated synaptic plasticity in the CNS. Annu Rev Neurosci 29:37–76
Chevaleyre V, Heifets BD, Kaeser PS et al. (2007) Endocannabinoid-mediated long-term plasticity requires cAMP/PKA signaling and RIM1α. Neuron 54:801–812
Crozier RA, Wang Y, Liu C-H et al. (2007) Deprivation-induced synaptic depression by distinct mechanisms in different layers of mouse visual cortex. Proc Natl Acad Sci USA 104:1383–1388
Dan Y, Poo M (2004) Spike timing-dependent plasticity of neural circuits. Neuron 44:23–30
Deadwyler SA, Hampson RE, Mu J et al. (1995) Cannabinoids modulate voltage sensitive potassium A-current in hippocampal neurons via a cAMP-dependent process. J Pharmacol Exp Ther 273:734–743
Di S, Malcher-Lopes R, Marcheselli VL et al. (2005) Rapid glucocorticoid-mediated endocannabinoid release and opposing regulation of glutamate and gamma-aminobutyric acid inputs to hypothalamic magnocellular neurons. Endocrinology 145:4292–4301
Diana MA, Marty A (2003) Characterization of depolarization-induced suppression of inhibition using paired interneuron – Purkinje cell recordings. J Neurosci 23:5906–5918
Dinh TP, Freund TF, Piomelli D (2002) A role for monoglyceride lipase in 2-arachidonoylglycerol inactivation. Chem Phys Lipids 121:149–158
Edwards DA, Kim J, Alger BE (2006) Multiple mechanisms of endocannabinoid response initiation in hippocampus. J Neurophysiol 95:67–75
Edwards DA, Zhang L, Alger BE (2008) Metaplastic control of the endocannabinoid system at inhibitory synapses in hippocampus. Proc Natl Acad Sci USA 105:8142–8147
Foldy C, Neu A, Jones MV et al. (2006) Presynaptic, activity-dependent modulation of cannabinoid type 1 receptor-mediated inhibition of GABA release. J Neurosci 26:1465–1469
Foldy C, Lee SY, Szabadics J et al. (2007) Cell type-specific gating of perisomatic inhibition by cholecystokinin. Nature Neurosci 10:1128–1130
Fortin DA, Levine ES (2007) Differential effects of endocannabinoids on glutamatergic and GABAergic inputs to layer 5 pyramidal neurons. Cereb Cortex 17:163–174
Fortin DA, Trettel J, Levine ES (2004) Brief trains of action potentials enhance pyramidal neuron excitability via endocannabinoid-mediated suppression of inhibition. J Neurophysiol 92:2105–2112

Freund TF, Katona I, Piomelli D (2003) Role of endogenous cannabinoids in synaptic signaling. Physiol Rev 83:1017–1066

Gerdeman G, Lovinger DM (2001) CB1 cannabinoid receptor inhibits synaptic release of glutamate in rat dorsolateral striatum. J Neurophysiol 85:468–471

Gerdeman GL, Ronesi J, Lovinger DM (2002) Postsynaptic endocannabinoid release is critical to long-term depression in the striatum. Nature Neurosci 5:446–451

Gibson HE, Edwards JG, Page RS et al. (2008) TRPV1 channels mediate long-term depression at synapses on hippocampal interneurons. Neuron 57:746–759

Giuffrida A, Parsons LH, Kerr TM et al. (1999) Dopamine activation of endogenous cannabinoid signaling in dorsal striatum. Nature Neurosci 2:358–363

Gubellini P, Pisani A, Centonze D et al. (2004) Metabotropic glutamate receptors and striatal synaptic plasticity: implications for neurological diseases. Prog Neurobiol 74:271–300

Hampson RE, Zhuang SY, Weiner JL et al. (2003) Functional significance of cannabinoid-mediated, depolarization induced suppression of inhibition (DSI) in the hippocampus. J Neurophysiol 90:55–64

Hashimotodani Y, Ohno-Shosaku T, Tsubokawa H et al. (2005) Phospholipase Cβ serves as a coincidence detector through its Ca2+ dependency for triggering retrograde endocannabinoid signal. Neuron 45:257–268

Hashimotodani Y, Ohno-Shosaku T, Kano M (2007) Presynaptic monoacylglycerol lipase activity determines basal endocannabinoid tone and terminates retrograde endocannabinoid signaling in the hippocampus. J Neurosci 27:1211–1219

Heinbockel T, Brager DH, Reich CG et al. (2005) Endocannabinoid signaling dynamics probed with optical tools. J Neurosci 25:9449–9459

Hentges ST, Low MJ, Williams JT (2005) Differential regulation of synaptic inputs by constitutively released endocannabinoids and exogenous cannabinoids. J Neurosci 25:9746–9751

Hirasawa M, Schwab Y, Natah S et al. (2004) Dendritically released transmitters cooperate via autocrine and retrograde actions to inhibit afferent excitation in rat brain. J Physiol (Lond) 559:611–624

Hoffman AF, Lupica CR (2000) Mechanisms of cannabinoid inhibition of $GABA_A$ synaptic transmission in the hippocampus. J Neurosci 20:2470–2479

Hoffman AF, Riegel AC, Lupica CR (2003) Functional localization of cannabinoid receptors and endogenous cannabinoid production in distinct neuron populations of the hippocampus. Eur J Neurosci 18:524–534

Ikeda SR (1991) Double-pulse calcium channel current facilitation in adult rat sympathetic neurones. J Physiol (Lond) 439:181–214

Isokawa M, Alger BE (2006) Ryanodine receptor regulates endogenous cannabinoid mobilization in the hippocampus. J Neurophysiol 95:3001–3011

Ito M (2001) Cerebellar long-term depression: characterization, signal transduction, and functional roles. Physiol Rev 81:1143–1195

Kawamura Y, Fukaya M, Maejima T et al. (2006) The CB1 cannabinoid receptor is the major cannabinoid receptor at excitatory presynaptic sites in the hippocampus and cerebellum. J Neurosci 26:2991–3001

Kim J, Isokawa M, Ledent C et al. (2002) Activation of muscarinic acetylcholine receptors enhances the release of endogenous cannabinoids in the hippocampus. J Neurosci 22:10182–10191

Klapstein GJ, Colmers WF (1992) 4-Aminopyridine and low Ca2+ differentiate presynaptic inhibition mediated by neuropeptide Y, baclofen and 2-chloroadenosine in rat hippocampal CA1 in vitro. Br J Pharmacol 105:470–474

Kombian SB, Mouginot D, Pittman QJ (1997) Dendritically released peptides act as retrograde modulators of afferent excitation in the supraoptic nucleus in vitro. Neuron 19:903–912

Kreitzer AC, Malenka RC (2005) Dopamine modulation of state-dependent endocannabinoid release and long-term depression in the striatum. J Neurosci 25:10537–10545

Kreitzer AC, Regehr WG (2001) Retrograde inhibition of presynaptic calcium influx by endogenous cannabinoids at excitatory synapses onto Purkinje cells. Neuron 29:717–727

Kreitzer AC, Carter AG, Regehr WG (2002) Inhibition of interneuron firing extends the spread of endocannabinoid signaling in the cerebellum. Neuron 34:787–796
Lenz RA, Wagner JJ, Alger BE (1998) N- and L-type calcium channel involvement in depolarization-induced suppression of inhibition in rat hippocampal CA1 cells. J Physiol (Lond) 512:61–73
Liu C-H, Heynen AJ, Schuler MG et al. (2008) Cannabinoid receptor blockade reveals parallel plasticity mechanisms in different layers of mouse visual cortex. Neuron 58:340–345
Llano I, Leresche N, Marty A (1991) Calcium entry increases the sensitivity of cerebellar Purkinje cells to applied GABA and decreases inhibitory synaptic currents. Neuron 6:565–574
Losonczy A, Biro AA, Nusser Z (2004) Persistently active cannabinoid receptors mute a subpopulation of hippocampal interneurons. Proc Natl Acad Sci USA 101:1362–1367
Ludányi A, Eross L, Czirják S, Vajda J, Halász P, Watanabe M, Palkovits M, Maglóczky Z, Freund TF, Katona I (2008) Downregulation of the CB1 cannabinoid receptor and related molecular elements of the endocannabinoid system in epileptic human hippocampus. J Neurosci 28:2976–2990
Maccarrone M, Rossi S, Bari M et al. (2008) Anandamide inhibits metabolism and physiological actions of 2-arachidonoyglycerol in the striatum. Nature Neurosci 11:152–159
Maejima T, Hasimoto K, Yoshida T et al. (2001) Presynaptic inhibition caused by retrograde signal from metabotropic glutamate to cannabinoid receptors. Neuron 31:463–475
Maejima T, Oka S, Hashimotodani Y et al. (2005) Synaptically driven endocannabinoid release requires Ca2+-assisted metabotropic glutamate receptor subtype 1 to phospholipase C β4 signaling cascade in the cerebellum. J Neurosci 25:6826–6835
Makara JK, Katona I, Nyiri G et al. (2007) Involvement of nitric oxide in depolarization-induced suppression of inhibition in hippocampal pyramidal cells during activation of cholinergic receptors. J Neurosci 27:10211–10222
Marcaggi P, Attwell D (2005) Endocannabinoid signaling depends on the spatial pattern of synapse activation. Nat Neurosci 8:776–781
Marsicano G, Wotjak CT, Azad SC et al. (2002) The endogenous cannabinoid system controls extinction of aversive memories. Nature 418:530–534
Marsicano G, Goodenough S, Monory K et al. (2003) CB1 cannabinoid receptors and on-demand defense against excitotoxicity. Science 302:84–88
Melis M, Perra S, Muntoni AL et al. (2004a) Prefrontal cortex stimulation induces 2-arachidonoyl-glycerol-mediated suppression of excitation in dopamine neurons. J Neurosci 24:10707–10715
Melis M, Pistis M, Perra S et al. (2004b) Endocannabinoids mediate presynaptic inhibition of glutamatergic transmission in rat ventral tegmental area dopamine neurons through activation of CB1 receptors. J Neurosci 24:53–62
Monory K, Massa F, Egertova M et al. (2006) The endocannabinoid system controls key epileptogenic circuits in the hippocampus. Neuron 51:455–466
Mu J, Zhuang S-Y, Hampson RE et al. (2000) Protein kinase-dependent phosphorylation and cannabinoid receptor modulation of potassium A current (IA) in cultured rat hippocampal neurons. Pflugers Arch 439:541–546
Mukhtarov M, Ragozzino D, Bregestovski P (2005) Dual Ca2+ modulation of glycinergic synaptic currents in rodent hypoglossal motoneurones. J Physiol (Lond) 569:817–831
Neu A, Foldy C, Soltesz I (2007) Postsynaptic origin of CB1-dependent tonic inhibition of GABA release at CCK-positive basket cell to pyramidal cell synapses in the CA1 region of the rat hippocampus. J Physiol (Lond) 578:233–247
Nevian T, Sakmann B (2006) Spine Ca2+ signalling in spike-timing-dependent plasticity. J Neurosci 26:11001–11013
Ohno-Shosaku T, Maejima T, Kano M (2001) Endogenous cannabinoids mediate retrograde signals from depolarized postsynaptic neurons to presynaptic terminals. Neuron 29:729–738
Ohno-Shosaku T, Tsubokawa H, Mizushima I et al. (2002) Presynaptic cannabinoid sensitivity is a major determinant of depolarization-induced retrograde suppression at hippocampal synapses. J Neurosci 22:3864–3872

Ohno-Shosaku T, Matsui M, Fukudome Y et al. (2003) Postsynaptic M1 and M3 receptors are responsible for the muscarinic enhancement of retrograde endocannabinoid signalling in the hippocampus. Eur J Neurosci 18:109–116
Ohno-Shosaku T, Hashimotodani Y, Ano M et al. (2007) Endocannabinoid signaling triggered by NMDA receptor-mediated calcium entry into rat hippocampal neurons. J Physiol (Lond) 584:407–418
Oliet SHR, Baimoukhametova DV, Piet R et al. (2007) Retrograde regulation of GABA transmission by the tonic release of oxytocin and endocannabinoids governs postsynaptic firing. J Neurosci 27:1325–1333
Pan B, Hillard CJ, Liu QS (2008) Endocannabinoid signaling mediates cocaine-induced inhibitory synaptic plasticity in midbrain dopamine neurons. J Neurosci 28:1385–1397
Pertwee RG (2005) Inverse agonism and neutral antagonism at cannabinoid CB1 receptors. Life Sci 76:1307–1324
Pitler TA, Alger BE (1992) Postsynaptic spike firing reduces synaptic GABAA responses in hippocampal pyramidal cells. J Neurosci 12:4122–4132
Reich CG, Karson MA, Karnup SV et al. (2005) Regulation of IPSP theta rhythms by muscarinic receptors and endocannabinoids in hippocampus. J Neurophysiol 94:4290–4299
Robbe D, Kopf M, Remaury A et al. (2002) Endogenous cannabinoids mediate long-term synaptic depression in the nucleus accumbens. Proc Natl Acad Sci USA 99:8384–8388
Robbe D, Montgomery SM, Thome A et al. (2006) Cannabinoids reveal importance of spike timing coordination in hippocampal function. Nat Neurosci 9:1526–1533
Ronesi J, Gerdeman GL, Lovinger DM (2004) Disruption of endocannabinoid release and striatal long-term depression by postsynaptic blockade of endocannabinoid membrane transport. J Neurosci 24:1673–1679
Ryberg E, Larsson N, Sjogren S et al. (2007) The orphan receptor GPR55 is a novel cannabinoid receptor. Br J Pharmacol 152:1092–1101
Safo PK, Regehr WG (2005) Endocannabinoids control the induction of cerebellar LTD. Neuron 48:647–659
Shin JH, Linden DJ (2005) An NMDA receptor/nitric oxide cascade is involved in cerebellar LTD but is not localized to the parallel fiber terminal. J Neurophysiol 94:4281–4289
Singla S, Kreitzer AC, Malenka RC (2007) Mechanisms for synapse specificity during striatal long-term depression. J Neurosci 27:5260–5264
Sjostrom PJ, Turrigiano GG, Nelson SB (2003) Neocortical LTD via coincident activation of presynaptic NMDA and cannabinoid receptors. Neuron 39:641–654
Sjostrom PJ, Turrigiano GG, Nelson SB (2004) Endocannabinoid-dependent neocortical layer-5 LTD in the absence of postsynaptic spiking. J Neurophysiol 92:3338–3343
Starowicz K, Nigam S, Di Marzo V (2007) Biochemistry and pharmacology of endovanilloids. Pharmacol Ther 114:13–33
Szabadits E, Cserep C, Ludanyi A et al. (2007) Hippocampal GABAergic synapses possess the molecular machinery for retrograde nitric oxide signaling. J Neurosci 27:8101–8111
Takahashi KA, Linden DJ (2000) Cannabinoid receptor modulation of synapses received by cerebellar Purkinje cells. J Neurophysiol 83:1167–1180
Trettel J, Levine ES (2002) Cannabinoids depress inhibitory synaptic inputs received by layer 2/3 pyramidal neurons of the neocortex. J Neurophysiol 88:534–539
Tzounopoulos T, Rubio ME, Keen JE et al. (2007) Coactivation of pre- and postsynaptic signaling mechanisms determines cell-specific spike-timing-dependent plasticity. Neuron 54:291–301
van Beugen BJ, Nagaraja RY, Hansel C (2006) Climbing fiber-evoked endocannabinoid signaling heterosynaptically suppresses presynaptic cerebellar long-term potentiation. J Neurosci 26:8289–8294
Varma N, Carlson GC, Ledent C et al. (2001) Metabotropic glutamate receptors drive the endocannabinoid system in hippocampus. J Neurosci 21(RC188):1–5

Varma N, Brager DH, Morishita W et al. (2002) Presynaptic factors in the regulation of DSI expression in hippocampus. Neuropharmacology 43:550–562
Vasquez C, Lewis DL (1999) The CB1 cannabinoid receptor can sequester G-proteins, making them unavailable to couple to other receptors. J Neurosci 19:9271–9280
Wagner JJ, Alger BE (1996) Increased neuronal excitability during depolarization-induced suppression of inhibition in rat hippocampus. J Physiol (Lond) 495:107–112
Wang J, Zucker RS (2001) Photolysis-induced suppression of inhibition in rat hippocampal CA1 pyramidal neurons. J Physiol (Lond) 533:757–763
Wilson RI, Nicoll RA (2001) Endogenous cannabinoids mediate retrograde signalling at hippocampal synapses. Nature 410:588–592
Wilson RI, Kunos G, Nicoll RA (2001) Presynaptic specificity of endocannabinoid signaling in the hippocampus. Neuron 31:453–462
Yamasaki M, Hashimoto K, Kano M (2006) Miniature synaptic events elicited by presynaptic Ca2+ rise are selectively suppressed by cannabinoid receptor activation in cerebellar Purkinje cells. J Neurosci 26:86–95
Yasuda H, Huang Y, Tsumoto T (2008) Regulation of excitability and plasticity by endocannabinoids and PKA in developing hippocampus. Proc Natl Acad Sci USA 105:3106–3111
Yin HH, Lovinger DM (2006) Frequency-specific and D2 receptor-mediated inhibition of glutamate release by retrograde endocannabinoid signaling. Proc Natl Acad Sci USA 103:8251–8256
Yin HH, Davis MI, Ronesi JA et al. (2006) The role of protein synthesis in striatal long-term depression. J Neurosci 26:11811–11820
Zhu PJ, Lovinger DM (2005) Retrograde endocannabinoid signaling in a postsynaptic neuron/synaptic bouton preparation from basolateral amygdala. J Neurosci 25:6199–6207
Zhu PJ, Lovinger DM (2007) Persistent synaptic activity produces long-lasting enhancement of endocannabinoid modulation and alters long-term synaptic plasticity. J Neurophysiol 97:4386–4389

Lessons from Nonmammalian Species

Ken Soderstrom

Contents

Abstract There is abundant evidence for the presence of endogenous cannabinoid signaling systems in many nonmammalian species, including several classes of invertebrates. Interest in the study of these animals largely relates to their production of distinct and measurable specialized behaviors. The ability to alter these behaviors through manipulation of cannabinoid signaling has provided important insight into both the phylogenetic history and physiological relevance of this essential neuromodulatory system.

This chapter presents a review of literature relevant to cannabinoid-altered behaviors in nonmammalian species from insects through advanced vocal learning avian species. Integration of findings supports a common role for endocannabinoid (ECB) modulation of ingestive and locomotor behaviors, with interesting contrasting agonist effects that distinguish vertebrate and invertebrate classes. Studies in amphibians and birds suggest that ECB signaling may function as a behavioral switch, allowing redirection from less- to more-essential behaviors in response to emergent environmental changes. Overall, the studies provide evidence for

---

K. Soderstrom
Department of Pharmacology and Toxicology, Brody School of Medicine, East Carolina University, Greenville, NC, 27834, USA
e-mail: soderstromk@ecu.edu
Supported by NIDA grant R01DA020109

D. Kendall and S. Alexander (eds.), *Behavioral Neurobiology of the Endocannabinoid System*, Current Topics in Behavioral Neurosciences 1,
DOI: 10.1007/978-3-540-88955-7_7,

cannabinoid modulation of aggression, emesis, feeding behavior, locomotor activity, reproductive behaviors, vocal learning, sensory perception and stress responses.

Keywords Invertebrate • Non-mammalian • Aggression • Emesis • Feeding • Locomotion • Reproduction • Sensory perception • Stress • Vocal learning

Abbreviations

2AG 2-Arachidonoylglycerol
AEA Anandamide
ECB Endocannabinoid

1 Invertebrates

The presence of behaviorally relevant cannabinoid signaling systems within invertebrate species has been controversial, probably due to difficulties isolating and functionally expressing cDNA encoding invertebrate cannabinoid receptors. Despite controversy, there is clear evidence for the presence of specific binding sites for cannabinoid receptor ligands within various invertebrate tissues (recently and thoroughly reviewed by McPartland et al. 2006). These binding data, combined with several in vivo studies, provide good evidence that cannabinoid ligands can influence invertebrate behavior. Whether these behavioral influences are produced through interaction with bona fide ECB signaling systems remains an open question depending upon the species involved.

1.1 Insects

Although concerted efforts have been made to identify specific cannabinoid binding sites in various insect species, positive evidence for the presence of cannabinoid receptors in insect tissues has yet to be obtained (McPartland et al. 2001). While this lack of evidence may indicate the absence of endogenous cannabinoid (ECB) signaling systems in insects, there is credible behavioral evidence indicative of Δ^9-tetrahydrocannabinol (THC)-altered behavior in ants.

1.1.1 Ants

A series of remarkably well-described studies investigating effects of psychoactive compounds on various ant behaviors were completed in the late 1970s at the

Institute of Pharmacology, University of Zurich. Compounds evaluated include amphetamine, D-lysergic acid (LSD) and THC. These agents were both microinjected at various dosages into the digestive tract and fed as solutions or suspensions prepared at various concentrations in sugar water.

The first study of this series focused on general behavioral effects including feeding, food preference, and "grouping" behavior; a tendency for ants to collect together within distinct regions of an open arena. Amphetamine was not consumed by ants at concentrations sufficient to produce altered behavior. LSD fed at a concentration of 100 μg ml^{-1} significantly impaired performance on each of the measured behaviors. Interestingly, THC fed at a concentration of 1 mg ml^{-1} had no measurable effect on feeding, food preference or grouping behaviors (Frischknecht and Waser 1978).

This group's second study focused on social behaviors of ant colonies, including social interactions between worker ants, adoption of new queen ants into colonies, and effects on general colony activity. Amphetamine was not used due to prior administration problems. As in the first study, LSD produced significant effects on each of the ant behaviors measured. Unlike the first study, an interesting behavioral effect of THC was measured. When fed 1 mg ml^{-1} THC in sugar water over 15 days, nest excursions (as ingeniously determined by activity detectors placed at nest entrances and exits) gradually increased to a significant degree (Fig. 1) (Frischknecht and Waser 1980). Upon cessation of THC feeding, activity levels gradually decreased over 15 days, suggesting a long duration of action consistent

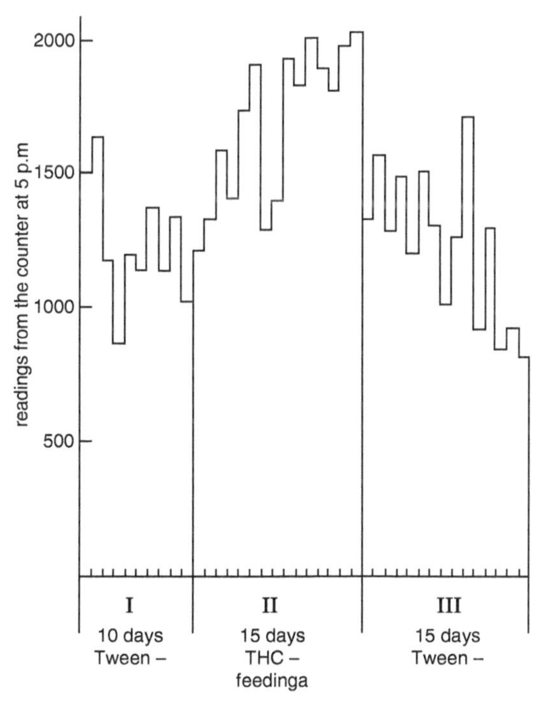

Fig. 1 Total daily ant colony nest excursions after feeding 1 mg ml^{-1} THC or vehicle (Tween) for periods of 10 or 15 days. From Frischknecht and Waser (1980) with permission from Elsevier

with the lipophilic nature of THC. Unfortunately, ant size, weight and individual consumption information was not reported and so accurate calculation of likely daily THC dosages is impossible. But with adoption of a series of assumptions, a crude estimation is possible: if we assume daily consumption of 1 μl of the 1 mg ml^{-1} THC solution, and an average ant weight of30 μg, we arrive at a crudely estimated daily dosage of 33 mg kg^{-1} THC. This dosage is in reasonable agreement with those employed to produce significant behavioral effects in mammalian species (e.g. 10 mg kg^{-1} twice daily to produce tolerance in mice, Bass and Martin 2000). Although the reported activity increases are not usually associated with cannabinoid effects, an interesting "triphasic" effect on Sprague–Dawley rat locomotor behavior has been described (Sanudo-Pena et al. 2000). As discussed below for planaria and other species, cannabinoid agonists clearly dose-dependently increase locomotor behavior in several animals, suggesting that perhaps inhibitory effects on locomotor activity is a later, vertebrate adaption to ECB signaling.

1.2 *Aquatic Invertebrates and Annelids*

1.2.1 Hydra (*Hydra Vulgaris*)

Hydra (Fig. 2) are small predatory aquatic organisms notable for primitive, but organized neural circuitry (Rawls et al. 2007). Nerve fibers coalesce to form a ring-like structure within the "head" of these animals to which a set of extensible

Fig. 2 Sessile hydra. Used with permission from Creative Commons

tentacles emerge that are used to capture prey. This ring-like neural structure is associated with control of a stereotypical, and measurable, feeding response that can be elicited by treatment with glutathione. This response is characterized by tentacle extension and opening of the mouth cavity that is surrounded by the neural ring. Interestingly, this feeding response is prolonged in duration by benzodiazepines and muscimol and inhibited by bicuculline, implicating involvement of GABAergic chloride channel signaling (Pierobon et al. 1995).

Using this animal model, Luciano De Petrocellis, Vincenzo Di Marzo and other members of the ECB Research Group in Pozzuoli, Italy, found that glutathione stimulation of hydra feeding behavior results in increased release of arachidonic acid and related compounds (Pierobon et al. 1997). Perhaps due to structural relationships between arachidonic acid and the endogenous cannabinoid compound anandamide, later experiments explored a possible role for cannabinoid signaling in modulation of hydra feeding behavior. Results of these behavioral experiments are summarized in Fig. 3 (De Petrocellis et al. 1999).

The ECB AEA was found to be an exceptionally potent inhibitor of the hydra feeding response, significantly reducing mouth opening duration at a concentration of only 10 nM. Because AEA inhibition was reversed by co-administration of the CB_1-selective antagonist, rimonabant, the effect appears to be caused by cannabinoid receptor activation. Strengthening this argument is the finding of specific and saturable binding of [^{3}H]-rimonabant to membranes prepared from hydra polyps. The affinity of this binding interaction was measured at ~1.9 nM, with a binding site density of ~27 fmol mg^{-1} protein. Although this receptor density is low relative to those found within vertebrate CNS (ranging from about 1,000–3,000 fmol mg^{-1} protein, Soderstrom and Johnson 2001), this is likely attributable to use of membranes prepared from whole animals, rather than receptor-enriched neuronal tissues.

The authors of these studies suggest that AEA and arachidonic acid signaling in hydra may function as a type of inhibitory feedback important in terminating feeding behavior. If this is true, then, as for cannabinoid-stimulated ant and planaria locomotor activity, inhibition of hydra feeding appears to contrast with orexigenic effects produced by cannabinoid agonists in vertebrate species (reviewed by Kirkham 2005 and in the chapter "Roles of the Endocannabinoid System in Learning and Memory" by Giovanni Marsicano and Pauline Lafenêtre, this volume).

1.2.2 Planaria

Planaria are well-studied animals due to their (1) status as a primitive example of bilateral body symmetry and (2) ability to regenerate excised tissue. They are free-swimming and engage in interesting and measurable exploratory locomotor behaviors (Buttarelli et al. 2008). These exploratory behaviors are characterized by purposeful swimming that, with increasing dosages of stimulant drugs, degrades into stereotyped, nonproductive "snake-like movements" and finally to "screw-like" hyperkinesia (Fig. 4). Consistent with effects in mammalian species, dopaminergic

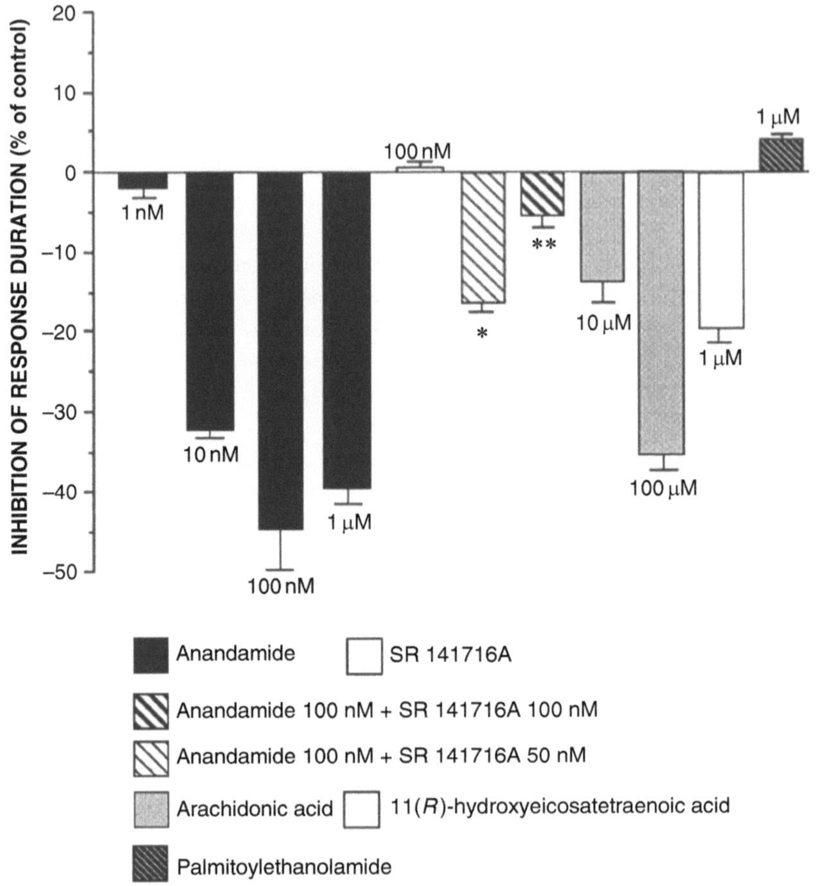

Fig. 3 Effects of AEA and related fatty acids on the hydra feeding response. Anandamide dose-dependently reduces the duration of the feeding response induced by 10 μM glutathione measured as a percentage of response duration in vehicle-treated animals (1,500 s at 10 μM glutathione). From De Petrocellis et al. (1999) with permission from Elsevier

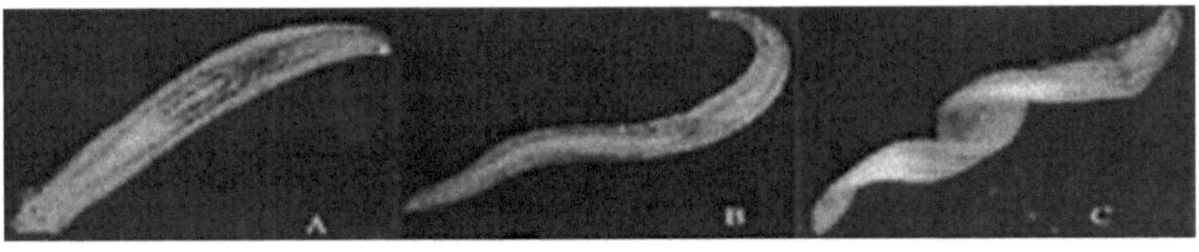

Fig. 4 Behavioral consequences of exposure to increasing dosages of stimulant drugs (in this case the cannabinoid agonist WIN55212-2) in planaria. Shown are: (**a**) normal behavior, (**b**) "snake-like" movement and (**c**) "screw-like" movement. From Buttarelli et al. (2002) used with permission from Elsevier

agonists dose-dependently stimulate these planarian locomotor behaviors in an antagonist-reversible manner (Venturini et al. 1989). Also consistent with mid-range dosages of cannabinoid agonists in mice (Sanudo-Pena et al. 2000), treatment of planaria with the synthetic cannabinoid agonist WIN55212-2 results in stimulation of locomotor activity that progresses to snake- and screw-like stereotypies (see Fig. 4 and Buttarelli et al. 2002). Cannabinoid-stimulated planarian locomotor activity was mitigated by pretreatment with the CB_1-receptor-selective antagonist rimonabant (also known as SR141716A), suggesting a receptor-mediated mechanism. Interestingly, cannabinoid stimulation was also reversed by the opioid receptor antagonist naloxone in a dose-dependent manner, suggesting convergence of cannabinoid and opioid signaling systems in this species.

Another research group has used planaria as a model system for investigating neural mechanisms underlying cannabinoid withdrawal. In these experiments, chronic administration of WIN55212-2 followed by abrupt cessation results in a withdrawal syndrome characterized by significantly reduced levels of spontaneous locomotor activity (Rawls et al. 2007). Treatment with LY 235959, an NMDA receptor antagonist, prevented manifestation of withdrawal symptoms, suggesting that cannabinoid dependence may involve glutamatergic signaling systems.

1.2.3 Leech

Leeches are carnivorous and often predatory annelids. There have been multiple studies detailing discovery of biochemical elements required to comprise an ECB signaling system within the central nervous systems of two leech species (*Theromyzon tessulatum* and *Hirudo medicinalis*).

The earliest of these reports demonstrated specific binding of [^{3}H]-AEA to leech neuronal membranes with an affinity of 32 nM and binding site density of 550 fmol mg^{-1} protein (Stefano et al. 1997). These binding data were accompanied by functional assays demonstrating potent AEA- and CP55940-induced release of nitric oxide from cultured leech neurons, effects that were blocked by the CB_1-selective antagonist rimonabant. The final component of this exceptionally complete study was the RT-PCR amplification of leech cDNA that, upon sequencing, demonstrated significant sequence homology to mammalian cannabinoid receptors. Additional analysis of this cDNA sequence indicated that it was likely a chimera consisting of part melanocortin receptor-like sequence, and part cannabinoid receptor-like sequence (Elphick 1998). In addition, the remarkably high (98%) sequence identity of portions of the leech-derived cDNA to that encoding bovine melanocortin receptors suggests that the leech sequence may be the result of horizontal gene transfer of genomic DNA derived from host blood. This fascinating possibility also suggests that the transcript is unlikely to be translated to a functional receptor, and so additional work will be required to identify the binding site associated with cannabinoid effects in leeches. Another possibility is that some contamination of neural tissue with digestive tract remains occurred during dissection – the power of the polymerase chain reaction is both a blessing and a curse, particularly when using degenerate primers.

A second report documented that all elements needed to constitute an ECB signaling system are present within the CNS of *H. medicinalis*. First, ECB ligands and their precursors were measured from extracts of leech neural tissue. AEA was found to be present at ~22 pmol g^{-1} wet weight, 2-arachidonoylglycerol (2AG) at ~147 pmol g^{-1} and *N*-arachidonoylphosphatidylethanolamine (a precursor in the formation of AEA) at ~17 pmol g^{-1}. This demonstrates that ECBs are produced within the leech CNS. Secondly, using an antibody directed against the conserved amidase domain of fatty acid amide hydrolase (FAAH, the enzyme responsible for AEA metabolism in mammalian species), the distribution of expression of this signal-terminating enzyme was determined. These immunohistochemistry experiments demonstrated an interesting pattern of expression within esophageal ganglia, suggesting a possible role in feeding behavior in this species (Fig. 5) (Matias et al. 2001). Finally, a series of adenylyl cyclase assays were done to examine the ability of various cannabinoids to alter cyclic AMP production within cultured leech ganglia. AEA was found to inhibit forskolin-stimulated cyclase activity. Interestingly, this AEA inhibition was reversed by pretreatment with the nitric oxide synthase inhibitor, L-NAME, implicating a role for formation of nitric oxide as an essential step in cannabinoid-induced cyclase inhibition.

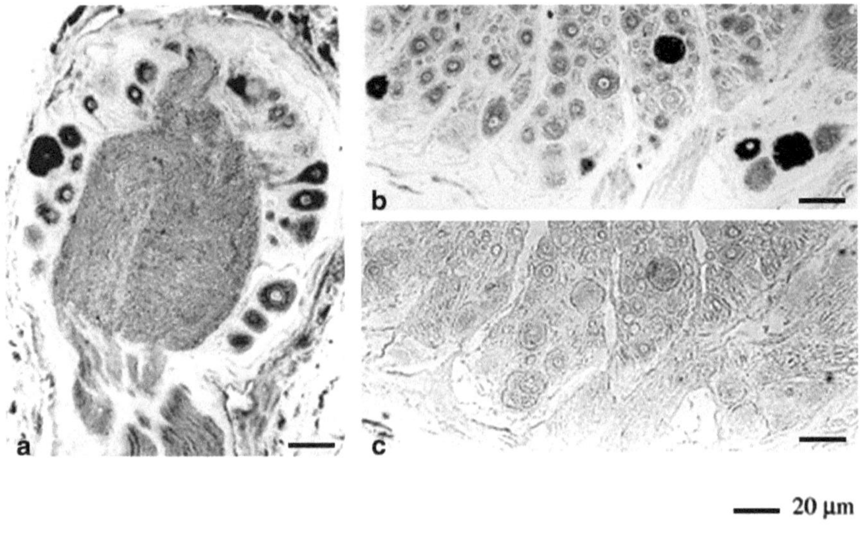

Fig. 5 Immunohistochemical detection (indirect peroxidase) of an amidase-like protein in frontal sections of *H. medicinalis*. (**a**) Immunoreactive neurons are detected in the supra-esophageal ganglia. (**b, c**) Adjacent sections of different compartments of the supra-esophageal ganglion treated either with anti-amidase serum (**b**) or with anti-amidase serum preadsorbed with the homologous antigen (**c**). Pre-adsorption of the serum with the immunizing peptide eliminated staining. From Matias et al. (2001) with permission

1.2.4 Snail

Snails comprise an exceptionally large group of molluscs that typically possess a conspicuous coiled shell. One of these animals, *Helix lucorum (L.),* has become distinctly useful for neurophysiological studies due to a large monosynaptic connection between sensory and premotor neurons responsible for initiating avoidance behavior (reviewed by Balaban 2002). This synapse is glutamatergic and application of high-frequency stimulation results in a persistent facilitation of transmission across it. Pretreatment of isolated neurons maintained in an organ bath with 10 μM AEA prevented this tetany-induced facilitation.

A strikingly different synaptic response is produced following lower frequency stimulation of transmission across this synapse, resulting in a short-term refractory period wherein excitatory postsynaptic potentials (EPSPs) are prevented from occurring for about 1 min. This low-frequency stimulation-induced refractory period is also modulated by cannabinoid exposure. Pretreatment of isolated neurons with 20 μM of the cannabinoid receptor antagonist AM251 eliminates the refractory period, suggesting that the temporary lack of responsiveness is due to release of an ECB agonist (Lemak et al. 2007).

1.2.5 Sea Urchin

Sea urchins (Fig. 6), while apparently sessile, are actually related to starfish and are similarly mobile through the use of thousands of tube feet. This mobility allows them to pursue a diet that consists of algae and an assortment of invertebrate animal species. Their spiny exterior protects an enclosed cavity containing the mouth, stomach, anus, and large, easily identified gonads that are considered delicacies in several cultures (e.g. Japanese uni sushi). Both eggs and sperm are easily harvested

Fig. 6 Sea urchin

and combined to produce larvae. Perhaps due to easy laboratory culture, these animals have become an important model for studying processes involved in sexual fertilization (reviewed by Neill and Vacquier 2004).

Sexual fertilization occurs through an acrosomal reaction (reviewed by Breitbart 2003). Within the tip of sperm cells is a peptide-hormone-containing granule, referred to as the acrosomal granule. Upon making contact with the jelly coat layer surrounding an appropriate egg, this granule fuses with the sperm membrane releasing signaling peptides. These peptides bind receptors on the egg membrane and initiate a biochemical cascade that ultimately results in fusion of egg and sperm membranes and creation of an embryo.

An interesting series of papers published from the late 1980s through the 1990s documents effects of THC and synthetic cannabinoid agonists to decrease viability of sea urchin sperm (cf. Berdyshev 1999; Schuel et al. 1987). Additional studies led to the discovery that cannabinoid signaling appears to play a key permissive role in allowing the acrosomal reaction to proceed. The naturally occurring cannabinoid agonist THC, synthetic agonists like CP55940 and ECB agonists including AEA have all been demonstrated to inhibit release of acrosomal granule contents, thereby impairing sperm–egg fusion. In the case of AEA and THC, potent effects are observed at concentrations of only 100 nM (Schuel et al. 1994).

1.2.6 Urochordates Including the Sea Squirt (*Ciona intestinalis)*

Urochordates like the sea squirt are of particular scientific interest because they exist on the evolutionary border separating vertebrate and invertebrate species. As larvae, these animals are free-swimming and possess a notochord: a hollow dorsal nerve column without vertebrae. This notochord is lost over development and adult animals attach to rocks and other underwater structures to adopt sessile lifestyles in maturity.

The genome sequence of *C. intestinalis* was published in late 2002 demonstrating that these animals express about 16,000 separate proteins, about half of the number expressed by later vertebrates (Dehal et al. 2002). Of the proteins expressed, many have human orthologs, including, as Maurice Elphick and Nori Satoh discovered, a G-protein-coupled receptor orthologous to CB_1 and CB_2 cannabinoid receptors (Elphick et al. 2003). Because *C. intestinalis* posseses a cannabinoid receptor ortholog and the sequenced genomes from the insect *Drosophila melanogaster* and worm *Caenorhabditis elegans* do not, Elphick and Satoh conclude that the ancestor of vertebrate CB_1 and CB_2 cannabinoid receptors must have originated in a deuterostomian invertebrate. This indicates that the cannabinoid-altered behaviors documented above to occur in more primitive invertebrate species must be due to either nonreceptor-mediated mechanisms, or perhaps more likely (given evidence for high-affinity specific binding of cannabinoid ligands to neuronal membranes across multiple invertebrate species), through activation of specific receptors that are unrelated phylogenetically to vertebrate CB_1 and CB_2. The issue of primitive cannabinoid receptors will undoubtedly be the subject of fascinating future studies.

Using the deduced amino acid sequence for the *C. intestinalis* cannabinoid receptor (CiCBR), Egertova and Elphick synthesized and purified an antibody probe directed against the C-terminal tail region of the receptor. Employing this antibody in immunohistochemistry experiments, they discovered distinct and dense expression within the cerebral ganglion (consistent with CB_1 expression throughout vertebrate forebrain). In addition to this "central" staining, and important in the context of behavioral effects of cannabinoid signaling in this species, distinct staining of neuronal projections to the oral and atrial siphons of the animal were also discovered (see Fig. 7 and Egertova and Elphick 2007).

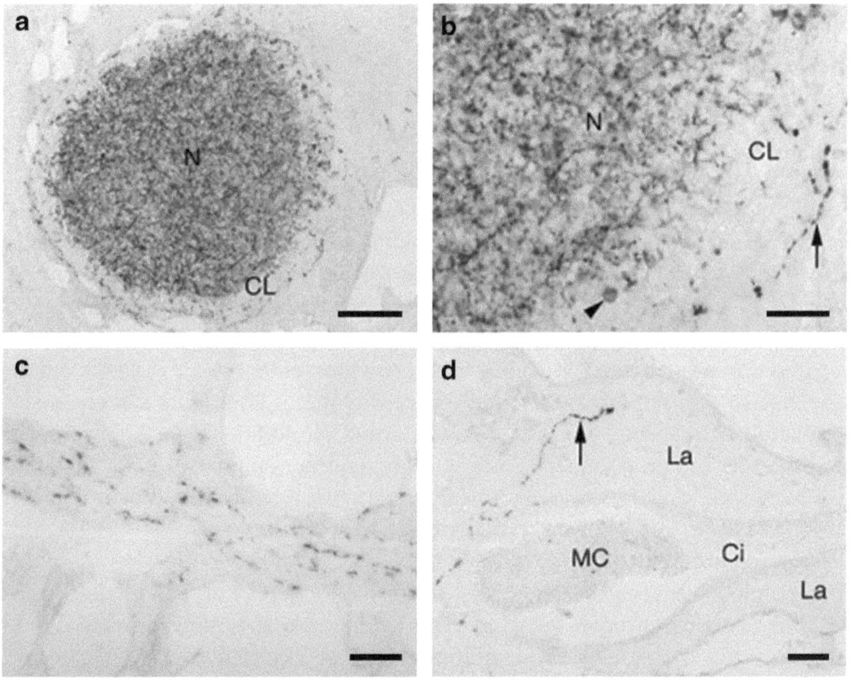

Fig. 7 CiCBR immunoreactivity in the central and peripheral nervous system of adult *Ciona* at high magnification. (**a**) CiCBR immunoreactivity is concentrated in fibers in the central neuropil (N) region of the cerebral ganglion. The cortical layer (CL) of the ganglion where most of the neuronal somata are located is largely unstained. (**b**) High-magnification image of the cerebral ganglion showing CiCBR immunoreactivity associated with fibers and axon varicosities in the neuropil (N) of the ganglion. The cortical layer (CL) is largely unstained. A single axon with stained varicosities can be seen emanating from the surface of the ganglion (arrow). (**c**) CiCBR-immunoreactive fibers in a branch of one of the nerves associated with the atrial siphon. (**d**) A single CiCBR-immunoreactive fiber (arrow) is associated with one of the languets (La). The languets collect food-laden sheets of mucus and move these as a rolled-up cord toward the esophagus. In this image, a mucus cord (MC) can be seen in contact with cilia (Ci) emanating from the epithelium of the languets. Scale bars = 100 μm in **a**; 50 μm in **b**; 25 μm in **c**,**d**. From Egertova and Elphick (2007) with permission from Wiley-Liss, Inc

Measurable behaviors displayed by adult *C. intestinalis* include oral siphon opening and closing. This behavior is both a defensive response, and important for feeding and respiration. Physical stimulation of the oral siphon results in its retraction and closing. The latency to reopening is measured. Following injection of various concentrations of the synthetic cannabinoid agonist, HU210, into the lumen of the oral siphon, a dose-dependent delay in reopening was observed. These delays were significant 2 h following injections of 133 and 300 nmol HU210, and the effect persisted for another hour after onset (Matias et al. 2005). The agonist-induced delay in siphon opening was reversed by pretreatment with a cocktail of 200 nmol each AM251 and AM630, CB_1- and CB_2-receptor-selective antagonists, respectively, implicating a receptor-mediated effect.

2 Nonmammalian Vertebrates

2.1 Fish

The puffer fish, *Fugu rubripes* (fugu), has become a notable animal model due to its distinctly compact genome. A lack of noncoding sequence relative to other vertebrate species made fugu an attractive candidate for early genome sequencing projects (Brenner et al. 1993). During low-stringency screening for dopamine receptor-encoding sequences of a library prepared from fugu genomic DNA (presumably prepared for sequencing), a cannabinoid receptor cDNA was isolated. Interestingly, use of this initial cannabinoid receptor clone in additional screens resulted in isolation of a second separate gene encoding an additional cannabinoid receptor. Both CB receptor sequences show high sequence similarity with mammalian CB_1 receptors, and much more distant relationship with CB_2 mammalian isoforms. These results indicate that fugu has evolved two separate CB_1 receptors whereas mammalian vertebrates have only one. Whether expression of multiple CB_1 receptor subtypes results in distinct behavioral phenotypes related to ECB signaling remains an open question. This unique property may provide opportunities to study consequences of multiple receptor expression.

In terms of the behavioral relevance of piscine cannabinoid signaling, distinct expression of CB_1 receptors within hypothalamic lobes of the teleost fish, *Pelvicachromis pulcher*, has led to the suggestion that cannabinoid signaling may play a role in feeding behavior (Cottone et al. 2005). This hypothesis was directly tested in the goldfish, *Carassius auratus* (Fig. 8) (Valenti et al. 2005).

Interestingly, a biphasic, dose-dependent feeding response was induced by the ECB, AEA. A potent stimulation of feeding was produced by low doses (e.g. 1 pg g^{-1}

Fig. 8 (**a**) Goldfish food intake (mg) after intraperitoneal administration of anandamide (1, 10 and 100 pg g^{-1} body weight). (**b**) Effect of intraperitoneal administration of anandamide (10 pg g^{-1}) and AM251 (1 μg g^{-1}) on goldfish total motor activity. (**c**) Goldfish plasma cortisol levels (μg dL^{-1}) at 2 h postinjection with vehicle (control) or anandamide (10 pg g^{-1})

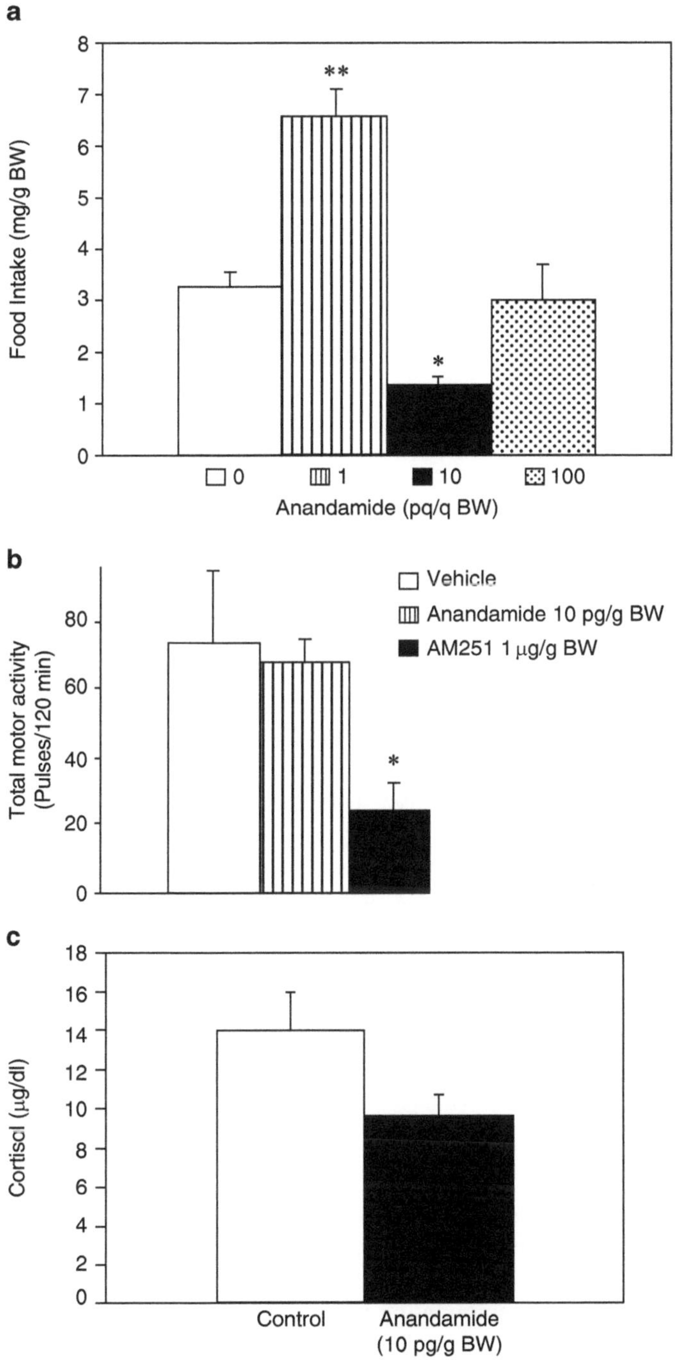
a
8
7
6
5
4
3
2
1
0
Food Intake (mg/g BW)
**
*
0
1
10
100
Anandamide (pq/q BW)
b
80
60
40
20
0
Total motor activity (Pulses/120 min)
*
Vehicle
Anandamide 10 pg/g BW
AM251 1 μg/g BW
c
18
16
14
12
10
8
6
4
2
0
Cortiscl (μg/dl)
Control
Anandamide (10 pg/g BW)

AEA, Fig. 8) while a tenfold higher dosage significantly reduced food intake. The inhibition of feeding produced by 10 pg g^{-1} anandamide was not accompanied by a general decrease in locomotor activity (Fig. 8b) or significant changes in serum cortisol (Fig. 8c), reducing the possibility of confounding motor or stress effects. Thus, it appears that goldfish feeding is distinctly sensitive to ECBs in a dose-dependent manner.

Goldfish are also an important model species for studying retinal neurobiology (Levine 2007). A distinct and dense population of CB_1 cannabinoid receptors is present within goldfish retina, particularly at the synaptic terminals of cone cells (Straiker et al. 1999). Activation of these receptors with the synthetic agonist WIN55212-2 produces interesting dose-dependent effects: low concentrations ($\leq 1\,\mu M$) cause increased conductance of K^+, Cl^- and Ca^{++}, while higher concentrations ($>1\,\mu M$) suppresses these currents. Both effects were mitigated by pretreatment with the CB_1-selective antagonist rimonabant. Each dosage was distinctly sensitive to pretreatment with cholera and pertussis toxins: low-dose stimulatory effects were prevented by cholera toxin (implicating Gs-mediated signaling), while higher-dose inhibition was reversed with pertussis toxin (implicating Gi/o, Fan and Yazulla 2003).

In a functional assay of goldfish retinal responsiveness, 10 μM WIN55212-2 speeds up recovery of cone cell responsiveness following light inactivation. This effect may lead to increased contrast sensitivity which may explain visual effects of systemic cannabinoid exposure (Struik et al. 2006).

The Siamese fighting fish (*Betta splendens*), typically referred to as "Betta", is a popular species among freshwater aquarium enthusiasts. In addition to a variety of attractive colors and fin morphologies, these fish are noted for their tendency toward aggressive territory defense. This characteristic territoriality has led to their use as a model for studying aggressive behavior. Remarkably, aggressive behavior in these fish persists even after orchiectomy (Weiss and Coughlin 1979). A particularly interesting study done with these animals involved adding a suspension of cannabis extract or purified THC to tank water. After 2 h of such exposure, time spent fighting over 5 min testing periods was significantly reduced (see Fig. 9 and Gonzalez et al. 1971). This is one of the earliest reports of the ability of cannabinoid agonist to effect a reduction in aggressive behavior.

2.2 Amphibians

Taricha granulosa, the roughskin newt, has been a productive animal model for understanding the hormonal control of neuronal signaling responsible for sex behaviors (reviewed by Rose and Moore 1999). These animals engage in a characteristic courtship clasping mating behavior. Male newts initiate this behavior by capturing a female in an amplectic clasp (Fig. 10) with all four limbs. This clasping position is easily identified and measured, and is maintained for long periods, often hours.

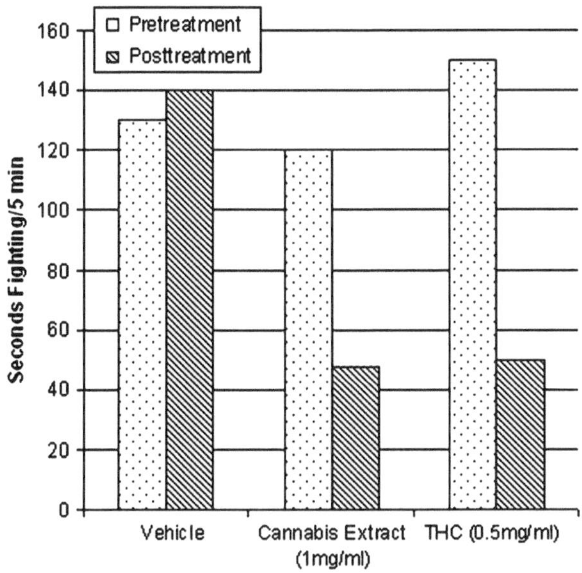

Fig. 9 Effects of various treatments added to tank water on *B. splendens* aggressive behavior. Pretreatment represents testing done 2 h prior to addition of treatments. Posttreatment represents tests done 2 h after addition of vehicle, cannabis extract or THC to tank water. Number of seconds spent fighting during a 5 min test were recorded. Adapted from Gonzalez et al. (1971) with permission from S. Karger AG

Fig. 10 Male (*top*) roughskin newt displaying courtship clasping of a female in an aquarium. The male's prominent cloaca (seen behind the rear limb) is in contact with the dorsal surface of the female. Mechanical stimulation of the cloaca triggers clasping and stimulation caused by movements of the female causes clasping to be intensified. Photograph generously provided by Prof. Frank L. Moore (see Rose and Moore 1999)

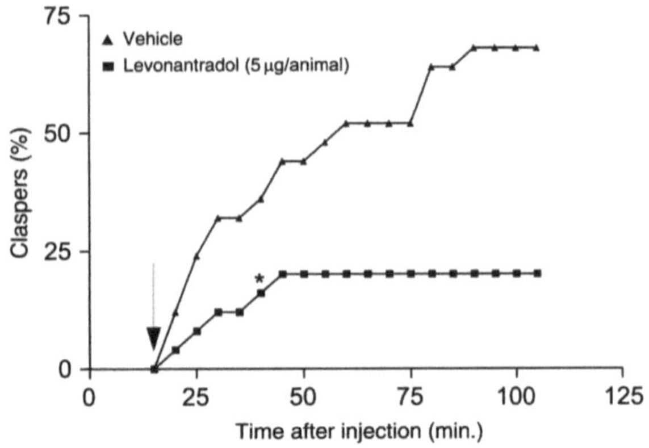

Fig. 11 Effects of the synthetic cannabinoid agonist levonantradol on incidence and latency to incidence of male clasping behavior in roughskin newts. The *arrow* indicates the time at which progesterone-primed females were added to testing buckets. Data are presented as cumulative percentage of clasping males at 5-min intervals. Males injected with levonantradol were significantly inhibited within 40 min: $*p < 0.05$ by Fisher's exact probability test. From Soderstrom et al. (2000), with permission from Blackwell Publishing

Use of this behavior allowed detection of rapid inhibitory effects of the adrenal steroid corticosterone, leading to identification of a membrane receptor-mediated mechanism of steroid action (Orchinik et al. 1991). We used the behavioral methods developed by Orchinik et al. to study effects of the synthetic cannabinoid agonist levonantradol on clasping and locomotor activity of these amphibians (Fig. 11) (Soderstrom et al. 2000).

Cannabinoid effects on newt clasping behavior were measured in the field at a lake in the coastal mountains near Benton County, Oregon. Female roughskin newts were captured before each experiment and taken to the laboratory where they were injected with 100 µg of progesterone the day before clasping behavior experiments and transported back to the field for use as clasping targets.

The day of experiments, male newts were collected from the lake and placed in plastic five-gallon buckets prepared with perforations that allowed free circulation of pond water. The perforated buckets were suspended from a raft floating in the lake. After a 1 h equilibration period following capture, males were treated with intraperitoneal levonantradol or a vehicle control. Fifteen minutes after treatments, primed female newts were introduced into testing buckets, and observation and recording of clasping behavior commenced. Numbers of males clasping and the latency to clasp were recorded.

Dose–response experiments demonstrated the ability of levonantradol to potently (IC_{50}=1.2 µg/animal [average newt weight~15 g]) inhibit clasping incidence. Injection of single dosages of levonantradol (5 µg/animal) significantly reduced the percentage of male newts engaging in clasping behavior, an effect that became significant 40 min following treatments (Fig. 10) (Soderstrom et al. 2000). It should be noted that treatment with 5 µg/animal levonantradol also significantly inhibited newt locomotor activity. Therefore, despite potent efficacy, levonantradol inhibition of clasping may be at least partially attributable to decreased motor activity.

A more direct investigation of the role for ECB signaling in the suppression of newt courtship clasping behavior has recently been reported by Emma Coddington working with Frank L. Moore at Oregon State University (Coddington et al. 2007). Using the antagonist AM281, they found that blocking cannabinoid receptor activation mitigates inhibitory effects of both corticosterone treatment and stress on newt clasping behavior. These results are interpreted to suggest that corticosterone and stress inhibit clasping through the release of ECBs (see Fig. 3 in Coddington et al. 2007). The authors also suggest that an ECB-mediated relationship between stress and inhibition of reproductive behaviors may indicate a role for cannabinoid signaling in adapting behavior to changing environmental conditions.

A role for cannabinoid signaling in behavioral responses to stress was further suggested by discovery of dense expression of CB_1-encoding mRNA within the bed nucleus of the stria terminalis in *Taricha* CNS (Hollis et al. 2006). This brain region appears to perform functions similar to those regulated by the same region of mammalian brain: involvement in autonomic and behavioral reactions to fearful stimuli (Nijsen et al. 2001).

In addition to roughskin newts, cannabinoid effects related to the behavior of multiple frog species have been reported. Early studies demonstrated that conduction at the frog neuromuscular junction is inhibited by both the naturally occurring agonist, THC (Turkanis and Karler 1986), and the ECB, AEA (Van der Kloot 1994), suggesting a potential locomotor role via alteration of signaling within the peripheral nervous system.

Using *Xenopus laevis* tissue and an antibody directed against the rat CB_1 receptor, interesting distinct labeling of olfactory structures and brain regions responsible for integration of sensory information was observed (Cesa et al. 2001). Less prominent labeling of analogous structures in mammalian brain suggests that ECB systems may have been more important for sensory processing in more primitive species, and that functional differences in the role for cannabinoid signaling may be present across phylogeny. Also interesting is a distinct CB_1 expression within the anterior lobe of *Xenopus* pituitary (Cesa et al. 2002). This expression is particularly prominent within lactotrophs and gonadotrophs, with a distinct absence in adrenocorticotropin-releasing cells, suggesting that ECB signaling may influence behaviors dependent upon pituitary secretion.

2.3 Birds

Pigeons have long been productively employed in stimulus discrimination studies of CNS active drugs (McMillan 1990). These animals are easily trained to perform in classic behavioral paradigms and were useful in several early studies investigating the mechanism of action of the naturally occurring cannabinoid, THC. Pigeon experiments showed that THC was not perceived to produce psychopharmacological effects similar to those produced by various opioid agonists, barbiturates, benzodiazepines, cholinergic antagonists, amphetamine or lysergic acid (Gouvier et al. 1984; Henriksson et al. 1975; Jarbe and Hiltunen 1988). Pigeons are also important as an emesis model (Earl et al. 1955) and have been useful for investigating the potent antiemetic effects of cannabinoids (Feigenbaum et al. 1989).

We found several years ago that CB_1 cannabinoid receptors are distinctly and densely expressed in almost all of the telencephalic brain regions known to be important for vocal learning and control of a songbird, the zebra finch (Soderstrom and Johnson 2000). A similar pattern of distinct song region expression of CB_1 receptors has now been reported in another vocal learning species, the budgerigar (Alonso-Ferrero et al. 2006).

Distinct song region receptor expression made us suspect that ECB signaling may be important to normal vocal development in these (and possibly other) vocal learning animals. Zebra finch song is a form of vocal communication learned during distinct stages of late postnatal development (Doupe and Kuhl 1999). Coordinated control of song learning, perception and production involves a discrete set of interconnected midbrain, thalamic and telencephalic brain regions (see Bottjer and Johnson 1997 for review). CB_1 cannabinoid receptors are densely and distinctly expressed in several of these song regions through adulthood (Fig. 12).

Important to our hypothesis that ECB signaling plays a role in normal vocal development, we have found that the density and pattern of distinct CB_1 expression in several telencephalic song regions notably waxes and wanes over the course of song learning. Song regions notable for particularly distinct changes in the density and pattern of receptor expression during vocal learning include the rostral telecephalic regions lMAN and Area X, and caudal regions HVC and RA (Soderstrom and Tian 2006).

Using zebra finches as a pharmacological model to study drug effects on learning during late-postnatal development, we have found that single daily treatments with a modest dosage (1 mg kg^{-1}) of the cannabinoid agonist WIN55212-2 from 50 to 100 days of age (the time-course of zebra finch postnatal development is similar to that of the rat, see Fig. 13a) alters vocal learning by reducing (1) the number of note-types produced and (2) song stereotypy (a measure of song quality developed by Scharff and Nottebohm 1991, see Fig. 14). Because these changes did not occur in adults administered the same treatment, the effect is restricted to periods of vocal development (Soderstrom and Johnson 2003). Further experiments have revealed that these effects on note number and stereotypy are produced independently: stereotypy is reduced by WIN55212-2 exposure from 50 to 75 days; while note numbers are altered by exposure from 75 to 100 days (Soderstrom and Tian 2004). Thus, there are apparently distinct periods of cannabinoid sensitivity during vocal development.

Currently we are working to identify physiological changes responsible for cannabinoid-altered vocal learning. Our hypothesis is that identification of these physiological changes will allow better appreciation of (1) neurobiology underlying normal vocal development and (2) mechanisms responsible for persistent behavioral effects caused by developmental exposure to CNS-active drugs.

Specific physiological differences that we hypothesize may be produced by developmental cannabinoid exposure include alterations in the expression of ECB

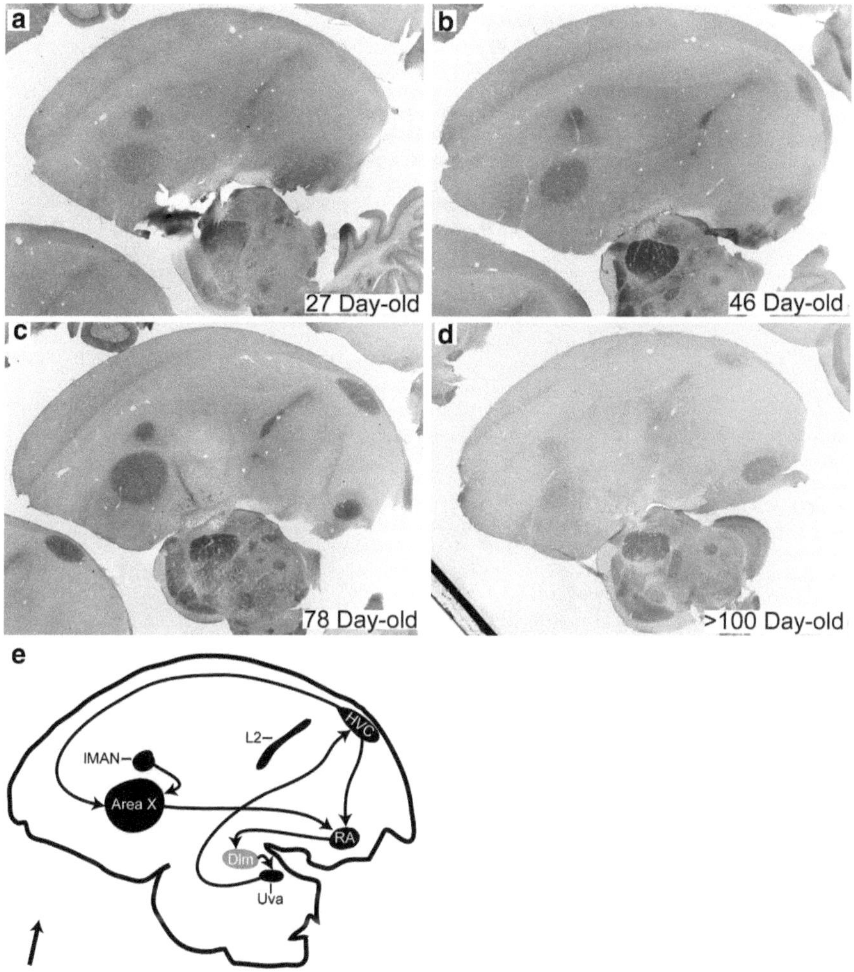

Fig. 12 12.5 × images of immunohistochemical staining of zebra finch brain with antizebra finch CB_1 receptor antibody. Sections shown were reacted together. A–D, Parasaggital sections represent planes about 2.5 mm lateral from the midline. Rostral is left. The *arrow* = 1 mm and points dorsal. E, Diagram based on panel C illustrates regions of distinct staining (*shown in black*) including song regions lMAN, Area X, L2, HVC and RA and the thalamic region nucleus uvaformis (Uva). Dorsal lateral nucleus of the medial thalamus (DLM), a thalamic region that is distinctly stained but not present in panel C, is shown diagrammatically in *gray*. *Arrows* between regions represent trajectories of some known interconnections (Bottjer and Johnson 1997). From Soderstrom and Tian (2006) reprinted with permission of Wiley-Liss, Inc

signaling elements; both in density and patterns of expression of CB_1 receptors and in levels of endogenous ligands (e.g. AEA and 2AG).

We have been fortunate to work with Vincenzo Di Marzo and his excellent ECB Research Group to develop methods to measure ECB levels from lipid extracts of zebra finch brain. While developing these methods using adult-derived tissue, we

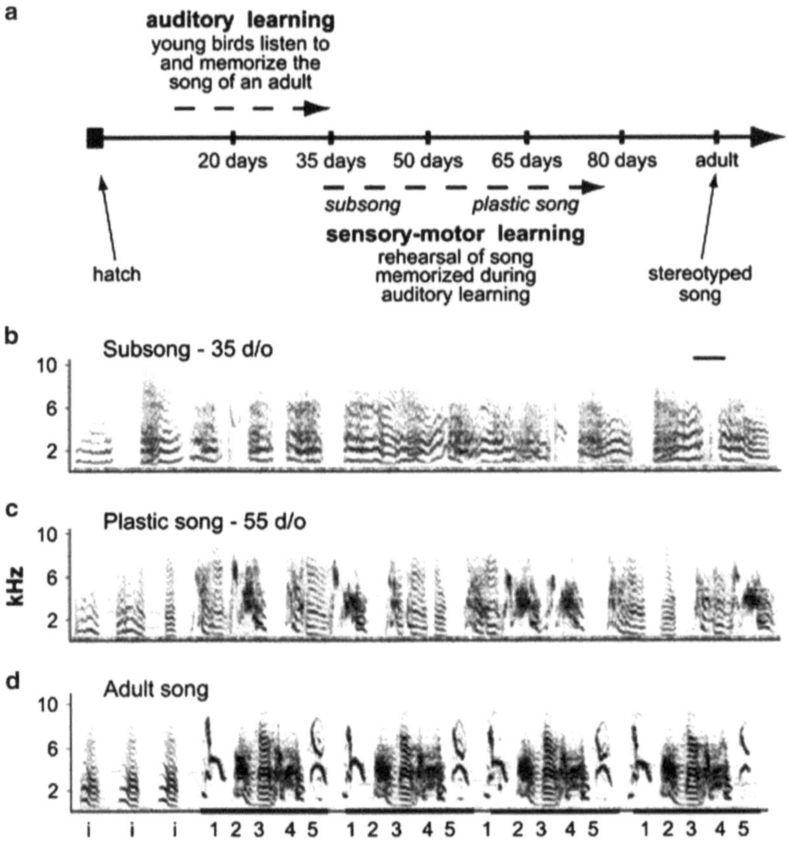

Fig. 13 Timing and sonogram examples of zebra finch vocal development. (**a**) Song learning occurs during at least two distinct stages termed auditory and sensory-motor learning. (**b**) Subsong is characterized by highly variable vocal output. (**c**) Plastic song is indicated by production of distinct note types. (**d**) Distinct note types are produced in distinct, stereotyped order at adulthood

became aware of an interesting study demonstrating that brief periods of food restriction reduces the amount that zebra finches sing (Johnson and Rashotte 2002). Because we had previously found that exogenous cannabinoid exposure also reduces song output (Soderstrom and Johnson 2001), combined with accumulating evidence of a role for ECB signaling in orexigenesis (Kirkham et al. 2002), we speculated that reduced vocal behavior caused by food restrictions might be due to ECB signaling. We discovered that under conditions of restricted food access (for 4 h periods daily, a manipulation that does not alter body weight), levels of the ECB 2AG were significantly elevated within zebra finch brain (Fig. 15) (Soderstrom et al. 2004). This result suggests an interesting relationship between feeding state and vocal behavior that is mediated by ECB signaling. This relationship suggests a potential role for cannabinoid signaling in the redirection of behavior from those

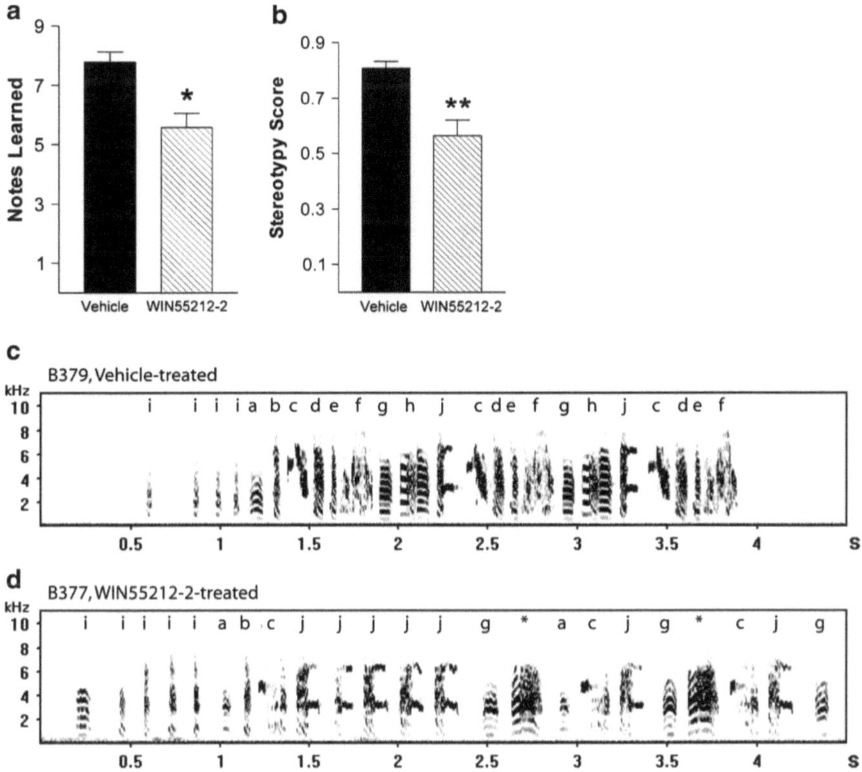

Fig. 14 Daily injections of a cannabinoid agonist (WIN55212-2, 1 mg kg^{-1} IM) during sensory-motor learning (50–100 days) (**a**) reduces number of note types learned and (**b**) decreases stereotypy (a measure of song quality. (**c** and **d**) Audiospectrograms indicate that cannabinoid treatment alters song patterns with repetitive note production (e.g. note j)

less to more critical. For example, a potentially life-threatening food shortage causes increased brain ECB levels which inhibit vocal behavior, allowing more urgent pursuit of food.

3 Lessons Learned

A summary of behavioral effects produced by activation of cannabinoid signaling systems in various nonmammalian species is presented in Table 1. From this survey, several patterns emerge. First, it is clear that similar behaviors are influenced by cannabinoid signaling across mammalian and nonmammalian species; particularly notable are involvement in feeding and locomotor behaviors. Within these nonmammalian behaviors, interesting contrasting effects are seen when comparing vertebrate and invertebrate species. Thus, it appears that the physiological roles played

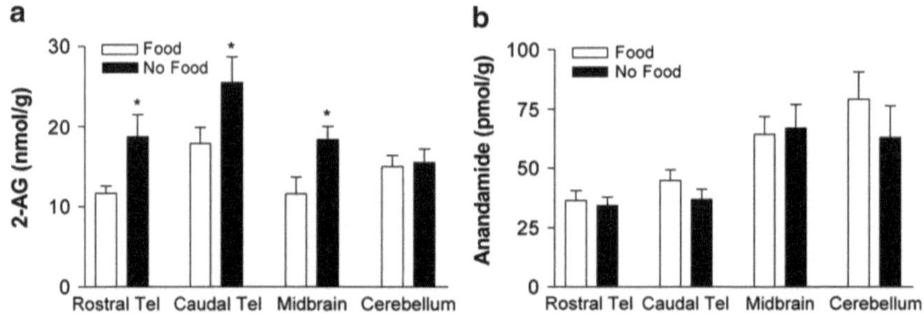

Fig. 15 Effect of food access on endogenous cannabinoid levels in zebra finch brain. Animals were either provided food ad libitum (food) or subjected to 4 h of limited food access (no food). Brains were rapidly dissected into rostral (Rostral Tel) and caudal (Caudal Tel) telencephalon, midbrain, and cerebellum. Lipids were extracted, spiked with deuterium-labeled internal standards, and subjected to LC-APCI-MS for quantitative analysis of 2AG (A) and AEA (B) content. ANOVA indicated a significant relationship between brain region and 2AG content ($p < 0.05$). Post hoc analysis revealed a significant increase in 2AG content in the midbrain and rostral and caudal telencephalon of animals subjected to limited food access (*$p < 0.05$). From Soderstrom et al. (2004) with permission from the Society for Neuroscience

Table 1 Summary of cannabinoid effects on non-mammalian behaviors

Behavior	Animal	Effect
Aggression	Betta	Inhibition
Emesis	Pigeon	Inhibition
Feeding	Goldfish	Biphasic low-dose stimulation, high-dose inhibition
Feeding	Hydra	Inhibition
Feeding	Leech	Unknown, amidase expression in esophageal ganglia
Feeding	Zebra finch	Stimulation, inhibition of vocal behavior
Locomotor activity	Ants	Stimulation
Locomotor activity	Newt	Inhibition
Locomotor activity	Planaria	Stimulation
Locomotor activity	Snail	Unknown, facilitation of premotor refractory period
Locomotor activity	Zebra finch	Inhibition
Reproduction	Newt	Inhibition of courtship clasping
Reproduction	Urchin	Inhibition of sperm acrosomal release
Sensory perception	Goldfish	Enhanced visual contrast
Stress responses	Newt	Inhibition of reproductive behavior
Vocal learning	Zebra finch	Inhibition

by cannabinoid signaling became transformed at the level of vertebrata, with stimulatory motor and inhibitory feeding effects of cannabinoids in invertebrates becoming reversed in vertebrates.

Because CB_1-like cannabinoid receptors do not appear to have evolved prior to the urochordates, mechanisms responsible for behavioral effects produced by cannabinoids in invertebrate species remain unclear. The potent behavioral effects of cannabinoids on hydra feeding and planarian locomotion, combined with high-affinity specific binding of cannabinoid radioligands to membranes prepared from these and other invertebrate species, suggest receptor-mediated mechanisms.

Identification of receptors responsible would represent a significant advance in understanding of the phylogenetic history of cannabinoid signaling. Perhaps such receptors will prove related to the postulated but as yet uncharacterized "CB_3" receptor in mammals (Fride et al. 2003).

Another interesting feature noted in studies of cannabinoid-altered feeding in both hydra and goldfish are biphasic, dose-dependent effects of the ECB, AEA. In both animals, low doses were more efficacious in altering feeding behavior than were higher dosages. In the case of hydra, 100 nM AEA produced greater inhibition of mouth opening than did 1 μM. In the case of goldfish, 1 pg g^{-1} stimulated feeding while 10 pg g^{-1} actually inhibited it. As ECBs have been shown to inhibit release of neurotransmitter at both excitatory (Takahashi and Castillo 2006) and inhibitory (Szabo et al. 2002) synapses, it is possible, if not likely, that differential ECB sensitivities of inhibitory and excitatory transmission may be responsible. This may be a profitable subject for future studies.

An interesting insight into potential behavioral relevance of ECB signaling systems is suggested by newt courtship clasping and zebra finch food restriction studies. In the case of the newt experiments, reversal of the inhibitory effects of stress and adrenal steroids by pretreatment with a CB_1 receptor antagonist suggests that ECB signaling may be important to allow redirection of behavior from one less (reproduction) to more critical (dealing with the emergent situation responsible for stress). A similar behavioral redirection is suggested by the finding of increased 2AG in the brains of food-deprived zebra finches. Increased ECB levels caused by food-deprivation are associated with reduced singing, suggesting that ECB signaling in the zebra finch may be important for allowing redirection of vocal behavior to a more critical search for food. It will be interesting to see if evidence accumulates supporting similar "behavior redirection" roles for ECB signaling in mammalian species.

References

Alonso-Ferrero ME, Paniagua MA, Mostany R et al. A (2006) Cannabinoid system in the budgerigar brain. Brain Res 1087:105–113

Balaban PM (2002) Cellular mechanisms of behavioral plasticity in terrestrial snail. Neurosci Biobehav Rev 26:597–630

Bass CE, Martin BR (2000) Time course for the induction and maintenance of tolerance to Delta (9)-tetrahydrocannabinol in mice. Drug Alcohol Depend 60:113–119

Berdyshev EV (1999) Inhibition of sea urchin fertilization by fatty acid ethanolamides and cannabinoids. Comp Biochem Physiol C Pharmacol Toxicol Endocrinol 122:327–330

Breitbart H (2003) Signaling pathways in sperm capacitation and acrosome reaction. Cell Mol Biol (Noisy-le-grand) 49:321–327

Brenner S, Elgar G, Sandford R et al. (1993) Characterization of the pufferfish (Fugu) genome as a compact model vertebrate genome. Nature 366:265–268

Bottjer SW, Johnson F (1997) Circuits, hormones, and learning: vocal behavior in songbirds. J Neurobiol 33:602–618

Buttarelli FR, Pontieri FE, Margotta V et al. (2002) Cannabinoid-induced stimulation of motor activity in planaria through an opioid receptor-mediated mechanism. Prog Neuropsychopharmacol Biol Psychiatry 26:65–68
Buttarelli FR, Pellicano C, Pontieri FE (2008) Neuropharmacology and behavior in planarians: translations to mammals. Comp Biochem Physiol C Toxicol Pharmacol 147(4):399–408
Cesa R, Mackie K, Beltramo M et al. (2001) Cannabinoid receptor CB1-like and glutamic acid decarboxylase-like immunoreactivities in the brain of *Xenopus laevis*. Cell Tissue Res 306:391–398
Cesa R, Guastalla A, Cottone E et al. (2002) Relationships between CB1 cannabinoid receptors and pituitary endocrine cells in *Xenopus laevis*: an immunohistochemical study. Gen Comp Endocrinol 125:17–24
Coddington E, Lewis C, Rose JD et al. (2007) Endocannabinoids mediate the effects of acute stress and corticosterone on sex behavior. Endocrinology 148:493–500
Cottone E, Forno S, Campantico E et al. (2005) Expression and distribution of CB1 cannabinoid receptors in the central nervous system of the African cichlid fish *Pelvicachromis pulcher*. J Comp Neurol 485:293–303
De Petrocellis L, Melck D, Bisogno T et al. (1999) Finding of the endocannabinoid signalling system in Hydra, a very primitive organism: possible role in the feeding response. Neuroscience 92:377–387
Dehal P, Satou Y, Campbell RK et al. (2002) The draft genome of *Ciona intestinalis*: insights into chordate and vertebrate origins. Science 298:2157–2167
Doupe AJ, Kuhl PK (1999) Birdsong and human speech: common themes and mechanisms. Annu Rev Neurosci 22:567–631
Earl AE, Winters RL, Schneider CM (1955) Assay of reserpine based on emesis in pigeons. J Pharmacol Exp Ther 115:55–60
Egertova M, Elphick MR (2007) Localization of CiCBR in the invertebrate chordate *Ciona intestinalis*: evidence of an ancient role for cannabinoid receptors as axonal regulators of neuronal signalling. J Comp Neurol 502:660–672
Elphick MR (1998) An invertebrate G-protein coupled receptor is a chimeric cannabinoid/melanocortin receptor. Brain Res 780:170–173
Elphick MR, Satou Y, Satoh N (2003) The invertebrate ancestry of endocannabinoid signalling: an orthologue of vertebrate cannabinoid receptors in the urochordate *Ciona intestinalis*. Gene 302:95–101
Fan SF, Yazulla S (2003) Biphasic modulation of voltage-dependent currents of retinal cones by cannabinoid CB1 receptor agonist WIN 55212-2. Vis Neurosci 20:177–188
Feigenbaum JJ, Richmond SA, Weissman Y et al. (1989) Inhibition of cisplatin-induced emesis in the pigeon by a non-psychotropic synthetic cannabinoid. Eur J Pharmacol 169:159–165
Fride E, Foox A, Rosenberg E et al. (2003) Milk intake and survival in newborn cannabinoid CB1 receptor knockout mice: evidence for a "CB3" receptor. Eur J Pharmacol 461:27–34
Frischknecht HR, Waser PG (1978) Actions of hallucinogens on ants (Formica pratensis)–II. Effects of amphetamine, LSD and delta-9-tetrahydrocannabinol. Gen Pharmacol 9:375–380
Frischknecht HR, Waser PG (1980) Actions of hallucinogens on ants (Formica pratensis) – III. Social behavior under the influence of LSD and tetrahydrocannabinol. Gen Pharmacol 11:97–106
Gonzalez SC, Matsudo VK, Carlini EA (1971) Effects of marihuana compounds on the fighting behavior of Siamese fighting fish (*Betta splendens*). Pharmacology 6:186–909
Gouvier WD, Akins FR, Trapold MA (1984) Assessment of drug state dimensionality via drug-drug training and stimulus generalization testing. Pharmacol Biochem Behav 21:687–693
Henriksson BG, Johansson JO, Jarbe TU (1975) Delta 9-tetrahydrocannabinol produced discrimination in pigeons. Pharmacol Biochem Behav 3:771–774
Hollis DM, Coddington EJ, Moore FL (2006) Neuroanatomical distribution of cannabinoid receptor gene expression in the brain of the rough-skinned newt, *Taricha granulosa*. Brain Behav Evol 67:135–149

Jarbe TU, Hiltunen AJ (1988) Limited stimulus generalization between delta 9-THC and diazepam in pigeons and gerbils. Psychopharmacology (Berl) 94:328–331

Johnson F, Rashotte M (2002) Food availability but not cold ambient temperature affects undirected singing in adult male zebra finches. Physiol Behav 76:9–20

Kirkham TC (2005) Endocannabinoids in the regulation of appetite and body weight. Behav Pharmacol 16:297–313

Kirkham TC, Williams CM, Fezza F et al. (2002) Endocannabinoid levels in rat limbic forebrain and hypothalamus in relation to fasting, feeding and satiation: stimulation of eating by 2-arachidonoyl glycerol. Br J Pharmacol 136:550–557

Lemak MS, Bravarenko NI, Bobrov MY et al. (2007) Cannabinoid regulation in identified synapse of terrestrial snail. Eur J NeuroSci 26:3207–3214

Levine MW (2007) Variability in the firing of retinal ganglion cells of goldfish: a review. Vis Neurosci 24:239–246

Matias I, Bisogno T, Melck D et al. (2001) Evidence for an endocannabinoid system in the central nervous system of the leech *Hirudo medicinalis*. Brain Res Mol Brain Res 87:145–159

Matias I, McPartland JM, Di Marzo V (2005) Occurrence and possible biological role of the endocannabinoid system in the sea squirt *Ciona intestinalis*. J Neurochem 93:1141–1156

McMillan DE (1990) The pigeon as a model for comparative behavioral pharmacology and toxicology. Neurotoxicol Teratol 12:523–529

McPartland J, Di Marzo V, De Petrocellis L et al. (2001) Cannabinoid receptors are absent in insects. J Comp Neurol 436:423–429

McPartland JM, Agraval J, Gleeson D et al. (2006) Cannabinoid receptors in invertebrates. J Evol Biol 19:366–373

Neill AT, Vacquier VD (2004) Ligands and receptors mediating signal transduction in sea urchin spermatozoa. Reproduction 127:141–149

Nijsen MJ, Croiset G, Diamant M et al. (2001) CRH signalling in the bed nucleus of the stria terminalis is involved in stress-induced cardiac vagal activation in conscious rats. Neuropsychopharmacology 24:1–10

Orchinik M, Murray TF, Moore FL (1991) A corticosteroid receptor in neuronal membranes. Science 252:1848–1851

Pierobon P, Concas A, Santoro G et al. (1995) Biochemical and functional identification of GABA receptors in *hydra vulgaris*. Life Sci 56:1485

Pierobon P, De Petrocellis L, Minei R et al. (1997) Arachidonic acid and eicosanoids in Hydra: possible endogenous signals involved in chemoreception and modulation of the feeding behaviour. Adv Exp Med Biol 433:363–366

Rawls SM, Gomez T, Raffa RB (2007) An NMDA antagonist (LY 235959) attenuates abstinence-induced withdrawal of planarians following acute exposure to a cannabinoid agonist (WIN 55212-2). Pharmacol Biochem Behav 86:499–504

Rose JD, Moore FL (1999) A neurobehavioral model for rapid actions of corticosterone on sensorimotor integration. Steroids 64:92

Sanudo-Pena MC, Romero J, Seale GE et al. (2000) Activational role of cannabinoids on movement. Eur J Pharmacol 391:269

Schuel H, Schuel R, Zimmerman AM et al. (1987) Cannabinoids reduce fertility of sea urchin sperm. Biochem Cell Biol 65:130–136

Schuel H, Goldstein E, Mechoulam R et al. (1994) Anandamide (arachidonylethanolamide), a brain cannabinoid receptor agonist, reduces sperm fertilizing capacity in sea urchins by inhibiting the acrosome reaction. Proc Natl Acad Sci USA 91:7678–7682

Scharff C, Nottebohm, F (1991) A comparative study of the behavioral deficits following lesions of various parts of the zebra finch song system: implications for vocal learning. J Neurosci 11:2896–2913

Soderstrom K, Johnson F (2000) CB1 cannabinoid receptor expression in brain regions associated with zebra finch song control. Brain Res 857:151–157

Soderstrom K, Johnson F (2001) The zebra finch CB1 cannabinoid receptor: pharmacology and in vivo and in vitro effects of activation. J Pharmacol Exp Ther 297:189–197
Soderstrom K, Tian Q (2006) Developmental pattern of CB1 cannabinoid receptor immunoreactivity in brain regions important to zebra finch (*Taeniopygia guttata*) song learning and control. J Comp Neurol 496:739–758
Soderstrom K, Leid M, Moore FL et al. (2000) Behavioral, pharmacological and molecular characterization of an amphibian cannabinoid receptor. J Neurochem 75:413–423
Soderstrom K, Tian Q, Valenti M et al. (2004) Endocannabinoids link feeding state and auditory perception-related gene expression. J Neurosci 24:10013–10021
Soderstrom K, Johnson F (2003) Cannabinoid exposure alters learning of zebra finch vocal patterns. Brain Res Dev Brain Res 142:215–217
Soderstrom K, Tian Q (2004) Distinct periods of cannabinoid sensitivity during zebra finch vocal development. Developmental Brain Research 153:225–232
Stefano GB, Salzet B, Salzet M (1997) Identification and characterization of the leech CNS cannabinoid receptor: coupling to nitric oxide release. Brain Res 753:219–224
Straiker A, Stella N, Piomelli D et al. (1999) Cannabinoid CB1 receptors and ligands in vertebrate retina: localization and function of an endogenous signaling system. Proc Natl Acad Sci USA 96:14565–14570
Struik ML, Yazulla S, Kamermans M (2006) Cannabinoid agonist WIN 55212-2 speeds up the cone response to light offset in goldfish retina. Vis Neurosci 23:285–293
Szabo B, Siemes S, Wallmichrath I (2002) Inhibition of GABAergic neurotransmission in the ventral tegmental area by cannabinoids. Eur J NeuroSci 15:2057–2061
Takahashi KA, Castillo PE (2006) The CB1 cannabinoid receptor mediates glutamatergic synaptic suppression in the hippocampus. Neuroscience 139:795–802
Turkanis SA, Karler R (1986) Effects of delta-9-tetrahydrocannabinol, 11-hydroxy-delta-9-tetrahydrocannabinol and cannabidiol on neuromuscular transmission in the frog. Neuropharmacology 25:1273–1278
Valenti M, Cottone E, Martinez R et al. (2005) The endocannabinoid system in the brain of *Carassius auratus* and its possible role in the control of food intake. J Neurochem 95:662–672
Van der Kloot W (1994) Anandamide, a naturally-occurring agonist of the cannabinoid receptor, blocks adenylate cyclase at the frog neuromuscular junction. Brain Res 649:181–184
Venturini G, Stocchi F, Margotta V et al. (1989) A pharmacological study of dopaminergic receptors in planaria. Neuropharmacology 28:1377–1382
Weiss CS, Coughlin JP (1979) Maintained aggressive behavior in gonadectomized male Siamese fighting fish (*Betta splendens*). Physiol Behav 23:173–177

Part II
Physiology and Pathophysiology

Roles of the Endocannabinoid System in Learning and Memory

Giovanni Marsicano and Pauline Lafenêtre

Contents

Abstract The endocannabinoid system (ECS) plays a central role in the regulation of learning and memory processes. The fine-tuned regulation of neural transmission by the system is likely to be the mechanism underlying this important function. In this chapter, we review the data in the literature showing the direct involvement of the physiological activation of cannabinoid receptors in the modulation of different forms of learning and memory. When possible, we also address the likely mechanisms of this involvement. Finally, given the apparent special role of the ECS in the extinction of fear, we propose a reasonable model to assess how neuronal networks could be influenced by the endocannabinoids in these processes. Overall, the data reviewed indicate that, despite the enormous progress of recent years, much is still to be done to fully elucidate the mechanisms of the ECS influence on learning and memory processes.

G. Marsicano (✉) and P. Lafenêtre
Group Molecular Mechanisms of Behavioural Adaptation, Research Centre INSERM U862 NeuroCentre Magendie Université Bordeaux 2, 146, rue Léo Saignat, 33077, Bordeaux, France
e-mail: marsicano@bordeaux.inserm.fr

D. Kendall and S. Alexander (eds.), *Behavioral Neurobiology of the Endocannabinoid System*, Current Topics in Behavioral Neurosciences 1,
DOI: 10.1007/978-3-540-88955-7_8,

Keywords Learning and memory • Endocannabinoid system • CB1 receptors • Extinction • Amygdala • Model

Abbreviations

AEA	Anandamide
BA	Basal nucleus of the amygdala
BLA	Basolateral amygdala
CaMKII	Calcium/calmodulin protein kinase II
CE	Central nucleus of the amygdale
CR	Conditioned response
CS	Conditioned stimulus
DSI	Depolarization-induced suppression of inhibition
ECS	Endocannabinoid system
FAAH	Fatty acid amide hydrolase
GPCR	G protein-coupled receptor
ICM	Intercalated cell masses
LA	Lateral nucleus of the amygdala
LTP	Long-term synaptic potentiation
PFC	Prefrontal cortex
PI3K	Phosphatidylinositol 3-kinase
THC	Δ^9-Tetrahydrocannabinol
US	Unconditioned stimulus

1 Introduction

Marijuana has been used for its psychotropic effects for centuries. It was, however, only during the last decades that this topic attracted scientific interest. The discovery of the active compound of the plant (Gaoni and Mechoulam 1964), the cloning of cannabinoid receptors (Matsuda et al. 1990), the synthesis of receptor antagonists (Rinaldi-Carmona et al. 1994), the identification of endogenously produced ligands of cannabinoid receptors (Devane et al. 1992; Mechoulam et al. 1995; Sugiura et al. 1995), the identification of endocannabinoids as retrograde synaptic signalling molecules (Maejima et al. 2001; Wilson and Nicoll 2001) and the first use of cannabinoid-interfering drugs in clinical trials (Pagotto et al. 2006; Van Gaal et al. 2005) represent scientific milestones in the study of cannabinoids over the last decades. Accordingly, Fig. 1 shows the evolution of the rate of published papers during the last 50 years in relation to these key discoveries, which have boosted new waves of interest. Nowadays, the number of scientific publications on the subject has achieved an exponential rate of growth, further strengthening the enormous

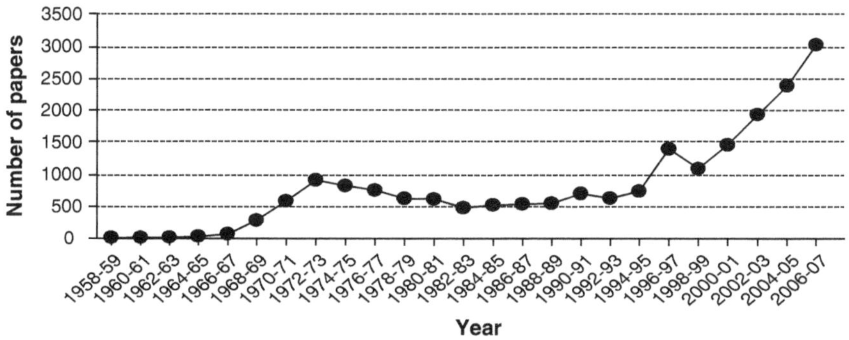

Fig. 1 Number of publications on cannabinoids during the last 50 years. Data obtained from PubMed (www.ncbi.nlm.nih.gov/pubmed/), using the following search keywords: cannabinoid OR marijuana OR marihuana OR tetrahydrocannabinol OR cannabis. Key discoveries are indicated: (1) Purification of Δ^9-tetrahydrocannabinol (THC) (Gaoni and Mechoulam 1964); (2) Cloning of CB_1 receptors (Matsuda et al. 1990); (3) Identification of anandamide (Devane et al. 1992); (4) Synthesis of the first CB_1 antagonist (Rinaldi-Carmona et al. 1994); (5) ECS and retrograde synaptic transmission (Maejima et al. 2001;Wilson and Nicoll 2001); (6) First clinical trial with cannabinoid antagonists (Van Gaal et al. 2005)

interest that the cannabinoid field elicits in the scientific community. The most important passages in this history were the ones related to the definition of the endocannabinoid system (ECS). The discovery of cannabinoid receptors, of their endogenous lipid ligands (endocannabinoids) and of the machinery to synthesise and degrade endocannabinoids led to the identification of an endogenous signalling system, which is emerging as a very important element in mammalian physiology and in learning and memory processes.

2 Expression of Cannabinoid Receptors in the Brain: Focus on CB_1

Evidence for the presence in neurons of a second cannabinoid receptor, CB_2, has been shown (Van Sickle et al. 2005) and other possible targets of endocannabinoids (e.g. vanilloid receptors TRPV1, GPR55, potassium channels TASKs) are very likely involved in the functions of the ECS brain in the brain (Di Marzo et al. 2002). However, most studies dealing with the roles of the ECS in learning and memory refer to the endogenous activation of the cannabinoid receptor type 1 (CB_1). Hence, this chapter will deal with the roles of the ECS in learning and memory with special emphasis on CB_1 receptors. CB_1 is a seven transmembrane G protein-coupled receptor (GPCR) that is expressed at very high levels in the brain. It has been calculated that the protein levels of CB_1 in the brain are a good deal higher than other GPCRs and that they are comparable to those of NMDA or $GABA_A$ receptors (Howlett et al. 2002). CB_1 is expressed in many different brain regions (including most of the ones classically involved in learning and memory processes), and it is present in different neuronal subpopulations (Marsicano and Lutz 1999). CB_1 is generally expressed presynaptically (Freund et al. 2003), although the evidence for its presence at the somatodendritic level is steadily growing (Bacci et al. 2004). Activation of CB_1 generally leads to a hyperpolarisation of neuronal membranes and to the stimulation of different intracellular signalling cascades (Straiker and Mackie 2007). Therefore, activation of CB_1 generally causes an inhibition of neurotransmitter release (Freund et al. 2003). Strong evidence exists that endocannabinoids released postsynaptically signal retrogradely to presynaptic CB_1 receptors, thereby providing one of the best described retrograde synaptic signalling systems in the CNS. First discovered in the hippocampus and cerebellum, this mode of endocannabinoid signalling has since been extended to the great majority of the brain regions, demonstrating the universality of the ECS as a modulatory system in the CNS. The complexity of the picture derives from the fact that the ECS can regulate, via $CB_{1,}$ the release of neurotransmitters that have very different and even opposite effects. The most striking example is the fact that the ECS can regulate both inhibitory GABAergic and excitatory glutamatergic transmission in the same brain regions. These diverse actions, together with the concept of an "on demand" synthesis and release of endocannabinoids (as described in other chapters), reveal a very finely regulated way of functioning of the ECS. In fact, it is believed that, during specific functions, the ECS might finely modulate several aspects of neuronal signalling at the same time.

In addition, it should be remembered that CB_1 is not only expressed in the central nervous system but also at lower levels, in the retina and peripheral tissues, including the peripheral nervous system, adipocytes, hepatocytes, pancreatic cells and the gastrointestinal tract (Pacher et al. 2006). Though these organs and tissues are not classically considered in learning and memory experiments, the possible confounding effects of CB_1 expression in these extra-CNS sites should be kept in mind. For instance, CB_1 present in the retina should be considered when visual stimuli are used to study learning processes. Similarly, the presence of CB_1 in the gastrointestinal tract might play a role, when conditioned taste aversion experiments are performed (see below). Nevertheless, CB_1 is the main cannabinoid receptor involved in the functions of the ECS in the brain and this chapter will focus almost exclusively on its endogenous activation by endocannabinoids during learning and memory processes.

3 The ECS and Learning and Memory

The ability to acquire, store, retrieve and modify information concerning previous experience is a crucial function for individuals to survive and reproduce. Neuronal adaptation has been correlated with the changes in behaviour observed in response to all stages of memory. In the light of the enormous theoretical and experimental progress in the understanding of these neuronal systems (Kandel et al. 2000; Squire et al. 2008), we will modestly limit this review to the specific role of the endocannabinoid system in learning and memory. We will consider learning in terms of perception of environmental changes and the subsequent changes in intercellular communication in the brain, and memory as the relative persistence of these changes. This review is directed towards the involvement of the ECS in these processes referring to a classical and certainly oversimplified classification of the time-dependent types of memories, i.e. short-term and long-term. Concerning the latter, we will follow the classical declarative-like, procedural and emotional classification with a special focus on the different phases of these processes: encoding of information (learning), consolidation, retrieval (or recall), re-consolidation and extinction. Considering the major role of the ECS in aversive memories, we will finally propose a theoretical model, which may explain the functions of the system in the processing and extinction of aversive memories.

4 Pharmacology of Exogenous Cannabinoid Agonists and Physiology of the ECS: Important Differences in Complex Functions

There is an important body of literature on the effects of natural or synthetic cannabinoid exogenous agonists (as THC, HU210, CP55940, WIN55, 212-2) on learning and memory processes. The administration of these ligands of the CB_1 receptors obviously disturbs the physiological activation of the ECS but does not always reflect a potentiation of the role of the endocannabinoids. There are indeed some physiological functions of the ECS where exposure to cannabinoid agonists simply exerts an additive effect on the physiological action of endocannabinoids, e.g. in peripheral inflammation (Massa et al. 2004). However, in complex brain processes, the effects of cannabinoid agonists rarely mimic the actions of the ECS in physiological conditions. For instance, cannabinoid agonists, including marijuana intoxication in humans and synthetic agonists in animals, are very well known to induce important alterations mainly in short-term episodic and working memory (Ranganathan and D'Souza 2006). In rats, working memory is deeply impaired by cannabinoid agonists but only sparse evidence reports that the ECS would physiologically negatively regulate working memory (Deadwyler et al. 2007; Deadwyler and Hampson 2008) by modulating the spontaneous release of endocannabinoids. However, Carter and Wang (2007b) recently suggested that the physiological activation of the ECS might even be beneficial to working memory mechanisms.

It is very difficult to account for the reasons for these apparent discrepancies between pharmacological treatments with cannabinoid agonists and the physiological roles of the ECS. Differences between pharmacological properties of exogenous cannabinoids and endocannabinoids could account for this. Indeed, the prototypical exogenous cannabinoid agonist Δ^9-tetrahydrocannabinol (THC) is only a partial agonist of CB_1 receptors in vitro (Pertwee 2008) and was recently proposed to act as an antagonist in vivo (Straiker and Mackie 2007). However, this possible explanation might be valid for THC, but it is more difficult to extend to other exogenous synthetic agonists, which are very powerful full agonists of CB_1 and share with THC many pharmacological effects on animals, including on learning and memory. It is, therefore, more likely that the discrepancy relies on the different spatial and temporal ways of activation of cannabinoid receptors by exogenously applied and endogenously released cannabinoids. Systemic or local intracerebral administration of agonists leads to a generalised (in the whole body or in a brain region, respectively) activation of cannabinoid receptors. The duration of this activation depends only on the pharmacokinetics of the drug and can last for hours or days. Conversely, the activation of the ECS, as described in other chapters in this volume, is tightly regulated in terms of space and time. It is believed that few cells might be under the control of the physiological activation of the ECS whereas the neighbouring ones might remain outside the influence of endocannabinoids. Moreover, the time of activation of the ECS can be very short (in the range of a few hundred milliseconds to a few minutes) thanks to very efficient systems for the degradation of endocannabinoids. It is, therefore, possible that, during specific ongoing neuronal activity such as the processing of memory traces, the ECS is activated in a limited number of brain regions (possibly only involving a few cells) within a very short time frame. This would explain why a generalised and long-lasting action of exogenously administered cannabinoid agonists does not usually reproduce the physiological functions of the ECS and can even exert *opposite* effects.

An example of this view can be found in a typical electrophysiological function of the ECS, the depolarization-induced suppression of inhibition (DSI) in the hippocampus, one of the most important brain regions for memory processing (Alger 2002). A short depolarisation of hippocampal pyramidal neurons induces the postsynaptic production and release of endocannabinoids, which retrogradely travel across the synaptic cleft and activate CB_1 receptors present on the presynaptic terminals of GABAergic inhibitory interneurons (Alger 2002; Freund et al. 2003). This activation induces, in turn, a short-lasting (from a few seconds to a few minutes) inhibition of GABA release, thereby temporarily releasing the pyramidal cell from the GABAergic inhibitory tone. Importantly, it was shown that such a short-term form of synaptic plasticity can facilitate the induction of long-term synaptic potentiation (LTP) of the pyramidal neurons (Carlson et al. 2002). As LTP is the best acknowledged cellular model of long-term storage of information in the brain, it is very likely that DSI might somehow participate in memory processing in the hippocampus. The first evidence that the ECS is the main system responsible for the retrograde signalling component of DSI in the hippocampus derives from the seminal and simultaneous experiments of two independent groups. In both of these series of experiments, DSI was shown to be blocked by the application of a CB_1 antagonist and to be "occluded" by the application of a CB_1 agonist (Maejima et al. 2001; Wilson and Nicoll 2001; Fig. 2). Whereas the blockade of a phenomenon by a receptor antagonist is self-explanatory, the concept of "occlusion" by an agonist is

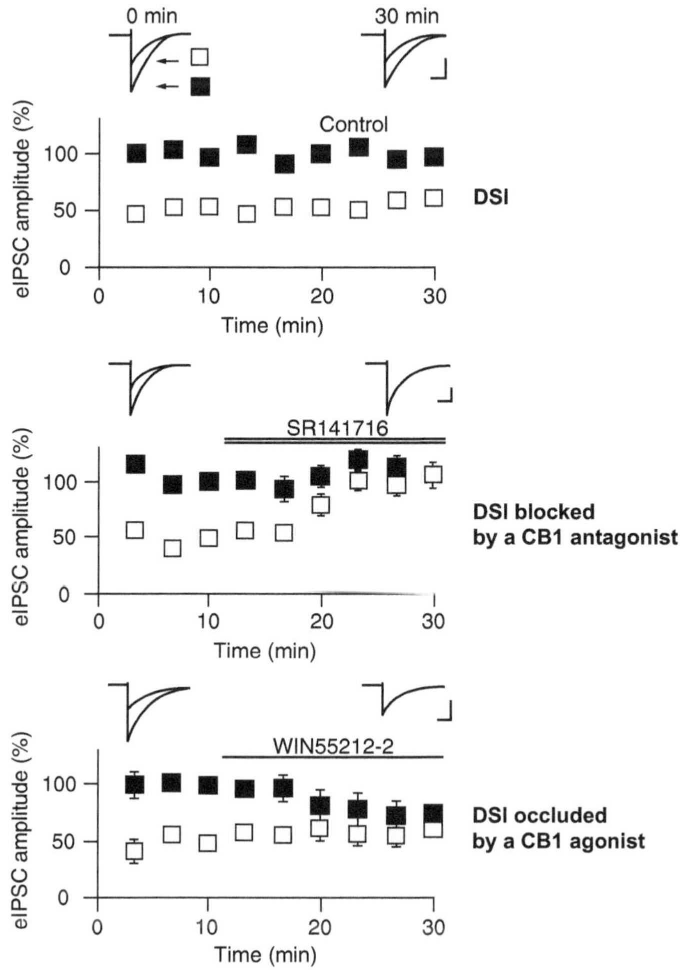

Fig. 2 Depolarisation-induced depression of inhibition (DSI). Endocannabinoids released from postsynaptic CA1 pyramidal hippocampal neurons induce a transient depression of GABAergic release from inhibitory interneurons. The specificity of the involvement of the ECS in this phenomenon was shown by (**a**) the blockade of DSI by a CB_1 antagonist and (**b**) its occlusion by a CB_1 agonist. These data indicate that the possible physiological effects of DSI in vivo are likely abolished both by antagonists and agonists treatments. Data reproduced with permission from Wilson and Nicoll (2001)

less intuitive. By "occlusion", it is meant that the application of an agonist compound, by occupying the available receptors, practically impedes the endogenously released endocannabinoids from binding to the same receptors. In other words, independently from the intrinsic effect of the agonist (which will induce a long-lasting decrease of GABAergic transmission), the spatially and temporally restricted action of endocannabinoids during DSI is abrogated in presence of exogenous CB_1

agonists (Fig. 2). We can assume that DSI does, indeed, participate in some forms of memory processes and its temporally and spatially restricted occurrence plays a central role in its function in living animals. Thus, if we treat animals with CB_1 agonists, the occurrence of DSI at the right place and at the right time will be impeded, in a very similar manner as with a CB_1 antagonist. It is, therefore, possible that pharmacological applications of agonists and antagonists result in very similar impairments of complex brain functions such as learning and memory.

In this chapter, we will omit the description of the effects of cannabinoid agonists on memory processes even though most of them were blocked by the administration of a CB_1 receptor antagonist. We direct interested readers to extensive reviews of the subject (Castellano et al. 2003; Fride 2005; Fujiwara and Egashira 2004; Iversen 2003a; Riedel and Davies 2005). Thus, we will focus our attention on experiments in which a clear physiological involvement of the ECS has been demonstrated. The direct approaches to showing such an involvement can be summarised in three experimental paradigms: the use of cannabinoid antagonists alone, the use of mutant animals bearing alterations in the expression of genes involved in the ECS (up to now, CB_1 and FAAH), and the use of drugs able to inhibit endocannabinoid degradation. In the latter case, the amount of endocannabinoids available to exert the functions of the ECS will be increased (as described in "The Life Cycle of the Endocannabinoids: Formation and Inactivation", the chapter by Stephen P.H. Alexander and David A. Kendall, this volume). This approach presents the advantage, as compared to the application of direct cannabinoid agonists to maintain the temporal and spatial range of activation of the ECS during physiological processes: only where and when endocannabinoids are synthesised to exert a specific function, are these drugs able to slightly increase and/or prolong their activity.

The important role of the ECS in learning and memory has been already assessed in a variety of behavioural tests. We will first mention again the role of the ECS in short-term working memory, then turn to its involvement in different kinds of long-term memories: declarative-like, procedural and emotional. Given the enormous amount of data present in the literature on the subject and the space limitation of this chapter, we apologise for possible omissions.

5 Involvement of the Endocannabinoid System in Learning and Memory

5.1 *The ECS and Working Memory*

Working memory is a form of memory that allows access to labile information for a short period of time. In humans, a trivial example of working memory is the ability to remember a phone number from the time of reading it in the directory to actually dialling it. Natural and synthetic cannabinoid agonists are known to negatively regulate working memory in humans (for reviews, see Iversen 2003b; Ranganathan and D'Souza 2006) and in animal models (for reviews, see Castellano et al. 2003; Riedel and Davies 2005). However, there is little evidence that the physiological action of endocannabinoids could induce such impairments. While Mallet and Beninger (1998) did not observe any effect of the CB_1 antagonist rimonabant on a non-match-to-position task in rats (Mallet and Beninger 1998), rats injected with rimonabant performed better in a delayed non-match-to-sample short-term memory task (Deadwyler et al. 2007) by reducing the influence of the biases encoded by the endocannabinoids in the hippocampus (Deadwyler and Hampson 2008). However, it has also been suggested that working memory would be enhanced by endocannabinoids in other tests. Indeed, Carter and Wang (2007) proposed that DSI can counteract the time-dependent decrease of accuracy of working memory in a model of spatial working memory (Carter and Wang 2007). Additional evidence comes from the use of FAAH-null mice lacking fatty acid amide hydrolase, the main enzyme involved in the degradation of the endocannabinoid anandamide (AEA). Indeed, these mice, which have tenfold increased brain levels of AEA and other fatty acid amide substrates (Cravatt et al. 2001) did not show any deficits in a working memory water maze task and even performed better on the first training session than their wild-type littermates (Cravatt et al. 2001; Varvel et al. 2006). This would argue that increased levels of endocannabinoids are not detrimental to working memory. In summary, existing studies strongly suggest that the ECS is somehow involved in working memory but it is still under discussion exactly how it affects this kind of processing.

As mentioned above, several phases of learning and memory processes can be identified. Classical theories propose that the acquisition of new information induces changes in neuronal connections that are supposed to encode a memory trace, somehow representing that information as subjectively perceived by the subject (Bailey et al. 2004; Barco et al. 2006; Dunning and During 2003; Kandel 2001). This memory trace is very labile (i.e. it can be easily disrupted through several treatments) at the beginning, and its strength increases through a process called consolidation (McGaugh 2000; Nader et al. 2000). Through consolidation, the memory is stored through long-lasting changes in neuronal connectivity and will be "retrieved" when the individual eventually needs it. More recent theories propose further complexity to the process. Every time consolidated memories are recalled, they probably switch again to a labile state. "Reactivated" traces can again be disrupted during this process. However, depending on the conditions of retrieval and on the strength of the original trace, "reactivated" memories can undergo two opposite processes. These are called "re-consolidation", when the conditions favour the permanence of the trace (Debiec et al. 2002; Duvarci and Nader 2004; Sara 2000), and "extinction", when the conditions indicate that the memory has no reason to persist any longer (Myers and Davis 2002; Quirk and Mueller 2008). Dissecting the neuronal mechanisms of these phases is one of the major goals of learning and memory research. In the specific case of the ECS, the tools described above (genetic and pharmacological) each present advantages and disadvantages. Genetic mutations are generally more specific than pharmacological tools, which are always liable to the risk of non-specific effects. However, any alteration in performance of mutant mice in test trials could be ascribed to a deficit during virtually any of the phases of learning and memory. On the other hand, pharmacological treatments could better differentiate these phases (acquisition, consolidation, retention, retrieval, and re-consolidation/extinction). These considerations should be kept in mind in the analysis of the different roles of the ECS in learning and memory processes.

5.2 *The ECS and Long-Term Memories*

5.2.1 The ECS and "Declarative"-Type Memories

Declarative memory refers to learned facts and information that are flexible and can be accessed consciously. It is also referred to as explicit memory. It is obviously quite difficult to measure "declarative" memory in animals. However, many scientists have developed different tasks in order to assess the "memory of facts" even in very simple organisms. In this chapter, we will review the role of the endocannabinoid system in two types of this form of memory: recognition memory and spatial memory.

5.2.2 The ECS and Recognition Memory

The first evidence of the involvement of the ECS in learning and memory comes from a non-procedural recognition memory task. Recognition memory is based on the innate preference of rodents for exploring novel places, objects or congeners instead of re-exploring something they have already been exposed to, referred to as *familiar*. The two most common protocols used in behavioural neurosciences are the object recognition and the social recognition tasks. The pharmacological blockade of CB_1 receptors resulted in a facilitation of short-term olfactory memory in a social recognition memory task (Terranova et al. 1996) and the administration of rimonabant reduced the deficits observed in aged rats and mice (Terranova et al. 1996). Two CB_1-null lines obtained from different genetic backgrounds were later examined at different time intervals according to the age of the animals. While CB_1-null mice performed better than the wild-type controls at younger age (6–8 weeks old), they were impaired during adulthood (3–5 months old), suggesting that the genetic blockade of CB_1 receptors *improves* social recognition memory in young animals, but it is deleterious at older ages (Bilkei-Gorzo et al. 2005). These data might set a link between age, memory performance and activity of the ECS. However, the absence of wild-type control littermates in this study did not permit the conclusion of whether the observed age-related changes were due to an intrinsic effect of CB_1 deletion in the tested mice or to altered embryonic and postnatal development due to the fact that the mothers of these animals were also full

CB_1-null mutants. Indeed, administration of rimonabant did not induce any effects in CD1 mice nor did AM251 in rats in the object recognition paradigm (Bura et al. 2007; Clarke et al. 2008).

Reibaud et al. (1999) extended these recognition memory results by assessing CB_1-null mice in an object recognition test. Mutant mice were able to retain the memory of a known object for at least 48 h whereas the wild-type controls did not remember the familiar object after 24 h. It was again suggested that this faculty could be age-dependent, because 1-month-old and 4-month-old CB_1-null mice explored selectively the novel object 24 h after the first exposure, but the difference from their wild-type controls was more obvious at the younger age (Maccarrone et al. 2002).

5.2.3 The ECS and Spatial Memory

The most frequently used cognitive tests for assessing declarative-like memories in rodents are spatial memory tasks. In a delay version of the radial-maze, in which rats have to remember and go to the only baited arm which was blocked on the previous trial, Aaron Lichtman found that the administration of the CB_1 antagonist rimonabant before the acquisition phase could improve the performance of the rats. Interestingly, no effect of the drug was observed when it was administered immediately after the acquisition phase nor before the retention test (Lichtman 2000), suggesting an involvement of the ECS in the learning phase of this task. The same group recently confirmed that rimonabant injection before, but not immediately after, acquisition, or before retrieval of the same task was able to decrease the number of errors of the rats even when the first exposition to the maze and the re-exposition were separated by a longer delay. This suggests a beneficial role of the ECS in choice accuracy. Moreover, the effect of the CB_1 antagonist was synergistic with that of an inhibitor of acetylcholinesterase, suggesting that the ECS and the cholinergic system may interact in this form of learning (Wise et al. 2007). Another version of the 8-arm-radial maze requires the rats to encode more information: during the test session, four arms that were baited during the acquisition phase are to be avoided, whereas the four other previously blocked arms are to be visited. In these conditions, the blockade of CB_1 receptors immediately after the acquisition phase could improve consolidation processes (Wolff and Leander 2003). The role of the ECS has repeatedly been assessed in the most commonly used behavioural paradigm to test spatial memory, the Morris Water Maze. In this test, rodents have to learn to find a hidden platform in a pool by using the surrounding spatial cues. While CB_1-null mice learned the task as well and as fast as the wild-type controls, they were impaired in the reversal learning phase where the platform had been moved to another location in the pool. Indeed, they repeatedly went to the previous location showing increased and non-adapted perseverance and a significant deficit in learning the new location, suggesting that the ECS is involved in extinction and/or forgetting processes (Varvel and Lichtman 2002). Consequently, the behaviour of rimonabant-treated mice and of CB_1-null mice was further analysed in two extinction procedures in order to differentiate these two processes. In the massed extinction procedure with many trials in a short period of time (20 in 5 days), no effect on the extinction was observed following either genetic or pharmacological blockade of the CB_1 receptor. On the contrary, neither the rimonabant-treated mice nor the CB_1-null mice exhibited a proper extinction in the spaced extinction procedure with five probe-tests over several months, suggesting a role for the ECS in the suppression of unadapted behaviours (Varvel et al. 2005). Robinson et al. (2008) found that intraperitoneal administration of rimonabant affected the learning abilities of rats, whereas the intrahippocampal infusion of the antagonist led to an enhanced acquisition but had no direct effect on consolidation processes, although they kept returning to the previous platform location up to seven days after reversal training. These data again confirmed an important involvement of the ECS in this form of declarative memory. Another way to better characterise the role of the ECS in spatial learning is by using FAAH-null mice that exhibit much higher levels of AEA and other fatty acid amides or by pharmacologically inhibiting this enzyme (e.g. with OL-135). However, no marked effects of these treatments were found in any versions of the task. FAAH-null mice even acquired the working memory procedure faster in the first session, whereas the administration of OL135 did not lead to a better acquisition but to a better extinction rate (Varvel et al. 2007). Additional evidence of the role of the ECS in spatial memory was observed in food-storing black-capped chickadees. When they were intrahippocampally infused with rimonabant, their long-term performance was improved but they could not recall the most recent reward location when the reward was moved (Shiflett et al. 2004).

5.3 The ECS and Procedural Memory

In contrast to declarative memory, procedural memory is an implicit memory that can be assessed by performance rather than by conscious recollection. It deals with the knowledge of "how" tasks should be performed, and it relies on processing like priming, conditioning (when subjects learn associations between two stimuli or a stimulus and an action) and skill learning (as riding a bicycle).

5.3.1 The ECS and Operant Conditioning

As the ECS is involved in functions other than learning and memory, such as emotions and reward, one should keep in mind the possible effects of the pharmacological and/or genetic treatments on locomotion and motivation of the animals to perform the task. This is particularly true in operant conditioning tests, where subjects associate a stimulus (e.g. a light) with a learned action (e.g. lever pressing or nose-poking) to obtain a reward (e.g. food). In order to improve learning rates and to overcome motivational problems, animals are generally food restricted or presented with palatable food as a reward, and this can obviously be a confounding factor, considering the important roles of the ECS in reward and energy balance (as described in other chapters of the present book). Indeed, in one of the earliest experiments aimed at studying the effects of the rimonabant on the performance of monkeys in an operant paradigm, the authors observed a decrease of overall response rate, but no changes in the percentage of errors, suggesting a predominant effect of the drug on the motivation to perform the task over the learning process itself (Winsauer et al. 1999). Mallet and Beninger had previously found no effect of this antagonist alone in a conditional discrimination task (Mallet and Beninger 1998). However, it was more recently shown that mature 3–5-month-old CB_1-null mice needed more time to learn the task as compared to younger CB_1-null mice (6–8 weeks), independently from the hedonic value of the reward (Bilkei-Gorzo et al. 2005) for which the motivational aspects were not altered (Soria et al. 2005). Mutant mice lacking the expression of CB_1 were also evaluated in another version of operant conditioning, the five-hole-choice task. Despite evidence for a lack of motivation, the mutant mice performed as well as their wild-type littermates. Moreover, in this task, both genotypes showed a similar decline when the reward was no longer delivered, suggesting that CB_1 receptors are not crucial for the extinction of appetitive tasks (Holter et al. 2005) in operant conditioning. Similarly, Niyuhire et al. (2007) recently demonstrated that there was no difference in the extinction rate of mice treated with either vehicle or rimonabant in similar tasks. However, the ones injected with the CB_1 antagonist lacked the "burst extinction" that refers to a strong response the first time the reward was not presented any more. The authors interpreted this effect as a possible impairment at the beginning of the extinction phase or in frustration-like behaviour without excluding a possible lack of motivation.

5.3.2 The ECS and Habits

When an action is repeated to reach a goal, it may become more automatic and not sensitive any more to the value of the outcome. In this case, a new habit has been formed but it is not easy for the observer to differentiate between goal-directed and habitual actions. For example, you can be assessed while typing on a French keyboard. If you are used to an English one, your "Q" will be an "A" but if you are still watching the letters, you will look for the proper one. In 2007, Hilário et al. used the operant chambers in order to assess this kind of procedural memory (Hilário et al. 2007). They first showed that an interval ratio procedure (where the reinforcer is delivered upon the first press after a delay of some seconds since the last reinforcer) led to habit actions in mice. Thus, CB_1-null mice and AM251-treated mice were trained in an interval ratio procedure and tested in a devaluation paradigm, i.e. after having eaten some "free" reinforcers. In this condition, the mice with pharmacological or genetic blockade of CB_1 receptors were sensitive to the devaluation procedure, indicating that they were still responding in a goal-directed manner. In this way, the authors emphasised the critical role of the endocannabinoid system in habit formation.

These results were recently complemented in a study in rats, suggesting a differential role of the dorsolateral and the hippocampal CB_1 receptors in the extinction of habits using a T-maze task (Rueda-Orozco et al. 2008a). Indeed, the focal administration of the antagonist AM251 either in the dorsolateral striatum or in the hippocampus respectively impaired or rather facilitated the extinction of procedural memories. These data suggest that endocannabinoids in the striatum are important for extinguishing previously relevant responses and that pharmacological blockade of CB_1 in the hippocampus would enhance learning in hippocampus-dependent tasks.

5.3.3 The ECS and Procedural Strategies in a Spatial Task

A similar dissociation between the hippocampal and the striatal CB_1 receptors was proposed to underlie the choice of strategy to perform a Barnes maze. In this test, rodents have to visit many holes at the periphery of a circular table in order to find the drop box through which they can escape. Individuals utilised different strategies to solve the task efficiently. They can preferentially use a spatial strategy using distal cues around the maze, or a serial strategy, visiting the holes in sequence and following one direction. This latter strategy is believed to rely on procedural memory (Packard and McGaugh 1992; White and McDonald 2002). According to the dark or light phases, rodents complete the task using preferentially and respectively serial and spatial strategies (Rueda-Orozco et al. 2008b). However, during the dark phase, rats that were injected with the CB_1 antagonist AM251 in the striatum at the end of the daily last trial used the serial strategy less (Rueda-Orozco et al. 2008b).

5.4 The ECS and Emotional Memory

5.4.1 The ECS and Aversive Memory

In order to survive and to adapt to their environment, animals easily learn to avoid insecure places and noxious food. Here we will review how the ECS has been involved in this kind of learning.

A possible role of the ECS was first suggested in aversive memory using the elevated T-maze. In this study, rats were placed in a starting closed arm and given the possibility to explore two open, aversive arms, which rats rapidly learn to avoid. However, rats that were injected with rimonabant acquired and consolidated this avoidance memory better (Takahashi et al. 2005).

The conditioned taste aversion (Garcia et al. 1955) was developed from the ethological observation that animals have a very good memory of noxious food, e.g. blue jays avoid monarch butterflies because of the toxins they contain (Brower and Glazier 1975). Indeed, animals learn to avoid eating or drinking a particular food or fluid because it has been previously associated with a noxious substance (e.g. lithium chloride, Welzl et al. 2001). It has been reported that the pharmacological blockade of the CB_1 receptors in the insular cortex promoted memory retention and blocked its extinction without affecting re-consolidation (Kobilo et al. 2007). In a similar way, chicks that show spontaneous pecking behaviour avoid this behaviour if they have previously pecked into a bitter bead. However, when they were injected with rimonabant before the retention test, they did not avoid the bead (Adam et al. 2008).

Another way of assessing aversive memory is the step-down passive avoidance test in which a rodent is placed on the top of a small platform and as soon as it steps down it receives an electric foot shock. Subsequently, it will stay longer on the platform to avoid the "dangerous" floor. Intrahippocampal injection of AM251 after the learning phase impaired the consolidation of this experience (de Oliveira et al. 2005), confirming an important role of the ECS in this kind of memory. More classically, passive avoidance memory in rodents is assessed by exploiting their innate behaviour to go towards unlit and closed environments. In a "shuttle box", made of one lit compartment and a dark one separated by a door, once the rodent has fled the illuminated compartment to reach the dark one, it receives an electric footshock. In the following trials, the rodents will avoid entering the dark compartment. It was recently reported that an i.p. injection of rimonabant did not induce any impairment in the acquisition of the task but the drug blocked the extinction of the avoidance response (Niyuhire et al. 2007). The lack of detectable acquisition impairment could be masked due to very strong conditioning conditions.

The "shuttle box" can be used also for active avoidance paradigms. In this case, the two compartments are identically shaped and illuminated and animals learn to flee to the other compartment when alerted (by a light or tone cue). CB_1-null mice were shown to learn this task better on the fifth day of training (Martin et al. 2002). However, Bura et al. (2007) were not able to replicate these data when they wanted to evaluate the interaction of the ECS and the cholinergic system in cognitive processes. Additional studies are therefore needed to better understand the role of the ECS in this task.

In summary, the ECS seems to be deeply implicated in the significance of an aversive stimulus and can attenuate overreactions that would not be suitable to the situation.

5.4.2 The ECS and Fear Memory

Classical fear conditioning is a common paradigm to assess implicit associative memory. However, because of the very high emotional component of the procedure, we next review studies using different versions of the test.

5.4.2.1 Acquisition of Fear Memories

In fear conditioning tests, subjects form associations between a previously neutral stimulus (conditioned stimulus, CS, e.g. a tone or a context) and an aversive unconditioned stimulus (US, e.g. a footshock). After conditioning (learning), the CS induces a fear response even in the absence of the US. This behaviour can be observed and quantified, providing an indirect measure of the strength of the learning (LeDoux 2000). The quantified behaviour can vary from the natural fear responses typical of a given species (e.g. "freezing" as the absence of any movement except for the ones for respiration in rodents) to increased "startle" response. In fear conditioning protocols, most of the data published indicate that the pharmacologic blockade or the genetic deletion of CB_1 receptors induces little or no effect on the acquisition of the task. In CB_1-null mice, cued fear conditioning, where the fear response (freezing) is induced by a tone previously paired with a footshock, acquisition is comparable to that of wild-type controls (Cannich et al. 2004; Kamprath et al. 2006; Marsicano et al. 2002). Similarly, pharmacological blockade of CB_1 was shown to have little effect on the acquisition phase of fear conditioning tasks (Kamprath et al. 2006; Marsicano et al. 2002).

However, in a context-dependent version of fear conditioning, Arenos et al. proposed that administration of a CB_1 antagonist prior to conditioning is able to disrupt learning of rats (Arenos et al. 2006). Similarly, CB_1-null mice did not show any fear response when re-exposed to a context in which they had previously been shocked, and mice treated with AM 251 showed a reduced peak of freezing, indicating that the ECS could play a role in the acquisition of hippocampus-dependent fear conditioning (Mikics et al. 2005). However, Suzuki et al. did not observe any disturbance of acquisition or early acquisition when they injected rimonabant before conditioning (Suzuki et al. 2004).

Moreover, Reich et al. (2008) have recently shown that administration of a CB_1 antagonist prior to conditioning enhanced the freezing response of mice in both "delay" and "trace" versions of a cued fear conditioning (in the former version the CS and US are terminated at the same time, whereas in the latter version, the CS and the US are temporally separated).

5.4.2.2 Consolidation, Re-consolidation and Extinction of Fear Memories

While the ECS may or may not be involved in the acquisition of fear memories, its role in fear extinction seems to be crucial. Indeed, it was first reported in 2002 that CB_1-null mice were able to learn a tone–footshock association but failed in adapting their fear response when exposed repeatedly or for a long time to the CS, whereas their wild type littermates showed a time-dependent reduction of freezing behaviour (Marsicano et al. 2002; Kamprath et al. 2006; Marsicano and Lutz 2006). The injection of rimonabant (3 mg kg^{-1}), before the extinction training in wild-type C57/Bl6N mice confirmed that endocannabinoids, through the activation of CB_1

receptors, play a major acute role in the extinction of cue-induced conditioned fear (Marsicano et al. 2002; Marsicano and Lutz 2006). Manipulating the levels of endocannabinoids by the administration of the inhibitor of endocannabinoid breakdown and uptake, AM404, Chhatwal et al. (2005) reported a dose-related enhancement of extinction. Conversely, the administration of rimonabant induced an impairment of the extinction of fear in rats (Chhatwal et al. 2005).

Besides, the infralimbic subregion of the medial prefrontal cortex seems to be a key region for the extinction of fear memories, as the focal infralimbic administration of AM 251 could block the diminution of the fear startle response in rats (Lin et al. 2009).

A precise pharmacological study was recently conducted to assess the role of the ECS at the different stages of memory (Reich et al. 2008). It was shown that the freezing response was enhanced during recall with the administration of AM251. However this was accompanied with enhanced generalised freezing. The extinction of fear memories was also further analysed and the authors suggested that the blockade of the CB_1 receptors impaired the extinction expression but not extinction learning (Reich et al. 2008).

Suzuki et al. (2004) further characterised this effect, showing that the pharmacological blockade of CB_1 receptors with rimonabant induces a specific impairment on extinction without affecting the consolidation or re-consolidation of fear memories. This contrasts with a later study which suggests that the endocannabinoids are involved in both consolidation and reactivation of aversive memories (Bucherelli et al. 2006). However, this discrepancy could be explained by the global (i.p. injections) vs. the local effects (local administration in the amygdala) and/or the nature of the CB_1 antagonists (rimonabant vs. AM251). In a subsequent study, Suzuki et al. (2008) proposed that the endocannabinoid system was important for the de-stabilisation of reactivated contextual fear memories. Indeed, while the pharmacological blockade of protein synthesis or the genetic disruption of CREB-dependent transcription interfered with memory re-establishment following reactivation, the prior blockade of the CB_1 receptors (and voltage-gated calcium channels) impeded this effect, indicating that a fear memory cannot be altered during re-stabilisation if it was not previously destabilised via the activation of the CB_1 receptors (Suzuki et al. 2008).

5.4.2.3 Intracellular Cascades

During the extinction of a fear response, the levels of endocannabinoids are increased in the basolateral amygdala (Marsicano et al. 2002). In order to better understand which signalling pathways could be subsequently triggered by the activation of the ECS, Cannich et al. (2004) further analysed the molecular and cellular signature of cued fear extinction. While fear extinction induces the activation of extracellular regulated kinases (Lu et al. 2001) in wild-type animals, this effect was strongly reduced in many brain regions of the CB_1-null mice, especially in the basolateral amygdala, the prefrontal cortex and the hippocampus that have been proposed to be involved in these processes (Cannich et al. 2004). Interestingly, CB_1 receptors also seem to control in the same regions the expression of calcineurin, a phosphatase which was proposed to play an important role in extinction (Lin et al. 2001, 2003a, b; Mansuy et al. 1998). Therefore, both intracellular phosphorylation and dephosphorylation processes can be modulated by the activation of the CB_1 receptors during fear extinction.

5.4.3 A Possible Mechanism for ECS-Dependent Extinction of Fear Memories in the Amygdala

Extinction of conditioned fear is a very complex behavioural process, and an exhaustive review of its neurobiological basis is beyond the scope of the present article. We refer the reader to excellent and extensive reviews for detailed analyses of the theoretical and experimental data present in the literature, e.g. Myers and Davis (2007). Here, we will propose a working hypothesis that might explain the neurobiological role of the endocannabinoid system in fear extinction and, at the same time, suggest a general mechanism of this important adaptive process. Extinction of fear is believed to be a behavioural process relying on multiple mechanisms (Myers and Davis 2007). For instance, both associative and non-associative learning events are probably involved in extinction (Myers and Davis 2007; Marks and Tobena 1990; Kamprath et al. 2006). Many different brain regions, including the amygdala, hippocampus, medial prefrontal cortex and ventral tegmental area have been proposed to contribute to fear extinction (Myers and Davis 2007; Pare et al. 2004; Pezze and Feldon 2004). Furthermore, several neurotransmitter systems, including GABA, glutamate, dopamine, acetycholine, ACTH, glucocorticoids, norepinephrine, brain-derived neurotrophic factor, and, as seen above, endocannabinoids, play important roles in several phases of fear extinction (reviewed in Myers and Davis 2007). Finally, fear extinction is known to involve a plethora of intracellular signalling messengers: pathways involving cAMP/PKA, ERKs, phosphatidylinositol 3-kinase (PI3K), calcium/calmodulin protein kinase II (CaMKII) and the phosphatase, calcineurin, were all proposed to be involved in the processing of fear extinction (reviewed in Myers and Davis 2007). Therefore, given the extreme complexity of the neurobiological processes involved in extinction of fear, proposing a mechanism with a general validity in different conditions might appear to be a kind of "mission impossible". Nevertheless, considering the involvement of the ECS in fear extinction, a reductive model limited to specific neurotransmitters in a specific brain region might represent a rationale for future investigations aimed at further understanding the neuronal mechanisms of extinction. As the amygdala and its connections with other brain areas (such as the medial prefrontal cortex and the hippocampus) appear to be some of the main loci where both fear acquisition and fear extinction occur (Myers and Davis 2007; Pare et al. 2004), we will limit our attention to this particular region, keeping in mind that, since the ECS is represented in many different brain areas, similar mechanisms as described below could theoretically also occur outside of the amygdala (e.g. in the medial prefrontal cortex, which is rich in CB_1 receptors and is centrally involved in extinction processes). For the sake of clarity and to limit the possible variables, our hypothesis will address specifically the modulatory role of the ECS on GABAergic and glutamatergic transmission within the amygdala. Again, the reader should keep in mind that the ECS has been shown to interact with all the neurotransmitter systems listed above that are involved in extinction and, therefore, should be aware that the reality of neuronal processes are surely much more complex than here proposed.

In many brain regions, including the amygdala, CB_1 receptor activation is able to modulate both glutamatergic and GABAergic neurotransmission and synaptic plasticity (Marsicano and Lutz 2006; Chevaleyre et al. 2006; Azad et al. 2003, 2004; Domenici et al. 2006). The relative importance of GABAergic vs. glutamatergic neurotransmission in the processing of fear extinction in different brain regions is still a matter of debate in the field (Harris and Westbrook 1998; McGaugh et al. 1990; Myers and Davis 2002, 2007). Indeed, there is evidence that both these neurotransmitter systems might play an important role during several phases of extinction. On the one hand, drugs interfering with $GABA_A$ receptors can differentially alter different phases of extinction (Harris and Westbrook 1998; McGaugh et al. 1990). On the other hand, there is compelling evidence that glutamatergic transmission, and in particular activation of NMDA receptors is not only necessary for fear acquisition, but also for fear extinction (Myers and Davis 2002, 2007). The amygdala is formed by different subnuclei, among which the lateral nucleus (LA), the basal nucleus (BA) and the central nucleus (CE) are well known to contain the neuronal circuitry in great part responsible for the processing of conditioned fear responses (LeDoux 2000). For the sake of brevity and simplicity, we will refer to the LA–BA complex as the basolateral complex (BLA) and we will omit the important functional distinctions between neurons belong to each of the individual nuclei (Phelps and LeDoux 2005). The BLA (mostly the LA) is the locus where different sensory inputs (i.e. the US and the CS) converge to form the fear memory, which is likely expressed by the potentiation of particular circuits' activity (Maren 1999). Anatomical and physiological data indicate that these circuits send information to the CE, which is the main output centre of the amygdala and is responsible, via several projections onto different areas of the brain, for the expression of the fear responses (e.g. freezing, increased startle or autonomic responses) (LeDoux 2000). Strategically located between the BLA and the CE, the intercalated cells (or intercalated cell masses, ICM) are projecting GABAergic neurons that receive excitatory inputs from glutamatergic BLA neurons and send inhibitory projections to CE neurons (Collins and Pare 1999a, b). Given their GABAergic nature, the activation of ICM neurons will lead to an inhibition of CE neurons, and this feature has been proposed to potentially participate in the extinction of fear, by the inhibition of the activity of the output nucleus of the amygdala (Pare et al. 2004). Consistently, a region that plays an important function in the storage of extinction memory, the medial prefrontal cortex (PFC) (Milad and Quirk 2002), is able to stimulate ICM neurons, e.g. as revealed by c-fos immunoreactivity (Berretta et al.

2005). In this view, fear acquisition would occur by the potentiation of the excitatory drive from BLA to CE, whereas extinction would be mediated by an inhibition of the CE, given by the GABAergic input of ICM neurons (Pare et al. 2004). Therefore, glutamatergic BLA neurons would mediate fear, whereas ICM neurons would mediate extinction by inhibiting CE outputs (Pare et al. 2004). A problematic issue with this proposal is that PFC stimulation is able not only to activate ICM neurons, but also to directly excite a certain proportion of glutamatergic BLA neurons (Likhtik et al. 2005), whereas other reports showed a PFC-mediated inhibition of BLA principal neurons, likely due to stimulation of inhibitory GABAergic interneurons (Rosenkranz and Grace 2002; Rosenkranz et al. 2003). These data suggest that, depending on the conditions, PFC can depress or activate glutamatergic neurons in the BLA. Whereas the former case fits easily with a scenario in which inhibition of glutamatergic neurons in the BLA mediates extinction, the possible excitation of the same neurons is more difficult to explain in such a scheme. Indeed, if PFC activity mediates extinction and BLA excitatory neurons mediate fear, one would expect that PFC stimulation should always inhibit, rather than excite BLA neurons. This apparent contradiction could be partially explained by the idea that a balanced activity of glutamatergic and GABAergic neurons within the amygdala could contribute to both acquisition and extinction of fear and that specific excitatory circuits exist to actively mediate fear and extinction of fear, respectively. A recent study confirmed this hypothesis. By means of single unit recordings in awake animals, Herry and colleagues (2008) showed that a certain proportion of pyramidal neurons in the basal amygdala are activated during fear acquisition, whereas another proportion is activated during fear extinction. In this balanced activity, the endocannabinoid system might play an important regulatory role. In a typical fear conditioning paradigm, an a priori neutral stimulus (often a tone, CS) is presented to the animals and immediately paired with an unpleasant stimulus (often a footshock, US). On successive presentations of the CS alone, animals will respond with increased fear reactions (often freezing behaviour, conditioned response, CR) (LeDoux 2000). Extinction occurs upon prolonged and/or repeated presentation of the CS in the absence of the US (i.e. in "non-reinforced" conditions), and it is defined as a continuous decrease of the CR (Myers and Davis 2007). The CS is generally considered neutral, because it normally does not elicit, in absence of a previously paired US, any observable fear reaction. However, the lack of an observable response is not per se an indication that the CS is not eliciting any neural response, but only that the stimulation induced is below the threshold to induce the response chosen as the "outcome" of the experiment. Indeed, many stimuli represent rodents' alarm signals. In other words, the auditory system is, for many species, an "alarm system", whose main function is to alert individuals concerning possible dangers present in the environment (Marks and Tobena 1990). This intuitive observation is corroborated by the experimental observation that "neutral" tones can elicit fear reactions simply by increasing their intensity (Marks and Tobena 1990; Kamprath et al. 2006). This is probably the reason why fear conditioning is such a robust protocol, which is easily learnt by many species without the need for intensive training: animals (and rodents in particular) are "prepared" to learn fear conditioning, especially when the modality of the CS involves the auditory system (Marks and Tobena 1990). On the other hand, fear reactions are generally characterised by an inhibition of normal activity, e.g. freezing or conditioned response inhibition (LeDoux 2000). Consequently, fear responses to stimuli should also be very flexible, in order to avoid the risk of a blockade of the common activities necessary for survival (e.g. feeding, reproduction, etc.). In this sense, extinction also appears as a primary function of neuronal circuits, allowing animals to maintain the right balance between "caution" and "activity" to guarantee proper survival. In our opinion, these concepts are important to keep in mind in the attempt to explain the neuronal events occurring both during acquisition and extinction of conditioned fear, and our hypothesis is that circuits in the amygdala might exist ready-made to generate behavioural "fear" and "no fear" reactions. The endocannabinoid system might be an important regulator of the relative activity of these pre-existing circuits. A simplified vision of the anatomical organisation of the neuronal circuits within the amygdaloid complex (Fig. 3), based on physiological and anatomical data (McDonald 1998; Collins and Pare 1999a;

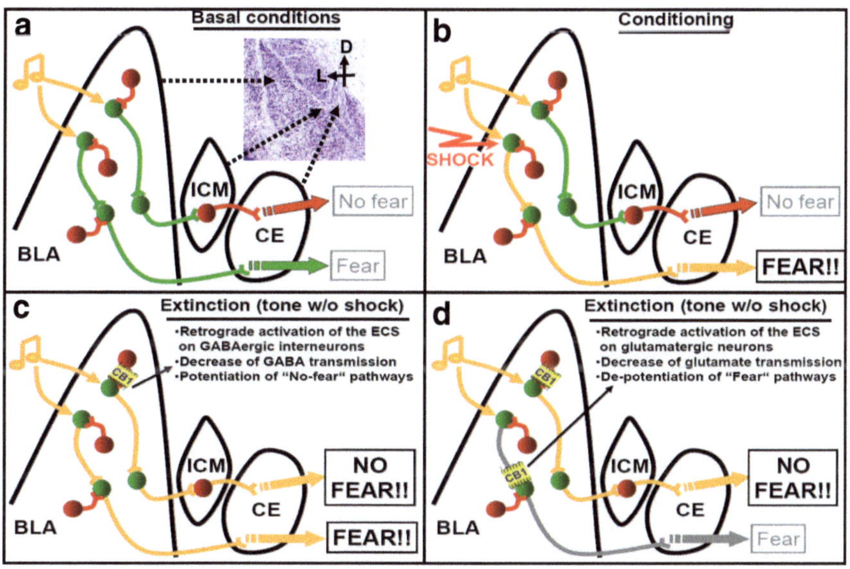

Fig. 3 Working hypothesis of extinction processing in the amygdala. (**a**) Schematic representation of the amygdala, with reference to subnuclei in a cresyl-violet staining of the region (inset, with orientation). Under basal conditions, a tone presentation is not sufficient to elicit a fear response. (**b**) During conditioning, the simultaneous presentation of the tone (CS) and the shock (US) potentiates "fear" circuits. (**c**) During extinction, non-reinforced and prolonged tone presentation might cause sustained stimulation of "no-fear" pathways, which might be "disinhibited" through ECS-mediated inhibition of GABAergic transmission. (**d**) Simultaneously, CB_1 on glutamatergic neurons might contribute to the depotentiation (habituation) of "fear" pathways. BLA, basolateral amygdala; CE, central nucleus of amygdala; ICM, intercalated cell masses. *Green*, glutamatergic pathways; *red*, GABAergic neurons; *orange*, potentiation; *grey*, depotentiation

Pare et al. 2004), suggests that two different glutamatergic circuits might exist within the BLA. One would directly send excitatory projections to the CE, whereas the second one would first activate ICM neurons, which, in turn, would inhibit CE neurons (Fig. 3a). Importantly, glutamatergic activity in the BLA is maintained under a strong control of local inhibition, which is mainly provided by GABAergic interneurons that are particularly active in this brain region (Bissiere et al. 2003) (Fig. 3a). Given this peculiar anatomical organisation, one could suggest that, in basal conditions (i.e. before conditioning), the neuronal activity conveying the information relative to the CS, deriving, in the case of a tone, from cortical and thalamic auditory areas (LeDoux 2000), might reach both "fear" and "no fear" circuits, eliciting a sub-threshold stimulation that is not sufficient to induce any relevant change in the behaviour of the animal (Fig. 3a). With the occurrence of the US, cortical sensory inputs to the BLA converge onto particular projecting neurons where the coincidence with the CS inputs elicits a long-term potentiation (LTP)-like plastic phenomenon, which is believed to mediate the acquisition of the fear conditioning (LeDoux 2000). In our model, this event would induce a plastic potentiation of the circuits mediating "fear" responses, whereas the hypothetical "no fear" circuits are left unchanged (Fig. 3b). Therefore, the successive presentation of the CS will find the "fear" pathway somehow potentiated, with the consequent stimulation of CE neurons and the behavioural and autonomic expression of the fear reaction (Fig. 3b). Sustained afferent stimulation at low intensity (900 pulses at 1 Hz) onto glutamatergic neurons in the BLA can induce a long-term depression of GABAergic inhibition onto the same neurons, which is mediated by a retrograde action of endocannabinoids likely released from the postsynaptic neurons and acting at presynaptic CB_1 receptors located on GABAergic terminals (LTDi; Azad et al. 2004; Marsicano et al. 2002). During successive "non-reinforced" (i. e. in absence of the US) presentations of the CS, the continuous stimulation of "no fear" neurons might, therefore, induce a similar form of "disinhibition" at specific circuits (Azad et al. 2004). If we assume that these circuits might be the "no fear" ones, an ICM-mediated inhibition of CE neurons will occur through their potentiation (Fig. 3c). Thus, during sustained or repeated non-reinforced CS presentation, a sort of unstable balance between two different and competitive kinds of circuits in the BLA would be generated, with the start of a decrease of the conditioned fear response. In this frame, within-session extinction (Myers and Davis 2002) would result from the contrast between two opposing "potentiations": the one of the "fear" pathways (potentiated by the previous conditioning), and the one of the "no fear" pathways (disinhibited by the CB_1-dependent retrograde decrease of inhibition). As CB_1 receptors can also strongly regulate glutamatergic transmission in the BLA (Azad et al. 2003), one could also hypothesize, although less electrophysiological evidence exists to support such a mechanism, that the sustained non-reinforced stimulation of glutamatergic neurons might somehow lead to a CB_1-dependent inhibition ("depotentiation") of excitatory synaptic strength in "fear" circuits, further pushing the response of the animal towards a "no fear" behavioural and autonomic reaction (Fig. 3d).

Importantly, given the short-lasting life of endocannabinoids (see the "on demand" activation of the ECS described in other chapters of the present book) these phenomena might be in part temporary (mostly the depotentiation of "fear" pathways), thereby justifying the observation that extinction is not "erasure" of the original fear conditioning and that different conditions (e.g. spontaneous recovery, reinstatement, renewal and others) can induce a rapid re-establishment of the original response (Myers and Davis 2007).

An important issue in modern theories of extinction is the associative or non-associative nature of this behavioural phenomenon. Extinction of the fear response might rely on associative processes ("new learning"), in which animals form a new inhibitory association between the CS and the lack of the US (Myers and Davis 2002, 2007) or on non-associative mechanisms, in which the subjects simply "habituate" to the aversive stimulus and thereby decrease their response (Marks and Tobena 1990; Kamprath and Wotjak 2004; Myers and Davis 2007). This discrimination is important also for the understanding of human anxiety-related pathologies, such as phobias or post-traumatic stress disorders, in which the impairment in the ability to extinguish fear might depend on alterations in cognitive (associative learning) or non-cognitive (habituation) processes (Marks and Tobena 1990). However, the distinction between these two processes is very difficult in animal experiments, because the experimenter can observe only the decrease of the CR, which is independent of the nature of the causative neuronal processes. However, there is evidence that the ECS might participate both in the associative component of extinction (Chhatwal et al. 2005; Azad et al. 2004) and the non-associative ones (Kamprath et al. 2006). The present model might account for both these functions of the ECS in extinction of fear. On the one hand, the endocannabinoid-mediated "disinhibition" of "no fear" pathways might be somehow regarded as the "associative" component of extinction learning, in the sense that it would be mediated by an active process of "potentiation" of certain neuronal circuits. On the other hand, the proposed CB_1-dependent decrease of glutamatergic transmission in the "fear" pathways would account for a form of "non-associative" habituation, in which the activation of the ECS would mediate a simple decrease in the stimulus-response reaction.

Furthermore, this model could also explain some apparently contradictory results concerning the CB_1-dependent activation of intracellular pathways in the BLA during extinction (Cannich et al. 2004). In fact, as seen before, CB_1 appears to be important during extinction (Cannich et al. 2004) both for the stimulation of the ERK pathway, which is believed to play a central role in synaptic potentiation (Sweatt 2001), and for the opposing action of calcineurin (phosphatase 2B, PP2B), believed to mediate certain forms of depotentiation (Mansuy 2003). The model described here would suggest that these events might be segregated in neurons specific to the "fear" and the "no fear" pathways, with ERK activation contributing to the potentiation of "no fear" neurons and calcineurin mediating the depotentiation of "fear" neurons, respectively.

As we have tried to argue, this model would present the advantage of reconciling many apparently discrepant observations present in the literature concerning the neuronal circuitry underlying extinction and, in particular, the role of the ECS in these processes. However, these ideas are still far from being proven and much experimental work is warranted to confirm or discard them and to fully understand the role of the ECS in the brain circuits mediating fear and anxiety responses.

6 General Conclusions

Many of the functions of the ECS observed in complex learning and memory processes could be explained by simple implicit forms of memory. Indeed, Kamprath et al. (2006) proposed that the deficits of the CB_1-null mice in the extinction of cued fear conditioning are due to habituation-like processes. In the recognition memory task (Bilkei-Gorzo et al. 2005), old CB_1-null mice or rimonabant-treated rodents could fail to habituate and would not make any distinction between familiar and novel stimuli, whereas their better performance at a younger age could be explained by a problem in habituating to the novel stimulus and they would thus remember longer. This is, however, in conflict with two studies that effectively reported that CB_1-null mice habituated to an open field much faster than the wild-type controls and exhibited respectively less locomotion or less rearing (Degroot et al. 2006; Thiemann et al. 2007). However, most of the studies did not report any difference in locomotor activity or rearing behaviour (de Oliveira et al. 2005). This could be due to the time needed to observe the effects of CB_1 blockade on habituation. However, it is also possible that the ECS participates in the regulation of typical associative processes of learning and memory tasks. For instance, as proposed above, the coordinate actions of the ECS on both habituation-like and associative processes could mediate the important role of endocannabinoids in extinction of aversive memories.

In attempting to understand the role of the ECS, it has been proposed that endocannabinoid signalling would be important only in situations where aversive stimuli were involved (Lutz 2002; Wotjak 2005). Indeed, the most prominent phenotype of the blockade of the ECS was obtained in the extinction of a fear response (Arenos et al. 2006; Cannich et al. 2004; Kamprath et al. 2006; Marsicano et al. 2002; Mikics et al. 2006; Suzuki et al. 2004, 2008) and in the reversal learning of a spatial task with an aversive component in the Morris Water Maze (Niyuhire et al. 2007; Robinson et al. 2008; Varvel and Lichtman 2002; Varvel et al. 2007). Moreover, it was found that the ECS was dispensable for the extinction in an appetitive task (Holter et al. 2005). In contrast with this theory, it has been very recently suggested that the ECS could also be involved in positively motivated behaviours (Hilário et al. 2007; Rueda-Orozco et al. 2008b). At this point, the ECS would rather have a major role in behavioural flexibility and/or in the ability to change attentional set, whereas it would have a limited function in initial learning. On the contrary, CB_1 receptor blockade induced reduced perseveration in a strategy-shift task, indicating that the switch of strategy was even facilitated (Hill et al. 2006). The endocannabinoid system could modulate many transmitter systems that have been involved in behavioural flexibility, such as the glutamatergic (Stefani and Moghaddam 2003; Stefani et al. 2003), serotoninergic (Clarke et al. 2008) and dopaminergic (Floresco et al. 2006; Floresco and Magyar 2006) systems in the prefrontal cortex, but further analyses are still needed to characterise these interactions.

The recent idea that the ECS could be involved in the "destabilisation" of memory traces (Suzuki et al. 2008) might be very important for the interpretation of the different (and sometimes apparently discrepant) results in the literature. Indeed, if the role of the ECS is to temporarily "weaken" acquired memory, this mechanism could be also involved in the prolongation of the time for which new memories are labile during acquisition in certain conditions. This could, theoretically, explain why, in different conditions, the ECS might influence different phases of learning and memory processing.

In conclusion, the roles of the ECS in learning and memory are on the way to be dissected and clarified. However, many aspects are still obscure and further studies are mandatory both to provide knowledge of the general mechanisms of learning and also to develop novel therapeutic tools to tackle diseases characterised by improper processing and storage of information in the brain.

References

Adam AS, Wenger T, Csillag A (2008) The cannabinoid CB1 receptor antagonist rimonabant dose-dependently inhibits memory recall in the passive avoidance task in domestic chicks (Gallus domesticus). Brain Res Bull 76:272–274

Alger E (2002) Retrograde signaling in the regulation of synaptic transmission: focus on endocannabinoids. Prog Neurobiol 68:247–286

Arenos JD, Musty RE, Bucci DJ (2006) Blockade of cannabinoid CB1 receptors alters contextual learning and memory. Eur J Pharmacol 539:177–183

Azad SC, Eder M, Marsicano G et al. (2003) Activation of the cannabinoid receptor type 1 decreases glutamatergic and GABAergic synaptic transmission in the lateral amygdala of the mouse. Learn Memory 10:116–128

Azad SC, Monory K, Marsicano G et al. (2004) Circuitry for associative plasticity in the amygdala involves endocannabinoid signaling. J Neurosci 24:9953–9961

Bacci A, Huguenard JR, Prince DA (2004) Long-lasting self-inhibition of neocortical interneurons mediated by endocannabinoids. Nature 431:312–316

Bailey CH, Kandel ER, Si K (2004) The persistence of long-term memory: a molecular approach to self-sustaining changes in learning-induced synaptic growth. Neuron 44:49–57

Barco A, Bailey CH, Kandel ER (2006) Common molecular mechanisms in explicit and implicit memory. J Neurochem 97:1520–1533

Berretta S, Pantazopoulos H, Caldera M et al. (2005) Infralimbic cortex activation increases c-Fos expression in intercalated neurons of the amygdala. Neuroscience 132:943–953

Bilkei-Gorzo A, Racz I, Valverde O et al. (2005) Early age-related cognitive impairment in mice lacking cannabinoid CB1 receptors. Proc Natl Acad Sci USA 102:15670–15675

Bissiere S, Humeau Y, Luthi A (2003) Dopamine gates LTP induction in lateral amygdala by suppressing feedforward inhibition. Nat Neurosci 6:587–592

Brower LP, Glazier SC (1975) Localization of heart poisons in the monarch butterfly. Science 188:19–25

Bucherelli C, Baldi E, Mariottini C et al. (2006) Aversive memory reactivation engages in the amygdala only some neurotransmitters involved in consolidation. Learn Memory 13:426–430
Bura SA, Castane A, Ledent C et al. (2007) Genetic and pharmacological approaches to evaluate the interaction between the cannabinoid and cholinergic systems in cognitive processes. Br J Pharmacol 150:758–765
Cannich A, Wotjak CT, Kamprath K et al. (2004) CB1 cannabinoid receptors modulate kinase and phosphatase activity during extinction of conditioned fear in mice. Learn Memory 11:625–632
Carlson G, Wang Y, Alger BE (2002) Endocannabinoids facilitate the induction of LTP in the hippocampus. Nat Neurosci 5:723–724
Carter E, Wang XJ (2007) Cannabinoid-mediated disinhibition and working memory: dynamical interplay of multiple feedback mechanisms in a continuous attractor model of prefrontal cortex. Cereb Cortex 17(Suppl 1):i16–i26
Castellano C, Rossi-Arnaud C, Cestari V et al. (2003) Cannabinoids and memory: animal studies. Curr Drug Targets CNS Neurol Disord 2:389–402
Chevaleyre V, Takahashi KA, Castillo PE (2006) Endocannabinoid-mediated synaptic plasticity in the CNS. Annu Rev Neurosci 29:37–75
Chhatwal JP, Davis M, Maguschak KA et al. (2005) Enhancing cannabinoid neurotransmission augments the extinction of conditioned fear. Neuropsychopharmacology 30:516–524
Clarke JR, Rossato JI, Monteiro S et al. (2008) Posttraining activation of CB1 cannabinoid receptors in the CA1 region of the dorsal hippocampus impairs object recognition long-term memory. Neurobiol Learn Memory 90(2):374–381
Collins DR, Pare D (1999a) Reciprocal changes in the firing probability of lateral and central medial amygdala neurons. J Neurosci 19:836–844
Collins DR, Pare D (1999b) Spontaneous and evoked activity of intercalated amygdala neurons. Eur J Neurosci 11:3441–3448
Cravatt BF, Demarest K, Patricelli MP et al. (2001) Supersensitivity to anandamide and enhanced endogenous cannabinoid signaling in mice lacking fatty acid amide hydrolase. Proc Natl Acad Sci USA 98:9371–9376
de Oliveira AL, de Oliveira LF, Camboim C et al. (2005) Amnestic effect of intrahippocampal AM251, a CB1-selective blocker, in the inhibitory avoidance, but not in the open field habituation task, in rats. Neurobiol Learn Memory 83:119–124
Deadwyler SA, Hampson RE (2008) Endocannabinoids modulate encoding of sequential memory in the rat hippocampus. Psychopharmacology (Berl) 198:577–586
Deadwyler SA, Goonawardena AV, Hampson RE (2007) Short-term memory is modulated by the spontaneous release of endocannabinoids: evidence from hippocampal population codes. Behav Pharmacol 18:571–580
Debiec J, Ledoux JE, Nader K (2002) Cellular and systems reconsolidation in the hippocampus. Neuron 36:527–538
Degroot A, Kofalvi A, Wade MR et al. (2006) CB1 receptor antagonism increases hippocampal acetylcholine release: site and mechanism of action. Mol Pharmacol 70:1236–1245
Devane WA, Hanus L, Breuer A et al. (1992) Isolation and structure of a brain constituent that binds to the cannabinoid receptor. Science 258:1946–1949
Di Marzo V, De Petrocellis L, Fezza F et al. (2002) Anandamide receptors. Prostag Leukot Ess Fatty Acids 66:377–391
Domenici MR, Azad SC, Marsicano G et al. (2006) Cannabinoid receptor type 1 located on presynaptic terminals of principal neurons in the forebrain controls glutamatergic synaptic transmission. J Neurosci 26:5794–5799
Dunning J, During MJ (2003) Molecular mechanisms of learning and memory. Expert Rev Mol Med 5:1–11
Duvarci S, Nader K (2004) Characterization of fear memory reconsolidation. J Neurosci 24:9269–9275
Floresco SB, Magyar O (2006) Mesocortical dopamine modulation of executive functions: beyond working memory. Psychopharmacology (Berl) 188:567–585

Floresco SB, Magyar O, Ghods-Sharifi S et al. (2006) Multiple dopamine receptor subtypes in the medial prefrontal cortex of the rat regulate set-shifting. Neuropsychopharmacology 31:297–309
Freund TF, Katona I, Piomelli D (2003) Role of endogenous cannabinoids in synaptic signaling. Physiol Rev 83:1017–1066
Fride E (2005) Endocannabinoids in the central nervous system: from neuronal networks to behavior. Curr Drug Targets CNS Neurol Disord 4:633–642
Fujiwara M, Egashira N (2004) New perspectives in the studies on endocannabinoid and cannabis: abnormal behaviors associate with CB1 cannabinoid receptor and development of therapeutic application. J Pharmacol Sci 96:362–366
Gaoni Y, Mechoulam R (1964) Isolation, structure and partial synthesis of an active constituent of hashish. J Am Chem Soc 86:1646–1647
Garcia J, Kimeldorf DJ, Koelling RA (1955) Conditioned aversion to saccharin resulting from exposure to gamma radiation. Science 122:157–158
Harris JA, Westbrook RF (1998) Evidence that GABA transmission mediates context-specific extinction of learned fear. Psychopharmacology (Berl) 140:105–115
Herry C, Ciocchi S, Senn V et al. (2008) Switching on and off fear by distinct neuronal circuits. Nature 454:600–606
Hilário MRF, Clouse E, Yin HH et al. (2007) Endocannabinoid signaling is critical for habit formation. Front Integr Neurosci 1:6
Hill MN, Froese LM, Morrish AC et al. (2006) Alterations in behavioral flexibility by cannabinoid CB1 receptor agonists and antagonists. Psychopharmacology (Berl) 187:245–259
Holter SM, Kallnik M, Wurst W et al. (2005) Cannabinoid CB1 receptor is dispensable for memory extinction in an appetitively-motivated learning task. Eur J Pharmacol 510:69–74
Howlett AC, Barth F, Bonner TI et al. (2002) International Union of Pharmacology. XXVII. Classification of cannabinoid receptors. Pharmacol Rev 54:161–202
Iversen L (2003a) Cannabis and the brain. Brain 126:1252–1270
Iversen L (2003b) Comparing cannabis with tobacco: arithmetic does not add up. Br Med J 327:165
Kamprath K, Wotjak CT (2004) Nonassociative learning processes determine expression and extinction of conditioned fear in mice. Learn Memory 11:770–786
Kamprath K, Marsicano G, Tang J et al. (2006) Cannabinoid CB1 receptor mediates fear extinction via habituation-like processes. J Neurosci 26:6677–6686
Kandel ER (2001) The molecular biology of memory storage: a dialog between genes and synapses. Biosci Rep 21:565–611
Kandel ER, Schwartz JH, Jessel TM (2000) Principles of neural science. McGraw-Hill, New York
Kobilo T, Hazvi S, Dudai Y (2007) Role of cortical cannabinoid CB1 receptor in conditioned taste aversion memory. Eur J Neurosci 25:3417–3421
LeDoux JE (2000) Emotion circuits in the brain. Annu Rev Neurosci 23:155–184
Lichtman AH (2000) SR 141716A enhances spatial memory as assessed in a radial-arm maze task in rats. Eur J Pharmacol 404:175–179
Likhtik E, Pelletier JG, Paz R et al. (2005) Prefrontal control of the amygdala. J Neurosci 25:7429–7437
Lin CH, Yeh SH, Lin CH et al. (2001) A role for the PI-3 kinase signaling pathway in fear conditioning and synaptic plasticity in the amygdala. Neuron 31:841–851
Lin CH, Lee CC, Gean PW (2003a) Involvement of a calcineurin cascade in amygdala depotentiation and quenching of fear memory. Mol Pharmacol 63:44–52
Lin CH, Yeh SH, Leu TH et al. (2003b) Identification of calcineurin as a key signal in the extinction of fear memory. J Neurosci 23:1574–1579
Lin HC, Mao SC, Su CL et al. (2009) The role of prefrontal cortex CB1 receptors in the modulation of fear memory. Cereb Cortex 19(1):165–175
Lu KT, Walker DL, Davis M (2001) Mitogen-activated protein kinase cascade in the basolateral nucleus of amygdala is involved in extinction of fear-potentiated startle. J Neurosci 21: RC162–RC166

Lutz B (2002) Molecular biology of cannabinoid receptors. Prostag Leukot Ess Fatty Acids 66:123–142

Maccarrone M, Valverde O, Barbaccia ML et al. (2002) Age-related changes of anandamide metabolism in CB1 cannabinoid receptor knockout mice: correlation with behaviour. Eur J Neurosci 15:1178–1186

Maejima T, Ohno-Shosaku T, Kano M (2001) Endogenous cannabinoid as a retrograde messenger from depolarized postsynaptic neurons to presynaptic terminals. Neurosci Res 40:205–210

Mallet PE, Beninger RJ (1998) The cannabinoid CB1 receptor antagonist SR141716A attenuates the memory impairment produced by delta9-tetrahydrocannabinol or anandamide. Psychopharmacology (Berl) 140:11–19

Mansuy IM (2003) Calcineurin in memory and bidirectional plasticity. Biochem Biophys Res Commun 311:1195–1208

Mansuy IM, Mayford M, Jacob B et al. (1998) Restricted and regulated overexpression reveals calcineurin as a key component in the transition from short-term to long-term memory. Cell 92:39–49

Maren S (1999) Long-term potentiation in the amygdala: a mechanism for emotional learning and memory. Trends Neurosci 22:561–567

Marks I, Tobena A (1990) Learning and unlearning fear: a clinical and evolutionary perspective. Neurosci Biobehav Rev 14:365–384

Marsicano G, Lutz B (1999) Expression of the cannabinoid receptor CB1 in distinct neuronal subpopulations in the adult mouse forebrain. Eur J Neurosci 11:4213–4225

Marsicano G, Lutz B (2006) Neuromodulatory functions of the endocannabinoid system. J Endocrinol Invest 29:27–46

Marsicano G, Wotjak CT, Azad SC et al. (2002) The endogenous cannabinoid system controls extinction of aversive memories. Nature 418:530–534

Martin M, Ledent C, Parmentier M et al. (2002) Involvement of CB1 cannabinoid receptors in emotional behaviour. Psychopharmacology (Berl) 159:379–387

Massa F, Marsicano G, Hermann H et al. (2004) The endogenous cannabinoid system protects against colonic inflammation. J Clin Invest 113:1202–1209

Matsuda LA, Lolait SJ, Brownstein MJ et al. (1990) Structure of a cannabinoid receptor and functional expression of the cloned cDNA. Nature 346:561–564

McDonald AJ (1998) Cortical pathways to the mammalian amygdala. Prog Neurobiol 55:257–332

McGaugh JL (2000) Memory – a century of consolidation. Science 287:248–251

McGaugh JL, Castellano C, Brioni J (1990) Picrotoxin enhances latent extinction of conditioned fear. Behav Neurosci 104:264–267

Mechoulam R, Ben Shabat S, Hanus L et al. (1995) Identification of an endogenous 2-monoglyceride, present in canine gut, that binds to cannabinoid receptors. Biochem Pharmacol 50:83–90

Mikics E, Barsy B, Barsvari B et al. (2005) Behavioral specificity of non-genomic glucocorticoid effects in rats: effects on risk assessment in the elevated plus-maze and the open-field. Horm Behav 48:152–162

Mikics E, Dombi T, Barsvari B et al. (2006) The effects of cannabinoids on contextual conditioned fear in CB1 knockout and CD1 mice. Behav Pharmacol 17:223–230

Milad MR, Quirk GJ (2002) Neurons in medial prefrontal cortex signal memory for fear extinction. Nature 420:70–74

Myers KM, Davis M (2002) Behavioral and neural analysis of extinction. Neuron 36:567–584

Myers KM, Davis M (2007) Mechanisms of fear extinction. Mol Psychiatry 12:120–150

Nader K, Schafe GE, Ledoux JE (2000) The labile nature of consolidation theory. Nat Rev Neurosci 1:216–219

Niyuhire F, Varvel SA, Thorpe AJ et al. (2007) The disruptive effects of the CB1 receptor antagonist rimonabant on extinction learning in mice are task-specific. Psychopharmacology (Berl) 191:223–231

Pacher P, Batkai S, Kunos G (2006) The endocannabinoid system as an emerging target of pharmacotherapy. Pharmacol Rev 58:389–462

Packard MG, McGaugh JL (1992) Double dissociation of fornix and caudate nucleus lesions on acquisition of two water maze tasks: further evidence for multiple memory systems. Behav Neurosci 106:439–446

Pagotto U, Cervino C, Vicennati V et al. (2006) How many sites of action for endocannabinoids to control energy metabolism? Int J Obes (Lond) 30(Suppl 1):S39–S43

Pare D, Quirk GJ, LeDoux JE (2004) New vistas on amygdala networks in conditioned fear. J Neurophysiol 92:1–9

Pertwee RG (2008) The diverse CB1 and CB2 receptor pharmacology of three plant cannabinoids: delta9-tetrahydrocannabinol, cannabidiol and delta9-tetrahydrocannabivarin. Br J Pharmacol 153:199–215

Pezze MA, Feldon J (2004) Mesolimbic dopaminergic pathways in fear conditioning. Prog Neurobiol 74:301–320

Phelps EA, LeDoux JE (2005) Contributions of the amygdala to emotion processing: from animal models to human behavior. Neuron 48:175–187

Quirk GJ, Mueller D (2008) Neural mechanisms of extinction learning and retrieval. Neuropsychopharmacology 33:56–72

Ranganathan M, D'Souza DC (2006) The acute effects of cannabinoids on memory in humans: a review. Psychopharmacology (Berl) 188:425–444

Reibaud M, Obinu MC, Ledent C et al. (1999) Enhancement of memory in cannabinoid CB1 receptor knock-out mice. Eur J Pharmacol 379:R1–R2

Reich CG, Mohammadi MH, Alger BE (2008) Endocannabinoid modulation of fear responses: learning and state-dependent performance effects. J Psychopharmacol. 22:761–768

Riedel G, Davies SN (2005) Cannabinoid function in learning, memory and plasticity. Handb Exp Pharmacol 168:445–477

Rinaldi-Carmona M, Barth F, Heaulme M et al. (1994) SR141716A, a potent and selective antagonist of the brain cannabinoid receptor. FEBS Lett 350:240–244

Robinson L, Killop-Smith S, Ross NL et al. (2008) Hippocampal endocannabinoids inhibit spatial learning and limit spatial memory in rats. Psychopharmacology (Berl) 198:551–563

Rosenkranz JA, Grace AA (2002) Cellular mechanisms of infralimbic and prelimbic prefrontal cortical inhibition and dopaminergic modulation of basolateral amygdala neurons in vivo. J Neurosci 22:324–337

Rosenkranz JA, Moore H, Grace AA (2003) The prefrontal cortex regulates lateral amygdala neuronal plasticity and responses to previously conditioned stimuli. J Neurosci 23:11054–11064

Rueda-Orozco PE, Montes-Rodriguez CJ, Soria-Gomez E et al. (2008a) Impairment of endocannabinoids activity in the dorsolateral striatum delays extinction of behavior in a procedural memory task in rats. Neuropharmacology 55(1):55–62

Rueda-Orozco PE, Soria-Gomez E, Montes-Rodriguez CJ et al. (2008b) A potential function of endocannabinoids in the selection of a navigation strategy by rats. Psychopharmacology (Berl) 198:565–576

Sara SJ (2000) Retrieval and reconsolidation: toward a neurobiology of remembering. Learn Memory 7:73–84

Shiflett MW, Rankin AZ, Tomaszycki ML et al. (2004) Cannabinoid inhibition improves memory in food-storing birds, but with a cost. Proc Biol Sci 271:2043–2048

Soria G, Mendizabal V, Tourino C et al. (2005) Lack of CB1 cannabinoid receptor impairs cocaine self-administration. Neuropsychopharmacology 30:1670–1680

Squire LR, Berq D, du Lac S et al. (2008) Fundamentals of Neuroscience, 3rd edn

Stefani MR, Moghaddam B (2003) Distinct contributions of glutamate receptor subtypes to cognitive set-shifting abilities in the rat. Ann N Y Acad Sci 1003:464–467

Stefani MR, Groth K, Moghaddam B (2003) Glutamate receptors in the rat medial prefrontal cortex regulate set-shifting ability. Behav Neurosci 117:728–737

Straiker A, Mackie K (2007) Metabotropic suppression of excitation in murine autaptic hippocampal neurons. J Physiol 578:773–785

Sugiura T, Kondo S, Sukagawa A et al. (1995) 2-Arachidonoylglycerol: a possible endogenous cannabinoid receptor ligand in brain. Biochem Biophys Res Commun 215:89–97
Suzuki A, Josselyn SA, Frankland PW et al. (2004) Memory reconsolidation and extinction have distinct temporal and biochemical signatures. J Neurosci 24:4787–4795
Suzuki A, Mukawa T, Tsukagoshi A et al. (2008) Activation of LVGCCs and CB1 receptors required for destabilization of reactivated contextual fear memories. Learn Memory 15:426–433
Sweatt JD (2001) The neuronal MAP kinase cascade: a biochemical signal integration system subserving synaptic plasticity and memory. J Neurochem 76:1–10
Takahashi RN, Pamplona FA, Fernandes MS (2005) The cannabinoid antagonist SR141716A facilitates memory acquisition and consolidation in the mouse elevated T-maze. Neurosci Lett 380:270–275
Terranova JP, Storme JJ, Lafon N et al. (1996) Improvement of memory in rodents by the selective CB1 cannabinoid receptor antagonist, SR 141,716. Psychopharmacology (Berl) 126:165–172
Thiemann G, Fletcher BC, Ledent C et al. (2007) The genetic versus pharmacological invalidation of the cannabinoid CB(1) receptor results in differential effects on 'non-associative' memory and forebrain monoamine concentrations in mice. Neurobiol Learn Memory 88:416–423
Van Gaal LF, Rissanen AM, Scheen AJ et al. (2005) Effects of the cannabinoid-1 receptor blocker rimonabant on weight reduction and cardiovascular risk factors in overweight patients: 1-year experience from the RIO-Europe study. Lancet 365:1389–1397
Van Sickle MD, Duncan M, Kingsley PJ et al. (2005) Identification and functional characterization of brainstem cannabinoid CB2 receptors. Science 310:329–332
Varvel SA, Lichtman AH (2002) Evaluation of CB(1) Receptor Knockout Mice in the Morris Water Maze. J Pharmacol Exp Ther 301:915–924
Varvel SA, Anum EA, Lichtman AH (2005) Disruption of CB(1) receptor signaling impairs extinction of spatial memory in mice. Psychopharmacology (Berl) 179:863–872
Varvel SA, Cravatt BF, Engram AE et al. (2006) Fatty acid amide hydrolase (-/-) mice exhibit an increased sensitivity to the disruptive effects of anandamide or oleamide in a working memory water maze task. J Pharmacol Exp Ther 317:251–257
Varvel SA, Wise LE, Niyuhire F et al. (2007) Inhibition of fatty-acid amide hydrolase accelerates acquisition and extinction rates in a spatial memory task. Neuropsychopharmacology 32:1032–1041
Welzl H, D'Adamo P, Lipp HP (2001) Conditioned taste aversion as a learning and memory paradigm. Behav Brain Res 125:205–213
White NM, McDonald RJ (2002) Multiple parallel memory systems in the brain of the rat. Neurobiol Learn Memory 77:125–184
Wilson RI, Nicoll RA (2001) Endogenous cannabinoids mediate retrograde signalling at hippocampal synapses. Nature 410:588–592
Winsauer PJ, Lambert P, Moerschbaecher JM (1999) Cannabinoid ligands and their effects on learning and performance in rhesus monkeys. Behav Pharmacol 10:497–511
Wise LE, Iredale PA, Stokes RJ et al. (2007) Combination of rimonabant and donepezil prolongs spatial memory duration. Neuropsychopharmacology 32:1805–1812
Wolff MC, Leander JD (2003) SR141716A, a cannabinoid CB1 receptor antagonist, improves memory in a delayed radial maze task. Eur J Pharmacol 477:213–217
Wotjak CT (2005) Role of endogenous cannabinoids in cognition and emotionality. Mini Rev Med Chem 5:659–670

Endocannabinoids and the Non-Homeostatic Control of Appetite

Tim C. Kirkham

Contents

Abstract The usual physiological perspective on appetite and food intake regards control of eating simplistically, as merely the reflexive behavioural component of a strict homeostatic regulatory system. Hunger is seen to arise in response to energy deficit; meal size is determined by the passage of nutrients into the gut and the stimulation of multiple satiety signals; and overall energy intake is modified to reflect the balance of fuel reserves and energy expenditure. But everyday experience shows that we rarely eat simply through need. Rather, food stimuli exert a powerful influence over consumption through their appeal to innate and learned appetites, generating the psychological experiences of hunger, craving and delight independently of energy status. That these important and influential subjective experiences are mediated through complex neurochemical processes is self-evident; but the chemical nature of our infatuation with, and subservience to, the motivating

T.C. Kirkham
School of Psychology, University of Liverpool, Eleanor Rathbone Building, Bedford Street South, Liverpool, L69 7ZA, England
e-mail: t.c.kirkham@liverpool.ac.uk

D. Kendall and S. Alexander (eds.), *Behavioral Neurobiology of the Endocannabinoid System*, Current Topics in Behavioral Neurosciences 1,
DOI: 10.1007/978-3-540-88955-7_9,

properties of foods are overshadowed by mechanistic, peripherally anchored models that take little account of psychological factors, and which consequently struggle to explain the phenomenon of obesity. This chapter discusses recent developments that suggest the endocannabinoids are key components of the central mechanisms that give rise to the emotional and motivational experiences that lead us to eat and to overconsume.

Keywords Appetite • Craving • Hyperphagia • Orexigenic

1 Obesity and the Problem of Homeostasis

When considering the orthodox position that energy intake and body weight are homeostatically regulated in reference to energy expenditure and levels of available fuel reserves, the early twenty-first century provides us with some problems. Specifically, if efficient homeostatic mechanisms have evolved to maintain energetic constancy, how do we explain the increasing prevalence of obesity? Are overconsumption and obesity a consequence of the failure of energy/body weight homeostasis due to powerful psychological factors that can override regulation for the sake of indulgence; or are they actually evidence of the lack of adequate regulatory systems that function to modulate appetite against energy reserves?

The past few decades have witnessed a dramatic increase in the number of central and peripheral chemical factors proposed to control food intake. But with the general acceptance of homeostatic models in this field, the predominant theme of research and theorising centres on the determination of the negative feedback signals that such a model requires to regulate feeding. Neurocrine, endocrine and adipokine messengers have been identified and proposed, through hypothalamic integration, to provide the neural brake to ingestion; arguably to prevent unnecessary overconsumption, and to maintain a stable internal energetic milieu and body weight constancy. The list of these "satiety" factors grows annually: most, by dint of their ability to suppress food intake (naturalistically or otherwise), are seen as key to the regulation of energy intake and to directly serve the homeostatic imperative (Konturek et al. 2005; Maljaars et al. 2007; Valassi et al. 2008; Wren and Bloom 2007). With such an armoury of inhibitory feedback signals it is surprising that overconsumption, let alone obesity, is ever expressed. Yet people continue to overconsume and to gain weight – and to become massively obese. And a typical, but less-than-satisfactory theoretical response is that negative regulation fails due to excessive demand on one or more of these multiple inhibitory systems (Scarpaci and Zhang 2007; Kalra 2008).

By contrast, the mechanisms that promote eating motivation, and the signals that may act to instigate feeding have received far less attention (Beck 2007). However, the systems that underlie hunger, and particularly food-craving and hedonic responses to food, represent considerably greater influences on the frequency, quantity and variety of consumption. Consequently, should we wish to pursue the totality of any homeostatic influences on feeding, or be set on the development of pharmaceutical interventions to restrict food intake, these positive motivational factors represent crucial targets for investigation. There is, however, an unfortunate tendency for researchers to ignore their own, common experience of the power of food to engender appetite, or to consider the possibility that this phenomenon represents more than an accident of psychology and an inconvenient interference to homeostatic regulation. This, in turn, leads to neglect of the evolutionary parsimony of the pre-eminence of the psychological processes that maintain positive energy balance and adequate nutritional status in the face of potentially unreliable nutrient sources.

The concept of satiation of a motivated behaviour was initially invoked to explain its eventual cessation. This conceptualization may, however, be distinguished from the existence of dynamic mechanisms that actively suppress the desire to eat. Satiation of eating may be a phenomenon that, acutely, is influenced by food intake and the bodily recognition that reserves are being restored or replenished. But, given the unpredictability of food availability in the world of our evolutionary antecedents, control of eating should not be wholly dependent upon such short-term considerations. An adaptive organism in an unpredictable food environment requires the control of eating motivation to be responsive to the immediate vagaries of that environment, and also to reflect the need for long-term energetic security. The central integration of oro-oesophageal and gastrointestinal sensory inputs resulting from the ingestion of food may act to dampen the urgency of eating motivation, but it is unlikely that the rate of passage of food through the gut during a meal can provide sufficiently rapid calculation of nutrient or utilisable energy content to accurately control meal size in relation to homeostatic constraints. Rather, physico-chemical stimuli have predictive, partly learned, qualities that signal later energy absorption (Gibson and Brunstrom 2007). These qualities are permissive in that they allow attention to be diverted to other aspects of the environment that may then gain motivational impact in relation to the satisfaction of other bodily needs. Satiety, then, need not be considered as a condition that will only wane with the passage of time and the expenditure of recently consumed calories, according to some immutable energetic equation – but is rather a reflection of the relative valence of competing stimuli for alternative motivational imperatives.

Indeed, appetite for food, itself, provides a prime example of such motivational lability: evident in the all too common experience whereby the sight or smell of food can provoke the desire to eat in the absence of any pre-existing, conscious yearning for food – or even, as with some tantalising dessert, when we may actually be at the point of physical discomfort from the recent consumption of a large meal, and should presumably already be subject to the inhibitory actions of manifold homeostatic satiety signals. Similarly, eating may be disrupted and terminated by the appearance of more urgent, environmental stimuli. The primacy and potency of external stimuli to promote eating should not be considered as incidental to the physiological investigation of how food intake is controlled.

From an evolutionary perspective, eating may be considered to be an opportunistic activity (albeit a pre-eminent factor for survival), arguably developed to suit an ancestral environment of periods of plenty alternating with limited food availability (Speakman 2007). We might perhaps retain a weak homeostatic model, whereby specific physiological influences reflecting nutritional state can operate on behaviour – in the short- or long-term – to affect the relative priority of food seeking and consumption. Examples of these would include responses to enforced fasting, maintaining sufficient energy/nutrient levels for fecundity, or the necessity to consume essential amino acids to allow the synthesis of certain neurotransmitters. However, observing modern people, there is less than compelling support for the short-term regulation of energy intake relative to short-term energy expenditure. Certainly, there appears to be no strong case that excessive energy intake manages to counteract eating motivation sufficiently to prevent obesity. Thus, Levitsky has discussed the apparent "non-regulation" of human eating behaviour, and the fact that "in the vast literature on human feeding behaviour, there is very little evidence eating behaviour is tightly coupled to biological mechanisms involved in energy balance" (Levitsky 2005). Indeed, his research provides convincing demonstrations of how energetic manipulations fail to elicit compensatory behavioural responses, contrary to the predictions of a homeostatic system that couples feeding to body weight regulation (Levitsky 2002).

1.1 Eating for Survival

To consider the governance of eating solely in terms of the instantaneous maintenance of finite levels of key factors such as glucose availability or glycogen stores, or longer-term adjustments correlated with adiposity levels, is to avoid the realities of the ancestral environment that shaped the evolution of the motivational and behavioural processes controlling ingestion. In the sense used by Cannon (1932), homeostasis as "a condition which may vary, but which is relatively constant" may be easily applied to the demonstrable physiochemical processes that maintain the immediate availability of sufficient energy and nutrient levels to the cells of the body. But long-term survival requires both the persistence of sufficient reserves to support the regulation of the milieu interne, and the development of behavioural processes that will maintain those reserves in anticipation of probable – but unpredictable – shortfall. Hence, maintaining too fine a balance of energy reserves against expenditure renders an organism unnecessarily subject to the consequences of privation.

Meal-taking represents a punctuation between our other daily activities, its regularity – or otherwise – reflecting food availability, lifestyle and a host of individual and environmental factors. For our early ancestors, meal frequency must have been largely determined by the relative availability and nutritional quality of food. An opportunistic grazing and scavenging style would have consumed much time and energy in relatively impoverished pre-agrarian environments;

even the acquisition of more sustaining, meat-containing diets through hunting would have entailed considerable energy expenditure and irregular meals. Only in relatively recent times, with the development of social structures built around agriculture and industrialisation, did predictable food supplies enable a formalised ritual of regular daily meals. These were coordinated to match the demands of laborious, energy-costly lifestyles, the expediencies of food preparation and the cycles of working and social life. For modern people, the occurrence of the psychological and visceral components of hunger at pre-determined meal times merely reflects conditioned, cephalic responses to the imminent, expected arrival of food: a bodily adaptation for maximising the efficient absorption and utilisation of a predictable supply of nutrients, rather than – except under extremis – a response to actual energetic deficit. In a healthy individual, hunger will eventually subside if a mealtime passes but no food is eaten. This motivational lapse does not appear to be the signature of a truly homeostatic system, responding on an instantaneous basis to the demands of finely tuned energetic requirements. Individuals with similar physiques, activities and energy requirements may have very disparate styles of eating, in terms of the distribution and size of meals. In fact, eating patterns may be seen to be largely subject to habit and conditional influences (Gibson and Brunstrom 2007). Moreover, dietary styles and patterns of consumption can shift very quickly, depending on factors such as the availability and palatability of supplies, and these changes will occur in the absence of altered energetic demands.

Of course, hunger and voracious eating may be engendered in response to weight loss under extreme environmental conditions, such as starvation (as is the predominant mode in animal research), or by dramatic experimental manipulations that reduce the availability of cellular fuels (but which barely reflect physiological conditions). Under these circumstances, appetite arises from a real and urgent energy deficit – or some neuroendocrine representation that simulates that deficit – and consequently provokes the behaviours necessary to defend critical adipose mass and fuel availability. But so much (if not all) of our eating arises in the absence of any such urgency: seemingly defying the long-term defence of a critical level of adipose stores, and overcoming the apparent redundancy of the putative inhibitory controls of meal size and meal frequency.

The pre-potency of external food-related stimuli in determining the recrudescence of the desire to eat and the provocation of overindulgence, particularly apparent in modern societies, should not be considered accidental. We have clearly evolved systems that are acutely sensitive to environmental signals and which, under modern conditions, allow us to feel hungry when replete, overconsume in response to variety, and be particularly desirous of high calorie, sweet, fatty foods (Yeomans 2007). As many of us know to our cost, regular consumption of the latter diet will easily produce substantial weight gain, clearly provoking deviations from hypothetical, finely gauged, homeostatic set-points – or even less-finite "settling" points – for adiposity or body weight. And, acutely, an individual snacking on relatively small quantities of junk foods can easily and quickly ingest calories far in excess of any immediate need – and do so long before any inhibitory, satiating, regulatory gut signals might be invoked. The extensive literature on sensory-specific satiety also indicates that subjective feelings of being replete can be overridden by the simple presentation of a new, palatable food (Rolls 1986).

Authors who view homeostasis more flexibly argue that feeding is not necessarily tied to the authority of negative feedback signals and is, in fact, not a regulated variable in the traditional homeostatic sense. Rather, to quote Woods and Ramsay, "eating is a behaviour that functions to *stabilise* adiposity over long intervals" (Woods and Ramsay 2007, p. 393). It is possible to further question the notion of even approximate long-term stabilisation of adiposity, at least in the sense that food intake might decline once a certain, surplus, level of fat storage has been achieved. Judging by the increasingly common examples of morbid obesity, any regulatory influences that might be predicated to counter "excess" fat accumulation may be considered ineffectual. Is this ineffectuality brought about by the extreme "obesogenic" environment in which we live, and a failure of normal feedback mechanisms (implicit, for example, in the notions of leptin resistance or leptin insufficiency syndrome; Kalra 2008) – or is there a more straightforward explanation?

1.2 Gluttony and Externality

Given the fact of obesity, and that – self-evidently – the obese do not appear to display significant automatic regulatory changes in eating behaviour to sufficiently counter the accrual of adipose stores, parsimony would indicate that there are simpler accounts. More specifically, we can argue that evolution has provided us with a "greedy" phenotype. The premise here is that, in all but a very few pathological instances, the obesity epidemic that confronts us reflects the fact that we are biologically programmed for gluttony. While the modern environment now provides us greater opportunity to indulge in the pleasure from varied, energy-dense foods, it is reasonable to assume that the underlying neural mechanisms mediating these cravings and delights represent a deep-time relationship between our world and the neural mechanisms that reinforce behaviours that lead to food acquisition (Sullivan et al. 2008). Eating is absolutely fundamental to survival, and mechanisms that encourage us to eat in order to maintain buffering reserves of fuel might therefore be considered essential. In reward systems, natural selection has seemingly given us the necessary apparatus to ensure that we respond positively to opportunities to acquire nutrients and supplement our fuel reserves. Some argue that our modern susceptibility to greed is more akin to the development of addictions subsumed by the same incentive and reward pathways that can drive food consumption (Lowe and Butryn 2007; Acosta et al. 2008). But as the hedonic processes that underlie the anticipation or experience of pleasure and satisfaction are such profound and fundamental components of general motivational systems, can we reasonably argue that efficient feeding control mechanisms could have evolved without their integration (Figlewicz Latteman et al. 2007)?

Should we not question, therefore, the notion that evolutionary pressures would have given us multiple mechanisms to prevent the storage of excess calories, or actively redress accidental overconsumption, when the most pertinent challenge to survival was the lack of food, not its surfeit? If our opportunistic susceptibility to overconsumption does constitute a component of a regulatory regime, then it may most easily be regarded as one that increases the likelihood that future energetic demands will be met – to support the truly homeostatic, moment-to-moment maintenance of cellular fuel availability. Overeating then is not counter-regulatory, but represents the most effective behavioural mechanism for ensuring that energy input can match future requirements. The ease with which appetite can be engendered, the hedonic processes that promote overconsumption, and the efficiency and remarkable capacity of fuel storage in adipose tissues all provide evidence that the ready development of obesity reflects the evolution of highly effective, persistent mechanisms to ensure positive energy balance. From this perspective, the development of cuisines and eating styles, and the overwhelming variety of foods provided through the industrialisation of food production are the consequence of our proclivity for enjoyment of tastes and flavours. Modern food manufacturers only exploit that proclivity, reflecting centuries of use of salt, herbs, spices and sugars invoked to overcome the bland monotony of our staple diets.

As described by Cannon in his *Wisdom of the Body* (1932), "... the person beset by an appetite is tempted, not driven, to action – he seeks satisfaction, not relief." He also observed, however, that "... the two motivating agencies – the pang and the pleasure" may not always be separable. Lowe has coined the term "hedonic hunger" to describe eating arising from a pre-occupation with "thoughts, feelings and urges about food in the absence of any short- or long-term energy deficit" (Lowe and Butryn 2007). But whatever the descriptor, the desire for food in the absence of energy deficit, and in particular the craving and enjoyment of the most palatable, highly calorific foods is a fundamental aspect of eating motivation – and may be viewed as a quintessential component of normal, non-pathological mechanisms that promote survival through the maintenance of positive energy balance. Modern overindulgence is merely a reflection of the indulgent nature of motivational processes, and the critical importance of hedonic factors in energising and guiding behaviour.

Notwithstanding the preceding arguments on the validity of homeostatic models in the control of eating, the contribution of the processes that give rise to appetite in overconsumption, overweight and obesity is all too evident. The overshadowing of appetitive processes by the overwhelming study of satiety mechanisms and lipostatic negative feedback is therefore clearly deserving of some redress. A greater knowledge of the neurochemical factors underlying the urge to eat, food anticipation, or the pleasure derived from eating will have crucial implications for understanding general motivational processes, as well as having far-reaching clinical implications. To this end, we now turn our discussion to the endogenous cannabinoids, and address the evidence for their role in the normal biopsychological mechanisms that create appetite and stimulate eating. More specifically, the following sections will address their contribution to incentive processes and the hedonic evaluation of food stimuli – or what Berridge (2000) has respectively described as "wanting" and "liking".

2 Endocannabinoids in Food Craving, Anticipation and Palatability

The appetite-stimulating action of the cannabis plant (*Cannabis sativa*) and its extracts has been documented for many centuries (Abel 1971; Kirkham and Williams 2001a). This effect is attributable to Δ^9-tetrahydrocannabinol (THC), one of a large group of "cannabinoid" molecules characterised in the 1960s (Gaoni and Mechoulam 1971). As described in other chapters, the actions of these phytocannabinoids in people were explained by the discovery of cannabinoid (CB) receptors and their endogenous ligands, the "endocannabinoids" (Pertwee 2008). The psychoactive effects of THC are thus explained by its ability to mimic the neural actions of the endogenous agonists at their receptors.

2.1 THC Hyperphagia

The earliest references to the hyperphagic actions of cannabis are found in the ancient Hindu writings of the *Rajanirghanta* (1,700 BP). Surprisingly, however, very little empirical research has been conducted in humans to clarify the plant's specific actions on appetite. The same limitation applies to exploration of the actions of THC or other phytocannabinoids, in people or in animal models. As a consequence of governmental paranoia about the use of these compounds, theorising about the psychological and behavioural activity of the cannabinoids has been overly reliant on anecdotal accounts from cannabis users and a limited number of, often poorly designed, animal studies. Even with the recent enthusiasm for endocannabinoid systems as therapeutic targets for obesity, metabolic syndrome and cardiovascular disease, relatively little work has been conducted to explore the critical actions of cannabinoid receptor agonists on appetite. Nevertheless, recent years have produced some important insights into the actions of THC and the endocannabinoids (Kirkham 2005). Here, we shall discuss the principal findings that guide us to the likely mechanisms whereby CB_1-specific drugs affect eating motivation and the role for endocannabinoid systems in the normal control of ingestive behaviour.

Much early research failed to obtain THC hyperphagia in rats, or reported only relatively weak effects. However, by adopting a pre-feed paradigm in which the rats were thoroughly sated before drug administration, we found that oral THC administration would reliably stimulate nocturnal feeding (Williams et al. 1998). We also demonstrated that THC hyperphagia is mediated by CB_1 receptors, being attenuated by rimonabant (SR141716), the selective CB_1 receptor antagonist, but not by SR144258, a selective antagonist of the CB_2 receptor (Williams and Kirkham 2002a).

One aspect of our data that deserves particular emphasis is the magnitude of the overconsumption induced by THC. Our animals were thoroughly satiated, having already eaten a quantity of a palatable wet mash diet equivalent in weight to their normal daily food intake. The substantial intake that followed THC treatment thus signifies that stimulation of CB1 receptors can provoke an exceptionally powerful stimulus to eat (and, arguably, overriding any supposed inhibitory feedback that arises from meal consumption). Moreover, the extent of THC-induced overeating easily matched that induced by central administration of neuropeptide Y, commonly regarded as one of the most potent orexigens (Chee and Colmers 2008). The remarkable hyperphagic potency of THC, and its mediation by CB_1 receptors, thus provided strong support for involvement of the endocannabinoid systems in the normal control of feeding.

2.2 *Orexigenic Actions of the Endocannabinoids*

This possibility necessitated the demonstration that the endocannabinoids, themselves, will exert hyperphagic actions. Using our pre-feed design, we found that systemic anandamide (AEA) administration significantly increased food intake (Williams and Kirkham 1999), although the degree of overeating was modest compared with the effects of THC. Moreover, AEA hyperphagia was blocked by rimonabant pre-treatment – but not by a CB_2 antagonist, again indicating that the overeating was specifically mediated by CB1 receptors. Subsequently, the feeding effects of systemic AEA were replicated in mice (Hao et al. 2000) and rats (Gómez et al. 2002). Later experiments also confirmed that 2AG can exert a potent stimulus to eat (Kirkham et al. 2002). More recently, a third endogenous CB_1 agonist, noladin ether, has been shown to increase food intake in food-restricted mice (Avraham et al. 2005) and free-feeding rats (Rogers and Kirkham, unpublished results). Remarkably, there has been very little subsequent investigation of the orexigenic actions of these compounds (with only five publications in a decade directly exploring their feeding effects).

The first published demonstration of a central site of action of endocannabinoids was by Jamshidi and Taylor (2001), who found that modest, rimonabant-reversible hyperphagia could be obtained by direct injection of AEA into the ventromedial hypothalamus (VMH). Subsequently, we demonstrated (so far uniquely) that administration of 2AG into the shell sub-region of the nucleus accumbens (AcbSh) exerts a potent, CB_1-selective, hyperphagic action (Kirkham et al. 2002). Similar effects have also been obtained with intra-accumbens AEA treatments (Mahler et al. 2007; Soria-Gomez et al. 2007). As we shall discuss below, the sensitivity of the AcbSh as a locus for endocannabinoid-induced feeding is critical to current hypotheses concerning their role in eating motivation. However other brain regions are also implicated in endocannabinoid effects. We have shown increases in food intake after AEA or 2AG administration into the lateral (LH) and paraventricular (PVN) nuclei of the hypothalamus (Kirkham and Williams 2001c), while others have obtained reliable hyperphagia following lateral ventricular administration of THC

(Koch and Matthews 2001). Circuits located in feeding-relevant hindbrain areas, such as the dorsal motor nucleus of the vagus (DMV) and the nucleus tractus solitarius (NTS), may also be subject to cannabinoid regulation, with the synthetic CB_1 agonist CP55940 reported to enhance milk intake when administered into the fourth ventricle (Miller et al. 2004).

An essential corollary to the agonist data, of course, is the extensive investigations of the feeding effects of CB_1 antagonists. Critically, these drugs have universally been found to suppress food intake (although debate persists about the actual specificity of the suppression to direct interruption of eating motivation (Tallett et al. 2007a, b). An anorectic action of rimonabant was first reported by Arnone and colleagues, providing evidence for tonic endocannabinoid activity in feeding-related systems (Arnone et al. 1997). Reliable anorectic actions of rimonabant, or its analogues (e.g. AM281, AM251, surinabant), have since been reported, following systemic or central administration in satiated or food-deprived animals, and after acute or chronic treatments (e.g. Colombo et al. 1998; Shearman et al. 2003; Werner and Koch 2003; Chen et al. 2004; Rutkowska 2004; Wiley et al. 2005; Rinaldi-Carmona et al. 2004). Several studies have also reported anorectic effects of CB_1 blockade in dietary (Hildebrandt et al. 2003; Ravinet Trillou et al. 2003) and genetic models of obesity, often with greater effects than in lean littermates (e.g. Vickers et al. 2003; Zhou and Shearman 2004).

3 Behavioural Characterization of Cannabinoid Hyperphagia: The Reward Hypothesis

These findings provide evidence for some role of endocannabinoids in feeding. However, increases or decreases in food intake alone tell us little about what aspects of eating motivation are altered to affect behavioural change. Indeed, given the wide pharmacological spectrum of cannabinoids, it is essential to demonstrate that these changes follow from a natural adjustment to feeding motivation, rather than some non-specific action. (Until recently, the behavioural pharmacology of cannabinoids was largely the study of high, non-selective, frankly sedative doses of agonists – and for many years, anorectic actions of THC were the most frequently reported in feeding experiments, leading to suggestions of "species differences" between animals and rodents to explain the consequences of inappropriate choice of dose.)

An hypothesis that gained early currency – on the basis of the anecdotal accounts of cannabis users described earlier (Tart 1970) – was that endocannabinoids may provoke overconsumption by amplifying the orosensory reward, or palatability, of foods (Arnone et al. 1997). Initially, this notion was supported by studies that were interpreted as indicating a more marked susceptibility of palatable foods to the stimulant effects of THC. Arnone and colleagues reported that in rats and marmosets rimonabant selectively attenuated the consumption of palatable ingesta

(Arnone et al. 1997; Simiand et al. 1998), while having little effect on intake of bland food. These workers suggested that such preferential effects of CB_1 blockade indicated important tonic endocannabinoid activity underlying food reward. Thus, cannabinoid agonists might increase food intake by rendering foods more palatable, while antagonists might tend to diminish the hedonic value of foods, and so reduce consumption. As we shall see, there are data that support this notion but the specifics are a little more complex.

With these antagonist data in mind, we began a series of studies to directly address the cannabinoid-reward hypothesis. Initially, we measured the effects of rimonabant on sucrose sham-feeding to explore drug effects on ingestion maintained entirely by palatability. In this model, rats are surgically implanted with a chronic gastric fistula and ingest palatable sucrose solutions, which are recovered within seconds directly from the stomach – thus avoiding the normal inhibitory consequences of gastric distension or of nutrient entry into the duodenum. Sham-feeding rats will consume many times the amount of sucrose solution ingested by intact, normally feeding rats. Moreover, the rate of sham-feeding is proportional to the palatability of the sucrose: the sweeter the solution, the more avid the ingestive response. Consequently, the model is particularly sensitive to manipulations that affect orosensory reward.

We hypothesised that, if endocannabinoids directly mediate food reward, sham-feeding should be disrupted by CB1 blockade. More specifically, we anticipated that suppression of sham-feeding by rimonabant would produce changes in behaviour which resemble the effect of diluting the sucrose solution (Kirkham 1990; Kirkham and Cooper 1988). A precedent for such an effect comes from our previous work with opioid antagonists. Opioids are heavily implicated in orosensory reward (see Sect. 4, below), and opioid receptor antagonists reduce sucrose sham-feeding in a manner which exactly mimics the changes in ingestion produced by mere sucrose dilution, and hence the palatability, of the sucrose. In addition, attenuation of sham-feeding by opioid antagonists can be reversed by increasing the palatability of the sucrose during a sham-feeding test (Kirkham and Cooper 1988; Kirkham 1990; Leventhal et al. 1995).

Contrary to our expectations, rimonabant failed to affect sucrose sham-feeding (Kirkham and Williams 2001b). Even doses of the drug ten times greater than those required to reverse cannabinoid-induced feeding (Williams and Kirkham 1999), or doses which suppress sucrose drinking in intact animals (Arnone et al. 1997), were ineffective. This failure argued against significant endogenous cannabinoid activity within the pathways that *maintain* sucrose ingestion. In other words, endocannabinoids did not seem to be primarily involved in food reward during ingestion, and might not be considered crucial to the pleasure derived from orosensory characteristics of food ("liking"); we shall return to this issue later. However, while our data did not support endocannabinoid mediation of the consummatory aspects of food reward, they did not preclude their involvement in some other aspect of feeding-related, emotional processes. For example, endocannabinoids could still be associated with appetitive, or incentive, aspects of feeding motivation, related to the anticipation of food or the desire to eat ("wanting").

3.1 Endocannabinoids and "Wanting": Primary Motivational Actions

Evidence for such a role of endocannabinoids comes from studies using progressive ratio paradigms as a model of craving. In these, rats are required to complete a progressively greater number of responses to obtain successive rewards of small amounts of liquid or food (typically sucrose solutions or pellets). The ratio at which animals cease to respond (the "break-point") is taken as an index of the degree of craving. Gallate and McGregor (1999) found that rimonabant dose-dependently reduced break-point, while after administration of the CB_1 agonist CP55,940 rats would work harder to obtain reinforcement, resulting in increased break-points (Gallate et al. 1999). THC will also produce rimonabant-reversible increases in responding for food (Solinas and Goldberg 2005). Recently, we have shown that systemic administration of a hyperphagic dose of noladin ether will also increase break-point (Rogers and Kirkham, unpublished data). These effects, which were reversed by the CB_1 antagonist surinabant, strongly implicate endocannabinoid systems in the processes underlying the motivation to obtain palatable ingesta. In line with these studies are reports that CB_1 knockout mice have reduced sensitivity to the motivating properties of food. Thus, $CB_1^{-/-}$ animals show lower levels of responding for sweet food and exhibit lower break-points than wild-type mice (Sanchis-Segura et al. 2004). Interestingly, however, antagonist effects on operant responding are also evident with more bland foods (Freedland et al. 2000; Pério et al. 2001), and rimonabant has proved equi-anorectic when tested with foods of differing macronutrient content and intrinsic palatability (Verty et al. 2004a, b). This generality of effect suggests that endocannabinoids modulate appetitive processes per se, to provide a general gain in the incentive value of all food (and is, incidentally, therefore supportive of the notion that common incentive processes are engaged by external food quality or internal deficit stimuli).

We have obtained further data to support endocannabinoid involvement in incentive motivation. Using an open-field apparatus, we observed the behaviour of pre-satiated rats following administration of THC or AEA. Under control conditions, rats generally displayed little motivation to eat: when eating did occur, it did so only after many minutes engaged in exploratory behaviours. By contrast, both exogenous and endogenous cannabinoid treatments stimulated feeding, dramatically reducing the latency to eat. Crucially, once initiated, the subsequent pattern of feeding behaviour displayed by THC- and AEA-treated rats in the open field was similar to that of untreated rats feeding freely in their home cages (Williams and Kirkham 2002b). These data are again compatible with an action of cannabinoids to increase the incentive value or salience of the food, and importantly, indicate that cannabinoids provoke feeding through adjustments to natural feeding control mechanisms.

We have also observed these effects using a more naturalistic, continuous meal pattern monitoring technique, where moment-to-moment feeding is monitored in animals' home cages. Under these circumstances, the latency to the first meal of pre-satiated or free-feeding rats is consistently reduced after central AEA and 2AG administration, often by more than an hour compared with the control condition (Kirkham and Williams 2001b). Indeed, in all experiments where we have measured the temporal distribution of feeding episodes, increases in total food intake over a finite test interval derive principally from the advance of eating onset. Together with the break-point data, such findings again imply that stimulation of CB_1 receptors increases the salience of food and hence the motivation to eat. We thus begin to see the development of a model which links endocannabinoids directly to the processes that lead to the initiation of feeding. Recalling the lack of effect of cannabinoid receptor blockade on the intra-meal palatability factors that maintain sham-feeding, most data support specific endocannabinoid involvement in the motivational processes that culminate in meal taking. The combination of CB_1 ligand effects on feeding microstructure and the motivation to work for food thus strongly implicate endocannabinoids in "wanting" processes.

In effect, the stimulatory actions of the cannabinoids on eating resemble the changes that occur with food deprivation, since both increase food salience, reduce eating latency and promote short-term hyperphagia (Marín Bivens et al. 1998). We might therefore expect to see activation of endocannabinoid systems under conditions where the incentive value of food is naturally raised. Some support for this comes from our finding that regional brain levels of AEA and 2AG increase after fasting (Kirkham et al. 2002), and that the anorectic action of rimonabant is significantly enhanced in food-deprived rats compared to non-deprived animals (Kirkham and Williams 2001b; Osei-Hyiamen et al. 2005). By contrast, levels of AEA or 2AG in animals re-feeding after fasting, or in rats eating a palatable diet, did not show any increase – again indicating only an indirect role for endocannabinoids in the maintenance of feeding once it has been initiated. It remains to be seen whether direct measures of brain endocannabinoids will support their activation by external food stimuli of high valence (such as the presence, or promise, of highly palatable foods).

In addition to the animal literature, there are some indications from recent human studies with a CB_1 agonist and antagonist that provide support for the role of endocannabinoids in appetitive processes. For example, we explored the acute effects of oro-mucosally administered THC on eating in healthy volunteers. In addition to a significant increase of energy intake, we found that one of the principal effects of the drug was a marked amplification of the normal pre-lunch rise in subjective hunger scores. Compared to placebo, hunger was seen to increase far earlier in the morning after THC, with significantly higher pre-prandial peak levels. This effect was associated with an earlier onset and increased incidence of snacking (Townson and Kirkham, unpublished results). Complementary data from CB_1 blockade are provided by a clinical trial with rimonabant. Over the course of 3 months of daily dosing, patients' appetite ratings were periodically assessed by laboratory test meals and home questionnaires (Blundell et al. 2006). Critically, rimonabant was found to lower hunger and desire to eat at the start of the meal, while having no effect on post-meal ratings of hunger or fullness. Significant reductions in hunger and the frequency and strength of food cravings were also detected over the course of the study, while rimonabant had no reliable effect on ratings of food pleasantness. These agonist and antagonist effects provide the most direct indications so far that CB_1 receptor ligands can specifically modulate "wanting" aspects of eating motivation, and are clearly worthy of further investigation.

Importantly, the apparent involvement of endocannabinoids in appetitive aspects of feeding is compatible with the known effects of CB_1 agonists and antagonists on mesolimbic dopaminergic neurons, arising in the ventral tegmental area (VTA) and projecting to the nucleus accumbens. This pathway is linked to incentive motivation, and the generation of emotional arousal and behavioural activation in response to stimuli that predict reward (Berridge 2007). Food stimuli cause dopamine release in the nucleus accumbens, especially after deprivation, or if the food is novel or palatable. There is growing support for an influence of endocannabinoids on mesolimbic dopaminergic activity. For example, CB_1 receptors are co-localised, and interact, with dopamine D1 and D2 receptors (Pickel et al. 2006; Meschler and Howlett 2001). Moreover, both THC and AEA have been reported to stimulate dopamine release in the nucleus accumbens (Gardner and Vorel 1998; Solinas et al. 2006). And Verty et al. (2004b) have shown that a behaviourally silent dose of the D1 antagonist SCH23390 prevents THC hyperphagia. It is therefore noteworthy that the accumbens dopamine release that is provoked by presentation of a novel, palatable food is blocked by rimonabant (Melis et al. 2007), suggesting that endocannabinoids normally facilitate the mesolimbic dopamine signalling that can give rise to appetite. This may occur in part through a disinhibitory action of endocannabinoids, whereby stimulation of accumbens CB_1 receptors suppresses glutaminergic activity and inhibits GABAergic medium spiny neurons that normally constrain the firing of VTA dopamine neurons (van der Stelt and Di Marzo 2003). Riegel and Lupica (2004) have suggested that, through independent pre- and post-synaptic mechanisms, endocannabinoids and dopamine in the VTA-accumbens pathways may co-operate to dynamically fine-tune the activity of incentive pathways in response to salient environmental stimuli. Endocannabinoids may thus be essential for the orientation to motivationally significant stimuli, the attribution of incentive salience and reward anticipation, and the elicitation of appropriate behavioural responses such as food seeking and eating initiation.

One might expect that endogenously entrained eating patterns (and perhaps also those arising from conditioned hunger) might show some linkage to changes in endocannabinoid activity. So far, there is little evidence for such specific associations, but circadian rhythms in endocannabinoid levels have been detected in the brains of rats. For example, Valenti et al. (2004) demonstrated clear diurnal variations, albeit with differential changes evident for AEA and 2AG. Thus, AEA levels were highest in the dark phase, when most eating occurs, while 2AG exhibited peak levels during daylight. These opposing variations (which may reflect changes in synthesis or metabolism of endocannabinoids) indicate the urgent necessity to explore the individual roles of each endocannabinoid in behavioural processes. It is also a priority to explore such changes with higher temporal resolution, to more precisely match behaviour to endocannabinoid activity.

3.2 *Endocannabinoids and "Liking": Secondary Motivational Actions*

The above findings link endocannabinoids to appetitive processes in feeding. However, despite our sham-feeding data, their role may also be extended to involvement in food "liking". As we have noted, such a role is clearly suggested by the anecdotal reports of cannabis users (Tart 1970), and recent animal studies have provided support for a specific interaction of endocannabinoids with food palatability.

In early reports, CB_1 receptor blockade was reported to preferentially attenuate the intake of palatable, sweet foods (Arnone et al. 1997; Simiand et al. 1998) and reduce operant responding for sweet food (Pério et al. 2001); while CB_1 knockout mice consume less sucrose than wild types (Poncelet et al. 2003). We have observed that central injection of endocannabinoids can induce modest increases in meal duration and, consequently, meal size. Although these effects are weak in relation to the effects on the latency noted above, they are compatible with an action of the agonists to enhance food palatability (i.e. eating may persist for longer as a consequence of increased food palatability/liking). More definitively, we examined the actions of CB_1 receptor ligands on the microstructure of sucrose drinking and found that alterations to licking behaviour induced by THC, AEA and 2AG are reminiscent of those observed in drug-free animals drinking more palatable solutions (Higgs et al. 2003). Conversely, rimonabant alters drinking in a way that is consistent with a reduction in the palatability of the sucrose solution. Additionally, $CB_1^{-/-}$ mice are less responsive to sweet taste, consistently drinking less of a range of sucrose solutions than the wild type (Sanchis-Segura et al. 2004). Moreover, these differences are abolished when sucrose solutions are adulterated with bitter quinine, indicating that they arise from differences in the rewarding consequences of palatable ingesta rather than from any sensory impairment.

Further support for an endocannabinoid role in palatability is provided by experiments employing a taste reactivity paradigm to gauge hedonic reactions to flavours by monitoring innate ingestive responses. Thus, Jarret and colleagues have reported that: THC produces rimonabant-reversible increases in ingestive "liking" responses to intra-oral delivery of sucrose solutions; THC reduces the rejection of a quinine solution – an effect blocked by the CB_1 antagonist AM251; AM251 alone both decreases hedonic reactions to sucrose, and increases aversive reactions to quinine solutions (Jarrett et al. 2005, 2007). These important findings thus support the hypothesis that endocannabinoid activity can contribute significantly to the hedonic evaluation of ingesta, and that CB_1 stimulation or blockade/deletion can respectively render food more or less pleasurable. That these effects on liking responses can be obtained under conditions where tastants are delivered directly into the mouth, independently of any volition on the part of the animal (and presumably also of activation of incentive mechanisms), suggest that endocannabinoid modulation of liking can occur separately from their effects on wanting processes.

Consistent with this notion is the fact that key components of the neural mechanisms underlying food palatability lie within the AcbSh (Stratford 2007) and, as already noted, 2AG administered into this site produces a profound hyperphagic response (Kirkham et al. 2002). AEA is also an effective orexigen in this region, as are agents that increase endocannabinoid levels by blocking their enzymatic breakdown or reuptake (Soria-Gomez et al. 2007). Moreover, Harrold and colleagues (2002) showed that accumbens CB_1 receptors are down-regulated in rats that overconsume palatable food supplements. This latter effect is consistent with increased activation of these receptors by endocannabinoids, and again suggests that they mediate the hedonic evaluation of palatable foods. That accumbens endocannabinoids can indeed enhance the hedonic impact of sweet taste is directly supported by the finding that intra-AcbSh administration of AEA specifically increases the number of positive ingestive responses to intra-oral infusions of sweet solutions in taste reactivity tests (Mahler et al. 2007).

4 Endocannabinoid–Opioid Interactions in Eating Motivation

Opioid receptor agonists and antagonists respectively increase or reduce food intake, and these effects have been shown to involve changes in the hedonic evaluation of foods (Cooper and Kirkham 1993; Bodnar 2004). For example, in people, opioid antagonists are reported to reduce the perceived palatability of previously preferred foods and fluids (Drewnowski et al. 1992; Yeomans and Gray 1996). There is now convincing evidence for interactions between endocannabinoids and opioids in relation to feeding, and that cannabinoids modulate the motivation to ingest via actions on both cannabinoid and opioid systems. For example, the hyperphagic action of THC is significantly attenuated by sub-anorectic doses of naloxone (Williams and Kirkham 2002a). Importantly, the facilitatory effects of a CB_1 agonist on responding for palatable solutions are reversed by both a CB_1 antagonist and naloxone (Gallate and McGregor 1999). Moreover, low doses of rimonabant and opioid antagonists that are behaviourally inactive when administered singly, combine synergistically to produce a profound anorectic action when co-administered – far outweighing the suppressive effects of even large doses of either drug given separately (Kirkham and Williams 2001b; Chen et al. 2004).

Given the established ability of opioid antagonists to reduce the hedonic evaluation of foods and to reverse CB_1 agonist-stimulated ingestion, the marked anorexia induced by combined CB_1 and opioid receptor blockade suggests that endocannabinoids also contribute to orosensory reward through the activation of opioid processes. Mesolimbic dopamine neurons synapse with accumbens enkephalinergic neurons that are critical to the expression of reward-related behaviours (Solinas et al. 2008), and there is ultrastructural evidence that cannabinoid–opioid interactions are mediated by activation of CB_1 and -opioid receptors within the same, or synaptically linked, reward-relevant neurons in the AcbSh (Pickel et al. 2004). Moreover, systemic administration of THC has been shown to stimulate β-endorphin release in the accumbens – a phenomenon that has been previously linked to consumption of palatable foods (Solinas et al. 2004). Importantly, as with AEA, administration of morphine into the AcbSh increases the liking of sweet solutions in taste reactivity tests, and it is notable that there is a very close correspondence between the opioid- and cannabinoid-sensitive sites (Pecina and Berridge 2000; Mahler et al. 2007).

Independent manipulations of endocannabinoid or opioid processes produce distinct behavioural/motivational consequences. As we have seen, cannabinoids principally reduce eating latency without dramatic effects on meal duration (i.e. primarily actions on appetitive processes); while opioids typically do not alter eating latency but extend meal duration by enhancing palatability (i.e. they primarily exert actions on consummatory processes). However, a closer temporal relationship, or merging, of direct cannabinoid influences on appetitive motivation and their secondary facilitation of opioid consummatory components may be revealed by a study by Solinas and Goldberg (2005). They reported that THC and morphine dose-dependently increased break-points for food reinforcement, while rimonabant and naloxone dose-dependently decreased break-points. Confirming our findings in free-feeding rats, THC effects on break-point were blocked by naloxone. But more surprisingly, morphine's effects were also blocked by rimonabant.

These data support interactive cannabinoid–opioid mediation of eating motivation, potentially linking the two systems in the reciprocal modulation of hedonic factors that control appetitive and consummatory behaviour. Certainly, the known effects of exogenously administered cannabinoids to promote activation (or disinhibition) of mesolimbic incentive circuits and activate nucleus accumbens circuits involved in hedonic evaluation could account for the heightened intensity of food craving and enhanced appreciation of food reported by cannabis users. It would therefore be extremely instructive to more fully explore the actions of THC or CB_1 antagonists on the subjective experience of hunger and appetite measures in people. The emerging evidence indicates that – as one might expect from personal subjective experience of food's attractiveness – the neurochemical systems that mediate the anticipation or actual experience of the pleasure derived from eating may interact in complex ways. However, it is not too radical to consider that expectation of food (or "hedonic hunger") should incorporate some neurophysiological representation of the prospective delights of consumption. The relationship between endocannabinoids and opioids in modulating activity of incentive-reward circuits may be key to this – reinforcing the primacy of pleasure (or its anticipation) as a key factor in the normal generation of eating motivation.

5 Endocannabinoids and Interactions with Other Orexigens

As might be expected, functional interactions between endocannabinoids and systems implicated in feeding are not restricted to the opioids. Indeed, there is a growing body of evidence for endocannabinoid interactions with a wide range of other factors that are currently implicated in the control of appetite, including putative orexigens (Cota et al. 2003; Matias et al. 2008). So far, much of this evidence is based on histological or in vitro studies, and research is heavily weighted to interactions with feeding-inhibitory agents; the few behavioural studies relevant to appetite-stimulation are outlined below.

In the first description of rimonabant's anorectic action, Arnone et al. (1997) also reported that the drug could block the ability of neuropeptide Y (NPY) to increase the intake of a palatable sucrose solution. Despite this finding, the considerable potency of NPY as an orexigen, and its key interactions with other feeding-related hypothalamic neuropeptides, subsequent analysis of potentially important relationships between this peptide and the endocannabinoids has been limited. Further investigation is warranted however, since Poncelet et al. (2003) reported that rimonabant can prevent NPY hyperphagia, and that the peptide's ability to stimulate feeding is abolished in $CB_1^{-/-}$ mice; although rimonabant is as effective in reducing food intake in NPY knockout mice as in wild-type (Di Marzo et al. 2001).

Very little is presently known about the interaction of the endocannabinoids with other neuropeptides implicated in stimulating food intake. However, Cota and colleagues (2003) have shown co-localization in the PVN of the CB_1 receptor with the orexigenic melanin-concentrating hormone (MCH). Importantly, the very potent and naturalistic hyperphagic actions of centrally administered MCH are sensitive to cannabinoid receptor blockade. Thus, we have found that the eating induced by intra-ventricular MCH is prevented by pre-treatment with behaviourally silent doses of the CB_1 antagonists rimonabant and surinabant (Cooper, Rogers, and Kirkham; unpublished data).

Evidence has also been obtained for significant interactions between endocannabinoids and the orexigenic peptide, and putative hunger signal, ghrelin. Ghrelin is a gut–brain peptide that is synthesised in gastric tissues and the hypothalamus. Circulating levels of gastric ghrelin are closely correlated with meal taking: rising in advance of meals and declining rapidly post-prandially. We found that feeding stimulated by intrahypothalamic (PVN) ghrelin injection is blocked by pre-treatment with sub-anorectic doses of rimonabant, suggesting that expression of ghrelin hyperphagia is dependent on an intact endocannabinoid system (Tucci et al. 2004). Additionally, plasma levels of ghrelin are suppressed by systemic rimonabant treatment (Cani et al. 2004). Importantly, ghrelin hyperphagia is abolished in CB_1 knockout mice, indicating that an intact cannabinoid signalling pathway is required for the peptide to exert its effects on food intake – possibly through the involvement of hypothalamic AMP-activated protein kinase, a key enzyme in the regulation of metabolism (Kola et al. 2008).

6 Conclusion

The preceding discussion has questioned whether homeostatic models of food intake control can adequately account for the actual nature of human appetite and eating behaviour, and particularly their susceptibility to external influences. It has been argued that overconsumption provoked by the anticipated and actual pleasures to be derived from food does not represent an aberrational, counter-regulatory phenomenon. Instead, such "hedonic" eating should be considered as a core component of specialised motivational mechanisms that have evolved to engage us in food-seeking, ensure consumption of the foods most likely to sustain us, and thereby to provide long-term energetic security. Current problems of obesity may in fact be seen to arise from the sheer effectiveness of the hedonic processes guiding appetitive and consummatory components of eating. Traditionally, motivation and emotion have been regarded as being intimately related – with desire, pleasure and satisfaction acting as guiding principles in the arousal and direction of behaviour. To view intake control in the absence of these factors, and solely in terms of instantaneous input:output calculations about energy balance regulation, denies the evidence of our own experience and the record of centuries. The contribution of hedonic factors in determining food intake needs to be more fully acknowledged by physiologists and neuroscientists, and must become the focus of a more concerted effort to determine their neurochemical underpinnings. The well-documented actions of cannabis on appetite must now be integrated with rigorous, detailed investigations in humans of the motivational and emotional actions of endocannabinoid receptor agonists and antagonists. Building on what we have seen in the animal data, the availability of selective CB_1 ligands provides us with an important opportunity to further our understanding of the neurochemical controls of eating behaviour in people, and particularly of the neural underpinnings of the psychological experience of hunger, food craving, eating pleasure and satisfaction.

References

Abel EL (1971) Effects of marijuana on the solution of anagrams, memory and appetite. Nature 231:260–261

Acosta MC, Manubay J, Levin FR (2008) Pediatric obesity: parallels with addiction and treatment recommendations. Harv Rev Psychiatry. 16:80–96

Arnone M, Maruani J, Chaperon F et al. (1997) Selective inhibition of sucrose and ethanol intake by SR 141716, an antagonist of central cannabinoid (CB1) receptors. Psychopharmacology (Berlin) 132:104–106

Avraham Y, Ben Menachem A, Okun A et al. (2005) Effects of the endocannabinoid noladin ether on body weight, food consumption, locomotor activity, and cognitive index in mice. Brain Res Bull 65:117–123

Beck B (2007) Hypothalamic neuropeptides and feeding regulation. In: Kirkham TC, Cooper SJ (eds) Appetite and body weight: integrative systems and the development of anti-obesity drugs. Academic Press, Burlington, MA

Berridge KC (2000) Measuring hedonic impact in animals and infants: microstructure of affective taste reactivity patterns. Neurosci Biobehav Rev 24:173–198

Berridge KC (2007) Brain reward systems for food incentives hedonics in normal appetite and eating disorders. In: Kirkham TC, Cooper SJ (eds) Appetite and body weight: integrative systems and the development of anti-obesity drugs. Academic Press, Burlington, MA

Blundell JE, Jebb SA, Stubbs RJ et al. (2006) Effect of rimonabant on energy intake, motivation to eat and body weight with or without hypocaloric diet: the REBA study. International Congress on Obesity, Sydney, Australia (poster)

Bodnar RJ (2004) Endogenous opioids and feeding behavior: a 30-year historical perspective. Peptides 25:697–725
Cannon WB (1932) The wisdom of the body. WW Norton & Company, New York
Cani PD, Montoya ML, Neyrinck AM, Delzenne NM, Lambert DM (2004) Potential modulation of plasma ghrelin and glucagon-like peptide-l by anorexigenic cannabinoid compounds, SR141716A (rimonabant) and oleoylethanolamide. Br J Nutr 92:757–761
Chee MJ, Colmers WF (2008) Y eat? Nutrition 24:869–877
Chen R, Huang R, Shen C et al. (2004) Synergistic effects of cannabinoid inverse agonist AM251 and opioid antagonist nalmefene on food intake in mice. Brain Res Brain Res Rev 999:227–230
Colombo G, Agabio R, Diaz G et al. (1998) Appetite suppression and weight loss after the cannabinoid antagonist SR 141716. Life Sci 63:PL113–PL117
Cooper SJ, Kirkham TC (1993) Opioid mechanisms in the control of food consumption and taste preferences. In: Herz A, Akil H, Simon EJ (eds) Handbook of experimental pharmacology, vol. 104, part II. Springer, Berlin
Cota D, Marsicano G, Tschop M et al. (2003) The endogenous cannabinoid system affects energy balance via central orexigenic drive and peripheral lipogenesis. J Clin Invest 112:423–431
Di Marzo V, Goparaju SK, Wang L et al. (2001) Leptin-regulated endocannabinoids are involved in maintaining food intake. Nature 410:822–825
Drewnowski A, Krahn DD, Demitrack MA et al. (1992) Taste responses and preferences for sweet high-fat foods: evidence for opioid involvement. Physiol Behav 51:371–379
Figlewicz Latteman D, Sanders NM, MacDonald A et al. (2007) Integration of peripheral adiposity signals and psychological controls of appetite. In: Kirkham TC, Cooper SJ (eds) Appetite and body weight: integrative systems and the development of anti-obesity drugs. Academic Press, Burlington, MA
Freedland CS, Poston JS, Porrino LJ (2000) Effects of SR141716A, a central cannabinoid receptor antagonist, on food-maintained responding. Pharmacol Biochem Behav 67:265–270
Gallate J, McGregor I (1999) The motivation for beer in rats: effects of ritanserin, naloxone and SR 141716. Psychopharmacology 142:302–308
Gallate J, Saharov T, Mallet P et al. (1999) Increased motivation for beer in rats following administration of a cannabinoid CB1 receptor agonist. Eur J Pharmacol 370:233–240
Gaoni Y, Mechoulam R (1971) The isolation and structure of delta-1-tetrahydrocannabinol and other neutral cannabinoids from hashish. J Am Chem Soc 93:217–224
Gardner E, Vorel S (1998) Cannabinoid transmission and reward-related events. Neurobiol Disease 5:502–533
Gibson EL, Brunstrom JM (2007) Learned influences on appetite, food choice, and intake: evidence in human beings. In: Kirkham TC, Cooper SJ (eds) Appetite and body weight: integrative systems and the development of anti-obesity drugs. Academic Press, Burlington, MA
Hao S, Avraham Y, Mechoulam R et al. (2000) Low dose anandamide affects food intake, cognitive function, neurotransmitter and corticosterone levels in diet-restricted mice. Eur J Pharmacol 392:147–156
Harrold J, Elliott C, King P et al. (2002) Down-regulation of cannabinoid-1 (CB-1) receptors in specific extrahypothalamic regions of rats with dietary obesity: a role for endogenous cannabinoids in driving appetite for palatable food? Brain Res 952:232–238
Higgs S, Williams CM, Kirkham TC (2003) Cannabinoid influences on palatability: microstructural analysis of sucrose drinking after delta(9)-tetrahydrocannabinol, anandamide, 2-arachidonoyl glycerol and SR141716. Psychopharmacology (Berl) 165:370–377
Hildebrandt AL, Kelly-Sullivan DM, Black SC (2003) Antiobesity effects of chronic cannabinoid CB1 receptor antagonist treatment in diet-induced obese mice. Eur J Pharmacol 462:125–132
Jamshidi N, Taylor D (2001) Anandamide administration into the ventromedial hypothalamus stimulates appetite in rats. Br J Pharmacol 134:1151–1154
Jarrett MM, Limebeer CL, Parker LA (2005) Effect of delta9-tetrahydrocannabinol on sucrose palatability as measured by the taste reactivity test. Physiol Behav 86:475–479

Jarrett MM, Scantlebury J, Parker LA (2007) Effect of delta(9)-tetrahydrocannabinol on quinine palatability and AM251 on sucrose and quinine palatability using the taste reactivity test. Physiol Behav 90:425–430
Kalra SP (2008) Central leptin insufficiency syndrome: an interactive etiology for obesity, metabolic and neural diseases and for designing new therapeutic interventions. Peptides 29:127–138
Kirkham TC (1990) Enhanced anorectic potency of naloxone in rats sham feeding 30% sucrose: reversal by repeated naloxone administration. Physiol Behav 47:419–426
Kirkham TC (2005) Endocannabinoids in the regulation of appetite and body weight. Behav Pharmacol 16:297–313
Kirkham TC, Cooper SJ (1988) Naloxone attenuation of sham feeding is modified by manipulation of sucrose concentration. Physiol Behav 44:491–494
Kirkham TC, Williams CM (2001a) Endogenous cannabinoids and appetite. Nutr Res Rev 14:65–86
Kirkham TC, Williams CM (2001b) Endocannabinoids: neuromodulators of food craving? In: Hetherington M (ed) Food cravings and addiction. Leatherhead Publishing, Leatherhead, UK
Kirkham TC, Williams CM (2001c) Synergistic effects of opioid and cannabinoid antagonists on food intake. Psychopharmacology (Berl) 153:267–270
Kirkham TC, Williams CM, Fezza F et al. (2002) Endocannabinoid levels in rat limbic forebrain and hypothalamus in relation to fasting, feeding and satiation: stimulation of eating by 2-arachidonoyl glycerol. Br J Pharmacol 136:550–557
Koch JE, Matthews SM (2001) Delta9-tetrahydrocannabinol stimulates palatable food intake in Lewis rats: effects of peripheral and central administration. Nutr Neurosci 4:179–187
Kola B, Farkas I, Christ-Crain M et al. (2008) The orexigenic effect of ghrelin is mediated through central activation of the endogenous cannabinoid system. PLoS ONE 3:e1797
Konturek PC, Konturek JW, Czesnikiewicz-Guzik et al. (2005) Neuro-hormonal control food intake; basic mechanisms and clinical implications. J Physiol Pharmacol 56(Suppl 6):5–25
Leventhal L, Kirkham TC, Cole JL et al. (1995) Selective actions of central mu and kappa opioid antagonists upon sucrose intake in sham-fed rats. Brain Res 685:205–210
Levitsky DA (2002) Putting behavior back into feeding behavior: a tribute to George Collier. Appetite 38:143–148
Levitsky DA (2005) The non-regulation of food intake in humans: Hope for reversing the epidemic. 86:623–632
Lowe MR, Butryn ML (2007) Hedonic hunger: a new dimension of appetite? Physiol Behav 91:432–439
Mahler SV, Smith KS, Berridge KC (2007) Endocannabinoid hedonic hotspot for sensory pleasure: anandamide in nucleus accumbens shell enhances 'liking' of a sweet reward. Neuropsychopharmacology 32:2267–2278
Maljaars J, Peters HP, Masclee AM (2007) The gastrointestinal tract: neuroendocrine regulation of satiety and food intake. Aliment Pharmacol Ther 26(Suppl 2):241–250
Marín Bivens CL, Thomas WJ, Stanley BG (1998) Similar feeding patterns are induced by perifornical neuropeptide Y injection and by food deprivation. Brain Res 782:271–280
Matias I, Cristino L, di Marzo V (2008) Endocannabinoids: some like it fat (and sweet too). J Neuroendocrinol 20(Suppl 1):100–109
Melis T, Succu S, Sanna F et al. (2007) The cannabinoid antagonist SR 141716A (Rimonabant) reduces the increase of extra-cellular dopamine release in the rat nucleus accumbens induced by a novel high palatable food. Neurosci Lett 419:231–235
Meschler JP, Howlett AC (2001) Signal transduction interactions between CB1 cannabinoid and dopamine receptors in the rat and a monkey striatum. Neuropharmacology 14:918–926
Miller CC, Murray TF, Freeman KG et al. (2004) Cannabinoid agonist, CP 55, 940, facilitates intake of palatable foods when injected into the hindbrain. Physiol Behav 80:611–616
Osei-Hyiamen D, DePetrillo M, Pacher P et al. (2005) Endocannabinoid activation at hepatic CB1 receptors stimulates fatty acid synthesis and contributes to diet-induced obesity. J Clin Invest 115:1298–1305

Pecina S, Berridge KC (2000) Opioid site in nucleus accumbens shell mediates eating and hedonic 'liking' for food: map based on microinjection Fos plumes. Brain Res 863:71–86
Pério A, Barnouin MC, Poncelet M et al. (2001) Activity of SR141716 on post-reinforcement pauses in operant responding for sucrose reward in rats. Behav Pharmacol 12:641–645
Pertwee RG (2008) Ligands that target cannabinoid receptors in the brain: from THC to anandamide and beyond. Addict Biol 13(2):147–159
Pickel VM, Chan J, Kash TL et al. (2004) Compartment-specific localization of cannabinoid (CB1) and mu-opioid receptors in rat nucleus accumbens. Neuroscience 127:101–112
Pickel VM, Chan J, Kearn et al. (2006) Targeting dopamine D2 and cannabinoid-1 (CB-1) in rat nucleus accumbens. J Comp Neurol 495:299–313
Poncelet M, Maruani J, Calassi R et al. (2003) Overeating, alcohol and sucrose consumption decrease in CB1 receptor deleted mice. Neurosci Lett 343:216–218
Ravinet Trillou C, Arnone M, Delgorge C et al. (2003) Anti-obesity effect of SR141716, a CB1 receptor antagonist, in diet-induced obese mice. Am J Physiol Regul Integr Comp Physiol 284: R345–R353
Riegel AC, Lupica CR (2004) Independent presynaptic and postsynaptic mechanisms regulate endocannabinoid signaling at multiple synapses in the ventral tegmental area. J Neurosci 24:11070–11078
Rinaldi-Carmona M, Barth F, Congy C et al. (2004) SR147778 [5-(4-Bromophenyl)-1–1(2, 4-dichlorophenyl)-4-ethyl -N-(1-piperidinyl)-1H-pyrazole-3-carboxamide], a new potent and selective antagonist of the CB1 cannabinoid receptor: biochemical and pharmacolo- gical characterization. J Pharmacol Exp Ther 310:905–914
Rolls BJ (1986) Sensory-specific satiety. Nutr Rev 44:93–101
Rutkowska M (2004) The effect of AM251, a cannabinoid CB1 receptor antagonist, on food intake in rats. Acta Pol Pharm 61:401–403
Sanchis-Segura C, Cline B, Marsicano G et al. (2004) Reduced sensitivity to reward in CB1 knockout mice. Psychopharmacology (Berl) 176:223–232
Scarpaci PJ, Zhang Y (2007) Elevated leptin: consequence or cause of obesity? Front Biosci 12:3531–3544
Shearman L, Rosko K, Fleischer R et al. (2003) Antidepressant-like and anorectic effects of the cannabinoid CB1 receptor inverse agonist AM251 in mice. Behav Pharmacol 14:573–582
Simiand J, Keane M, Keane P et al. (1998) SR 141716, a CB1 cannabinoid receptor antagonist, selectively reduces sweet food intake in marmoset. Behav Pharmacol 9:179–181
Solinas M, Goldberg SR (2005) Motivational effects of cannabinoids and opioids on food reinforcement depend on simultaneous activation of cannabinoid and opioid systems. Neuropsychopharmacology 30:2035–2045
Solinas M, Justinova Z, Goldberg SR et al. (2006) Anandamide administration alone and after inhibition of fatty acid amide hydrolase (FAAH) increases dopamine levels in the nuclear accumbens shell in rats. J. Neurochem 98:408–419
Solinas M, Goldberg SR, Piomelli D (2008) The endocannabinoid system in brain reward processes. Br J Pharmacol 154:369–383
Solinas M, Zangen A, Thiriet N, Goldberg SR (2004) β-endorphin elevations in the ventral tegmental area regulate the discriminative effects of Δ9-tetrahydrocannabinol. Eur J Neurosci 19:3183–3192
Soria-Gomez E, Matias I et al. (2007) Pharmacological enhancement of the endocannabinoid system in the nucleus accumbens shell stimulates food intake and increases c-Fos expression in the hypothalamus. Br J Pharmacol 151:1109–1116
Speakman JR (2007) Gene environment interactions and the origin of the modern obesity epidemic: a novel "nonadaptive drift" scenario. In: Kirkham TC, Cooper SJ (eds) Appetite and body weight: integrative systems and the development of anti-obesity drugs. Academic Press, Burlington, MA

Stratford TR (2007) The nucleus accumbens shell as a model of integrative subcortical forebrain systems regulating food intake. In: Kirkham TC, Cooper SJ (eds) Appetite and body weight: integrative systems and the development of anti-obesity drugs. Academic Press, Burlington, MA

Sullivan RJ, Hagen EH, Hammerstein P (2008) Revealing the paradox of drug reward in human evolution. Proc Biol Sci 275:1231–1241

Tallett AJ, Blundell JE et al. (2007a) Grooming, scratching and feeding: role of response competition in acute anorectic response to rimonabant in male rats. Psychopharmacology (Berl) 195:27–39

Tallett AJ, Blundell JE et al. (2007b) Acute anorectic response to cannabinoid CB1 receptor antagonist/inverse agonist AM 251 in rats: indirect behavioural mediation. Behav Pharmacol 18:591–600

Tart C (1970) Marijuana intoxication: common experiences. Nature 226:701–704

Tucci SA, Rogers EK, Korbonits M, Kirkham TC (2004) The cannabinoid CBI receptor antagonist SR141716 blocks the orexigenic effects of intrahypothalamic ghrelin. Br J Pharmacol 143:520–523

Valassi E, Scacchi M, Cavagnini F (2008) Neuroendocrine control of food intake. Nutr Metab Cardiovascular Dis 18:158–168

Valenti M, Viganòb D, Cascioa MG et al. (2004) Differential diurnal variations of anandamide and 2 arachidonoyl-glycerol levels in rat brain. Cell Mol Life Sci 61:945–950

van der Stelt M, Di Marzo V (2003) The endocannabinoid system in the basal ganglia and in the mesolimbic reward system: implications for neurological and psychiatric disorders. Eur J Pharmacol 480:133–150

Verty A, McGregor I, Mallet P (2004a) The dopamine receptor antagonist SCH 23390 attenuates feeding induced by Delta9-tetrahydrocannabinol. Brain Res 1020:188–195

Verty AN, McGregor IS, Mallet PE (2004b) Consumption of high carbohydrate, high fat, and normal chow is equally suppressed by a cannabinoid receptor antagonist in non-deprived rats. Neurosci Lett 354:217–220

Vickers S, Webster L, Wyatt A et al. (2003) Preferential effects of the cannabinoid CB1 receptor antagonist, SR 141716, on food intake and body weight gain of obese (fa/fa) compared to lean Zucker rats. Psychopharmacology (Berl) 167:103–111

Werner N, Koch J (2003) Effects of the cannabinoid antagonists AM281 and AM630 on deprivation-induced intake in Lewis rats. Brain Res 967:290–292

Wiley JL, Burston JJ, Leggett DC et al. (2005) CB(1) cannabinoid receptor-mediated modulation of food intake in mice. Br J Pharmacol 145:293–300

Williams CM, Kirkham TC (1999) Anandamide induces overeating: mediation by central cannabinoid (CB1) receptors. Psychopharmacology (Berl) 143:315–317

Williams CM, Kirkham TC (2002a) Reversal of delta 9-THC hyperphagia by SR141716 and naloxone but not dexfenfluramine. Pharmacol Biochem Behav 71:333–340

Williams CM, Kirkham TC (2002b) Observational analysis of feeding induced by Delta9-THC and anandamide. Physiol Behav 76:241–250

Williams CM, Rogers PJ, Kirkham TC (1998) Hyperphagia in pre-fed rats following oral delta9 THC. Physiol Behav 65:343–346

Woods SC, Ramsay DS (2007) Homeostasis: beyond Curt Richter. Appetite 49(2):388–398

Wren AM, Bloom SR (2007) Gut hormones and appetite control. Gastroenterology 132:2116–2130

Yeomans MR (2007) The role of palatability in control of human appetite: implications for understanding and treating obesity. In: Kirkham TC, Cooper SJ (eds) Appetite and body weight: integrative systems and the development of anti-obesity drugs. Academic Press, Burlington, MA

Yeomans MR, Gray RW (1996) Selective effects of naltrexone on food pleasantness and intake. Physiol Behav 60:439–446

Zhou D, Shearman L (2004) Voluntary exercise augments acute effects of CB1-receptor inverse agonist on body weight loss in obese and lean mice. Pharmacol Biochem Behav 77:117–125

Cannabinoid/Endocannabinoid Signaling Impact on Early Pregnancy Events

Xiaofei Sun and Sudhansu K. Dey

Contents

Abstract It has been known for decades that marijuana and its major psychoactive component Δ^9-tetrahydrocannabinol (THC) alter both male and female reproductive functions in humans and laboratory animals. The discovery of cannabinoid-like molecules (endocannabinoids), anandamide (AEA) and 2-arachidonylglycerol (2AG), as well as G-protein-coupled cannabinoid/endocannabinoid receptors CB_1 and CB_2, created an opportunity to study the adverse and beneficial effects of cannabinoids/endocannabinoids on fertility using molecular, physiological and genetic approaches. In fact, studies to explore the significance of cannabinoid/endocannabinoid signaling in reproduction have revealed some intriguing physiological roles in early pregnant events. This review summarizes some aspects of these signaling molecules in preimplantation and implantation biology utilizing genetically engineered mouse models.

X. Sun and S.K. Dey (✉)
Reproductive Sciences, Cincinnati Children's Hospital Research Foundation, Cincinnati, OH, 45229, USA
e-mail: sk.dey@cchmc.org

D. Kendall and S. Alexander (eds.), *Behavioral Neurobiology of the Endocannabinoid System*, Current Topics in Behavioral Neurosciences 1,
DOI: 10.1007/978-3-540-88955-7_10,

Keywords Endocannaibnoids • CB1 • CB2 • Pregnancy • Implantation • Uterus • Oviduct

1 Introduction

Although the human population is growing rapidly, 15% of couples worldwide are infertile (Abma et al. 1997; Thonneau et al. 1991), with infertility defined as the inability to conceive after one year of regular sexual intercourse. Infertility is still a worldwide social and economic concern. Early pregnancy loss in humans often happens due to defects that occur before, during or immediately after implantation. Although in vitro fertilization and embryo transfer (IVF-ET) approaches have overcome several barriers of human infertility, the implantation rate still remains disappointingly low. Therefore, studying physiological, genetic and molecular bases of implantation is important. However, it is difficult to define the hierarchical landscape of molecular pathways during human pregnancy because of experimental difficulties and ethical restrictions on research with human embryos. It is hoped that experiments in mice and other animal models combined with feasible experiments in humans will generate meaningful information to address this critical issue. Although details of many of the molecular interactions during the peri-implantation events have not yet been defined, increasing evidence from gene expression and transgenic mouse studies reveals that synchronous development of the preimplantation embryo to the blastocyst stage and differentiation of the uterus to the receptive stage are prerequisites for the initiation of implantation (Dey et al. 2004; Paria et al. 2002; Wang and Dey 2006).

Over the past several years, molecular and genetic studies have provided evidence that lipid mediators are critical signaling molecules in coordinating events of early pregnancy (Shah and Catt 2005; Song et al. 2002; Wang and Dey 2005; Ye et al. 2005). Among these signaling pathways, endocannabinoid signaling has recently been highlighted as an important player in directing preimplantation development of embryos and their timely homing into the receptive uterus for implantation. This review highlights various aspects of the endocannabinoid system in female fertility. It is hoped that a deeper insight will lead to potential clinical applications, perhaps targeting the endocannabinoid signaling pathway to correct infertility and improve women's reproductive health. This article reviews the endocannabinoid system and its roles in peri-implantation biology primarily in genetically engineered mouse models.

2 Endocannabinoid Systems

Marijuana, derived from the plant *Cannabis sativa*, is widely used for its psychoactive effects, including euphoria and analgesia. Although it has been used recreationally for thousands of years, studies regarding the chemistry of *Cannabis* were initiated just decades ago (Mechoulam and Hanus 2000). In 1964, Δ^9-tetrahydrocannabinol (Δ^9-THC) was identified as the major active component of marijuana (Gaoni and Mechoulam 1964), stimulating research on marijuana. In the early 1990s, research on marijuana was further boosted by the discovery and cloning of two cannabinoid receptors, brain-type (CB_1) (Devane et al. 1988; Matsuda et al. 1990) and spleen-type (CB_2) (Munro et al. 1993). At around the same time, several endogenous ligands were identified that target CB_1 and CB_2, and subsequently they were termed endocannabinoids. The two most studied endocannabinoids are *N*-arachidonoylethanolamine, commonly known as anandamide (AEA) (Devane et al. 1992) and 2-arachidonoylglycerol (2AG) (Mechoulam et al. 1995; Sugiura et al. 1995). Also see "The Life Cycle of the Endocannabinoids: Formation and Inactivation" in the chapter by Stephen P.H. Alexander and David A. Kendall, this volume, for a more detailed description.

2.1 AEA Synthesis and Degradation

It is widely accepted that AEA is derived from the precursor *N*-arachidonoylphosphatidylethanolamine (NAPE) through its reaction with NAPE-hydrolyzing phospholipase D (NAPE-PLD) (Natarajan et al. 1982, 1984), a member of the metallo-lactamase family with Ca^{2+}–sensitive enzyme activity (Okamoto et al. 2004; Ueda et al. 2001). However, unaltered polyunsaturated NAE (N-acyl-ethanolamine) levels in NAPE-PLD deficient mice suggest that other AEA synthetic pathways also contribute to levels of AEA (Leung et al. 2006). Recently, two other enzymatic routes were identified: (1) double deacylation of NAPE by a phospholipase/lysophospholipase B, α/β-hydrolase 4 (Abh4), to generate glycerophospho-NAE (GP-NAE) which is then cleaved by a phosphodiesterase to liberate AEA (Simon and Cravatt 2006), and (2) cleavage of NAPE by phospholipase C to generate phosphor-AEA (pAEA) which is subsequently dephosphorylated by a protein tyrosine phosphatase (PTPN22) to release AEA (Liu et al. 2006). Although these pathways are found in both the CNS and peripheral tissues, mechanism(s) by which these pathways are regulated and affect each other are still unknown.

AEA signaling through CB receptors occurs through a two-step process. AEA is first taken up by the cell through an AEA membrane transporter (AMT) and then degraded intracellularly by the fatty acid amide hydrolase (FAAH) (Cravatt et al. 1996; Giang and Cravatt 1997). It is to be noted, however, that the existence of endocannabinoid transporters is still under debate (Glaser et al. 2005; Mechoulam and Deutsch 2005). The current models suggest that enzymes for the synthesis and degradation of endocannabinoids are localized within the cell. This means that stimulation of cannabinoid receptors by endocannabinoids from the extracellular component requires them to cross the cell membrane twice. This concept is controversial, based on research on the transporter. In fact, the uptake of AEA has features similar to facilitated transport, dependent on concentration, time, and temperature, and independent of external Na^+ ions or ATP hydrolysis (Mechoulam and Deutsch 2005). The development of new drugs that inhibit AMT selectively without affecting FAAH corroborates this speculation (Ortega-Gutierrez 2005). However, FAAH may not need a transporter to contact AEA for its degradation (Bracey et al. 2002). It is suggested that AEA uptake instead is driven by non-protein mediated diffusion and is regulated by its degree of hydrolysis by FAAH (Kaczocha et al. 2006). Along this same tenet, it is thought that the target of some of these recently developed transport inhibitors is an uncharacterized intracellular component that delivers AEA to FAAH (Kaczocha et al. 2006).

After AEA is accumulated within the cell, it is degraded to ethanolamine and arachidonic acid (AA) by FAAH (Cravatt et al. 1996; Giang and Cravatt 1997). Mammalian FAAH is a membrane-bound enzyme with a globular shape. It has 28 α-helices and 11 β-sheets, which account for approximately 53 and 13% of the whole protein structure, respectively (Bracey et al. 2002). This enzyme uses an unusual serine–serine–lysine (S241–S217–K142) catalytic triad (McKinney and Cravatt 2005). FAAH can also hydrolyze other endocannabinoids including 2AG and the sleep-inducing substance, oleamide (McKinney and Cravatt 2005). FAAH has also been shown to be critical for regulating both the magnitude and duration of AEA and other fatty acid amide signaling (Cravatt and Lichtman 2002). Recently, a second membrane-associated fatty acid amide hydrolase was found in human and other primate genomes but not in that of rodents (Wei et al. 2006).

2.2 2AG Synthesis and Degradation

2AG was discovered by two independent groups, with one group identifying it in the canine gut and the other in the rat brain (Mechoulam et al. 1995; Sugiura et al. 1995). 2AG is derived from the precursor diacylglycerol by a membrane-bound sn1-diacylglycerol lipase (DGL) (Moriyama et al. 1999). To date, two isoforms of DGL have been cloned: DGLα and DGLβ. The α and β isoforms have molecular masses of 120 and 70 kDa, respectively, with four transmembrane domains, and they are members of the serine lipase family with serine and aspartic acid (S443–D495) participating in the catalytic triad. DGLα is mainly expressed in the adult brain, whereas DGLβ is expressed in the developing brain (Bisogno et al. 2003). Like AEA, 2AG is produced as necessary, but these two endocannabinoids differ in that AEA often acts only as a partial agonist of cannabinoid receptors, while 2AG acts as a full agonist. Interestingly, the binding affinity of 2AG to cannabinoid receptors is approximately 24 times less than that of AEA, but under most physiological conditions, 2AG levels are much higher than AEA (Sugiura et al. 2006). It still remains to be determined, therefore, how only a small percentage of 2AG (10–20%) crosses the plasma membrane to interact with cannabinoid receptors (Bisogno et al. 1997).

Like AEA, the termination of pharmacological effects of 2AG requires it to be transported into the intracellular compartment. It is proposed that the 2AG membrane transporter is the same as AMT (Beltramo and Piomelli 2000). In fact, 2AG accumulation is directly reduced by an AMT inhibitor, AM404, and indirectly reduced by high concentrations of AA (Beltramo and Piomelli 2000). After 2AG accumulates in cells, it can then be degraded by either FAAH or a serine hydrolase, monoacylglycerol lipase (MGL) (Goparaju et al. 1999). MGL, a 33-kD protein, has been isolated, cloned and characterized in both rats and humans (Dinh et al. 2002; Goparaju et al. 1999; Ho et al. 2002). Unlike FAAH, MGL is localized primarily in the cytosol, but not on the plasma membrane. Recently, Muccioli et al. identified a novel protein in a mouse microglial cell line that has MGL activity and regulates 2AG levels (Muccioli et al. 2007).

2.3 Cannabinoid Receptors

Endocannabinoids, as well as plant-derived and synthetic cannabinoids, target cannabinoid receptors CB_1 and CB_2. They are G protein-coupled receptors with seven transmembrane domains. CB_1 is present mostly in the central nervous system and in some peripheral tissues including heart, testis, liver, small intestine and uterus, while CB_2 is abundantly expressed in astrocytes, spleen and several immune cells (Howlett et al. 2002; McAllister and Glass 2002; Pertwee and Ross 2002). CB_1 and CB_2 show 44% overall identity and both are coupled with G proteins in the $G_{i/o}$ and G_q families. Activation of each CB receptor subtype has different biological effects with most being cell-type dependent. Signal transduction pathways regulated by CB receptors include inhibition of adenylyl cyclase (Matsuda et al. 1990; Paria et al. 1995), regulation of Ca^{2+} channels (Caulfield and Brown 1992; Gebremedhin et al. 1999; Lauckner et al. 2005; Mackie and Hille 1992; Wang et al. 2003), activation of phospholipase C (Zoratti et al. 2003) and stimulation of mitogen-activated protein kinases (MAPKs) including ERK, JNK and p38 (Bouaboula et al. 1995; Murphy and Blenis 2006; Wang et al. 2003).

Some evidence indicates the existence of other putative cannabinoid receptors in addition to CB_1 and CB_2 (Baker et al. 2006). For example, it was shown that AEA can protect murine neuroblastoma cells subjected to low serum-induced apoptosis by non-CB_1, non-CB_2 receptors (Matas et al. 2007). Furthermore, a novel cannabinoid receptor 3 (GPR55) has been reported (McPartland et al. 2006; Sawzdargo et al. 1999), which, as yet, is a G protein-coupled orphan receptor. However, the physiological role of this receptor is not clearly understood.

AEA, but not 2AG, can also activate receptors other than CB_1 and CB_2. One receptor that AEA activates is the transient receptor potential vanilloid 1 (TRPV1) (Van Der Stelt and Di Marzo 2004), a ligand-gated non-selective cationic channel. TRPV1 can also be activated by molecules derived from plants, such as capsaicin (the pungent component of "hot" red peppers) and resinferatoxin, and also by stimuli including heat and low PH (Protons). Some recent studies suggest a physiological role for AEA as a TRPV1 agonist. The binding of AEA to the cytosolic binding site of TRPV1 triggers Ca^{2+} influx and eventual cytochrome c release (De Petrocellis et al. 2001; Maccarrone and Finazzi-Agro 2003).

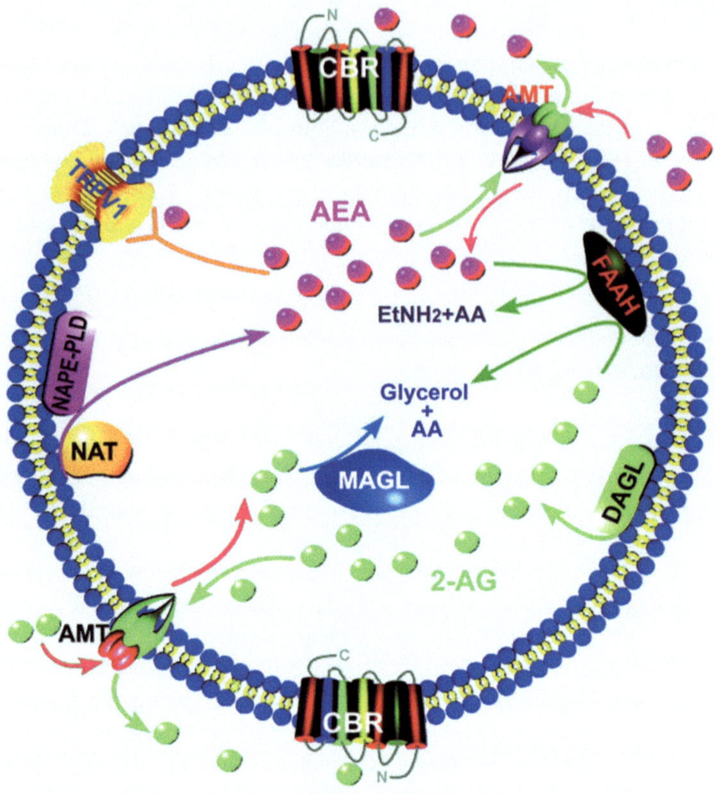

Fig. 1 The endocannabinoid system. Synthesis of AEA from membrane *N*-arachidonoylphosphatidylethanolamines is catalyzed by sequential activities of *N*-acyltransferase (NAT) and NAPE-PLD, which releases AEA and phosphatidic acid. AEA is transported in both directions through the cell membrane by a selective AMT and, once taken up, is hydrolyzed by FAAH to ethanolamine ($EtNH_2$) and AA. The main targets of AEA are CB_1 and CB_2 receptors (CBR) with extracellular binding sites, and type-1 vanilloid receptors (TRPV1) with intracellular binding sites. 2AG is also released from membrane lipids through the activity of DGL. 2AG can also be hydrolyzed by FAAH or more importantly by MGL, releasing glycerol and AA. The transport of 2AG across the cell membrane may be mediated by AMT or a related transporter, and CBR (but not TRPV1) is the target of this endocannabinoid. This figure is adapted from Wang et al. (2006a)

Endocannabinoid receptors and their ligands together with the synthesis and degradation enzymes collectively constitute the endocannabinoid system (Fig. 1).

3 Peri-implantation Events

Life begins with the fusion of two haploid gametes, an egg and a sperm (Evans and Florman 2002; Wassarman et al. 2001). The one-cell fertilized zygote, now termed an embryo, undergoes several mitotic cell divisions, eventually forming the blastocyst.

The blastocyst is comprised of two distinct cell populations, the inner cell mass (ICM) and an outer layer of trophectoderm cells (Rossant and Tam 2004; Wang and Dey 2006; Zernicka-Goetz 2005). The embryo proper is derived exclusively from the ICM, whereas the placenta and extraembryonic membranes are generated from cells contributed by the trophectoderm (Cross et al. 1994; Rossant 2004). During early pregnancy, another critical event occurs in parallel with preimplantation embryonic development – the embryos' timely transport from the oviduct into the uterus. In mice, embryos at the late morula or early blastocyst stage enter the uterus, where they develop and differentiate to the late blastocyst stage. A two-way interaction between the blastocyst and maternal uterine luminal epithelium initiates the process of implantation (Dey et al. 2004; Paria et al. 2002; Wang and Dey 2006). Although the precise sequence and details of the molecular interactions involved in these processes are not clearly understood, increasing evidence from gene expression and transgenic mouse studies during the last two decades shows that coordinated integration of a range of paracrine, autocrine, and/or juxtacrine signaling pathways participates in embryo–uterine dialog during implantation (Carson et al. 2000; Dey et al. 2004; Paria et al. 2002; Red-Horse et al. 2004; Wang and Dey 2006). Among these, endocannabinoid signaling has recently been highlighted as an important player in directing preimplantation embryo development, the timely homing of embryos into a receptive uterus, and coordinating blastocyst activation and uterine receptivity for implantation.

3.1 Preimplantation Embryo Development

Development of preimplantation embryos to blastocysts is critical for achieving implantation competency. Their delayed development causes defective or failure of implantation, leading to compromised pregnancy (Wang and Dey 2006).

Endocannabinoid signaling occurs in preimplantation embryos, the oviduct and uteri. Both CB_1 and CB_2 are present in preimplantation embryos (Das et al. 1995; Paria et al. 1995, 2001; Wang et al. 2004), while only CB_1 is expressed in the oviduct and uterus. While CB_1 mRNA is detected from the 4-cell through the blastocyst stages, CB_2 is present from the 1-cell through the blastocyst stages (Paria et al. 1995). AEA binding sites are also evident in embryos at these stages. Notably, these binding sites are primarily located in outer cells of embryos at 8-cell, morula, and blastocyst stages. Dey's group has shown that AEA binds to a single class of high-affinity receptors on blastocysts. The presence of CB_1 mRNA correlates with CB_1 protein as detected by immunocytochemistry (Paria et al. 2001; Yang et al. 1996). Moreover, blastocyst CB_1 is biologically active, since both THC and AEA inhibit forskolin-stimulated cAMP formation in the embryo, and this inhibition is prevented by pertussis toxin pretreatment (Das et al. 1995; Paria et al. 1995). Recent observations of expression of CB_2 in early embryos and embryonic stem cells by microarray analysis (Sharov et al. 2003), and the absence of its expression in trophoblast stem cells derived from preimplantation blastocysts, suggests that CB_2 expression is restricted to the ICM of blastocysts (Hamatani et al. 2004). Thus, while the role of CB_2 in early embryos remains unknown, the presence of functional CB_1 suggests that mouse embryos are potential targets of endocannabinoids and natural cannabinoids.

Embryos exposed to high levels of endocannabinoids, plant-derived and/or synthetic cannabinoids show retarded development. For example, high levels of AEA causes blastocysts to have a reduced number of trophectoderm cells and decreases the rate of zona-hatching (Schmid et al. 1997; Yang et al. 1996). Furthermore, AEA, 2AG, THC or WIN55212-2 (a synthetic cannabinoid agonist) arrests the development of two-cell embryos to blastocysts (Paria et al. 1995, 1998b). This developmental defect, however, is rescued by SR141716A or AM251 (synthetic CB_1-selective antagonists), but not by SR144528 (a CB_2-selective antagonist). Furthermore, a CB_2 agonist, AM663, fails to influence embryo development (Paria et al. 1998b). These studies collectively provide evidence that endocannabinoids or cannabinoids mediate their effects on preimplantation embryos via CB_1 (Fig. 2).

The availability of gene targeted *cnr1* and *cnr2* mouse models has greatly expanded the field of endocannabinoid research. It was observed that $CB_1^{-/-}$ and $CB_1^{-/-}/CB_2^{-/-}$ embryos recovered from oviducts (day 3) and uteri (day 4) of pregnant mice show asynchronous development compared with wild-type embryos (Paria et al. 2001). Interestingly, heterozygous embryos recovered from $CB_1^{-/-}$

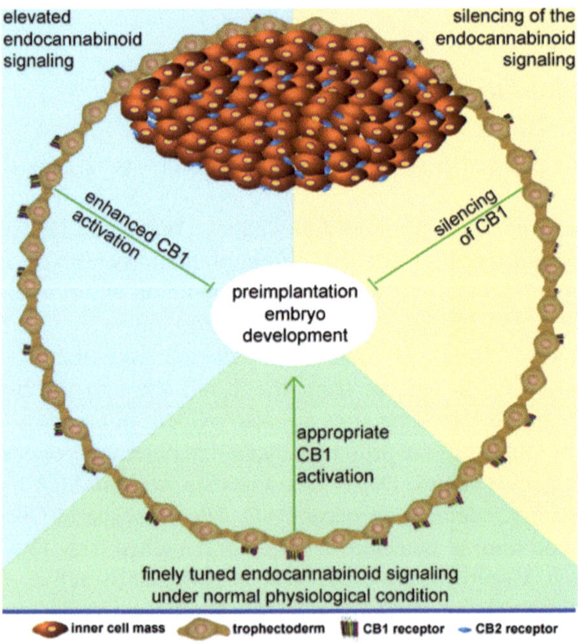

Fig. 2 Cannabinoid signaling in preimplantation embryo development. Both exaggerated or absent cannabinoid/endocannabinoid signaling mediated by CB_1 leads to aberrant preimplantation embryo development. This figure is adapted from Sun and Dey (2008)

females mated with wild-type males showed normal embryo development (Wang et al. 2004). These findings also imply that embryonic CB_1 receptors, but not oviductal (maternal) CB_1 receptors, direct appropriate early embryonic development (Wang et al. 2006a). Furthermore, normal development of heterozygous null embryos suggests that even one copy of CB_1 is sufficient for normal development. These findings prompted the hypothesis that appropriate endocannabinoid signaling is necessary for embryo development.

In vitro embryo culture experiments showed that most 2-cell wild-type embryos fail to develop to the blastocyst stage in the presence of excess AEA. However, low levels of AEA (7 nM) promoted trophoblast differentiation and growth, while higher levels (28 nM) inhibited such development (Wang et al. 1999). In contrast, more than 80% of $CB_1^{-/-}$ or $CB_1^{-/-}/CB_2^{-/-}$ double mutant embryos develop into blastocysts in the presence of similar levels of AEA. Interestingly, in vitro development of $CB_2^{-/-}$ embryos, like wild-type embryos, was severely compromised in the presence of AEA (Paria et al. 2001). These results lend genetic support that CB_1, but not CB_2, responds to cannabinoids to govern embryonic development.

Interestingly, $CB_2^{-/-}$ or $CB_1^{-/-}/CB_2^{-/-}$ embryos collected from the oviduct on day 3 and uterus on day 4 also show asynchronous development (Paria et al. 2001), indicating that CB_2 apparently has some role in preimplantation embryo development. The significance of this finding is not fully understood. However, recent observations of CB_2 expression in embryonic stem cells by microarray analysis (Sharov et al. 2003) together with its absence in trophoblast stem cells (Wang and Dey, unpublished data) suggest that CB_2 expression is restricted to the inner cell mass (ICM), pointing toward a role of CB_2 in ICM cell development and thus development of the embryo proper. Collectively, cannabinoid signaling can regulate preimplantation embryo development, with the current model implicating its effects mediated via CB_1 receptors. However, the role of CB_2 receptors in embryo development remains puzzling.

3.2 Oviductal–Uterine Embryo Transport

In parallel with preimplantation embryo development, embryos transit from the oviduct to the uterus. The oviduct consists of an ampulla and isthmus and is connected to the uterus through the utero-tubal junction. The ampulla is lined with many more ciliated cells than the isthmus, while the isthmus possesses a thicker muscular layer because of their distinct functions (Gaddum-Rosse and Blandau 1976). In mice, embryos transit rapidly though the oviduct ampulla due to the forward-moving beating of the cilia present on the epithelial cell surface. Once they reach the ampulla–isthmus junction, they reside at the isthmus for approximately 3 days. Then the embryos are propelled through the utero-tubal junction by a wave of regulated contraction and relaxation of the isthmus smooth muscle (Halbert et al. 1976). Embryos enter the uterus at the late morula stage, and coincident with this transport a cavity appears in the embryo, marking the early blastocyst stage. The embryo only achieves implantation competency at the blastocyst stage. Thus, a successful implantation depends on normal and timely transport of embryos from the oviduct to the uterus. Although there is no evidence for implantation of embryos in the mouse oviduct, human embryos can implant in the human oviduct (Fallopian tube). A dysfunctional regulation of oviductal–uterine transport results in oviductal retention of embryos, and thus can lead to ectopic pregnancy in women (Farquhar 2005; Pisarska et al. 1998).

In the mouse oviduct, CB_1, not CB_2, is detected (Das et al. 1995; Wang et al. 2004). Both NAPE-PLD and FAAH are also present in the oviduct. NAPE-PLD levels are higher in the isthmus compared to the ampullary region, whereas FAAH shows the reverse pattern, being higher in the ampullary region (Guo et al. 2005; Wang et al. 2006b). This spatially different expression pattern of key enzymes in AEA regulation suggests that endocannabinoid signaling has a physiological role in the oviduct.

Studies from our group have shown that almost half of $CB_1^{-/-}$ mice show pregnancy loss. These mice, however, have normal ovulation and fertilization (Paria et al. 2001; Wang et al. 2004). It was initially thought that asynchronous embryo development is a contributor to this pregnancy loss. Based on this assumption, normal pregnancy in $CB_1^{-/-}$ mice would be restored by mating mutant females with wild-type males to generate all heterozygous embryos with normal preimplantation growth, since $CB_1^{+/-}$ embryos have normal preimplantation development in $CB_1^{-/-}$ oviducts. However, almost half of $CB_1^{-/-}$ mothers still showed pregnancy loss (Paria et al. 2001; Wang et al. 2004). This suggested that maternal CB_1, and not embryonic CB_1, is the cause for pregnancy failure.

Further investigation found that pregnancy failure in $CB_1^{-/-}$ females was attributed to oviductal retention of embryos (Fig. 3a). $CB_1^{-/-}$/ $CB_2^{-/-}$ mice also show oviductal retention, but wild-type and $CB_2^{-/-}$ mice do not, suggesting that oviductal retention results from the lack of CB_1. This is consistent with the expression pattern of CB receptors in the oviduct, in that CB_1 is present in murine oviducts, but not CB_2. This same study showed that all the trapped embryos in the oviduct are morphologically and physiologically healthy, because they can implant when transferred into day 4 pseudopregnant uteri, again confirming that oviductal retention is due to lack of maternal CB_1. This was further confirmed by reciprocal embryo transfer between $CB_1^{-/-}$ and wild-type female mice. Only $CB_1^{-/-}$ recipients displayed oviductal retention of embryos, irrespective of embryonic genotypes (Wang et al. 2004). In addition, wild-type mice with pharmacologically inhibited CB_1, but not CB_2, also show high rate of embryo retention in the oviduct. Notably, $FAAH^{-/-}$ mice, which have higher oviductal AEA levels, and wild-type mice exposed to THC or meth-AEA (a stable AEA analog) also show oviductal retention of embryos (Wang et al. 2006b). All these observations suggest that the regulation of oviduct–uterine transport is not simply an up or down regulation of endocannabinoid signaling. Instead, it suggests that a finely regulated endocannabinoid tone mediated by CB_1 in the oviduct regulates normal embryo transport through the oviduct (Fig. 3b).

It is known that the transport of embryos through the oviduct is aided by a wave of movements in the oviduct muscle that is controlled by the sympathetic nervous

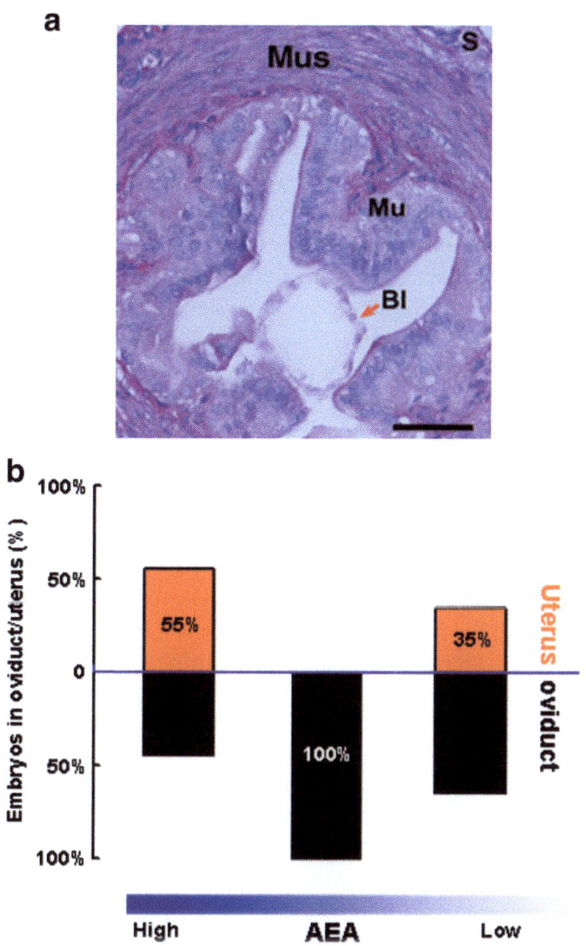

Fig. 3 Impaired oviductal embryo transport. (**a**) A representative histological section of a day 7 pregnant *Cnr1*$^{-/-}$ oviduct showing a trapped blastocyst (arrow) at the oviduct isthmus. Bl, blastocyst; Mus, muscularis; Mu, mucosa; S, serosa. Bar, 100 μm. (**b**) Percentage of embryos recovered from oviducts or uteri at different AEA levels. Panel **a** is adapted from Wang et al. (2004); and panel **b** is reproduced from Wang et al. (2004, 2006b)

system (Heilman et al. 1976). Stimulation of β_2-adrenoceptors (β_2-AR) causes muscle relaxation, whereas stimulation of α_1-AR confers muscle contraction. It has been shown that reciprocal stimulation of these two receptors causes a wave of contractility and relaxation, which is conducive to the passage of embryos from the oviduct to the uterus (Heilman et al. 1976; Howe and Black 1973). In this respect, exposure of wild-type oviducts to either an α_1-AR agonist or a β_2-AR antagonist leads to oviductal retention of embryos. In addition, CB_1 expression in the muscularis of the oviduct is colocalized with α_1- and β_2-adrenoceptors, and $CB_1^{-/-}$ oviducts show increased release of norepinephrine (NE) (Wang et al. 2004).

These observations provide evidence that CB_1-mediated endocannabinoid signaling is coupled to adrenergic signaling to regulate oviductal motility, and that the oviductal muscularis is predominantly in a contraction phase in the absence of CB_1. In contrast, heightened endocannabinoid signaling, in either $FAAH^{-/-}$ mice with naturally higher AEA levels or wild-type mice exposed to excessive natural or synthetic cannabinoid ligands, cause the oviductal muscularis to shift to a relaxation phase, thus impairing oviductal embryo transport to the uterus.

In conclusion, the spatiotemporal expression of NAPE-PLD and FAAH in the oviduct creates an appropriate endocannabinoid tone, executed by CB_1 receptors to regulate the release of NE. Silencing or enhanced endocannabinoid/cannabinoid signaling impedes the highly coordinated oviductal smooth muscle contraction and relaxation through the sympathetic nervous system, consequently regulating the transit of embryos from the oviduct to the uterine lumen.

3.3 Implantation

Attachment of the embryo to the luminal epithelium of the uterus is a crucial step in mammalian reproduction. As the embryo travels into the uterus and differentiates into a blastocyst, the uterine cells undergo proliferation and differentiation to achieve a receptive state to accept the blastocyst for implantation. It is thought that blastocyst activation (implantation competency) and uterine receptivity are two distinct events in the process of implantation (Paria et al. 1993). The attainment of implantation competency of the blastocyst and uterine receptivity are primarily coordinated by the ovarian steroid hormones, estrogen and progesterone (Paria et al. 1998a). Progesterone has been shown to be essential for implantation and pregnancy maintenance in all mammals studied, whereas the requirement for ovarian estrogen is species-specific. In mice, under progesterone priming, closure of the uterine lumen occurs and coincides with the escape of the blastocyst from the zona pellucida, bringing the blastocyst trophectoderm into close contact with the uterine luminal epithelium. Superimposition of the progesterone-primed uterus with preimplantation ovarian estrogen and its catechol metabolite, 4-hydroxy-17β-estradiol (4-OH-E_2) differentially regulate uterine preparation and blastocyst activation, respectively. Estrogen, via its interaction with nuclear estrogen receptors, participates in the preparation of the progesterone-primed uterus to the receptive state in an endocrine manner, whereas its metabolite, 4-OH-E_2, mediates blastocyst activation for implantation in a paracrine manner (Paria et al. 1998a). These coordinated actions of progesterone and estrogen are crucial for the regulation of the window of implantation.

One major step in the process of implantation is the attachment of the blastocyst trophectoderm with the uterine luminal epithelium. This occurs within a narrow time frame concurrent with an intimate two-way dialog that occurs between the implantation-competent blastocyst and the receptive uterus. In mice, this attachment reaction is initiated around midnight on day 4 of pregnancy (Das et al. 1994).

However, elimination of preimplantation estrogen secretion by ovariectomy on the morning of day 4 results in implantation failure with blastocyst dormancy within the quiescent uterine lumen (McLaren 1971; Yoshinaga and Adams 1966). This condition is referred to as delayed implantation and can be maintained for many days by continued progesterone treatment. However, implantation with blastocyst activation is rapidly initiated by a single injection of estrogen (McLaren 1971; Yoshinaga and Adams 1966). This physiologically relevant delayed implantation model has been widely used to identify signaling pathways mediating embryo–uterine cross-talk during implantation. Endocannabinoid signaling has also recently been shown to participate in embryo–uterine interactions during implantation.

Our group has found that lower levels of AEA and CB_1 receptors are beneficial for implantation. AEA levels have been measured in both receptive and nonreceptive uteri, with the former having lower levels of AEA compared with the latter (Schmid et al. 1997). In vitro experiments also show that natural, synthetic or endogenous cannabinoids inhibit preimplantation embryo development and blastocyst zona-hatching in culture, whereas blastocysts exposed to low levels of AEA show accelerated trophoblast differentiation and outgrowth (Paria et al. 1995, 1998b; Schmid et al. 1997). In vivo experiments show that wild-type blastocysts collected from the uterus on the early morning of day 4 of pregnancy have higher levels of AEA binding, and this binding remarkably declines in blastocysts recovered on the evening of day 4, prior to implantation. These observations suggest that implantation competency requires downregulation of AEA binding to the blastocyst (Paria et al. 2001). Immunostaining confirmed that CB_1 is lower in activated blastocysts compared to dormant blastocysts (Paria et al. 2001; Wang et al. 2003). Collectively, these results show that the coordinated down-regulation of blastocyst CB_1 and uterine AEA levels are critical in regulating the "window" of implantation by synchronizing trophoblast differentiation and uterine preparation to the receptive state.

Concurrent with this tenet, higher levels of *nape-pld* mRNA and NAPE-PLD activity are found in nonreceptive uteri and in interimplantation sites, compared to implantation sites and receptive uteri (Guo et al. 2005; Wang et al. 2007). It is interesting that FAAH expression and activity show the inverse relationship: higher FAAH expression and activity are observed at implantation sites and in receptive uteri. Some evidence points to the possibility that the implanting blastocyst exerts an inhibitory effect on uterine *nape-pld* expression, and upregulates uterine FAAH activity by releasing a lipid "FAAH activator" (Guo et al. 2005; Maccarrone et al. 2004). These observations suggest a potential role of the implanting embryo in regulating uterine AEA levels, perhaps to serve as a protective mechanism against exposure to detrimental levels of AEA. Regardless of its control, it is obvious that tight regulation of AEA plays an important role in implantation.

Because of the biphasic action of AEA during embryo implantation, several studies have begun to unravel the underlying mechanism(s) for this by delineating potential signaling pathways coupled with CB_1 (Fig. 4). These studies have found that, under different AEA concentrations, endocannabinoid signaling mediated by embryonic CB_1 is coupled with specific downstream signaling pathways.

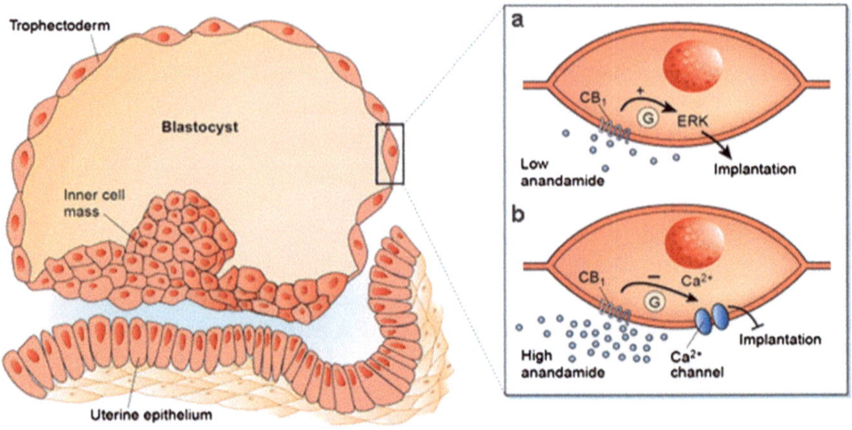

Fig. 4 Influence of cannabinoid/endocannabinoid signaling in embryo implantation in mice. (**a**) At low concentrations, anandamide activates CB_1 displayed on the surface of trophectoderm cells, stimulating ERK and facilitating implantation; (**b**) at higher concentrations, anandamide engages a second CB_1-dependent pathway, which inhibits the activity of voltage-operated N-type calcium channels, reduces calcium entry, and blocks implantation. CB_1 indicates brain-type cannabinoid receptor. This figure is adapted from Piomelli (2004)

For instance, AEA induces stimulatory and inhibitory effects on blastocyst function through ERK and Ca^{2+} signaling pathways, respectively. While AEA at a low concentration (7 nM) activates ERK signaling via CB_1, higher AEA levels (28 nM) fail to activate ERK, but instead inhibit Ca^{2+} mobilization (Wang et al. 2003).

4 Conclusion

In this review, we present molecular, genetic, physiological, and pharmacological studies describing roles of cannabinoid/endocannabinoid signaling that is operative during early pregnancy events. Studies in mouse models demonstrate that under normal physiological conditions, endocannabinoid signaling through CB_1 is crucial to development of embryos and their oviductal transport, as well as their homing and implantation in the receptive uterus. Either silenced or overwhelming endocannabinoid signaling derails these processes. A considerable amount of early pregnancy loss occurs due to either preimplantation embryonic death or implantation failure resulting from asynchronous embryonic development and failure of the uterus to differentiate to the receptive stage (Wilcox et al. 1988). Therefore, our findings in mice raise concerns not only for women of reproductive age who chronically abuse marijuana, but for those who use marijuana or other endocannabinoid system-oriented drugs for medicinal purposes. In addition, studies described here raise caution against the use of CB_1 antagonists to treat obesity in humans, since there is evidence that women with elevated peripheral AEA levels have spontaneous pregnancy loss (Maccarrone et al. 2000, 2002). Future studies need to be directed towards endocannabinoids' roles in placentation and parturition, since early pregnancy often influences the later developmental processes, ultimately determining the success of pregnancy.

Acknowledgements The authors thank Susanne Tranguch for her critical reading of the manuscript. This work was supported in part by NIH grants DA06668, HD12304, and P01-CA-77,839. S. K. Dey is a recipient of Method to Extend Research in Time (MERIT) Awards from the National Institute on Drug Abuse (NIDA) and the National Institute of Child Health and Human Development (NICHD).

References

Abma JC, Chandra A, Mosher WD et al. (1997) Fertility, family planning, and women's health: new data from the 1995 National Survey of Family Growth. Vital Health Stat 19:1–114

Baker D, Pryce G, Davies WL et al. (2006) In silico patent searching reveals a new cannabinoid receptor. Trends Pharmacol Sci 27:1–4

Beltramo M, Piomelli D (2000) Carrier-mediated transport and enzymatic hydrolysis of the endogenous cannabinoid 2-arachidonylglycerol. Neuroreport 11:1231–1235

Bisogno T, Sepe N, Melck D et al. (1997) Biosynthesis, release and degradation of the novel endogenous cannabimimetic metabolite 2-arachidonoylglycerol in mouse neuroblastoma cells. Biochem J 322(Pt 2):671–677

Bisogno T, Howell F, Williams G et al. (2003) Cloning of the first sn1-DAG lipases points to the spatial and temporal regulation of endocannabinoid signaling in the brain. J Cell Biol 163:463–468

Bouaboula M, Poinot-Chazel C, Bourrie B et al. (1995) Activation of mitogen-activated protein kinases by stimulation of the central cannabinoid receptor CB1. Biochem J 312(Pt 2):637–641

Bracey MH, Hanson MA, Masuda KR et al. (2002) Structural adaptations in a membrane enzyme that terminates endocannabinoid signaling. Science (New York, NY) 298:1793–1796

Carson DD, Bagchi I, Dey SK et al. (2000) Embryo implantation. Dev Biol 223:217–237

Caulfield MP, Brown DA (1992) Cannabinoid receptor agonists inhibit Ca current in NG108–15 neuroblastoma cells via a pertussis toxin-sensitive mechanism. Br J Pharmacol 106:231–232

Cravatt BF, Lichtman AH (2002) The enzymatic inactivation of the fatty acid amide class of signaling lipids. Chem Phys Lipids 121:135–148

Cravatt BF, Giang DK, Mayfield SP et al. (1996) Molecular characterization of an enzyme that degrades neuromodulatory fatty-acid amides. Nature 384:83–87

Cross JC, Werb Z, Fisher SJ (1994) Implantation and the placenta: key pieces of the development puzzle. Science (New York, NY) 266:1508–1518

Das SK, Wang XN, Paria BC et al. (1994) Heparin-binding EGF-like growth factor gene is induced in the mouse uterus temporally by the blastocyst solely at the site of its apposition: a possible ligand for interaction with blastocyst EGF-receptor in implantation. Development (Cambridge, England) 120:1071–1083

Das SK, Paria BC, Chakraborty I et al. (1995) Cannabinoid ligand–receptor signaling in the mouse uterus. Proc Natl Acad Sci USA 92:4332–4336

De Petrocellis L, Bisogno T, Maccarrone M et al. (2001) The activity of anandamide at vanilloid VR1 receptors requires facilitated transport across the cell membrane and is limited by intracellular metabolism. J Biol Chem 276:12856–12863

Devane WA, Dysarz FA 3rd, Johnson MR et al. (1988) Determination and characterization of a cannabinoid receptor in rat brain. Mol Pharmacol 34:605–613

Devane WA, Hanus L, Breuer A et al. (1992) Isolation and structure of a brain constituent that binds to the cannabinoid receptor. Science (New York, NY) 258:1946–1949

Dey SK, Lim H, Das SK et al. (2004) Molecular cues to implantation. Endocr Rev 25:341–373

Dinh TP, Carpenter D, Leslie FM et al. (2002) Brain monoglyceride lipase participating in endocannabinoid inactivation. Proc Natl Acad Sci U S A 99:10819–10824

Evans JP, Florman HM (2002) The state of the union: the cell biology of fertilization. Nature Cell Biol 4(Suppl):s57–s63

Farquhar CM (2005) Ectopic pregnancy. Lancet 366:583–591

Gaddum-Rosse P, Blandau RJ (1976) Comparative observations on ciliary currents in mammalian oviducts. Biol Reprod 14:605–609

Gaoni Y, Mechoulam R (1964) Isolation, structure, and partial synthesis of an active constituent of Hashish. J Am Chem Soc 86:1646–1647

Gebremedhin D, Lange AR, Campbell WB et al. (1999) Cannabinoid CB1 receptor of cat cerebral arterial muscle functions to inhibit L-type Ca2+ channel current. Am J Physiol 276: H2085–H2093

Giang DK, Cravatt BF (1997) Molecular characterization of human and mouse fatty acid amide hydrolases. Proc Natl Acad Sci USA 94:2238–2242

Glaser ST, Kaczocha M, Deutsch DG (2005) Anandamide transport: a critical review. Life Sci 77:1584–1604

Goparaju SK, Ueda N, Taniguchi K et al. (1999) Enzymes of porcine brain hydrolyzing 2-arachidonoylglycerol, an endogenous ligand of cannabinoid receptors. Biochem Pharmacol 57:417–423

Guo Y, Wang H, Okamoto Y et al. (2005) N-acylphosphatidylethanolamine-hydrolyzing phospholipase D is an important determinant of uterine anandamide levels during implantation. J Biol Chem 280:23429–23432

Halbert SA, Tam PY, Blandau RJ (1976) Egg transport in the rabbit oviduct: the roles of cilia and muscle. Science (New York, NY) 191:1052–1053

Hamatani T, Daikoku T, Wang H et al. (2004) Global gene expression analysis identifies molecular pathways distinguishing blastocyst dormancy and activation. Proc Natl Acad Sci U S A 101:10326–10331

Heilman RD, Reo RR, Hahn DW (1976) Changes in the sensitivity of adrenergic receptors in the oviduct during early gestation in the rabbit. Fertil Steril 27:426–430

Ho SY, Delgado L, Storch J (2002) Monoacylglycerol metabolism in human intestinal Caco-2 cells: evidence for metabolic compartmentation and hydrolysis. J Biol Chem 277:1816–1823

Howe GR, Black DL (1973) Autonomic nervous system and oviduct function in the rabbit. I. Hormones and contraction. Journal Reprod Fertil 33:425–430

Howlett AC, Barth F, Bonner TI et al. (2002) International Union of Pharmacology. XXVII. Classification of cannabinoid receptors. Pharmacol Rev 54:161–202

Kaczocha M, Hermann A, Glaser ST et al. (2006) Anandamide uptake is consistent with rate-limited diffusion and is regulated by the degree of its hydrolysis by fatty acid amide hydrolase. J Biol Chem 281:9066–9075

Lauckner JE, Hille B, Mackie K (2005) The cannabinoid agonist WIN55, 212-2 increases intracellular calcium via CB1 receptor coupling to Gq/11 G proteins. Proc Natl Acad Sci USA 102:19144–19149

Leung D, Saghatelian A, Simon GM et al. (2006) Inactivation of N-acyl phosphatidylethanolamine phospholipase D reveals multiple mechanisms for the biosynthesis of endocannabinoids. Biochemistry 45:4720–4726

Liu J, Wang L, Harvey-White J et al. (2006) A biosynthetic pathway for anandamide. Proc Natl Acad Sci USA 103:13345–13350

Maccarrone M, Finazzi-Agro A (2003) The endocannabinoid system, anandamide and the regulation of mammalian cell apoptosis. Cell Death Differ 10:946–955

Maccarrone M, Valensise H, Bari M et al. (2000) Relation between decreased anandamide hydrolase concentrations in human lymphocytes and miscarriage. Lancet 355:1326–1329
Maccarrone M, Bisogno T, Valensise H et al. (2002) Low fatty acid amide hydrolase and high anandamide levels are associated with failure to achieve an ongoing pregnancy after IVF and embryo transfer. Molecular Human Reprod 8:188–195
Maccarrone M, DeFelici M, Klinger FG et al. (2004) Mouse blastocysts release a lipid which activates anandamide hydrolase in intact uterus. Molecular Human Reprod 10:215–221
Mackie K, Hille B (1992) Cannabinoids inhibit N-type calcium channels in neuroblastoma-glioma cells. Proc Natl Acad Sci USA 89:3825–3829
Matas D, Juknat A, Pietr M et al. (2007) Anandamide protects from low serum-induced apoptosis via its degradation to ethanolamine. J Biol Chem 282:7885–7892
Matsuda LA, Lolait SJ, Brownstein MJ et al. (1990) Structure of a cannabinoid receptor and functional expression of the cloned cDNA. Nature 346:561–564
McAllister SD, Glass M (2002) CB(1) and CB(2) receptor-mediated signalling: a focus on endocannabinoids. Prostaglandins Leukot Essent Fatty Acids 66:161–171
McKinney MK, Cravatt BF (2005) Structure and function of fatty acid amide hydrolase. Annu Rev Biochem 74:411–432
McLaren A (1971) Blastocysts in the mouse uterus: the effect of ovariectomy, progesterone and oestrogen. J Endocrinol 50:515–526
McPartland JM, Matias I, Di Marzo V et al. (2006) Evolutionary origins of the endocannabinoid system. Gene 370:64–74
Mechoulam R, Deutsch DG (2005) Toward an anandamide transporter. Proc Natl Acad Sci U S A 102:17541–17542
Mechoulam R, Hanus L (2000) A historical overview of chemical research on cannabinoids. Chem Phys Lipids 108:1–13
Mechoulam R, Ben-Shabat S, Hanus L et al. (1995) Identification of an endogenous 2-monoglyceride, present in canine gut, that binds to cannabinoid receptors. Biochem Pharmacol 50:83–90
Moriyama T, Urade R, Kito M (1999) Purification and characterization of diacylglycerol lipase from human platelets. J Biochem 125:1077–1085
Muccioli GG, Xu C, Odah E et al. (2007) Identification of a novel endocannabinoid-hydrolyzing enzyme expressed by microglial cells. J Neurosci 27:2883–2889
Munro S, Thomas KL, Abu-Shaar M (1993) Molecular characterization of a peripheral receptor for cannabinoids. Nature 365:61–65
Murphy LO, Blenis J (2006) MAPK signal specificity: the right place at the right time. Trends Biochem Sci 31:268–275
Natarajan V, Reddy PV, Schmid PC et al. (1982) N-Acylation of ethanolamine phospholipids in canine myocardium. Biochim Biophys Acta 712:342–355
Natarajan V, Schmid PC, Reddy PV et al. (1984) Catabolism of N-acylethanolamine phospholipids by dog brain preparations. J Neurochem 42:1613–1619
Okamoto Y, Morishita J, Tsuboi K et al. (2004) Molecular characterization of a phospholipase D generating anandamide and its congeners. J Biol Chem 279:5298–5305
Ortega-Gutierrez S (2005) Therapeutic perspectives of inhibitors of endocannabinoid degradation. Curr Drug Targets 4:697–707
Paria BC, Huet-Hudson YM, Dey SK (1993) Blastocyst's state of activity determines the "window" of implantation in the receptive mouse uterus. Proc Natl Acad Sci USA 90:10159–10162
Paria BC, Das SK, Dey SK (1995) The preimplantation mouse embryo is a target for cannabinoid ligand-receptor signaling. Proc Natl Acad Sci USA 92:9460–9464
Paria BC, Lim H, Wang XN et al. (1998a) Coordination of differential effects of primary estrogen and catecholestrogen on two distinct targets mediates embryo implantation in the mouse. Endocrinology 139:5235–5246
Paria BC, Ma W, Andrenyak DM et al. (1998b) Effects of cannabinoids on preimplantation mouse embryo development and implantation are mediated by brain-type cannabinoid receptors. Biol Reprod 58:1490–1495

Paria BC, Song H, Wang X et al. (2001) Dysregulated cannabinoid signaling disrupts uterine receptivity for embryo implantation. J Biol Chem 276:20523–20528
Paria BC, Reese J, Das SK et al. (2002) Deciphering the cross-talk of implantation: advances and challenges. Science (New York, NY) 296:2185–2188
Pertwee RG, Ross RA (2002) Cannabinoid receptors and their ligands. Prostaglandins Leukot Essent Fatty Acids 66:101–121
Piomelli D (2004) THC: moderation during implantation. Nat Med 10:19–20
Pisarska MD, Carson SA, Buster JE (1998) Ectopic pregnancy. Lancet 351:1115–1120
Red-Horse K, Zhou Y, Genbacev O et al. (2004) Trophoblast differentiation during embryo implantation and formation of the maternal–fetal interface. J Clin Invest 114:744–754
Rossant J (2004) Lineage development and polar asymmetries in the peri-implantation mouse blastocyst. Semin Cell Develop Biol 15:573–581
Rossant J, Tam PP (2004) Emerging asymmetry and embryonic patterning in early mouse development. Develop Cell 7:155–164
Sawzdargo M, Nguyen T, Lee DK et al. (1999) Identification and cloning of three novel human G protein-coupled receptor genes GPR52, PsiGPR53 and GPR55: GPR55 is extensively expressed in human brain. Brain Res Mol Brain Res 64:193–198
Schmid PC, Paria BC, Krebsbach RJ et al. (1997) Changes in anandamide levels in mouse uterus are associated with uterine receptivity for embryo implantation. Proc Natl Acad Sci USA 94:4188–4192
Shah BH, Catt KJ (2005) Roles of LPA3 and COX-2 in implantation. Trends Endocrinol Metab 16:397–399
Sharov AA, Piao Y, Matoba R et al. (2003) Transcriptome analysis of mouse stem cells and early embryos. PLoS Biol 1:E74
Simon GM, Cravatt BF (2006) Endocannabinoid biosynthesis proceeding through glycerophospho-N-acyl ethanolamine and a role for alpha/beta-hydrolase 4 in this pathway. J Biol Chem 281:26465–26472
Song H, Lim H, Paria BC et al. (2002) Cytosolic phospholipase A2alpha is crucial [correction of A2alpha deficiency is crucial] for 'on-time' embryo implantation that directs subsequent development. Development (Cambridge, England) 129:2879–2889
Sugiura T, Kudo N, Ojima T et al. (1995) Coenzyme A-dependent cleavage of membrane phospholipids in several rat tissues: ATP-independent acyl-CoA synthesis and the generation of lysophospholipids. Biochim Biophys Acta 1255:167–176
Sugiura T, Kishimoto S, Oka S et al. (2006) Biochemistry, pharmacology and physiology of 2-arachidonoylglycerol, an endogenous cannabinoid receptor ligand. Prog Lipid Res 45:405–446
Sun X, Dey SK (2008) Aspects of endocannabinoid signaling in periimplantation biology. Mol Cell Endocrinol 286(1–2):S3–S11
Thonneau P, Marchand S, Tallec A et al. (1991) Incidence and main causes of infertility in a resident population (1,850,000) of three French regions (1988–1989). Human Reprod (Oxford, England) 6:811–816
Ueda N, Liu Q, Yamanaka K (2001) Marked activation of the N-acylphosphatidylethanolamine-hydrolyzing phosphodiesterase by divalent cations. Biochim Biophys Acta 1532:121–127
Van Der Stelt M, Di Marzo V (2004) Endovanilloids. Putative endogenous ligands of transient receptor potential vanilloid 1 channels. Eur J Biochem/FEBS 271:1827–1834
Wang H, Dey SK. Unpublished data
Wang H, Dey SK (2005) Lipid signaling in embryo implantation. Prostag Oth Lipid M 77:84–102
Wang H, Dey SK (2006) Roadmap to embryo implantation: clues from mouse models. Nature Rev 7:185–199
Wang J, Paria BC, Dey SK et al. (1999) Stage-specific excitation of cannabinoid receptor exhibits differential effects on mouse embryonic development. Biol Reprod 60:839–844
Wang H, Matsumoto H, Guo Y et al. (2003) Differential G protein-coupled cannabinoid receptor signaling by anandamide directs blastocyst activation for implantation. Proc Natl Acad Sci U S A 100:14914–14919

Wang H, Guo Y, Wang D et al. (2004) Aberrant cannabinoid signaling impairs oviductal transport of embryos. Nat Med 10:1074–1080
Wang H, Dey SK, Maccarrone M (2006a) Jekyll and hyde: two faces of cannabinoid signaling in male and female fertility. Endocr Rev 27:427–448
Wang H, Xie H, Guo Y et al. (2006b) Fatty acid amide hydrolase deficiency limits early pregnancy events. J Clin Invest 116:2122–2131
Wang H, Xie H, Sun X et al. (2007) Differential regulation of endocannabinoid synthesis and degradation in the uterus during embryo implantation. Prostag Oth Lipid M 83:62–74
Wassarman PM, Jovine L, Litscher ES (2001) A profile of fertilization in mammals. Nat Cell Biol 3:E59–E64
Wei BQ, Mikkelsen TS, McKinney MK et al. (2006) A second fatty acid amide hydrolase with variable distribution among placental mammals. J Biol Chem 281:36569–36578
Wilcox AJ, Weinberg CR, O'Connor JF et al. (1988) Incidence of early loss of pregnancy. N Engl J Med 319:189–194
Yang ZM, Paria BC, Dey SK (1996) Activation of brain-type cannabinoid receptors interferes with preimplantation mouse embryo development. Biol Reprod 55:756–761
Ye X, Hama K, Contos JJ et al (2005) LPA3-mediated lysophosphatidic acid signalling in embryo implantation and spacing. Nature 435:104–108
Yoshinaga K, Adams CE (1966) Delayed implantation in the spayed, progesterone treated adult mouse. J Reprod Fertil 12:593–595
Zernicka-Goetz M (2005) Cleavage pattern and emerging asymmetry of the mouse embryo. Nat Rev Mol Cell Biol 6:919–928
Zoratti C, Kipmen-Korgun D, Osibow K et al. (2003) Anandamide initiates Ca(2+) signaling via CB2 receptor linked to phospholipase C in calf pulmonary endothelial cells. Br J Pharmacol 140:1351–1362

Targeting the Cannabinoid System to Produce Analgesia

Devi Rani Sagar, Maulik Jhaveri, and Victoria Chapman

Contents

Abstract Cannabinoid receptors are present at key sites involved in the relay and modulation of nociceptive responses. The analgesic effects of the cannabinoid CB_1 receptor are well described. The widespread distribution of these receptors in the brain does, however, also explain the side-effects associated with CB_1 receptor agonists. The cannabinoid CB_2 receptor also produces analgesic effects in models of acute, inflammatory and neuropathic pain. The sites and mechanisms of CB_2 receptor-mediated analgesia are described herein. In addition to targeting cannabinoid receptors directly, protection of endocannabinoids (eCBs) from metabolism also produces analgesic effects. Indeed, reports that noxious stimulation elevates levels of eCBs in the spinal cord and brain provide further rationale for this approach. The effects of inhibition of fatty acid amide hydrolase (FAAH) on nociceptive responses in models of inflammatory and neuropathic pain are discussed.

D.R. Sagar (✉), M. Jhaverl and V. Chapman
School of Biomedical Sciences, University of Nottingham, Nottingham, NG7 2UH, UK
e-mail: Devi.sagar@nottingham.ac.uk

D. Kendall and S. Alexander (eds.), *Behavioral Neurobiology of the Endocannabinoid System*, Current Topics in Behavioral Neurosciences 1,
DOI: 10.1007/978-3-540-88955-7_11,

Keywords Inflammation • Neuropathic • Fatty acid amide hydrolase • Endocannabinoid • Arthritis

Abbreviations

2AG	2-Arachidonoylglycerol
AEA	*N*-Arachidonoylethanolamine; Anandamide
CB_1	Cannabinoid-1 receptor
CB_2	Cannabinoid-2 receptor
CCI	Chronic constriction injury
CFA	Complete Freund's adjuvant
COX-2	Cyclooxygenase type 2
DGL	Diacylglycerol lipase
DRG	Dorsal root ganglion
FAAH	Fatty acid amide hydrolase
i.p.	Intraperitoneal administration
i.pl.	Intraplantar administration
MAPK	Mitogen activated protein kinase
MGL	Monoacylglycerol lipase
NAAA	*N*-Acylethanolamine hydrolysing acid amidase
NADA	*N*-Arachidonoyl dopamine
NAE	*N*-Acylethanolamines
NAPE	*N*-Acylphosphatidylethanolamine
OA	Osteoarthritis
OEA	*N*-Oleoylethanolamine
PAG	Periaqueductal grey
PEA	*N*-Palmitoyl ethanolamine
PLC	Phospholipase C
PLD	Phospholipase D
p.o.	Oral administration
RA	Rheumatoid arthritis
SNL	Spinal nerve ligation
Δ^9-THC	Δ^9-Tetrahydrocannabinol
TRPV1	Transient receptor potential vanilloid type 1

1 Introduction

The anti-nociceptive effects of cannabinoids are well documented (Hohmann 2002; Iversen and Chapman 2002; Jhaveri et al. 2007a, b; Pertwee 2001; Rice et al. 2002; Walker and Huang 2002). The analgesic effects of cannabinoids are, however, often limited by psychoactive side-effects. In the last decade, rapid scientific progress has revealed an endogenous cannabinoid system, which consists of cannabinoid receptors, endogenous cannabinoid ligands and their synthesizing and metabolising enzymes. This progress has led to the investigation of the individual components of the cannabinoid system as targets for producing analgesia and other medicinal effects, with minimal side-effects. This chapter will discuss the application of these approaches to the development of novel analgesics.

2 The Endogenous Cannabinoid System

Two cannabinoid receptors, the cannabinoid-1 (CB_1) and cannabinoid-2 (CB_2) have been identified, cloned and pharmacologically characterised (see Mackie 2006). Both receptors are $G_{i/o}$-protein coupled receptors negatively coupled to adenylyl cyclase and positively coupled to mitogen activated protein kinase (MAPK). A third receptor, GPR55, binds a number of cannabinoid ligands and therefore has been proposed as a member of the cannabinoid receptor family (Brown 2007; Johns et al. 2007; Lauckner et al. 2008; Ryberg et al. 2007).

CB_1 receptors are associated with neuronal tissue, with high density in the central, peripheral and autonomic nervous system (Egertova and Elphick 2000; Herkenham et al. 1991; Tsou et al. 1998). CB_1 receptors are also present at lower densities in the heart, lung, testis, ovary, bone marrow, thymus, uterus and immune cells (Galiegue et al. 1995). CB_1 receptor density is moderate to high in regions involved in pain transmission and modulation, such as the dorsal root ganglion (DRG), spinal cord, thalamus, periaqueductal grey (PAG), amygdala and rostroventromedial medulla (Tsou et al. 1998). The effects of cannabinoid agonists on brain function have been investigated with functional magnetic resonance imaging studies. Systemic administration of a non-selective CB_1/CB_2 agonist increased regional cerebral blood flow, an indirect index of brain activity, in cortical regions, hippocampus, PAG, nucleus accumbens and striatum (Chin et al. 2008). Thus, the brain regions activated by the cannabinoid ligand correspond well to those regions identified by autoradiographic approaches.

Given that the bulk of the unwanted effects of cannabinoids arises due to activation of CB_1 receptors, recent research has focused on the potential for CB_2 receptor agonists as analgesics. At high densities, CB_2 receptors are primarily on immune tissues. Nevertheless, a putative role of the CB_2 receptor in the nervous system is becoming apparent. Although early studies indicated an absence of CB_2 receptors in the central nervous system, recent work has reported the presence of CB_2 mRNA in the spinal cord of control rats (Beltramo et al. 2006) and brain tissue (Gong et al. 2006; Van Sickle et al. 2005). The functional role of CB_2 receptors in the CNS is unclear. A functional imaging study demonstrated that CB_2 receptor antagonism did not alter brain activation evoked by systemic administration of a non-selective cannabinoid agonist (Chin et al. 2008). These data suggest there is little CB_2-mediated cannabinoid-induced brain activity under control conditions.

2.1 Endocannabinoids

Currently, five endogenous cannabinoid receptor ligands or eCBs have been discovered, of which anandamide (AEA) was the first to be identified (Devane et al. 1992). Since then, 2-arachidonoylglycerol (2AG; Mechoulam et al. 1995), noladin ether (Hanus et al. 2001), virodhamine (Porter et al. 2002) and *N*-arachidonoyldopamine (NADA; Huang et al. 2002) have been identified. ECBs are synthesized de novo and their actions are rapidly terminated by being taken up into cells where they are metabolised by enzymatic hydrolysis.

2.2 Endocannabinoid Synthesis

Several different pathways have been suggested to contribute to the synthesis of the *N*-acylethanolamines (NAEs) AEA, *N*-oleoylethanolamine (OEA) and *N*-palmitoylethanolamine (PEA) from *N*-acylphosphatidylethanolamine (NAPE). In addition to the established NAPE-phospholipase D (PLD) pathway, two alternative pathways via phospholipase C (PLC)-PTPN22 (Liu et al. 2008) and αβ hydrolase (αβH4)-GDE1 (Simon and Cravatt 2008) are able to generate NAEs. These multiple pathways may subserve differential synthesis of NAEs, as it has been suggested that NAPE-PLD mainly generates saturated *N*-acylethanolamines (NAEs) such as PEA (Leung et al. 2006). Tissue distribution of these synthetic enzyme pathways may also vary. AEA and PEA biosynthesis in the CNS is suggested to be predominantly via the αβH4-GDE1 pathway (Simon and Cravatt 2008). In macrophages, PLC-mediated cleavage of NAPE to phosphoanandamide prior to PTPN22-mediated dephosphorylation to NAE has been described (Liu et al. 2006).

2.3 Endocannabinoid Metabolism

AEA and other NAEs are mainly hydrolysed by FAAH (Cravatt et al. 1996; Deutsch and Chin 1993) whilst 2AG is mainly metabolised by monoacylglycerol lipase (MGL; Dinh et al. 2002). Although FAAH and MGL are the main enzymes for metabolism of AEA and 2AG, enzymes such as cyclooxygenase type 2 (COX-2; for review see Fowler 2007) also metabolise AEA and 2AG. In addition, *N*-acylethanolamine hydrolysing acid amidase (NAAA) can also metabolise AEA and PEA (Tsuboi et al. 2007).

3 Endocannabinoids and Pain Processing

The anti-nociceptive effects of eCBs are well described. We have demonstrated anti-nociceptive effects of AEA when administered spinally (Harris et al. 2000) and peripherally (Sokal et al. 2003) using extracellular recordings of dorsal horn neurones in carrageenan-inflamed rats. AEA is also anti-nociceptive in behavioural models of acute and chronic pain (for review see Pertwee 2001). Similarly, 2AG reduces pain behaviour in tail-flick (Mechoulam et al. 1995) and formalin tests (Guindon et al. 2007).

AEA and 2AG are present in key regions involved in the detection, relay and integration of nociceptive inputs, including the skin, DRG, spinal cord, PAG and rostral ventromedial medulla. There is good evidence that eCBs tonically inhibit pain responses and contribute to the setting of nociceptive thresholds. Indeed, spinal administration of selective CB_1 receptor antagonists increases evoked-firing of dorsal horn neurones and thermal hyperalgesia (Chapman 1999). In turn, levels of eCBs are altered under pathological conditions such as inflammation and neuropathic pain. We have demonstrated a significant reduction in levels of AEA and PEA in the hindpaw of rats with a carrageenan-induced inflammation (Jhaveri et al. 2008b). Similarly, levels of AEA, 2AG and PEA were decreased in the hindpaw following intraplantar injection of formalin (Maione et al. 2007). By contrast, Beaulieu et al. reported no significant alteration in levels of AEA, 2AG and PEA in the hindpaw of formalin-treated rats (Beaulieu et al. 2000). In addition to altering levels of eCBs at the site of injury, noxious stimulation such as formalin-evoked hindpaw inflammation increases levels of eCBs at other targets in the nociceptive pathway such as the periaqueductal grey, indicating a role for eCBs in descending control of pain processing (Walker et al. 1999).

Alterations in the levels of eCBs and NAEs under various pathological conditions may occur as a result of either enhanced synthesis or decreased catabolism. ECB levels increased in the spinal cord (Petrosino et al. 2007) and dorsal root ganglia (Mitrirattanakul et al. 2006) following peripheral nerve injury, a model of neuropathic pain. We have shown that levels of AEA are increased, whereas levels of PEA are decreased, in the spinal cord (unpublished observations) in a model of neuropathic pain. These data suggest there is differential synthesis, or catabolism, of AEA and PEA in the spinal cord of neuropathic rats. How these findings relate to the presence of additional cell types, such as activated microglia, which will contribute to the synthesis and catabolism of eCBs, and the emerging evidence for multiple cell related synthesis pathways, is unknown.

4 CB Receptor-Mediated Analgesia

The analgesic effects produced by activation of CB_1 receptors are well described and extensively reviewed (for reviews see Iversen and Chapman 2002; Pertwee 2001; Walker and Huang 2002). Activation of CB_1 receptors in the spinal cord (Hohmann et al. 1998; Kelly and Chapman 2001, 2003) and in the periphery (Kelly et al. 2003) attenuates nociceptive responses of dorsal horn neurones in naive rats. Supra-spinal CB_1 receptors in a number of discrete brain regions also contribute to the anti-nociceptive effects of cannabinoids in models of acute/tonic pain (Finn et al. 2003; Lichtman et al. 1996; Martin et al. 1999; Meng et al. 1998;

Welch et al. 1998; Welch and Stevens 1992). The broad distribution of CB_1 receptors in the brain results in both therapeutic effects, such as analgesia, as well as their side-effects. To avoid these psychoactive side-effects, the analgesic potential of selective activation of peripheral and spinal CB_1 receptors has been studied. Anti-nociceptive effects of CB_1 receptor agonist were substantially reduced in mice with CB_1 receptor gene deletion in the peripheral nociceptors (Agarwal et al. 2007). Thus it appears that CB_1 receptor agonists which do not cross the blood–brain barrier, and thus selectively activate peripheral CB_1 receptors, may be a promising analgesic strategy. This concept is supported by earlier work demonstrating that hindpaw injection of CB_1 receptor agonists produces anti-nociceptive effects in models of inflammatory and chronic pain (Clayton et al. 2002; Elmes et al. 2005; Kelly and Chapman 2002, 2003; Kelly et al. 2003; Richardson et al. 1998; Scott et al. 2004).

Recently, a number of studies have demonstrated analgesic effects of CB_2 agonist receptor agonists in models of acute and chronic pain (reviewed elsewhere by Guindon and Hohmann 2008; Jhaveri et al. 2007b). Administration of CB_2 agonists systemically (Ibrahim et al. 2006; Malan et al. 2001; Valenzano et al. 2005) or locally into the hindpaw (Elmes et al. 2004; Malan et al. 2001) attenuates nociceptive responses in naïve rats. CB_2 receptors are present in the skin and their activation is reported to release endorphins from keratinocytes which act via μ opioid receptors to produce analgesia (Ibrahim et al. 2005). There is little evidence that spinal (Sagar et al. 2005) or supra-spinal (Jhaveri et al. 2008a) CB_2 receptors modulate nociceptive responses in naive rats, despite the description of supra-spinal CB_2 receptors (see earlier). There is, however, evidence for a functional role of CB_2 receptors in the spinal cord (Romero-Sandoval et al. 2008; Sagar et al. 2005; Yamamoto et al. 2008) and thalamus (Jhaveri et al. 2008a) of neuropathic rats. Importantly, CB_2 receptor agonists are devoid of CNS-mediated side-effects (Malan et al. 2003).

5 Attenuation of Endocannabinoid Catabolism Produces Analgesia

As mentioned earlier, the beneficial and analgesic effects of eCBs are hampered by their short duration of action. In order to prolong these effects, research has investigated the effects of inhibiting the breakdown of eCBs. One of the benefits of inhibiting the catabolism of eCBs is that regions with elevated levels of eCBs, for example as a result of noxious stimulation, are targeted as opposed to the global effects of receptor agonists.

The important role of FAAH in metabolism of eCBs has been demonstrated in mice lacking FAAH, which exhibit 15-fold elevated levels of AEA compared to wild-type mice, and display phenotypic hypoalgesia in models of acute and inflammatory pain (Cravatt et al. 2001; Lichtman et al. 2004b), but not neuropathic pain

(Lichtman et al. 2004b). Inhibitors of FAAH, such as URB597 and OL135, are antinociceptive in models of acute and inflammatory pain (Chang et al. 2006; Fegley et al. 2005; Jayamanne et al. 2006; Kathuria et al. 2003; Lichtman et al. 2004a; Russo et al. 2007 (Table 1). A single systemic injection of URB597 significantly reduced thermal allodynia and mechanical hyperalgesia in the Complete Freund's adjuvant (CFA) model of inflammation (Jayamanne et al. 2006). In the carrageenan model of inflammation, we reported that intraplantar injection of URB597 increased levels of AEA and 2AG in hindpaw skin and reduced carrageenan hyperalgesia (Jhaveri et al. 2008b).

The effects of inhibition of FAAH on neuropathic pain behaviour are less consistent than those reported for inflammatory pain states. Acute systemic injection of URB597 (0.3 mg kg^{-1} i.p.) did not alter mechanical allodynia in a model of peripheral neuropathy (Jayamanne et al. 2006). Similarly, a single oral dose of URB597 (10 mg kg^{-1} p.o.) had limited effects on mechanical hyperalgesia in the chronic constriction injury model of peripheral neuropathy (Russo et al. 2007). Repeated oral dosing of URB597 (10 mg kg^{-1} for 4 days p.o.) significantly reduced thermal and mechanical hyperalgesia (Russo et al. 2007) and a far higher dose of OL135 (ED_{50} 9 mg kg^{-1} i.p.) reduced mechanical allodynia (Chang et al. 2006) in neuropathic rodents. These data suggest that there is an alteration in synthesis/metabolism of eCBs and eCB-like compounds or their receptor function following peripheral neuropathy, which is also supported by data from our electrophysiological experiments. Peripheral injection of URB597, at a dose (25 μg in 50 μl) effective in reducing mechanically evoked responses of spinal cord dorsal horn neurones in sham-operated rats, did not alter responses in neuropathic rats (Jhaveri et al. 2006). A fourfold higher dose (100 μg in 50 μl, i.pl.) of URB597 did, however, reduce mechanically evoked responses in these animals (Jhaveri et al. 2006). In the same study, spinal administration of URB597 (10–50 μg in 50 μl) was equi-effective at reducing mechanically evoked responses of dorsal horn neurones in neuropathic and sham-operated rats, suggesting that alteration in the synthesis/metabolism of endocannabinoids and related molecules is focal and not global (Jhaveri et al. 2006).

6 Arthritis – A Therapeutic Target for Cannabinoids?

One of the groups of pain patients in which clinical effectiveness of cannabis-based medicines has been shown is arthritis. Anecdotal evidence indicates the effectiveness of cannabis in arthritis patients (Wright et al. 2006) and the cannabis-based drug Sativex produced significant analgesia in a double-blind multicentre group comparison study of patients with arthritis (Blake et al. 2006) (Table 1).

Recently, we have demonstrated the expression of cannabinoid CB_1 and CB_2 receptors in the synovial tissue of patients with rheumatoid arthritis (RA) and osteoarthritis (OA). In addition, we reported the presence of AEA and 2AG in the synovial fluid of OA and RA patients, neither of which were detected in samples

Table 1 Comparison of effects of FAAH inhibitors in models of neuropathic and inflammatory pain

Species	Route	Neuropathic pain	Inflammatory pain
Rat	Peripheral (i.pl.)	URB597 inhibited mechanically evoked responses following SNL (Jhaveri et al. 2006)	URB597 inhibited carrageenan-evoked changes in weight-bearing (Jhaveri et al. 2008b)
Rat	(i.pl.)		*N*-Arachidonoyl-serotonin attenuates formalin-evoked hyperalgesia (Maione et al. 2007)
Rat	Spinal	URB597 inhibited mechanically evoked responses following SNL (Jhaveri et al. 2006)	
Rat	(i.p.)	*N*-Arachidonoyl-serotonin inhibited thermal hyperalgesia and mechanical allodynia following CCI (Maione et al. 2007)	*N*-Arachidonoyl-serotonin inhibited formalin-evoked hyperalgesia (Maione et al. 2007)
Rat	(i.p.)	URB597 had no effect on mechanical allodynia (Jayamanne et al. 2006)	URB597 reduced CFA-induced allodynia and thermal hyperalgesia (Jayamanne et al. 2006)
Rat	(i.p.)		URB597 decreased carrageenan-evoked paw oedema (Holt et al. 2005)
Mouse	(i.p.)	OL135 inhibited mechanical allodynia (Chang et al. 2006)	
Mouse	(p.o.)	Chronic dosing of URB597 decreased mechanical allodynia following partial nerve ligation (Russo et al. 2007)	

from normal volunteers (Richardson et al. 2008). Whether the peripheral cannabinoid receptors present in the synovium are able to modulate arthritis-induced pain remains unknown. Inhibitory effects of cannabinoids have, however, been demonstrated in animal models of both OA and RA. Systemic administration of Δ^9-tetrahydrocannabinol (Δ^9-THC) was anti-nociceptive in adjuvant-induced arthritis (Cox et al. 2007a, b; Cox and Welch 2004; Smith et al. 1998), an effect mediated by CB_2 receptors (Cox et al. 2007a) and which involved interaction with the opioid receptor system (Cox et al. 2007b; Cox and Welch 2004). Furthermore, application of the CB_1 receptor agonist ACEA onto joint primary afferent fibres decreased arthritis-induced increases in frequency of firing of joint primary afferent fibres in rats (Schuelert and McDougall 2008). AEA also produces analgesia in models of arthritis; these effects did not appear to be CB_1 mediated, but like the effects of Δ^9-THC were attenuated by the opioid receptor antagonist naloxone, implicating a role for the opioid system (Smith et al. 1998). It appears however that the dose of AEA and route of administration used is key, since close arterial injection of AEA produced excitation of nociceptive fibres in knee joints of arthritic rats and normal control rats which was mediated via the transient receptor potential vanilloid 1 (TRPV1) receptor (Gauldie et al. 2001).

A further advantage of cannabinoid-based medicines as candidates for the treatment of rheumatic conditions is their effects on bone metabolism. AEA and 2AG have been identified in bone at levels similar to that in the brain (Bab et al. 2008) and their synthesis is reported in both osteoclasts and osteoblasts in vitro (Tam et al. 2006). The expression of the synthetic enzyme for 2AG, diacylglycerol lipase (DGL), in osteoblasts, osteocytes and bone-lining cells and the presence of FAAH in bone cells (Bab et al. 2008) collectively indicate the role of the cannabinoid system in bone turnover. Indeed, both CB_1 and CB_2 receptors contribute to the regulation of bone mass and the CB_2 receptor is a putative target for osteoporosis and other bone diseases (Bab et al. 2008; Karsak et al. 2005; Ofek et al. 2006).

In conclusion, the analgesic effects of cannabinoid-based medicines acting at CB_1 receptors are well described, but limited by adverse side-effect profiles. The identification of alternative cannabinoid entities, such as the CB_2 receptor and enzymes engaged in the catabolism of eCBs, offers further opportunity for the development of novel cannabinoid based analgesics with an improved side-effect profile.

Acknowledgements We would like to thank the Wellcome Trust, Medical Research Council and GlaxoSmithKline for financial support towards the original research discussed in this review.

References

Agarwal N, Pacher P, Tegeder I et al. (2007) Cannabinoids mediate analgesia largely via peripheral type 1 cannabinoid receptors in nociceptors. Nat Neurosci 10:870–879

Bab I, Ofek O, Tam J et al. (2008) Endocannabinoids and the regulation of bone metabolism. J Neuroendocrinol 20(Suppl 1):69–74

Beaulieu P, Bisogno T, Punwar S et al. (2000) Role of the endogenous cannabinoid system in the formalin test of persistent pain in the rat. Eur J Pharmacol 396:85–92

Beltramo M, Bernardini N, Bertorelli R et al. (2006) CB2 receptor-mediated antihyperalgesia: possible direct involvement of neural mechanisms. Eur J Neurosci 23:1530–1538

Blake DR, Robson P, Ho M et al. (2006) Preliminary assessment of the efficacy, tolerability and safety of a cannabis-based medicine (Sativex) in the treatment of pain caused by rheumatoid arthritis. Rheumatology (Oxford) 45:50–52

Brown AJ (2007) Novel cannabinoid receptors. Br J Pharmacol 152:567–575

Chang L, Luo L, Palmer JA et al. (2006) Inhibition of fatty acid amide hydrolase produces analgesia by multiple mechanisms. Br J Pharmacol 148:102–113

Chapman V (1999) The cannabinoid CB1 receptor antagonist, SR141716A, selectively facilitates nociceptive responses of dorsal horn neurones in the rat. Br J Pharmacol 127:1765–1767

Chin CL, Tovcimak AE, Hradil VP et al. (2008) Differential effects of cannabinoid receptor agonists on regional brain activity using pharmacological MRI. Br J Pharmacol 153:367–379

Clayton N, Marshall FH, Bountra C et al. (2002) CB1 and CB2 cannabinoid receptors are implicated in inflammatory pain. Pain 96:253–260

Cox ML, Welch SP (2004) The antinociceptive effect of Delta9-tetrahydrocannabinol in the arthritic rat. Eur J Pharmacol 493:65–74

Cox ML, Haller VL, Welch SP (2007a) The antinociceptive effect of Delta9-tetrahydrocannabinol in the arthritic rat involves the CB(2) cannabinoid receptor. Eur J Pharmacol 570:50–56

Cox ML, Haller VL, Welch SP (2007b) Synergy between delta9-tetrahydrocannabinol and morphine in the arthritic rat. Eur J Pharmacol 567:125–130

Cravatt BF, Giang DK, Mayfield SP et al. (1996) Molecular characterization of an enzyme that degrades neuromodulatory fatty-acid amides. Nature 384:83–87
Cravatt BF, Demarest K, Patricelli MP et al. (2001) Supersensitivity to anandamide and enhanced endogenous cannabinoid signaling in mice lacking fatty acid amide hydrolase. Proc Natl Acad Sci USA 98:9371–9376
Deutsch DG, Chin SA (1993) Enzymatic synthesis and degradation of anandamide, a cannabinoid receptor agonist. Biochem Pharmacol 46:791–796
Devane WA, Hanus L, Breuer A et al. (1992) Isolation and structure of a brain constituent that binds to the cannabinoid receptor. Science 258:1946–1949
Dinh TP, Carpenter D, Leslie FM et al. (2002) Brain monoglyceride lipase participating in endocannabinoid inactivation. Proc Natl Acad Sci USA 99:10819–10824
Egertova M, Elphick MR (2000) Localisation of cannabinoid receptors in the rat brain using antibodies to the intracellular C-terminal tail of CB. J Comp Neurol 422:159–171
Elmes SJ, Jhaveri MD, Smart D et al. (2004) Cannabinoid CB2 receptor activation inhibits mechanically evoked responses of wide dynamic range dorsal horn neurons in naive rats and in rat models of inflammatory and neuropathic pain. Eur J Neurosci 20:2311–2320
Elmes SJ, Winyard LA, Medhurst SJ et al. (2005) Activation of CB1 and CB2 receptors attenuates the induction and maintenance of inflammatory pain in the rat. Pain 118:327–335
Fegley D, Gaetani S, Duranti A et al. (2005) Characterization of the fatty acid amide hydrolase inhibitor cyclohexyl carbamic acid 3′-carbamoyl-biphenyl-3-yl ester (URB597): effects on anandamide and oleoylethanolamide deactivation. J Pharmacol Exp Ther 313:352–358
Finn DP, Jhaveri MD, Beckett SR et al. (2003) Effects of direct periaqueductal grey administration of a cannabinoid receptor agonist on nociceptive and aversive responses in rats. Neuropharmacology 45:594–604
Fowler CJ (2007) The contribution of cyclooxygenase-2 to endocannabinoid metabolism and action. Br J Pharmacol 152:594–601
Galiegue S, Mary S, Marchand J et al. (1995) Expression of central and peripheral cannabinoid receptors in human immune tissues and leukocyte subpopulations. Eur J Biochem 232:54–61
Gauldie SD, McQueen DS, Pertwee R et al. (2001) Anandamide activates peripheral nociceptors in normal and arthritic rat knee joints. Br J Pharmacol 132:617–621
Gong JP, Onaivi ES, Ishiguro H et al. (2006) Cannabinoid CB2 receptors: immunohistochemical localization in rat brain. Brain Res 1071:10–23
Guindon J, Hohmann AG (2008) Cannabinoid CB2 receptors: a therapeutic target for the treatment of inflammatory and neuropathic pain. Br J Pharmacol 153:319–334
Guindon J, Desroches J, Beaulieu P (2007) The antinociceptive effects of intraplantar injections of 2-arachidonoyl glycerol are mediated by cannabinoid CB2 receptors. Br J Pharmacol 150:693–701
Hanus L, Abu-Lafi S, Fride E et al. (2001) 2-Arachidonyl glyceryl ether, an endogenous agonist of the cannabinoid CB1 receptor. Proc Natl Acad Sci USA 98:3662–3665
Harris J, Drew LJ, Chapman V (2000) Spinal anandamide inhibits nociceptive transmission via cannabinoid receptor activation in vivo. NeuroReport 11:2817–2819
Herkenham M, Groen BG, Lynn AB et al. (1991) Neuronal localization of cannabinoid receptors and second messengers in mutant mouse cerebellum. Brain Res 552:301–310
Hohmann AG (2002) Spinal and peripheral mechanisms of cannabinoid antinociception: behavioral, neurophysiological and neuroanatomical perspectives. Chem Phys Lipids 121:173–190
Hohmann AG, Tsou K, Walker JM (1998) Cannabinoid modulation of wide dynamic range neurons in the lumbar dorsal horn of the rat by spinally administered WIN55, 212-2. Neurosci Lett 257:119–122
Holt S, Comelli F, Costa B et al. (2005) Inhibitors of fatty acid amide hydrolase reduce carrageenan-induced hind paw inflammation in pentobarbital-treated mice: comparison with indomethacin and possible involvement of cannabinoid receptors. Br J Pharmacol 146:467–476

Huang SM, Bisogno T, Trevisani M et al. (2002) An endogenous capsaicin-like substance with high potency at recombinant and native vanilloid VR1 receptors. Proc Natl Acad Sci USA 99:8400–8405

Ibrahim MM, Porreca F, Lai J et al. (2005) CB2 cannabinoid receptor activation produces antinociception by stimulating peripheral release of endogenous opioids. Proc Natl Acad Sci USA 102:3093–3098

Ibrahim MM, Rude ML, Stagg NJ et al. (2006) CB2 cannabinoid receptor mediation of antinociception. Pain 122:36–42

Iversen L, Chapman V (2002) Cannabinoids: a real prospect for pain relief? Curr Opin Pharmacol 2:50–55

Jayamanne A, Greenwood R, Mitchell VA et al. (2006) Actions of the FAAH inhibitor URB597 in neuropathic and inflammatory chronic pain models. Br J Pharmacol 147:281–288

Jhaveri MD, Richardson D, Kendall DA et al. (2006) Analgesic effects of fatty acid amide hydrolase inhibition in a rat model of neuropathic pain. J Neurosci 26:13318–13327

Jhaveri MD, Richardson D, Chapman V (2007a) Endocannabinoid metabolism and uptake: novel targets for neuropathic and inflammatory pain. Br J Pharmacol 152:624–632

Jhaveri MD, Sagar DR, Elmes SJ et al. (2007b) Cannabinoid CB(2) receptor-mediated antinociception in models of acute and chronic pain. Mol Neurobiol 36:26–35

Jhaveri MD, Elmes SJ, Richardson D et al. (2008a) Evidence for a novel functional role of cannabinoid CB receptors in the thalamus of neuropathic rats. Eur J Neurosci 27:1722–1730

Jhaveri MD, Richardson D, Robinson I et al. (2008b) Inhibition of fatty acid amide hydrolase and cyclooxygenase-2 increases levels of endocannabinoid related molecules and produces analgesia via peroxisome proliferator-activated receptor-alpha in a model of inflammatory pain. Neuropharmacology 55:85–93

Johns DG, Behm DJ, Walker DJ et al. (2007) The novel endocannabinoid receptor GPR55 is activated by atypical cannabinoids but does not mediate their vasodilator effects. Br J Pharmacol 152:825–831

Karsak M, Cohen-Solal M, Freudenberg J et al. (2005) Cannabinoid receptor type 2 gene is associated with human osteoporosis. Hum Mol Genet 14:3389–3396

Kathuria S, Gaetani S, Fegley D et al. (2003) Modulation of anxiety through blockade of anandamide hydrolysis. Nat Med 9:76–81

Kelly S, Chapman V (2001) Selective cannabinoid CB1 receptor activation inhibits spinal nociceptive transmission in vivo. J Neurophysiol 86:3061–3064

Kelly S, Chapman V (2002) Spinal administration of capsazepine inhibits noxious evoked responses of dorsal horn neurons in non-inflamed and carrageenan inflamed rats. Brain Res 935:103–108

Kelly S, Chapman V (2003) Cannabinoid CB(1) receptor inhibition of mechanically evoked responses of spinal neurones in control rats, but not in rats with hindpaw inflammation. Eur J Pharmacol 474:209–216

Kelly S, Jhaveri MD, Sagar DR et al. (2003) Activation of peripheral cannabinoid CB1 receptors inhibits mechanically evoked responses of spinal neurons in noninflamed rats and rats with hindpaw inflammation. Eur J Neurosci 18:2239–2243

Lauckner JE, Jensen JB, Chen HY et al. (2008) GPR55 is a cannabinoid receptor that increases intracellular calcium and inhibits M current. Proc Natl Acad Sci USA 105:2699–2704

Leung D, Saghatelian A, Simon GM et al. (2006) Inactivation of N-acyl phosphatidylethanolamine phospholipase D reveals multiple mechanisms for the biosynthesis of endocannabinoids. Biochemistry 45:4720–4726

Lichtman AH, Cook SA, Martin BR (1996) Investigation of brain sites mediating cannabinoid-induced antinociception in rats: evidence supporting periaqueductal gray involvement. J Pharmacol Exp Ther 276:585–593

Lichtman AH, Leung D, Shelton CC et al. (2004a) Reversible inhibitors of fatty acid amide hydrolase that promote analgesia: evidence for an unprecedented combination of potency and selectivity. J Pharmacol Exp Ther 311:441–448

Lichtman AH, Shelton CC, Advani T et al. (2004b) Mice lacking fatty acid amide hydrolase exhibit a cannabinoid receptor-mediated phenotypic hypoalgesia. Pain 109:319–327
Liu J, Wang L, Harvey-White J et al. (2006) A biosynthetic pathway for anandamide. Proc Natl Acad Sci USA 103:13345–13350
Liu J, Wang L, Harvey-White J et al. (2008) Multiple pathways involved in the biosynthesis of anandamide. Neuropharmacology 54:1–7
Mackie K (2006) Cannabinoid receptors as therapeutic targets. Annu Rev Pharmacol Toxicol 46:101–122
Maione S, De Petrocellis L, de Novellis V et al. (2007) Analgesic actions of N-arachidonoyl-serotonin, a fatty acid amide hydrolase inhibitor with antagonistic activity at vanilloid TRPV1 receptors. Br J Pharmacol 150:766–781
Malan TP Jr, Ibrahim MM, Deng H et al. (2001) CB2 cannabinoid receptor-mediated peripheral antinociception. Pain 93:239–245
Malan TP Jr, Ibrahim MM, Lai J et al. (2003) CB2 cannabinoid receptor agonists: pain relief without psychoactive effects? Curr Opin Pharmacol 3:62–67
Martin WJ, Coffin PO, Attias E et al. (1999) Anatomical basis for cannabinoid-induced antinociception as revealed by intracerebral microinjections. Brain Res 822:237–242
Mechoulam R, Ben-Shabat S, Hanus L et al. (1995) Identification of an endogenous 2-monoglyceride, present in canine gut, that binds to cannabinoid receptors. Biochem Pharmacol 50:83–90
Meng ID, Manning BH, Martin WJ et al. (1998) An analgesia circuit activated by cannabinoids. Nature 395:381–383
Mitrirattanakul S, Ramakul N, Guerrero AV et al. (2006) Site-specific increases in peripheral cannabinoid receptors and their endogenous ligands in a model of neuropathic pain. Pain 126:102–114
Ofek O, Karsak M, Leclerc N et al. (2006) Peripheral cannabinoid receptor, CB2, regulates bone mass. Proc Natl Acad Sci USA 103:696–701
Pertwee RG (2001) Cannabinoid receptors and pain. Prog Neurobiol 63:569–611
Petrosino S, Palazzo E, de Novellis V et al. (2007) Changes in spinal and supraspinal endocannabinoid levels in neuropathic rats. Neuropharmacology 52:415–422
Porter AC, Sauer JM, Knierman MD et al. (2002) Characterization of a novel endocannabinoid, virodhamine, with antagonist activity at the CB1 receptor. J Pharmacol Exp Ther 301:1020–1024
Rice AS, Farquhar-Smith WP, Nagy I (2002) Endocannabinoids and pain: spinal and peripheral analgesia in inflammation and neuropathy. Prostaglandins Leukot Essent Fatty Acids 66:243–256
Richardson JD, Kilo S, Hargreaves KM (1998) Cannabinoids reduce hyperalgesia and inflammation via interaction with peripheral CB1 receptors. Pain 75:111–119
Richardson D, Pearson RG, Kurian N et al. (2008) Characterisation of the cannabinoid receptor system in synovial tissue and fluid in patients with osteoarthritis and rheumatoid arthritis. Arthritis Res Ther 10:R43
Romero-Sandoval A, Nutile-McMenemy N, DeLeo JA (2008) Spinal microglial and perivascular cell cannabinoid receptor type 2 activation reduces behavioral hypersensitivity without tolerance after peripheral nerve injury. Anesthesiology 108:722–734
Russo R, Loverme J, La Rana G et al. (2007) The fatty-acid amide hydrolase inhibitor URB597 (cyclohexyl carbamic acid 3′-carbamoyl-biphenyl-3-yl ester) reduces neuropathic pain after oral administration in mice. J Pharmacol Exp Ther 322:236–242
Ryberg E, Larsson N, Sjogren S et al. (2007) The orphan receptor GPR55 is a novel cannabinoid receptor. Br J Pharmacol 152:1092–1101
Sagar DR, Kelly S, Millns PJ et al. (2005) Inhibitory effects of CB1 and CB2 receptor agonists on responses of DRG neurons and dorsal horn neurons in neuropathic rats. Eur J Neurosci 22:371–379
Schuelert N, McDougall JJ (2008) Cannabinoid-mediated antinociception is enhanced in rat osteoarthritic knees. Arthritis Rheum 58:145–153
Scott DA, Wright CE, Angus JA (2004) Evidence that CB-1 and CB-2 cannabinoid receptors mediate antinociception in neuropathic pain in the rat. Pain 109:124–131

Simon GM, Cravatt BF (2008) Anandamide biosynthesis catalyzed by the phosphodiesterase GDE1 and detection of glycerophospho-N-acyl ethanolamine precursors in mouse brain. J Biol Chem 283:9341–9349
Smith FL, Fujimori K, Lowe J et al. (1998) Characterization of delta9-tetrahydrocannabinol and anandamide antinociception in nonarthritic and arthritic rats. Pharmacol Biochem Behav 60:183–191
Sokal DM, Elmes SJ, Kendall DA et al. (2003) Intraplantar injection of anandamide inhibits mechanically-evoked responses of spinal neurones via activation of CB2 receptors in anaesthetised rats. Neuropharmacology 45:404–411
Tam J, Ofek O, Fride E et al. (2006) Involvement of neuronal cannabinoid receptor CB1 in regulation of bone mass and bone remodeling. Mol Pharmacol 70:786–792
Tsou K, Brown S, Sanudo-Pena MC et al. (1998) Immunohistochemical distribution of cannabinoid CB1 receptors in the rat central nervous system. Neuroscience 83:393–411
Tsuboi K, Takezaki N, Ueda N (2007) The N-acylethanolamine-hydrolyzing acid amidase (NAAA). Chem Biodivers 4:1914–1925
Valenzano KJ, Tafesse L, Lee G et al. (2005) Pharmacological and pharmacokinetic characterization of the cannabinoid receptor 2 agonist, GW405833, utilizing rodent models of acute and chronic pain, anxiety, ataxia and catalepsy. Neuropharmacology 48:658–672
Van Sickle MD, Duncan M, Kingsley PJ et al. (2005) Identification and functional characterization of brainstem cannabinoid CB2 receptors. Science 310:329–332
Walker JM, Huang SM (2002) Cannabinoid analgesia. Pharmacol Ther 95:127–135
Walker JM, Huang SM, Strangman NM et al. (1999) Pain modulation by release of the endogenous cannabinoid anandamide. Proc Natl Acad Sci USA 96:12198–12203
Welch SP, Stevens DL (1992) Antinociceptive activity of intrathecally administered cannabinoids alone, and in combination with morphine, in mice. J Pharmacol Exp Ther 262:10–18
Welch SP, Huffman JW, Lowe J (1998) Differential blockade of the antinociceptive effects of centrally administered cannabinoids by SR141716A. J Pharmacol Exp Ther 286:1301–1308
Wright S, Ware M, Guy G (2006) The use of a cannabis-based medicine (Sativex) in the treatment of pain caused by rheumatoid arthritis. Rheumatology (Oxford) 45, 781; author reply 781–782
Yamamoto W, Mikami T, Iwamura H (2008) Involvement of central cannabinoid CB2 receptor in reducing mechanical allodynia in a mouse model of neuropathic pain. Eur J Pharmacol 583:56–61

Integration of Endocannabinoid Signaling into the Neural Network Regulating Stress-Induced Activation of the Hypothalamic–Pituitary–Adrenal Axis

Boris B. Gorzalka and Matthew N. Hill

Contents

Abstract The evidence that has been gathered to date strongly argues for an inhibitory role of endocannabinoid (ECB) signaling in regulating HPA axis activity. Under basal conditions, ECB signaling appears to be a driving force in the maintenance of low HPA axis activity, as disruption of CB_1 receptor activity results in basal hyperactivity of the HPA axis. Under conditions of acute stress, ECB signaling likewise appears to constrain activation of the HPA axis, possibly via both distal regulation of incoming amygdalar inputs and local regulation of excitatory input to CRF neurosecretory cells in the PVN. ECB neurotransmission is, in turn, modulated by stress, possibly acting as either a "gatekeeper" of the HPA axis, or a recovery system aimed at limiting HPA axis activity. Consistently, pharmacological enhancement of ECB signaling attenuates stress-induced HPA axis activity while impairment of CB_1 receptor signaling results in an exaggerated cellular and neuroendocrine response to stress. Additionally, under conditions of repeated stress,

B.B Gorzalka (✉) and M.N Hill
Department of Psychology, University of British Columbia, Vancouver, BC, Canada
e-mail: bgorzalka@psych.ubc.ca

D. Kendall and S. Alexander (eds.), *Behavioral Neurobiology of the Endocannabinoid System*, Current Topics in Behavioral Neurosciences 1,
DOI: 10.1007/978-3-540-88955-7_12, © Springer-Verlag Berlin Heidelberg 2009

a progressive increase in limbic 2AG/CB_1 receptor signaling contributes to the development and expression of neuroendocrine habituation.

Ultimately, these data demonstrate that the ECB system is likely to be an integral player in the neuronal response and plasticity to stress. The relevance of this relationship has not been fully explored with respect to both normal homeostasis and pathological states characterized by alterations in HPA axis function, but will be a focus of future research.

Keywords Cannabinoid • Anandamide • 2-AG • FAAH • Stress • PVN • HPA axis • Corticosterone • Amygdala • Adrenal

Abbreviations

2AG	2-Arachidonoylglycerol
ACTH	Adrenocorticotropic hormone
AEA	Anandamide
BLA	Basolateral nucleus of the amygdala
CeA	Central nucleus of the amygdala
CRF	Corticotrophin-releasing factor
ECB	Endocannabinoid
FAAH	Fatty acid amide hydrolase
GH	Growth hormone
HPA	Hypothalamic–pituitary–adrenal
MeA	Medial amygdala
MGL	Monoacylglycerol lipase
mPFC	Medial prefrontal cortex
PAG	Periaqueductal gray
PVN	Paraventricular nucleus of the hypothalamus

1 Stress and the Hypothalamic–Pituitary–Adrenal Axis

The hypothalamic–pituitary–adrenal (HPA) axis is a neuroendocrine system that responds in a coordinated manner to aversive stimuli. The integration center for the HPA axis is located within the paraventricular nucleus of the hypothalamus (PVN), where a dense population of neurons secreting corticotrophin-releasing factor (CRF) are located. The activity of these neurons is regulated by both excitatory and inhibitory signals arising from local hypothalamic nuclei such as the dorsomedial hypothalamic nucleus and the medial preoptic area, limbic structures such as the amygdala and prefrontal cortex, as well as monoaminergic cell bodies found in the midbrain and brainstem (Herman et al. 2002, 2003). Exposure of an organism to aversive stimuli activates a characteristic neural network which encodes both the salience and the threat of the stimuli, and ultimately culminates in the activation of the CRF neurosecretory cells in the PVN. Upon activation, these neurons release CRF into the portal blood, where it is carried to the anterior pituitary. In the pituitary, CRF acts as a secretagogue and stimulates the release of adrenocorticotropic hormone (ACTH) into the general circulation, where it travels to the adrenal cortex to stimulate the secretion of glucocorticoids and other adrenal steroids. Blood-borne glucocorticoids, such as cortisol and corticosterone, perform a variety of functions. In particular, these hormones promote the reallocation of energy resources to deal appropriately with a potential threat, including the redirection of glucose from adipose to muscle tissue, heightened vigilance to contextual stimuli and a suppression of motivated and neurovegetative behaviors (Pecoraro et al. 2006). Under acute conditions, these effects of glucocorticoids can be beneficial to survival of the organism; however, long-term hypersecretion of glucocorticoids can lead to a plethora of psychological, metabolic, cardiovascular and immune dysfunctions (McEwen 2005; Pecoraro et al. 2006). Thus, regulation of the HPA axis is extremely important for maintaining optimal functioning of an organism.

In the short term, glucocorticoids regulate their own secretion through several, distinct negative feedback processes (Dallman et al. 1994). Within adrenal, pituitary, hypothalamic and extrahypothalamic tissues, glucocorticoids act to inhibit their own release and/or synthesis. Within the hypothalamus, this negative feedback process can take the form of both fast feedback and delayed feedback. Glucocorticoid fast feedback occurs very rapidly within the PVN, where glucocorticoids appear to act through a nongenomic mechanism to inhibit glutamate release onto CRF neurosecretory cells (Dallman et al. 1994; Di et al. 2003). Delayed feedback elicited by glucocorticoids in the PVN occurs through a genomic mechanism in which glucocorticoids act to inhibit transcription of CRF (Dallman et al. 1994; Schulkin et al. 1998).

In the long term, adaptation to chronic stress is essential to prevent the detrimental effects of glucocorticoid hypersecretion. If an organism is exposed to the same stressor multiple times, habituation will occur even if the stressor is aversive (see Armario 2006). For example, physical restraint is an aversive, psychological stressor that will elicit a robust activation of the HPA axis during initial exposure in rats. However, following repeated restraint sessions, the magnitude of the HPA axis activation response is significantly reduced (Cole et al. 2000; Jaferi and Bhatnagar 2006; Viau and Sawchenko 2002). This plasticity of the HPA axis response is likely mediated by alterations in neuronal activity within extrahypothalamic limbic structures which regulate activation of CRF neurosecretory cells in the PVN, as lesions of specific neuroanatomical sites within the stress neural circuit can abrogate the development of habituation (Bhatnagar et al. 2002; Carter et al. 2004). In line with this hypothesis, activation of the neural stress circuit, as indicated via induction of immediate early genes in response to neuronal depolarization, declines in a similar fashion to adrenocortical secretion following repeated exposure to a common stressor (Melia et al. 1994; Patel et al. 2005; Viau and Sawchenko 2002; Watanabe et al. 1994). The flexibility of an organism to respond to disturbances of homeostasis in the short term, but to adapt to long-term exposure, is a process that is critical to survival in a constantly changing environment. The importance of adaptation to stress in humans is emphasized by the association between maladaptive responses to stress and the development of neuropsychiatric disorders such as depression or post-traumatic stress disorder (Korte et al. 2005).

2 The Endocannabinoid System

Cannabis has been used recreationally for centuries by various cultures around the world, in part due to its relaxing and stress-alleviating properties. The psychoactive constituent of cannabis that is predominately responsible for eliciting most of these emotional alterations is Δ^9-tetrahydrocannabinol (THC; Isbell et al. 1967). The actions of THC are mediated by its ability to interact with specific cannabinoid receptors throughout the brain and periphery. Two receptors have been characterized to date. The CB_1 cannabinoid receptor is the predominant central cannabinoid receptor and exhibits widespread distribution in the brain (Herkenham et al. 1991; Moldrich and Wenger 2000; Tsou et al. 1998), and, at lower expression levels, in peripheral tissue, such as blood vessels, immune cells and reproductive tissues (Gorzalka and Hill 2006; Hillard 2000; Parolaro 1999). The CB_2 cannabinoid receptor is located mainly in cells of the circulating immune system, such as macrophages (Munro et al. 1993; Parolaro 1999), as well as resident immune cells, including microglia (Cabral and Marciano-Cabral 2005; Carrier et al. 2004). Recent evidence suggests that the CB_2 receptor may also be expressed by neurons in some species (Gong et al. 2006; Van Sickle et al. 2005). Both cannabinoid receptors are G-protein-coupled receptors; these receptors activate $G\alpha_{i/o}$ proteins resulting in inhibition of adenylyl cyclase activity and inhibition of calcium channel activation by depolarization (Felder and Glass 1998; Howlett and Mukhopadhyay 2000; Piomelli 2003). The CB_1 receptor is present at high densities on presynaptic axon terminals, where it functions to inhibit neurotransmitter release (Schlicker and Kathmann 2001; Vaughan and Christie 2005). The CB_1 receptor is expressed by sub-populations of glutamate, gamma-aminobutyric acid (GABA), acetylcholine, serotonin and noradrenergic neurons (Nakazi et al. 2000; Ohno-Shosaku et al. 2001; Schlicker and Kathmann 2001), indicating that cannabinoids possess the ability to suppress release of many neurotransmitters and neuromodulators. This distribution often complicates the interpretation of in vivo pharmacological data. In addition to these known targets for cannabinoid agents, there is also increasing, but not definitive, evidence that cannabinoids may also exhibit affinity for other receptor subtypes, such as vanilloid receptors (TRPV1; Ross 2003), peroxisome-proliferator-activated receptors (PPAR; Sun et al. 2006) and non-CB_1/CB_2 G-protein coupled receptors, such as GPR55 (Ryberg et al. 2007).

The endogenous ligands of the cannabinoid receptors have been called ECBs. The two primary molecules that have been functionally identified as ECBs are the arachidonate-derived lipophilic molecules *N*-arachidonoylethanolamine (anandamide; AEA; Devane et al. 1992) and 2-arachidonoylglycerol (2AG; Sugiura et al.

1995). Several other lipid molecules, including virodhamine (Porter et al. 2002), noladin ether (Hanus et al. 2001) and *N*-arachidonoyldopamine (Bisogno et al. 2000) have been identified as putative ECB ligands; however, far less is known about these compounds than is known about AEA and 2AG. The ECBs are not typical neurotransmitters but are synthesized post-synaptically by activity-dependent cleavage of phospholipid head-groups by activation of specific enzymes. The ECB molecules are not released vesicularly; instead, these molecules are thought to be synthesized "on demand" in response to increased neuronal excitation and/or increased intracellular calcium. The ECBs are released into the synapse, where they act in a retrograde manner to activate the presynaptically located CB_1 receptor and inhibit neurotransmitter release (Bisogno et al. 2005; Schlicker and Kathmann 2001; Wilson and Nicoll 2002). Regulation of ECB content is also maintained by degradative enzymes such as fatty acid amide hydrolase (FAAH), which is the primary enzyme class capable of AEA hydrolysis, and monoacylglyceride lipase (MGL), which is the primary, but not exclusive, catabolic enzyme for 2AG (Deutsch et al. 2002; Dinh et al. 2002; Ueda 2002). A more detailed examination of the biochemical and pharmacological properties of the ECB system can be found in earlier chapters in this book.

3 Endocannabinoid-Mediated Regulation of the HPA Axis

3.1 Endocannabinoid Signaling within the HPA Axis

As cannabis is known typically to evoke stress-reducing and relaxing effects, it is not surprising that increasing evidence has indicated that the ECB system may contribute to regulation of the HPA axis. In terms of functional expression, all the major players of the ECB system are widely distributed throughout the central stress circuit as well as within peripheral endocrine tissue. Within the brain, the CB_1 receptor and both ECB ligands are found throughout all of the extrahypothalamic sites that regulate PVN neuronal activation, such as the hippocampus, prefrontal cortex, amygdala, bed nucleus of the stria terminalis and midbrain monoaminergic nuclei such as the locus coeruleus and dorsal raphe (Bisogno et al. 1999; Cadas et al. 1997; Herkenham et al. 1991; Tsou et al. 1998). Given that the ECB system modulates synaptic transmission via effects on neurotransmitter release (Freund et al. 2003), it is well-suited to regulate neuronal activation in stress-sensitive anatomical circuits.

The ECB system is strategically located both externally at sites regulating the HPA axis as well as internally throughout the HPA axis in a fashion conducive to an integral regulator. Particularly, the CB_1 receptor has been characterized on glutamatergic afferents within the PVN, and activation of this receptor attenuates excitatory activation of CRF neurosecretory cells (as well as oxytocinergic and vasopressinergic; Di et al. 2003). Microdialysis studies have revealed that antagonism of the CB_1 receptor in the PVN increases excitatory amino acid release and subsequent activation of neuropeptidergic cells (Succu et al. 2006). Thus, as well as regulating activity within extrahypothalamic structures that regulate the PVN, ECB signaling within the PVN proper can also potentially modulate activation of the HPA axis.

Within the periphery, both the CB_1 receptor and ECB ligands have been identified within the pituitary gland (Gonzalez et al. 1999; Pagotto et al. 2001; Wenger et al. 1999). Co-expression analysis in human pituitaries has identified that CB_1 receptor mRNA is only seen in the anterior lobe of the pituitary, and particularly within cells synthesizing ACTH, GH or prolactin (Pagotto et al. 2001). Finally, at the termination point of the HPA axis, CB_1 receptor expression has been documented within the adrenal gland (Galiegue et al. 1995). Collectively, these data indicate that the ECB system is present in all the major structures integrated into the HPA axis, making it an ideal candidate for regulating stress responsivity.

3.2 *Endocannabinoid Signaling Inhibits HPA Axis Activity*

Genetic and pharmacological studies have revealed a critical role of ECB signaling as a negative regulator of the HPA axis. Under non-stress conditions, deletion of the CB_1 receptor results in an increase in adrenocortical secretion, particularly during the daily peak of the circadian rhythm (Barna et al. 2004; Cota et al. 2007; Haller et al. 2004; Steiner et al. 2008). Similarly, those transgenic animals lacking the CB_1 receptor also exhibit an increase in the expression of CRF within the PVN under basal conditions (Cota et al. 2003, 2007). It should be noted, however, that there are also reports of either no change or even a reduction in basal corticosterone (Aso et al. 2008; Fride et al. 2005; Uriguen et al. 2004; Wade et al. 2006). These findings do appear to be strain-specific, and are likely a compensatory response that occurs in some strains, as pharmacological studies in mice and rats have supported the hypothesis that ECB activity suppresses basal HPA axis activity. Specifically, both acute and chronic administration of CB_1 receptor antagonists have reliably been found to elevate basal levels of ACTH and corticosterone (Doyon et al. 2006; Hill et al. 2007; Lamota et al. 2008; Manzaneres et al. 1999; Patel et al. 2004; Steiner et al. 2008; Wade et al. 2006). These data would suggest that ECB signaling dampens HPA axis activity under non-stress conditions.

While somewhat equivocal outcomes occur under basal conditions, under stress conditions the ECB system appears consistently to constrain activation of the HPA axis. Deletion of the CB_1 receptor results in a robust increase in corticosterone and/or ACTH secretion in response to restraint stress (Uriguen et al. 2004), tail suspension stress (Aso et al. 2008), forced swim stress (Steiner et al. 2008) and novelty stress (Barna et al. 2004; Haller et al. 2004), but not audiogenic stress (Fride et al. 2005). Similarly, pharmacological antagonism of the CB_1 receptor potentiates stress-induced glucocorticoid secretion and neuronal activation within the PVN (Evanson et al. 2007; Ginsberg et al. 2006; Patel et al. 2004; Steiner et al. 2008). Similarly, antagonism of the CB_1 receptor is capable of reversing the ability of chronic antidepressant treatment to dampen stress-induced neuronal activation within the PVN and corticosterone secretion (Hill et al. 2006). Consistent with the hypothesis that ECB signaling constrains stress-induced activation of the HPA axis, pharmacological inhibition of ECB uptake or FAAH activity attenuates corticosterone secretion in response to restraint stress (Patel et al. 2004). It should be noted that administration of the ECB AEA has been found to activate the HPA axis (Hao et al. 2000; Weidenfeld et al. 1994; Wenger et al. 1997, 2003); however, this effect is not mediated by the CB_1 receptor (Wenger et al. 1997, 2003), but is likely mediated by the rapid metabolism of AEA by FAAH and subsequent synthesis of prostaglandins, which themselves are potent activators of the HPA axis (Malcher-Lopes et al. 2008), and accordingly cannot be assumed to represent an ECB effect.

Despite some discrepancies, most of which appear to be a function of methodological issues, the bulk of current evidence argues that ECB signaling negatively regulates HPA axis activity, both under basal conditions and following exposure to acute stress.

3.3 At What Sites of Action Does Endocannabinoid Signaling Modulate HPA Axis Activity?

Through a collection of lesion and immediate early gene studies, several critical brain structures have been highlighted as exerting regulation of the HPA axis (Herman et al. 2003, 2005). Thus, it is possible that the inhibitory effects of ECB signaling are due not to local actions within the HPA axis, but to modulation of neuronal circuits which subserve activation of the HPA axis. Given the inhibitory nature of ECB neurotransmission, these effects would likely be localized to structures which promote HPA axis activation, and not to those which inhibit it.

The prefrontal cortex is an important site for coordinating incoming stimuli and behavioral/visceral responses. Lesion studies have highlighted that dorsal regions of the prefrontal cortex (such as the anterior cingulate cortex and prelimbic cortex) function to inhibit HPA axis activation, while ventral regions of the prefrontal cortex (such as the infralimbic cortex) appear to facilitate HPA axis activity (Diorio et al. 1993; Radley et al. 2006). We have recently examined the effects of a local infusion of a CB_1 receptor agonist or antagonist into the infralimbic cortex on stress-induced adrenocortical secretion (Hill et al. 2009a). Neither of these manipulations modulated basal glucocorticoid levels or the increase in plasma corticosterone following restraint stress (Hill et al. 2009a), suggesting that ECB signaling in the prefrontal cortex is not relevant for regulation of the HPA axis.

The amygdala is one of the main limbic structures which promotes HPA axis activity (Herman et al. 2003, 2005). Incoming sensory stimuli activate neurons within the basolateral nucleus of the amygdala (BLA), which in turn activates the central nucleus (CeA) or medial amygdala (MeA), which projects both directly and indirectly to the PVN to modulate HPA axis activation (Herman et al. 2003, 2005). We have recently found that infusion of a CB_1 receptor agonist into the BLA prior to restraint stress significantly attenuated the subsequent increase in plasma corticosterone (Hill et al. 2009a). In line with this, infusion of a CB_1 receptor antagonist into the BLA increased HPA axis output (Hill et al. 2009a). These effects were relatively specific to the BLA, as local administration of CB_1 receptor ligands into either the CeA or the MeA did not exert these effects (Hill et al. 2009a). Thus, it would appear that ECB activity within the amygdala, and particularly the BLA, may act to curb HPA axis activation.

In addition to these effects in the amygdala, there also appears to be a role for local ECB signaling within the PVN to modulate HPA axis activity. As mentioned, CB_1 receptors can attenuate glutamatergic activation of CRF neurosecretory cells in the PVN, likely via an inhibition of glutamate release (Di et al. 2003). This hypothesis is consistent with both the basal increase in CRF in the PVN seen in CB_1 receptor knockout mice (Cota et al. 2003, 2007) and the finding that systemic administration of a CB_1 receptor antagonist increases excitatory amino acid release within the PVN (Succu et al. 2006). As such, it appears quite plausible that ECB signaling directly inhibits HPA axis activation by suppressing incoming excitatory input onto CRF neurons in the PVN.

The presence of ECB activity in the pituitary suggests that this endocrine tissue may also be a locus of action; however, it should be noted that intracerebroventricular administration of a CB_1 receptor antagonist increases HPA axis activity, suggesting that this effect is mediated by a central site of action (Manzaneres et al. 1999). An initial study demonstrated that pituitary function of CB_1 receptor knockout mice is unaltered, in that both basal and CRF-stimulated ACTH secretion were not different between mutant and wild-type mice (Barna et al. 2004). However, a more recent study has revealed that while basal ACTH secretion is unchanged in pituitary cells derived from transgenic mice lacking the CB_1 receptor, both forskolin- and CRF-stimulated ACTH secretion is increased significantly (Cota et al. 2007). It is not clear why these differences exist between these studies, but it does appear that CB_1 signaling within the pituitary may contribute to the inhibitory effects of ECBs on the HPA axis. The same argument cannot be made for the adrenal gland, as no hypertrophy or gross morphological changes have been found within the adrenal glands of mice lacking the CB_1 receptor (Barna et al. 2004; Cota et al. 2007).

Ultimately, the current data would suggest that the ability of ECB signaling to negatively regulate both basal and stress-induced HPA axis activity is mediated through a coordinated suppression of incoming amygdalar input and local excitatory tone within the PVN following activation of the CB_1 receptor in the BLA and PVN, respectively. However, ongoing research is seeking to determine if other neuroanatomical sites are relevant for these effects.

4 Acute Stress-Induced Modulation of Endocannabinoid Signaling: Contributions to Glucocorticoid Negative Feedback

While ECB signaling may attenuate the neuroendocrine responses to stress, the question remains as to whether ECB activity itself is modulated by stress. The first evidence suggesting that the ECB system could be modulated by stress emerged from a series of in vitro studies employing glucocorticoid treatment to hypothalamic tissue. Specifically, bath application of glucocorticoids (such as corticosterone and dexamethasone) to slices of hypothalamic tissue, largely composed of the PVN, resulted in a rapid and significant elevation in both AEA and 2AG tissue contents (Di et al. 2005; Malcher-Lopes et al. 2006). These effects were mediated by the activation of a putative membrane-bound glucocorticoid G-protein coupled receptor that enhanced intracellular cAMP-protein kinase A signaling and resulted in rapid synthesis of ECB ligands (Malcher-Lopes et al. 2006).

These in vitro data lead to the hypothesis that stress, via elevation of glucocorticoids, could result in a rapid induction of ECB signaling. However, the first in vivo study published on this topic actually demonstrated the converse. Specifically, Patel and colleagues (2004) exposed male mice to 30 min of restraint stress, after which they were immediately sacrificed and ECB contents were measured in the hypothalamus. 2AG content was found to be significantly reduced, while AEA content was not modified by stress (Patel et al. 2004). This group reported subsequently that 2AG levels did not change in the forebrain, amygdala, cerebellum (Patel et al. 2005; Rademacher et al. 2008), hippocampus (S. Patel and C.J. Hillard, unpublished findings), ventral striatum or medial prefrontal cortex (mPFC; Rademacher et al. 2008) of mice exposed to 30 min restraint stress. Intriguingly, preliminary data from our laboratory has found that, unlike mice, in rats 30 min of restraint stress resulted in a significant increase in 2AG content in both the mPFC and the hypothalamus, but not the amygdala (Hill et al. 2009a). While this finding does support the hypothesis that stress-induced glucocorticoid secretion may induce ECB synthesis, these increases in 2AG did not correlate with the magnitude of the adrenocortical response to stress (Hill et al. 2009a). Similarly, a second study employing rats has demonstrated an immediate and transient increase in 2AG content in the periaqueductal gray (PAG) of rats as early as 2 min following exposure of the rats to foot shock stress (Hohmann et al. 2005). Collectively, these data do illustrate the potential for dramatic species differences in the effects of stress and/or glucocorticoids on 2AG synthesis (increase in rats, decrease in mice), especially since all of the in vitro studies examining glucocorticoid-induced ECB synthesis have been performed in tissue harvested from rats and not mice (Di et al. 2003, 2005; Malcher-Lopes et al. 2006). This species difference is not unprecedented. Restraint stress has previously been shown to have opposite effects on neurogenesis in mice and rats (Bain et al. 2004) and other stressors have been shown to have opposite behavioral effects in mice and rats (De Catanzaro and Gorzalka 1979). With regards to relevance to humans, it should be noted that we have recently determined that circulating 2AG levels rapidly elevate in response to stress (Hill et al. 2009b), similar to what is seen in central tissue in rats.

On the other hand, exposure of mice to restraint stress resulted in significant reductions in amygdalar (Patel et al. 2005; Rademacher et al. 2008) and hippocampal (S. Patel and C.J. Hillard, unpublished findings) AEA contents, while AEA content in the forebrain, cerebellum, ventral striatum and mPFC did not change (Patel et al. 2005; Rademacher et al. 2008). Unlike the discrepancy that was seen with 2AG, we have recently found a significant reduction in amygdalar, but not prefrontal cortical or hypothalamic, AEA content following 30 min restraint stress in rats (Hill et al. 2009a). Similarly, hippocampal AEA content is reduced following acute administration of corticosterone to male rats (Hill et al. 2008). Thus, while stress and/or glucocorticoids appear to have differential effects on 2AG synthesis in mice and rats, in both species a reliable reduction in AEA content, specifically in the amygdala and hippocampus, is seen following these treatments. The effects of acute stress and/or glucocorticoid treatment on ECB content can be seen in Table 1.

As discussed previously, ECB signaling constrains activation of the HPA axis, and has been hypothesized to function as a stress-recovery system (Di Marzo et al.

Table 1 The effects of acute stress (or glucocorticoid treatment) on the tissue content of the endocannabinoid ligands anandamide (AEA) and 2-arachidonylglycerol (2AG) in discrete brain regions

Species/strain	Stressor	Brain region	AEA	2AG	Reference
ICR mice	30 min restraint stress	Hypothalamus	NC	–	Patel et al. (2004)
ICR mice	30 min restrain stress	Forebrain	NC	NC	Patel et al. (2005)
		Amygdala	–	NC	
		Cerebellum	NC	NC	
ICR mice	30 min restraint stress	Hippocampus	–	NC	Patel and Hillard, unpublished data
ICR mice	30 min restraint stress	Prefrontal cortex	NC	NC	Rademacher et al. (2008)
		Amygdala	–	NC	
		Ventral striatum	NC	NC	
Sprague–Dawley rats	30 min restraint stress	Prefrontal cortex	NC	+	Hill et al. (2009a)
		Amygdala	–	NC	
		Hypothalamus	NC	+	
Sprague–Dawley rats	3 min foot shock	Periaqueductal gray (PAG)	+	+	Hohmann et al. (2005)
		Occipital cortex	NC	NC	
Long Evans rats	Single injection corticosterone (20 mg kg^{-1})	Hippocampus	–	NC	Hill et al. (2008)

NC = no change; – = significant reduction; + = significant increase

1998; Gorzalka et al. 2008). With regards to data obtained from rats, this hypothesis seems plausible given that hypothalamic tissue from rats exhibits an increase in 2AG synthesis following glucocorticoid exposure (Di et al. 2005; Malcher-Lopes et al. 2006), and in vivo rats exposed to 30 min restraint exhibit increased hypothalamic levels of 2AG (Hill et al. 2009a). Accordingly, it can be predicted that stress-induced glucocorticoid secretion may induce 2AG synthesis within the PVN, in which 2AG binding to presynaptic CB_1 receptors will reduce the excitatory drive on CRF neurosecretory cells and aid in shutting down the HPA axis. However, in mice the biochemical studies discussed above indicate that hypothalamic ECBs are reduced following stress, not increased (Patel et al. 2004). Given that pharmacological modulation of the ECB system exhibits comparable regulation of the HPA axis in both mice and rats, to reconcile this apparent discrepancy in hypothalamic 2AG responses to stress, a "gatekeeper" hypothesis has been proposed (Patel et al. 2004). According to this hypothesis, ECB content within the PVN of the hypothalamus is high during the basal (non-stressed) state, resulting in a tonic inhibition of excitatory inputs to the HPA axis. Upon exposure to stress, 2AG levels rapidly decline, through an undetermined mechanism, resulting in a disinhibition of glutamatergic projections to the PVN and allowing activation of the HPA axis (Patel et al. 2004). If ECB levels are maintained at a high level (through administration of an ECB uptake inhibitor, for example [Patel et al. 2004]) then this inhibition is maintained and HPA axis activation is attenuated. If this inhibition is disrupted by the administration of a CB_1 receptor antagonist or genetic deletion of the CB_1 receptor, then an exaggerated activation of the HPA axis occurs. The fact that the CB_1 receptor appears to constrain HPA axis activation under non-stress conditions in mice (Cota et al. 2007; Patel et al. 2004) attests to the "gatekeeper" theory of ECB regulation of the HPA axis. The same argument can theoretically be proposed for ECB signaling within the amygdala. As discussed, activation of amygdalar CB_1 receptors can inhibit activation of the HPA axis and disruption of amygdalar CB_1 receptor activity can promote HPA axis function (Hill et al. 2009a). Across both rats and mice, stress evoked a reduction in amygdalar AEA content; in rats, this reduction has been found to negatively correlate with the magnitude of the adrenocortical response, such that larger reductions in amygdalar AEA equated to larger increases in plasma corticosterone (Hill et al. 2009a). Thus, AEA/CB_1 receptor coupling in the amygdala, particularly the BLA, may provide a similar "gatekeeper" function as described for 2AG within the mouse hypothalamus (Patel et al. 2004). Taken together, the available data indicate that the ECB/CB_1 receptor signaling within the PVN of the hypothalamus, and possibly the amygdala, is likely a critical player in the regulation of the HPA axis under conditions of acute stress. While the mechanism by which ECBs fine tune HPA axis activity may slightly differ between rats and mice, the end result appears to be the same.

In addition to these hypotheses, it has also been suggested that ECB signaling may modify HPA axis activity through a modulation of glucocorticoid-mediated negative feedback. The in vitro studies examining the effects of glucocorticoid activity on ECBs revealed that the induction of ECB synthesis mediated glucocorticoid-suppression of PVN neuronal activation, or "glucocorticoid fast-feedback"

(Di et al. 2003). Accordingly, ECBs provided the missing link through which glucocorticoids rapidly suppressed CRF neural activity to inhibit subsequent activation of the HPA axis (Di et al. 2003). This phenomenon appears to be specific to the PVN, however, as deletion of the CB_1 receptor does not impair glucocorticoid-mediated suppression of ACTH secretion from pituitary cells (Barna et al. 2004), nor does it affect CRF expression in other brain regions such as the amygdala and prefrontal cortex (Cota et al. 2007). In line with a hypothetical role of ECBs in glucocorticoid feedback, in vivo glucocorticoid administration prior to stress induction can attenuate ACTH secretion, a phenomenon which has been found to be sensitive to genetic or pharmacological blockade of the CB_1 receptor (Cota et al. 2007; Evanson et al. 2007). Thus, the hyperactivity of the HPA axis that occurs in response to disruption of CB_1 receptor signaling may occur via a deficiency in glucocorticoid feedback that promotes a state of glucocorticoid hypersecretion.

5 Chronic Stress-Induced Regulation of Endocannabinoid Signaling: A Driving Force for Stress Habituation

The picture to emerge thus far indicates that stress modulates ECB signaling in limbic structures, which in turn appears to contribute to the activation, and subsequent suppression, of the HPA axis. Habituation to stress is a progressive reduction in neuronal and neuroendocrine responses to exposure to an aversive stimulus, such as restraint stress (Bhatnagar et al. 2002; Cole et al. 2000; Patel et al. 2005; Viau and Sawchenko 2002). Given the neuroanatomical distribution of the ECB system, as well as its robust ability to modulate HPA axis activation under conditions of acute stress, it is reasonable to hypothesize that the ECB system may be important for regulation of the HPA axis under conditions of chronic stress.

The effects of repeated, homotypic stress exposure on the ECB system has been examined in male mice exposed to 30 min of restraint sessions for 5–10 consecutive days. Repeated exposure to homotypic stress does not change CB_1 receptor density (Rademacher et al. 2008). Repeated restraint stress has no effect on hypothalamic AEA content, but 2AG content is significantly increased compared to a control condition and compared to a single restraint exposure (Patel et al. 2004). 2AG content is also significantly elevated in both the amygdala and limbic forebrain as a whole, and the mPFC in particular, following consecutive stress exposures (Patel et al. 2005; Rademacher et al. 2008). Similar to what was seen under acute stress conditions, AEA content in the amygdala is significantly reduced following 5, 7 or 10 days of repeated stress, as is AEA content in the mPFC following 7 or 10 days of repeated restraint (Patel et al. 2005; Rademacher et al. 2008). However, AEA content is not changed if the tissue piece includes the entire limbic forebrain, demonstrating the AEA changes within forebrain structures may be limited to specific structures such as the mPFC (Patel et al. 2005). Thus, repeated homotypic stress is associated with an increase in limbic 2AG/CB_1 receptor signaling and a concomitant reduction in AEA/CB_1 receptor coupling.

Differences in the synaptic functions of AEA and 2AG have been suggested; one hypothesis being that 2AG/CB_1 receptor activity represents a rapid, burst-like signal while AEA/CB_1 receptor coupling may provide more of a tonic level of CB_1 receptor activation (Gorzalka et al. 2008). In line with this hypothesis, it can be predicted that repeated restraint stress is associated with a reduction in the tone of limbic ECB signaling, but an increase in the rapid, intense signal. This increase in 2AG may be relevant for the suppression of neuronal activation within the stress circuit during the development of habituation; there are several lines of evidence to support this hypothesis. First, acute treatment with a CB_1 receptor antagonist to mice repeatedly exposed to restraint stress reverses the habituation of stress-induced neuronal activation within limbic structures (as assessed by immediate early gene induction; Patel et al. 2005). Similarly, administration of a CB_1 receptor antagonist, either on the final day of stress or daily throughout the stress exposure, to rats repeatedly exposed to restraint stress reversed neuroendocrine habituation to stress (Hill et al. 2007). Second, 2AG signaling is known to suppress excitatory neurotransmission (Uchigashima et al. 2007; Yoshida et al. 2006); thus a repeated stress-induced increase in 2AG release may function to attenuate excitatory activation of the neuronal stress circuit, thus contributing to the manifestation of stress habituation. Third, we have found that potentiation of AEA/CB_1 signaling (via pharmacological inhibition of FAAH) does not modulate habituation to stress (Hill et al. 2007), indicating that 2AG is likely the signaling molecule at the CB_1 receptor driving the habituation response. Fourth, ECB signaling promotes other forms of habituation to aversive stimuli, such as behavioral inhibition in response to shock (Kamprath et al. 2006) and audiogenic stress (Fride et al. 2005). Taken together, these data collectively argue that ECB signaling is recruited under conditions of repeated stress to suppress neuronal activation in stress circuits, thereby facilitating cellular, neuroendocrine and behavioral adaptations to stress.

6 Conclusion

The aim of the current chapter was to review the current literature examining the interactions between ECBs and the HPA axis. The evidence that has been gathered to date strongly argues for an inhibitory role of ECB signaling in regulating HPA axis activity. Under basal conditions, ECB signaling appears to be a driving force in the maintenance of low HPA axis activity, as disruption of CB_1 receptor activity results in basal hyperactivity of the HPA axis. Under conditions of acute stress, ECB signaling likewise appears to constrain activation of the HPA axis, possibly both via distal regulation of incoming amygdalar inputs and local regulation of excitatory input to CRF neurosecretory cells in the PVN. ECB neurotransmission is, in turn, modulated by stress, possibly acting as either a "gatekeeper" of the HPA axis, or a recovery system aimed at limiting HPA axis activity. Consistently, pharmacological enhancement of ECB signaling attenuates stress-induced HPA axis activity while impairment in CB_1 receptor signaling results in an exaggerated cellular and neuroendocrine response to stress. Additionally, under conditions of repeated stress, a progressive increase in limbic $2AG/CB_1$ receptor signaling contributes to the development and expression of neuroendocrine habituation.

Ultimately, these data demonstrate that the ECB system is likely an integral player in the neuronal response and plasticity to stress. The relevance of this relationship has not been fully explored with respect to both normal homeostasis and pathological states characterized by alterations in HPA axis function, but will be a focus of future research.

References

Armario A. The hypothalamic–pituitary–adrenal axis: what can it tell us about stressors? CNS Neurol Disord Drug Target. 2006;5:485–501.

Aso E, Ozaita A, Valdizan EM et al. BDNF impairment in the hippocampus is related to enhanced despair behavior in CB(1) knockout mice. J Neurochem 2008;105:565–572

Bain MJ, Dwyer SM, Rusak B. Restraint stress affects hippocampal cell proliferation differently in rats and mice. Neurosci Lett. 2004;368:7–10.

Barna I, Zelena D, Arszovszki AC et al. The role of endogenous cannabinoids in the hypothalamo-pituitary-adrenal axis regulation: in vivo and in vitro studies in CB1 receptor knockout mice. Life Sci. 2004;75:2959–2970.

Bhatnagar S, Huber R, Nowak N et al. Lesions of the posterior paraventricular thalamus block habituation of hypothalamic–pituitary–adrenal responses to repeated restraint. J Neuroendocrinol. 2002;14:403–410.

Bisogno T, Berrendero F, Ambrasino G et al. Brain regional distribution of endocannabinoids: Implications for their biosynthesis and biological functions. Biochem Biophys Res Commun. 1999;256:377–380.

Bisogno T, Melck D, Bobrov MY et al. N-acyl-dopamines: novel synthetic CB(1) cannabinoid-receptor ligands and inhibitors of anandamide inactivation with cannabimimetic activity in vitro and in vivo. Biochem J. 2000;351:817–824.

Bisogno T, Ligresti A, Di Marzo V. The endocannabinoid signalling system: biochemical aspects. Pharmacol Biochem Behav. 2005;81:224–238.

Cabral GA, Marciano-Cabral F. Cannabinoid receptors in microglia of the central nervous system: immune functional relevance. J Leukoc Biol. 2005;78:1192–1197.

Cadas H, di Tomaso E, Piomelli D. Occurrence and biosynthesis of endogenous cannabinoid precursor, N-arachidonoyl phosphatidylethanolamine, in rat brain. J Neurosci. 1997;17:1226–1242.

Carrier EJ, Kearn CS, Barkmeier AJ et al. Cultured rat microglial cells synthesize the endocannabinoid 2-arachidonylglycerol, which increases proliferation via a CB2 receptor-dependent mechanism. Mol Pharmacol. 2004;65:999–1007.

Carter RN, Pinnock SB, Herbert J. Does the amygdala modulate adaptation to repeated stress? Neuroscience. 2004;126:9–19.

Cole MA, Kalman BA, Pace TW et al. Selective blockade of the mineralocorticoid receptor impairs hypothalamic–pituitary–adrenal axis expression of habituation. J Neuroendocrinol. 2000;12:1034–1042.

Cota D, Marsicano G, Tschop M et al. The endogenous cannabinoid system affects energy balance via central orexigenic drive and peripheral lipogenesis. J Clin Invest. 2003;112:423–431.

Cota D, Steiner MA, Marsicano G et al. Requirement of cannabinoid receptor type 1 for the basal modulation of hypothalamic–pituitary–adrenal axis function. Endocrinology. 2007;148:1574–1581.

Dallman MF, Akana SF, Levin N et al. Corticosteroids and the control of function in the hypothalamo–pituitary–adrenal (HPA) axis. Ann NY Acad Sci. 1994;746:22–31.
De Catanzaro D, Gorzalka BB. Postpubertal social isolation and male sexual behavior in rodents: Facilitation or inhibition is species-dependent. Anim Learn Behav. 1979;7:555–561.
Deutsch DG, Ueda N, Yamamoto S. The fatty acid amide hydrolase (FAAH). Prostag Leukotr Ess Fatty Acids. 2002;66:201–210.
Devane WA, Hanus L, Breuer A et al. Isolation and structure of a brain constituent that binds to the cannabinoid receptor. Science. 1992;258:1946–1949.
Di S, Malcher-Lopes R, Halmos KC et al. Nongenomic glucocorticoid inhibition via endocannabinoid release in the hypothalamus: a fast feedback mechanism. J Neurosci. 2003;23:4850–4857.
Di S, Malcher-Lopes R, Marcheselli VL et al. Rapid glucocorticoid-mediated endocannabinoid release and opposing regulation of glutamate and gamma-aminobutyric acid inputs to hypothalamic magnocellular neurons. Endocrinology. 2005;146:4292–4301.
Di Marzo V, Melck D, Bisogno T et al. Endocannabinoids: endogenous cannabinoid receptor ligands with neuromodulatory action. Trends Neurosci. 1998;21:521–528.
Dinh TP, Carpenter D, Leslie FM et al. Brain monoglyceride lipase participating in endocannabinoid inactivation. Proc Natl Acad Sci USA. 2002;99:10819–10824.
Diorio D, Viau V, Meaney MJ. The role of the medial prefrontal cortex (cingulate gyrus) in the regulation of hypothalamic–pituitary–adrenal responses to stress. J Neurosci. 1993;13:3839–3847.
Doyon C, Denis RG, Baraboi ED et al. Effects of rimonabant (SR141716) on fasting-induced hypothalamic–pituitary–adrenal axis and neuronal activation in lean and obese Zucker rats. Diabetes. 2006;55:3403–3410.
Evanson NK, Ulrich-Lai YM, Furay AR et al. (2007) Hypothalamic paraventricular cannabinoid receptor signaling in fast feedback inhibition of the hypothalamic–pituitary–adreanl response to acute restraint stress. Soc Neurosci. Abstr 197.5
Felder CC, Glass M. Cannabinoid receptors and their endogenous agonists. Annu Rev Pharmacol Toxicol. 1998;38:179–200.
Freund TF, Katona I, Piomelli D. Role of endogenous cannabinoids in synaptic signaling. Physiol Rev. 2003;83:1017–1066.
Fride E, Suris R, Weidenfeld J et al. Differential response to acute and repeated stress in cannabinoid CB1 receptor knockout newborn and adult mice. Behav Pharmacol. 2005;16:431–440.
Galiegue S, Mary S, Marchand J et al. Expression of central and peripheral cannabinoid receptors in human immune tissues and leukocyte subpopulations. Eur J Biochem. 1995;232:54–61.
Ginsberg AB, Pecoraro NC, van den Heuvel JK et al. (2006) Cannabinoid CB_1 receptor blockade rapidly enhances the hormonal response to restraint. Soc Neurosci. Abstr 563.8
Gong JP, Onaivi ES, Ishiguro H et al. Cannabinoid CB2 receptors: immunohistochemical localization in rat brain. Brain Res. 2006;1071:10–23.
Gonzalez S, Manzanares J, Berrendero F et al. Identification of endocannabinoids and cannabinoid CB(1) receptor mRNA in the pituitary gland. Neuroendocrinology. 1999;70:137–145.
Gorzalka BB, Hill MN. Cannabinoids, reproduction and sexual behavior. Annu Rev Sex Res. 2006;17:132–161.
Gorzalka BB, Hill MN, Hillard CJ. Regulation of endocannabinoid signaling by stress: Implications for stress-related affective disorders. Neurosci Biobehav Rev 2008;32:1152–1160.
Haller J, Varga B, Ledent C et al. Context-dependent effects of CB1 cannabinoid gene disruption on anxiety-like and social behaviour in mice. Eur J NeuroSci. 2004;19:1906–1912.
Hanus L, Abu-Lafi S, Fride E et al. 2-Arachidonyl glyceryl ether, an endogenous agonist of the cannabinoid CB1 receptor. Proc Natl Acad Sci U S A. 2001;98:3662–3665.
Hao S, Avraham Y, Mechoulam R et al. Low dose anandamide affects food intake, cognitive function, neurotransmitter and corticosterone levels in diet-restricted mice. Eur J Pharmacol. 2000;392:147–156.
Herkenham M, Lynn AB, Johnson MR et al. Characterization and localization of cannabinoid receptors in rat brain: a quantitative in vitro autoradiographic study. J Neurosci. 1991;11:563–583.

Herman JP, Tasker JG, Ziegler DR et al. Local circuit regulation of paraventricular nucleus stress integration: glutamate-GABA connections. Pharmacol Biochem Behav. 2002;71:457–468.
Herman JP, Figueiredo H, Mueller NK et al. Central mechanisms of stress integration: hierarchical circuitry controlling hypothalamo–pituitary–adrenocortical responsiveness. Front Neuroendocrinol. 2003;24:151–180.
Herman JP, Ostrander MM, Mueller NK et al. Limbic system mechanisms of stress regulation: hypothalamo-pituitary-adrenocortical axis. Prog Neuropsychopharmacol Biol Psychiatry. 2005;29:1201–1213.
Hill MN, Ho WS, Sinopoli KJ et al. Involvement of the endocannabinoid system in the ability of long-term tricyclic antidepressant treatment to suppress stress-induced activation of the hypothalamic–pituitary–adrenal axis. Neuropsychopharmacology. 2006;31:2591–2599.
Hill MN, Morrish AC, McLaughlin RJ et al. (2007) Endogenous cannabinoids regulate habituation to stress. Soc Neurosci. Abstr 197.4
Hill MN, Carrier EJ, Ho WS et al. Prolonged glucocorticoid treatment decreases cannabinoid CB_1 receptor binding in the hippocampus. Hippocampus. 2008;18:221–226.
Hill MN, McLaughlin RJ, Morrish AC et al (2009a) Suppression of amygdalr endocannabinoid signaling by stress contributes to activation of the hypothalamic-pituitary-adrenal axis. Neuropsychopharmacology (in press).
Hill MN, Miller GE, Carrier EJ et al (2009b) Circulating endocannabinoids and N-acyl ethanolamines are differentially regulated in major depression and following exposure to social stress. Psychoneuroendocrinology (in press).
Hillard CJ. Endocannabinoids and vascular function. J Pharmacol Exp Ther. 2000;294:27–32.
Hohmann AG, Suplita RL, Bolton NM et al. An endocannabinoid mechanism for stress-induced analgesia. Nature. 2005;435:1108–1112.
Howlett AC, Mukhopadhyay S. Cellular signal transduction by anandamide and 2-arachidonoylglycerol. Chem Phys Lipids. 2000;108:53–70.
Isbell H, Gorodetzsky CW, Jasinski D et al. Effects of (–) delta-9-tetrahydrocannabinol in man. Psychopharmacology. 1967;11:184–188.
Jaferi A, Bhatnagar S. Corticosterone can act at the posterior paraventricular thalamus to inhibit hypothalamic–pituitary–adrenal activity in animals that habituate to repeated stress. Endocrinology. 2006;147:4917–4930.
Kamprath K, Marsicano G, Tang J et al. Cannabinoid CB1 receptor mediates fear extinction via habituation-like processes. J Neurosci. 2006;26:6677–6686.
Korte SM, Koolhaas JM, Wingfield JC et al. The Darwinian concept of stress: benefits of allostasis and costs of allostatic load and the trade-offs in health and disease. Neurosci Biobehav Rev. 2005;29:3–38.
Lamota L, Bermudez-Silva FJ, Marco EM et al. Effects of adolescent nicotine and SR 147778 (Surinabant) administration on food intake, somatic growth and metabolic parameters in rats. Neuropharmacology. 2008;54:194–205.
Malcher-Lopes R, Di S, Marcheselli VS et al. Opposing crosstalk between leptin and glucocorticoids rapidly modulates synaptic excitation via endocannabinoid release. J Neurosci. 2006;26:6643–6650.
Malcher-Lopes R, Franco A, Tasker JG. Glucocorticoids shift arachidonic acid metabolism toward endocannabinoid synthesis: A non-genomic anti-inflammatory switch. Eur J Pharmacol. 2008;583:322–339.
Manzaneres J, Corchero J, Fuentes JA. Opioid and cannabinoid receptor-mediated regulation of the increase in adrenocorticotropin hormone and corticosterone plasma concentrations induced by central administration of delta(9)-tetrahydrocannabinol in rats. Brain Res. 1999;839:173–179.
McEwen BS. Glucocorticoids, depression, and mood disorders: structural remodeling in the brain. Metabolism. 2005;54:20–23.
Melia KR, Ryabinin AE, Schroeder R et al. Induction and habituation of immediate early gene expression in rat brain by acute and repeated restraint stress. J Neurosci. 1994;14:5929–5938.

Moldrich G, Wenger T. Localization of the CB1 cannabinoid receptor in the rat brain. An immunohistochemical study. Peptides. 2000;21:1735–1742.
Munro S, Thomas KL, Abu-Shaar M. Molecular characterization of a peripheral receptor for cannabinoids. Nature. 1993;365:61–65.
Nakazi M, Bauer U, Nickel T et al. Inhibition of serotonin release in the mouse brain via presynaptic cannabinoid CB1 receptors. Naunyn Schmiedebergs Arch Pharmacol. 2000;361:80–84.
Ohno-Shosaku T, Maejima T, Kano M. Endogenous cannabinoids mediate retrograde signals from depolarized postsynaptic neurons to presynaptic terminals. Neuron. 2001;29:729–738.
Pagotto U, Marsicano G, Fezza F et al. Normal human pituitary gland and pituitary adenomas express cannabinoid receptor type 1 and synthesize endogenous cannabinoids: first evidence for a direct role of cannabinoids on hormone modulation at the human pituitary level. J Clin Endocrinol Metab. 2001;86:2687–2696.
Parolaro D. Presence and functional regulation of cannabinoid receptors in immune cells. Life Sci. 1999;65:637–644.
Patel S, Roelke CT, Rademacher DJ et al. Endocannabinoid signaling negatively modulates stress-induced activation of the hypothalamic–pituitary–adrenal axis. Endocrinology. 2004;145:5431–5438.
Patel S, Roelke CT, Rademacher DJ et al. Inhibition of restraint stress-induced neural and behavioural activation by endogenous cannabinoid signalling. Eur J NeuroSci. 2005;21:1057–1069.
Pecoraro N, Dallman MF, Warne JP et al. From Malthus to motive: how the HPA axis engineers the phenotype, yoking needs to wants. Prog Neurobiol. 2006;79:247–340.
Piomelli D. The molecular logic of endocannabinoid signalling. Nat Rev Neurosci. 2003;4:873–884.
Porter AC, Sauer JM, Knierman MD et al. Characterization of a novel endocannabinoid, virodhamine, with antagonist activity at the CB1 receptor. J Pharmacol Exp Ther. 2002;301:1020–1024.
Rademacher DJ, Meier SE, Shi L et al. Effects of acute and repeated restraint stress on endocannabinoid content in the amygdala, ventral striatum, and medial prefrontal cortex in mice. Neuropharmacology. 2008;54:108–116.
Radley JJ, Arias CM, Sawchenko PE. Regional differentiation of the medial prefrontal cortex in regulating adaptive responses to acute emotional stress. J Neurosci. 2006;26:12967–12976.
Ross RA. Anandamide and vanilloid TRPV1 receptors. Br J Pharmacol. 2003;140:790–801.
Ryberg E, Larsson N, Sjogren S et al. The orphan receptor GPR55 is a novel cannabinoid receptor. Br J Pharmacol. 2007;152:984–986.
Schlicker E, Kathmann M. Modulation of transmitter release via presynaptic cannabinoid receptors. Trends Pharmacol Sci. 2001;22:565–572.
Schulkin J, Gold PW, McEwen BS. Induction of corticotropin-releasing hormone gene expression by glucocorticoids: Implication for understanding the states of fear and anxiety and allostatic load. Psychoneuroendocrinology. 1998;23:219–243.
Steiner MA, Marsicano G, Nestler EJ et al. Antidepressant-like behavioral effects of impaired cannabinoid receptor type 1 signaling coincide with exaggerated corticosterone secretion in mice. Psychoneuroendocrinology. 2008;33:54–67.
Succu S, Mascia MS, Sanna F et al. The cannabinoid CB1 receptor antagonist SR 141716A induces penile erection by increasing extra-cellular glutamic acid in the paraventricular nucleus of male rats. Behav Brain Res. 2006;169:274–281.
Sugiura T, Kondo S, Sukagawa A et al. 2-Arachidonoylglycerol: a possible endogenous cannabinoid receptor ligand in brain. Biochem Biophys Res Commun. 1995;215:89–97.
Sun Y, Alexander SP, Kendall DA et al. Cannabinoids and PPARalpha signalling. Biochem Soc Trans. 2006;34:1095–1097.
Tsou K, Brown S, Sanudo-Pena MC et al. Immunohistochemical distribution of cannabinoid CB1 receptors in the rat central nervous system. Neuroscience. 1998;83:393–411.
Uchigashima M, Narushima M, Fukaya M et al. Subcellular arrangement of molecules for 2-arachidonoyl-glycerol-mediated retrograde signaling and its physiological contribution to synaptic modulation in the striatum. J Neurosci. 2007;27:3663–3676.
Ueda N. Endocannabinoid hydrolases. Prostaglandins Oth Lipid M. 2002;68–69:521–534.

Uriguen L, Perez-Rial S, Ledent C et al. Impaired action of anxiolytic drugs in mice deficient in cannabinoid CB1 receptors. Neuropharmacology. 2004;46:966–973.
Van Sickle MD, Duncan M, Kingsley PJ et al. Identification and functional characterization of brainstem cannabinoid CB2 receptors. Science. 2005;310:329–332.
Vaughan CW, Christie MJ. Retrograde signalling by endocannabinoids. Handbook Exp Pharmacol. 2005;168:367–383.
Viau V, Sawchenko PE. Hypophysiotropic neurons of the paraventricular nucleus respond in spatially, temporally, and phenotypically differentiated manners to acute vs. repeated restraint stress. J Comp Neurol. 2002;445:293–307.
Wade MR, Degroot A, Nomikos GG. Cannabinoid CB1 receptor antagonism modulates plasma corticosterone in rodents. Eur J Pharmacol. 2006;551:162–167.
Watanabe Y, Stone E, McEwen BS. Induction and habituation of c-fos and zif/268 by acute and repeated stressors. NeuroReport. 1994;5:1321–1324.
Weidenfeld J, Feldman S, Mechoulam R. Effect of the brain constituent anandamide, a cannabinoid receptor agonist, on the hypothalamo–pituitary–adrenal axis in the rat. Neuroendocrinology. 1994;59:110–112.
Wenger T, Jamali KA, Juaneda C et al. Arachidonyl ethanolamide (anandamide) activates the parvocellular part of hypothalamic paraventricular nucleus. Biochem Biophys Res Commun. 1997;237:724–728.
Wenger T, Fernandez-Ruiz JJ, Ramos JA. Immunocytochemical demonstration of CB1 cannabinoid receptors in the anterior lobe of the pituitary gland. J Neuroendocrinol. 1999;11:873–878.
Wenger T, Ledent C, Tramu G. The endogenous cannabinoid, anandamide, activates the hypothalamo–pituitary–adrenal axis in CB1 cannabinoid receptor knockout mice. Neuroendocrinology. 2003;78:294–300.
Wilson RI, Nicoll RA. Endocannabinoid signaling in the brain. Science. 2002;296:678–682.
Yoshida T, Fukaya M, Uchigashima M et al. Localization of diacylglycerol lipase-alpha around postsynaptic spine suggests close proximity between production site of an endocannabinoid, 2-arachidonoyl-glycerol, and presynaptic cannabinoid CB1 receptor. J Neurosci. 2006;26:4740–4751.

Part III
Pathology

Drug Addiction

Zuzana Justinova, Leigh V. Panlilio, and Steven R. Goldberg

Contents

Abstract Many drugs of abuse, including cannabinoids, opioids, alcohol and nicotine, can alter the levels of endocannabinoids in the brain. Recent studies show that release of endocannabinoids in the ventral tegmental area can modulate the reward-related effects of dopamine and might therefore be an important neurobiological mechanism underlying drug addiction. There is strong evidence that the endocannabinoid system is involved in drug-seeking behavior (especially behavior that is reinforced by drug-related cues), as well as in the mechanisms that underlie relapse to drug use. The cannabinoid CB_1 antagonist/inverse agonist rimonabant

Z. Justinova, L.V. Panlilio and S.R. Goldberg (✉)
Department of Health and Human Services, Preclinical Pharmacology Section, Behavioral Neuroscience Research Branch, Intramural Research Program, National Institute on Drug Abuse, National Institutes of Health, Baltimore, MD 21224, USA
e-mail: sgoldber@intra.nida.nih.gov

D. Kendall and S. Alexander (eds.), *Behavioral Neurobiology of the Endocannabinoid System*, Current Topics in Behavioral Neurosciences 1,
DOI: 10.1007/978-3-540-88955-7_13,

has been shown to reduce the behavioral effects of stimuli associated with drugs of abuse, including nicotine, alcohol, cocaine, and marijuana. Thus, the endocannabinoid system represents a promising target for development of new treatments for drug addiction.

Keywords Drug addiction • Cannabinoids • Endocannabinoids • Self-administration • Relapse • Reward • THC

Abbreviations

2-AG	2-Arachidonoylglycerol
AEA	Anandamide
VTA	ventral tegmental area
DAT	dopamine transporter
THC	delta-9-tetrahydrocannabinol
FAAH	fatty acid amide hydrolase

1 Introduction

1.1 Drug Addiction

The abuse of drugs and alcohol is a major problem worldwide, costing 250 billion dollars annually due to premature deaths, healthcare expenditures, reduction of productivity, lost earnings and drug-related crime in the United States alone (estimated by US National Institute on Drug Abuse and National Institute on Alcohol Abuse and Alcoholism). Drug addiction is considered to be a chronic, relapsing disorder characterized by compulsive drug-seeking, by continued use despite serious negative socioeconomic and health consequences, and by loss of control over drug use (Cami and Farre 2003). The World Health Organization and the American Psychiatric Association use the term "substance dependence" rather than "drug addiction." Both terms are used interchangeably in the literature, but the latter term is less likely to be confused with physical dependence and emphasizes the behavioral component of the process. According to the DSM-IV (American Psychiatric Association 1994), three or more of the following must be present in order to diagnose substance dependence: (a) symptoms of tolerance, (b) symptoms of withdrawal, (c) large amounts of drug taken, (d) unsuccessful attempts or desire to control use, (e) considerable time spent obtaining the substance, (f) reduction of social and occupational activities due to abuse, (g) continued use of a substance despite physical or psychological problems.

Repeated drug use arises from the drug's neurochemical actions that produce positive reinforcing effects, progressively leading to neurobiological changes in the brain reward circuits and behaviors characteristic of addiction: tolerance, sensitization, dependence, withdrawal and craving (Kreek et al. 2002). The transition from casual drug use to drug addiction might also involve an additional source of reinforcement, such as the reduction of a negative emotional state during acute abstinence (Koob et al. 1998). The combination of positive (e.g., euphoria) and negative (e.g., alleviation of dysphoria or withdrawal symptoms) reinforcement may provide a powerful motivational force for compulsive drug taking. Associated neurobiological changes and behavioral abnormalities and deficits in cognitive function may persist for months or years after discontinuation of drug use (Cami and Farre 2003).

1.2 Endocannabinoid System in Brain Reward Circuitry

The initial events that lead to drug addiction involve acute effects at the specific sites of action of the abused drug. These sites of action (e.g., G-protein coupled receptors and ligand-gated ion channels) typically activate neural circuits associated with positive reinforcement/reward, particularly the mesocorticolimbic dopaminergic system. This system, originating in the ventral tegmental area (VTA) and projecting to the nucleus accumbens, olfactory tubercle, frontal cortex, and amygdala (Wise 2004), interacts with glutamatergic projections from the cerebral cortex, hippocampus, and amygdala, and thus regulates responses to natural reinforcers such as food, drink, social interactions or sex (Kauer 2004). The mesocorticolimbic dopaminergic system is part of a brain reward circuit that has long been thought to play a major role in mediating the reinforcing/rewarding effects of drugs of abuse (Di Chiara et al. 1999; Koob 1992; Wise and Bozarth 1987). Abused drugs (like opioids, cannabinoids, psychostimulants, alcohol, nicotine, sedative-hypnotics, anxiolytics, and anesthetics) directly or indirectly elevate extracellular levels of dopamine in the shell of the nucleus accumbens (Brodie et al. 1999; Chen et al. 1990; Masuzawa et al. 2003; Pontieri et al. 1995, 1996; Spyraki and Fibiger 1988; van der Laan et al. 1992).

In the striatum, cannabinoid CB_1 receptors are localized presynaptically in GABAergic and glutamatergic nerve terminals, but also postsynaptically in the dendritic shafts and spines of both enkephalinergic and dynorphinergic GABAergic efferent neurons (Fusco et al. 2004; Hohmann and Herkenham 2000; Kofalvi et al. 2005; Pickel et al. 2004, 2006). When these cells are depolarized, endocannabinoids can be released in the nucleus accumbens (Robbe et al. 2001) and VTA (Melis et al. 2004; Riegel and Lupica 2004), where they modulate the excitatory (glutamatergic) and inhibitory (GABAergic) inputs that control dopaminergic neurons of the mesocorticolimbic pathway by acting as retrograde messengers on CB_1 receptors. Endocannabinoids are also involved in synaptic plasticity in the mesolimbic system – please see the chapter, "Endocannabinoid Signaling in Neural Plasticity."

The dopaminergic system has a well-established role in the reinforcing effects of drugs of abuse. It has become increasingly clear that the endocannabinoid system can modulate dopaminergic reward circuits, which suggests that endocannabinoids also play a major role in the mechanisms underlying drug addiction.

1.3 Release of Endocannabinoids by Abused Drugs

The endocannabinoid system can modulate the primary rewarding effects of non-cannabinoid drugs of abuse, and this ability appears to depend on endocannabinoid release in the VTA (Lupica and Riegel 2005). This hypothesis is consistent with evidence that repeated non-contingent drug administration alters levels of the endocannabinoids anandamide (AEA) and 2-arachidonoylglycerol (2-AG).

Analysis of AEA and 2-AG levels in brains of animals treated chronically with cocaine, nicotine, or ethanol showed that chronic cocaine administration produced a modest but significant decrease in the content of 2-AG in the limbic forebrain (Gonzalez et al. 2002). In contrast, chronic ethanol and nicotine exposure produced an increase in AEA content in this area. Chronic ethanol administration caused a decrease in the contents of both AEA and 2-AG in the midbrain. Chronic nicotine exposure increased both AEA and 2-AG in the brainstem and decreased their content in the hippocampus, striatum and cerebral cortex. It appears that the most consistent finding with these drugs of abuse is that chronic administration led to an elevation in endocannabinoid levels in the limbic system. This observation is consistent with the notion that endocannabinoids enhance the reinforcing effects of addictive drugs by increasing dopamine release via the inhibition of GABA release in the limbic system. Chronic administration of Δ^9-tetrahydrocannabinol (THC) itself decreases the levels of AEA and 2-AG in the striatum (Di Marzo et al. 2000). Chronic morphine administration decreases 2-AG levels in the striatum without altering AEA levels (Gonzalez et al. 2003; Vigano et al. 2003). Acute morphine administration, on the other hand, increased AEA levels and decreased 2-AG levels in the striatum (Vigano et al. 2004). Thus, the selection of the time point for endocannabinoid analysis is critical for determination of the nature of alterations in endocannabinoid levels.

It should be noted that, in the studies just described, endocannabinoid levels were measured in postmortem rat brain tissue, usually at a single time-point after chronic administration of the drugs. Therefore, it is not clear whether these findings reflect sustained changes in the brain endocannabinoid levels, or acute alterations of endocannabinoid formation. Another potential problem with these results is that endocannabinoid levels in brain tissue are affected by rapid postmortem increases in endocannabinoid formation (Bazinet et al. 2005; Patel et al. 2005). In addition, these studies have used non-contingent drug administration, which can produce neurochemical, proteomic, and genomic effects that differ substantially from those induced by free-choice self-administration (Jacobs et al. 2003). Thus, further research is required to determine whether endocannabinoid levels would be affected in the same way if the drugs were self-administered.

In vivo microdialysis techniques offer an effective means of studying levels of neurotransmitters during drug self-administration. However, microdialysis is difficult to perform with AEA and 2-AG due to their highly lipophilic nature and instability. The first study in which endocannabinoid levels were measured by microdialysis showed that local intrastriatal administration of the dopamine D_2 agonist quinpirole elevated levels of AEA, but not 2-AG (Giuffrida et al. 1999). Caille and colleagues (Caille et al. 2007) were the first to measure changes in endocannabinoid levels during self-administration of a drug of abuse. They found that self-administration of cocaine did not alter either AEA or 2-AG levels in the nucleus accumbens, but self-administration of heroin increased AEA and decreased 2-AG levels, and self-administration of ethanol increased 2-AG without altering AEA levels. These data provide in vivo evidence for an endocannabinoid involvement in the motivational effects of ethanol and heroin but not cocaine.

An exciting new analytical method, combining online in vivo brain microdialysis with solid-phase extraction–liquid chromatography–tandem mass spectrometry, allows real-time detection of trace amounts of endocannabinoids in the extracellular fluid. This technique has been used to show that the CB_1 receptor antagonist/inverse agonist rimonabant increased, whereas the CB_1 receptor agonist WIN 55,212-2 decreased, AEA release in the rat hypothalamus (Bequet et al. 2007). Interestingly, the same treatments induced opposite changes in 2-AG release. The same study also shows that inhibition of fatty acid amide hydrolase (FAAH), the primary enzyme responsible for AEA degradation, induced an increase in outflow of AEA, but not 2-AG. In this study, FAAH was inhibited by systemic administration of URB597, a selective FAAH inhibitor now entering clinical trial.

1.4 Endocannabinoids in Drug-Seeking and Relapse

Relapse to drug use, even after a long period of forced or voluntary withdrawal and detoxification, is one of the defining features of addiction, and perhaps the most important impediment to effective treatment (American Psychiatric Association 1994; O'Brien 2001). Reinstatement of drug-seeking behavior in laboratory animals is an experimental procedure that is used to model relapse. In this model, animals are initially trained to self-administer drugs intravenously by making an operant response (e.g., pressing a lever). Subsequently, the drug-reinforced behavior is extinguished by replacing the self-administered drug solution with saline or by disconnecting the infusion pump. After extinction of the drug-reinforced behavior, reinstatement of drug-seeking behavior can be tested by one of several procedures that mirror the triggers that induce relapse in humans. These include non-contingent injection of a drug (drug-induced reinstatement), presentation of visual or auditory stimuli that had previously signaled availability or delivery of the drug (cue-induced reinstatement), or exposure to a brief period of intermittent foot-shock in the self-administration chamber (stress-induced reinstatement).

The first evidence for involvement of the endocannabinoid system in reinstatement of extinguished drug-seeking behavior was provided by De Vries and colleagues (De Vries et al. 2001). They found that the synthetic CB_1 receptor agonist HU210 could reinstate cocaine-seeking behavior. The CB_1 antagonist/inverse agonist rimonabant blocked this effect and also reduced cocaine- and cue-induced reinstatement, but not stress-induced reinstatement. Another CB_1 antagonist/inverse agonist, AM251, was later found to block cocaine-induced reinstatement of drug-seeking behavior (Xi et al. 2006). Rimonabant has since been found to reduce reinstatement induced by heroin, methamphetamine, nicotine, WIN 55, 212-2 and ethanol, and to attenuate or block cue-induced reinstatement of the seeking of cocaine, nicotine, heroin, methamphetamine, and alcohol in rodents (De Vries and Schoffelmeer 2005; Fattore et al. 2007). Rimonabant blocks both THC-induced and cue-induced reinstatement of THC-seeking behavior in non-human primates (Justinova et al. 2008b). At present, studies investigating a potential ability of CB_1 receptor blockade to alter reinstatement have been described as indicating that rimonabant is unable to affect stress-induced relapse to cocaine or ethanol seeking (Fattore et al. 2007). Although stress-induced relapse has received less attention than drug- and cue-induced relapse, the evidence accumulated to date indicates that endocannabinoid signaling might not be involved in stress-induced reinstatement.

Early clinical trials examining the effectiveness of rimonabant as an aid in smoking cessation and obesity treatment were very promising. Rimonabant has been approved and marketed as an anti-obesity treatment, but not an anti-smoking treatment, in the European Union and in a number of other countries. However, in 2006 the American Food and Drug Administration (FDA) declined to approve rimonabant for smoking cessation and required further studies before final approval for weight management. This was due to concerns over possible depression-like side effects. It has been suggested that CB_1 neutral antagonists may be devoid of the side effects produced by CB_1 antagonists/inverse agonists, and neutral antagonists are now being tested in animal studies (Salamone et al. 2007; Sink et al. 2008). Thus, while manipulations of the endocannabinoid system show promise for the treatment of addiction, there is not yet a specific clinically tested compound that has been shown to be both effective and safe.

2 Cannabinoids

Cannabinoids, usually abused by humans in the form of marijuana, have become the most frequently abused illicit class of drugs in the United States. There is ample evidence that most of the centrally mediated effects of cannabinoids occur through the endocannabinoid system. The effects of marijuana in humans are quite complex and highly variable depending on the dose of the drug, environment and expectations of the user. The subjective effects may include excitement and dissociation of ideas, quickening of mental associations, heightened perception, distortion of the sense of time, irresistible impulses and illusions accompanied by decrease of psychomotor activity (Dewey 1986). In addition to its euphorigenic properties (Haney et al. 1997), marijuana can produce anxiety, analgesia, hypothermia, increased appetite, anti-emetic effects, vasorelaxation, and alterations in cognition and memory (Hollister 1986). A controlled study in healthy cannabis users showed that the intoxicant effects are clearly mediated by CB_1 receptors (Huestis et al. 2001). In rodents, administration of cannabis or its major psychoactive ingredient, THC, produces a characteristic combination of four signs, analgesia, hypoactivity, catalepsy and hypothermia, associated with anxiogenesis, memory changes and cardiovascular changes (Chaperon and Thiebot 1999). The tetrad of behavioral and physiological assays (motor activity, ring catalepsy, body temperature and analgesia tests) was developed to assess in vivo activity of cannabinoid analogs in mice (Martin et al. 1991). In monkeys, disruption of behavior and static ataxia have also been observed (Branch et al. 1980).

Although cannabis dependence is often considered to be a less serious problem than dependence on other drugs, the number of people seeking treatment for cannabis use in the US is higher than the number seeking treatment for cocaine use (CEWG 2007; Substance Abuse and Mental Health Services Administration 2007). Cannabis produces clear subjective motivational responses in humans, leading to drug-seeking and drug-taking behavior (Maldonado 2002). Many different animal models are used to elucidate the consequences of chronic exposure to cannabinoids and to predict their abuse liability. Tolerance and withdrawal syndromes provide only a partial correlate of their addictive properties. The reinforcement-related motivational properties of drugs, including cannabinoids, can be evaluated using several different behavioral models: drug self-administration, conditioned-place preference/aversion and drug-discrimination paradigms.

2.1 *Self-Administration of Cannabinoids*

2.1.1 Drug Self-Administration Paradigm

Drug self-administration behavior is one of the most direct and productive approaches for studying the reinforcing effects of psychoactive drugs, which are critical in determining their abuse potential (Johanson and Balster 1978). Even in a non-dependent state, animals and humans will readily self-administer drugs of abuse. Allowing limited access to drugs provides a reliable model for their acute reinforcing effects and a means for exploring the neuropharmacological mechanisms involved in these effects (Koob and Weiss 1990). Reliable and persistent self-administration behavior has now been demonstrated in laboratory animals for almost all drugs abused by humans, including psychostimulants, opiates, ethanol, nicotine (Collins et al. 1984; Goldberg et al. 1981; Koob and Weiss 1990; Yokel 1987; Young and Herling 1986), and recently marijuana (THC) (Justinova et al. 2003, 2008b; Tanda et al. 2000).

During the intravenous self-administration procedure, animals are allowed to self-administer a drug by making an operant response, such as pressing a lever or inserting their nose into a hole (a "nose-poking" response), which activates a pump to intravenously deliver the drug. Subjects are prepared with intravenous catheters. Primates often wear a vest to protect the catheter. In many studies, a certain number of responses is required for each injection, in a procedure known as a fixed-ratio schedule. The behavioral measures in drug self-administration studies include the rate of responding and the number of drug injections delivered. Although there are many variations of the self-administration paradigm, usually the reinforcing efficacy of a tested drug is compared to a standard drug of known abuse potential from a similar pharmacological class and also to the drug's vehicle in the same subject (Bergman and Johanson 1985; Tanda et al. 2000; Young and Woods 1981). These studies are often performed in rhesus (*Macaca mulatta*) or squirrel (*Saimiri sciureus*) monkeys, which have learned to self-administer a drug, for example cocaine, under a schedule requiring a certain number of responses to obtain each injection (e.g., ten-response, fixed-ratio schedule of drug injection) (Goldberg et al. 1971). The drugs being tested are then substituted for the training drug and evaluated for their ability to maintain a level of responding resulting in their frequent injection. It is important to mention that the functional state of the brain reward circuits in naïve versus experienced drug self-administering animals is different, and neurobiological adaptations might predispose to or limit subsequent self-administration (Young et al. 1981).

To study the relative reinforcing efficacy of drugs and compare the effects of pharmacological treatments, progressive-ratio schedules of drug self-administration are often used, in which the number of responses required for each injection increases progressively within a session until the drug-seeking response ceases (Arnold and Roberts 1997). Progressive-ratio schedules allow an estimation of the maximal effort an individual will put forth under a specified set of conditions to obtain a particular reinforcement. The behavioral measure obtained is the maximal number of responses an animal will perform for a single drug injection, often called the "break-point," which is taken as a measure of the motivational strength of the reinforcing event and predicts rewarding efficacy of drugs (Hodos 1961). Many different and more complicated schedules of reinforcement also exist and are used to focus on various aspects of addiction. One of these variations, the second order schedule, is discussed in detail in Sect. 2.1.3.

2.1.2 Fixed-Ratio Schedule

THC and Synthetic Cannabinoids

During the last three decades, many attempts to demonstrate intravenous self-administration of THC or of synthetic CB_1 receptor agonists by experimental animals were relatively unsuccessful (for review see Justinova et al. 2005a; Tanda and Goldberg 2003). In none of these studies were THC or synthetic cannabinoids clearly shown to maintain self-administration behavior that was persistent, dose-related and susceptible to vehicle extinction and subsequent reinstatement. Only a few studies reported self-administration of THC at levels higher than vehicle controls. In one of these studies (Kaymakcalan 1973), two monkeys out of six acquired THC self-administration behavior, but only after withdrawal from forced automatic intravenous injections of THC, when overt signs of physical dependence occurred.

Although THC has not been found to maintain persistent intravenous self-administration in mice or rats, it has been reported to be self-administered intracerebroventricularly (Braida et al. 2004) and into the VTA and the shell of the nucleus accumbens (Zangen et al. 2006) in rats. There have also been several reports of intravenous and intracerebroventricular self-administration of synthetic CB_1 receptor agonists in rodents: WIN 55, 212-2 (Ledent et al. 1999; Martellotta et al. 1998), CP55940 (Braida et al. 2001b; Navarro et al. 2001) and HU210 (Navarro et al. 2001) in mice and WIN 55, 212-2 in rats (Fattore et al. 2001; Spano et al. 2004). However, the experimental procedures employed in some of these studies limit the scope and generality of the findings. For example, the studies with mice (Ledent et al. 1999; Martellotta et al. 1998; Navarro et al. 2001) employed 1-day experimental tests during which mice were restrained for acute intravenous administration through the tail vein. This procedure provides little information about acquisition, extinction and relapse to self-administration behavior. In contrast, the study by Fattore and colleagues (Fattore et al. 2001) utilized unrestrained, freely moving rats as subjects that were allowed to self-administer WIN 55, 212-2 over repeated daily sessions. Spano and colleagues (Spano et al. 2004) used the same model to provide the first evidence of drug-induced reinstatement of cannabinoid-seeking behavior.

It is important to note that chronic diet restriction (rats were maintained at 80% of their normal body weight) was a necessary condition in the study by Fattore and colleagues (Fattore et al. 2001), since rats on an unrestricted diet did not acquire cannabinoid self-administration behavior. Diet restriction has been repeatedly shown to increase a wide variety of appetitive behaviors, including self-administration of drugs from each of the major classes of abused drugs (Carroll and Meisch 1984). Thus, the need for food restriction may simply indicate that cannabinoid agonists are only weakly reinforcing in rats, or that they may have aversive effects that can counteract their reinforcing effects. In another series of studies, food was not only restricted in the rats, but delivered during the THC self-administration sessions (Takahashi and Singer 1979, 1980). In those studies, THC self-administration behavior above placebo levels was found in diet-restricted rats (maintained at 80% of normal body weight), under conditions where a food pellet was automatically delivered once every minute. However, this self-administration behavior immediately decreased to placebo levels when food restriction was discontinued, suggesting that this was probably a schedule-induced adjunctive behavior, rather than a case of drug reinforcement, per se.

Robust, dose-related, intravenous self-administration of THC by animals was first demonstrated under a fixed-ratio schedule in squirrel monkeys (Tanda et al. 2000). This study utilized monkeys with cocaine self-administration experience that were not food-deprived and had access to THC only after at least 1 week of saline extinction. The dose range of THC in this study (1–8 g kg^{-1} per injection) was lower than that previously used in THC self-administration studies and comparable to that received from smoking a marijuana cigarette (Agurell et al. 1986; Tanda et al. 2000). Under these conditions, monkeys readily acquired THC self-administration behavior. Once acquired, self-administration behavior could be rapidly extinguished by substituting vehicle for THC or by administering the CB_1 antagonist/inverse agonist, rimonabant, suggesting that the behavior was mediated by CB_1 receptors. The opioid-receptor antagonist naltrexone can also partially reduce THC self-administration (Justinova et al. 2004).

Although earlier attempts to obtain THC self-administration behavior in monkeys that had prior experience with cocaine self-administration were unsuccessful – even when THC was directly substituted for cocaine with no intervening vehicle extinction (Harris et al. 1974) – the fact that the monkeys in the study by Tanda and colleagues (Tanda et al. 2000) had prior experience with cocaine raised the possibility that cocaine might induce persistent neurobiological adaptations that subsequently predispose animals to self-administer THC (Maldonado 2002). However, such adaptations are clearly not a necessary condition, since further experiments established that drug-naïve squirrel monkeys readily acquired THC self-administration behavior (Justinova et al. 2003).

The ability of THC to maintain drug-taking behavior in monkeys without a history of exposure to other drugs shows that this drug possesses reinforcing properties of its own that are not dependent on prior self-administration of other drugs. These findings support the previous conclusion that, under certain experimental conditions, THC has a pronounced abuse liability that is comparable to that of other drugs of abuse (Justinova et al. 2005a). Self-administration of THC by squirrel monkeys provides the most reliable animal model for human marijuana abuse available to date. This animal model now makes it possible to study the relative abuse liability of other natural and synthetic cannabinoids and to preclinically assess new therapeutic strategies for the treatment or prevention of marijuana abuse in humans.

AEA and Methanandamide

Building on the procedures that were successfully used to obtain THC self-administration in squirrel monkeys, it was shown that the endocannabinoid AEA is also self-administered by squirrel monkeys, with or without previous exposure to other drugs (Justinova et al. 2005b). This study also showed that methanandamide, a longer-lasting synthetic analog of AEA, can serve as an effective reinforcer of drug-taking behavior when self-administered intravenously by squirrel monkeys. The reinforcing effects of both AEA and methanandamide in squirrel monkeys appear to be mediated by CB_1 receptors, because pre-session treatment with the CB_1 antagonist/inverse agonist rimonabant dramatically decreased self-administration behavior for both cannabinoids. The fact that the endocannabinoid AEA is self-administered is consistent with the hypothesis that the release of endogenous cannabinoids is involved in brain reward processes and that activation of CB_1 receptors by AEA is part of the signaling of natural rewarding events (Solinas et al. 2007d, 2008). As discussed below, intravenous self-administration of AEA by squirrel monkeys provides a procedure for studying the potential abuse liability of medications that activate the endogenous cannabinoid system by interfering with inactivation of endocannabinoids and for investigating mechanisms involved in the reinforcing effects of endocannabinoids.

Fatty Acid Amide Hydrolase (FAAH) Inhibitors and AEA Transport Inhibitors

URB597

Interest in the development of medications that enhance endocannabinoid signaling in the brain without inducing the psychotropic side effects associated with systemic administration of direct acting CB_1 receptor agonists (like THC) led our laboratory to study the selective FAAH inhibitor URB597, focusing on evaluating its abuse liability and measuring its effects on endocannabinoid levels in the brain. We found (Justinova et al. 2007, 2008a) that URB597 suppresses FAAH activity and increases AEA levels in regions of the squirrel monkey brain that participate in motivational, cognitive and emotional functions. This effect is accompanied by a marked decrease in the levels of 2-AG, which would presumably have major effects on endocannabinoid signaling in the brain. This was surprising, because URB597 does not affect activity of the 2-AG-metabolizing enzyme; it may be due to enhanced levels of AEA causing a compensatory up-regulation in 2-AG mobilization.

We further observed that, over a broad range of experimental conditions, URB597 does not display overt reinforcing properties in monkeys. Indeed, the drug did not have reinforcing effects (i.e., was not self-administered more than vehicle) even when its cumulative intake exceeded by several fold a fully effective dose for FAAH inhibition. Furthermore, neither previous cocaine nor THC exposure predisposed monkeys to self-administer URB597. Indeed, even monkeys that had previously self-administered AEA at very high rates failed to self-administer the FAAH inhibitor. Finally, URB597 did not alter the reinforcing effects of THC or cocaine, and did not reinstate extinguished drug-seeking behavior in monkeys that had previously self-administered THC or cocaine. These results indicate that the potentiation of endogenous AEA-mediated transmission produced by URB597 is insufficient per se to produce reinforcing effects. Our findings further imply that FAAH inhibitors such as URB597 – which have demonstrated analgesic, anxiolytic, antidepressant and antihypertensive properties in rodents (Gobbi et al. 2005; Jayamanne et al. 2006; Kathuria et al. 2003) – may be used in humans without anticipated risk of inducing abuse or provoking relapse to drug use in abstinent individuals.

AM404

Another mechanism by which brain levels of AEA can be increased is by inhibition of AEA transport into neurons. The most studied drug of this class is AM404, which has been found to exert anxiolytic effects (Bortolato et al. 2006), serve as a reinforcer of intravenous drug-taking behavior in squirrel monkeys, and reinstate extinguished drug-seeking behavior (Justinova and Goldberg 2004). These findings are consistent with the fact that AM404 produces conditioned place preference, an indication of rewarding properties, in rats housed under enriched conditions (Bortolato et al. 2006). On the other hand, place preference does not develop with URB597 under the same conditions (Gobbi et al. 2005). Although both AM404 and URB597 do not have THC-like discriminative or neurochemical effects in rodents (Gobbi et al. 2005; Solinas et al. 2006a, 2007c), only AM404 has motivational effects in rodents and primates that suggest the potential for abuse.

2.1.3 Second-Order Schedule and Drug Seeking

The second-order schedule of drug self-administration (Goldberg et al. 1975; Schindler et al. 2002) has been strongly advocated as an animal model that can be used to focus on drug-seeking behavior, as opposed to drug taking (Arroyo et al. 1998; Everitt and Robbins 2000). This drug-seeking schedule incorporates drug-related stimuli that model the environmental cues that maintain drug seeking and induce drug craving and relapse in humans. Unlike fixed-ratio schedules of drug self-administration, which can only be used to evaluate the ability of a treatment to block the effects of the abused drug after it has been self-administered, second-order schedules can be used to evaluate treatments that target drug seeking, per se, before the abused drug is received. This is important because treatments that reduce drug seeking might provide an especially effective means of achieving and maintaining drug abstinence. In addition to studying the effects of treatments on drug seeking, the second-order schedule can also be used to study relapse induced by drug-related cues, as well as relapse caused by re-exposure to the abused drug or exposure to other drugs. The study by Justinova and colleagues (Justinova et al. 2008b), described in detail below, took advantage of all of these features of a second-order schedule to study the effects of treatments on the maintenance of and relapse to THC seeking.

In the first study utilizing a second-order schedule of THC self-administration (Justinova et al. 2008b), squirrel monkeys' lever-pressing responses intermittently produced brief presentations of a visual stimulus (a colored light). This drug-seeking response produced only the stimulus until the end of the 30-min session, when the last response of the session produced both the stimulus and intravenous delivery of THC. Monkeys' THC-seeking behavior occurred at a high rate even though the drug was not delivered until the end of the session. This behavior depended on both the delivery of THC and the response-contingent presentations of the drug-paired stimulus. That is, when the brief light stimulus was not presented during the session, THC seeking decreased abruptly and continued to occur at a low rate even when THC was still delivered paired with the stimulus at the end of each session. When both the stimulus and THC delivery were discontinued, responding ceased, but it was immediately reinstated when stimulus presentations were reinstituted. Thus, like re-exposure to the drug, re-exposure to THC-associated stimuli (cues) was a highly effective trigger for relapse following a period of abstinence.

When the THC-seeking procedure was used to evaluate the effects of potential therapeutic treatments, it was found that the CB_1 antagonist/inverse agonist, rimonabant, was highly effective in reducing the drug-seeking response. Importantly, treatment with rimonabant produced an immediate decrease in THC seeking, indicating that rimonabant blocked the ability of the stimulus to maintain THC seeking. This finding is consistent with a number of studies showing that rimonabant can reduce the behavioral effects of stimuli associated with other drugs of abuse, including nicotine, alcohol, cocaine and heroin (Cohen et al. 2005; De Vries and Schoffelmeer 2005; Fattore et al. 2007; Le Foll and Goldberg 2005; Maldonado et al. 2006), as well as the effects of similar cues under second-order food-seeking procedures (Evenden and Ko 2007; Thornton-Jones et al. 2005). Thus, this effect of rimonabant on responding maintained by drug-paired cues appears to be a general effect, unlike its ability to reduce drug-taking behavior, which seems to be limited to specific drugs (De Vries and Schoffelmeer 2005). This suggests that the ability to block both drug seeking (behavior reinforced by drug-related cues) and drug taking (behavior reinforced directly by the drug) might make rimonabant and similar drugs especially useful for treating cannabinoid use disorders.

In contrast with rimonabant, treatment with the opioid antagonist, naltrexone, had a more limited effect under the second-order schedule. In another study by Justinova and colleagues (Justinova et al. 2004), naltrexone produced a partial reduction in THC taking under a fixed-ratio schedule over most of a 5-day course of treatment. However, under the second-order schedule, naltrexone only decreased THC seeking during the first 2 days of treatment. These results might suggest that, like rimonabant, naltrexone can alter both THC seeking and THC taking, but that naltrexone only partially blocks the reinforcing effects of THC. This finding is consistent with the many studies showing functional interactions between the cannabinoid and opioid systems, but it appears that an opioid antagonist alone might not provide significant protection against drug seeking induced by THC-related environmental cues.

During reinstatement testing with the second-order schedule, it was also found that THC seeking was reinstated when the monkeys were passively exposed to THC, AEA, methanandamide, or the AEA transport inhibitor AM404. Also consistent with evidence for functional links between the cannabinoid and opioid systems (see Sect. 3 for more details), passive exposure to morphine reinstated THC seeking. Although it has been shown that passive cannabinoid exposure can reinstate cocaine seeking in rats (De Vries et al. 2001; Xi et al. 2006), cocaine did not reinstate THC seeking in the second-order study. This finding is consistent with those of Spano and colleagues (Spano et al. 2004), who found that the cannabinoid agonist, WIN 55, 212-2, or heroin, reinstated seeking of WIN 55, 212-2 in rats, but cocaine did not.

Rimonabant and naltrexone were tested to determine whether they could block the reinstating effects of passive exposure to THC or morphine. The cannabinoid antagonist/inverse agonist only blocked the effects of the cannabinoid agonist, and the opioid antagonist only blocked the effects of the opioid agonist. These findings contrast with those of Spano and colleagues (Spano et al. 2004) that rimonabant and the opioid antagonist, naloxone, were both capable of preventing WIN 55, 212-2-induced as well as heroin-induced reinstatement of WIN 55, 212-2 seeking in rats. This discrepancy could be due to differences between rats and monkeys, or due to differences between THC and WIN 55, 212-2, which show different profiles of non-cannabinoid receptor binding.

2.2 Conditioned Place Preference and Aversion with Cannabinoids

2.2.1 THC and Synthetic Cannabinoids

An alternative way to assess the rewarding effects of cannabinoids in experimental animals is to study cannabinoid-induced conditioned place preference. Although methodological details differ among laboratories, a typical place-conditioning experiment involves differentially pairing a distinct set of environmental (contextual) cues with the effects of a drug. This occurs in a training chamber with two compartments. During the conditioning procedure, the animal receives the drug in one compartment and receives vehicle in the other. These pairings are repeated several times over a number of days. Following conditioning, a choice test is conducted in which a door is opened between the two compartments, and the animal is allowed unrestricted access to both contexts in the absence of the drug. An increase in time spent in the drug-paired context relative to a control value is taken as evidence that the drug has rewarding effects. On the other hand, a decrease in time spent in the drug-paired context is taken as evidence that the drug has aversive effects.

Unfortunately, the results of conditioned place preference studies with cannabinoid agonists have ranged from positive place preference to no effect to place aversion. THC, as well as synthetic cannabinoid agonists like CP55940 (McGregor et al. 1996), WIN 55, 212-2 (Chaperon et al. 1998) and HU210 (Cheer et al. 2000), can induce conditioned place aversion in rats (Hutcheson et al. 1998; Mallet and Beninger 1998; Parker and Gillies 1995; Sanudo-Pena et al. 1997) and mice (Valjent and Maldonado 2000). THC-induced conditioned place preferences have been reported within limited dose ranges and under restricted experimental conditions in rats and in mice (Braida et al. 2004; Ghozland et al. 2002; Le Foll et al. 2006; Lepore et al. 1995; Valjent and Maldonado 2000). CP55940-induced conditioned place preference have been reported in rats (Braida et al. 2001a). Interestingly, THC microinjections into the VTA or the shell of the nucleus accumbens can produce conditioned place preference in rats (Zangen et al. 2006). Because of the unresolved inconsistencies in this area of research, it is difficult to draw general conclusions on whether cannabinoids have rewarding or aversive effects in this paradigm.

One explanation for this inconsistency might be that THC's rewarding effects in place-conditioning procedures are often masked or reversed by its aversive effects. Differences are reported to exist in the rewarding and aversive effects of cannabinoids in rats and mice in a measure of anxiety. Cannabinoid agonists produced predominantly anxiolytic effects in mice, but predominantly anxiogenic effects in rats (Haller et al. 2007). There also seem to be different mechanisms involved in the THC-induced conditioned place preference compared to aversion. It was found that aversions in mice depend on kappa-opioid receptors (Cheng et al. 2004; Ghozland et al. 2002) and endogenous dynorphin (Zimmer et al. 2001), while preference depends on mu-opioid receptors (Ghozland et al. 2002).

Given the difficulty in obtaining cannabinoid self-administration in rodents, place-preference procedures will likely remain a valuable alternative for studying the abuse-related effects of cannabinoid agonists, despite the fact that results have so far been inconsistent. One approach that does not appear to have been attempted is to use cannabinoid-induced place preference to study reinstatement, as has been done with morphine and other drugs (Parker and Mcdonald 2000).

2.2.2 AEA

There are only two studies to date that evaluated rewarding or aversive effects of AEA in a place-conditioning procedure. First, Mallet and colleagues (Mallet and Beninger 1998) compared effects of THC and AEA. Rats in this study received injections of the potent, but non-selective, FAAH inhibitor phenylmethylsulfonyl fluoride (PMSF) prior to AEA injections in order to prolong its half-life. The study showed that THC, but not AEA, induced significant conditioned place aversion. Second, Scherma and colleagues (Scherma et al. 2008a) found that AEA alone had no effects on place conditioning, but it induced conditioned place aversion when its metabolism was inhibited by the selective FAAH inhibitor URB597, which by itself does not produce conditioned place preference or aversion (Gobbi et al. 2005; Kathuria et al. 2003). The latter study by Scherma and colleagues used intravenous catheters for AEA delivery, while in the former study AEA was injected intraperitoneally. It is possible that, when injected intraperitoneally, AEA availability was not sufficient to produce effects in the place preference procedure because of hepatic first-passage metabolism, which does not favor rapid entry of AEA into the brain.

2.3 Discriminative-Stimulus Effects of Cannabinoids

Drug discrimination is a powerful behavioral assay for discerning similarities and differences among drugs active in the central nervous system (CNS). The subjective and perceptible CNS effects of a compound can be evaluated in this paradigm by training subjects to respond differently when these effects are present versus when they are absent. During drug-discrimination training, the interoceptive effects of a training drug (e.g., THC) are established as a cue for performing a specific operant response (e.g., lever pressing reinforced by food). One of the widely used protocols is the two-lever choice drug-discrimination procedure. Pressing one lever is reinforced during sessions when the training drug has been injected, and pressing on a second lever is reinforced during sessions when vehicle has been injected. Lever choice during test sessions can be used as an indication of whether a novel drug has effects similar to the training drug, or whether a potential therapeutic alters the effects of the training drug (Solinas et al. 2006b). The range of effects measured by drug discrimination is wider than those of direct measures of reward and reinforcement and can include aversive, anxiogenic or anxiolytic effects (Colpaert 1999).

Discriminative-stimulus effects of CB_1 agonists (like THC) in animals show a high degree of pharmacological specificity. Generally, only CB_1 agonists produce discriminative-stimulus effects similar to THC, and only CB_1 antagonists block them (Jarbe et al. 2001; Solinas et al. 2004, 2007c; Wiley et al. 1995a, b). Among non-cannabinoid drugs, only pentobarbital and diazepam have been found to produce partial generalization to a cannabinoid cue (Barrett et al. 1995; Mokler et al. 1986; Wiley and Martin 1999). The effect of diazepam was not blocked by the CB_1 antagonist/inverse agonist rimonabant, suggesting that this effect is mediated by an interaction with the GABAergic system (Wiley and Martin 1999).

Several studies have investigated whether endogenous cannabinoid ligands produce THC-like discriminative stimulus effects when they are systemically administered. AEA does not generally produce THC-like responding in monkeys and rats in drug-discrimination studies or does so only at very high doses that also dramatically depress rates of responding (Burkey and Nation 1997; Jarbe et al. 2001; Wiley et al. 1997, 1998). However, metabolically stable, synthetic analogs of AEA, methanandamide, O1812 and AM1346, did induce THC-like responding (Alici and Appel 2004; Burkey and Nation 1997; Jarbe et al. 2006; Wiley et al. 2004). Thus, AEA's fast metabolic inactivation is likely responsible for its observed weak THC-like discriminative-stimulus effects.

When metabolic inactivation of AEA via FAAH was blocked by the FAAH inhibitor URB597, AEA produced dose-related THC-like discriminative-stimulus effects (Solinas et al. 2007c). URB597 alone did not produce any THC-like effects, even at doses several times higher than those that potentiated the effects of AEA (Gobbi et al. 2005). Another compound interfering with AEA inactivation, AM404, which is thought to inhibit the transport of AEA into neurons, produced no THC-like effects itself, but also did not potentiate the THC-like effects of AEA (Solinas et al. 2007c). These different effects of FAAH blockade and blockade of AEA transport on THC-like discriminative effects of AEA suggest that membrane transport is not the main mechanism for AEA inactivation in the brain regions mediating the discriminative-stimulus effects of THC. Interestingly, nicotine was shown to produce THC-like discriminative effects after FAAH inhibition with URB597 (Solinas et al. 2007b), which implicates nicotine-induced increases in the release of endocannabinoids in another effect observed in the study, the ability of nicotine to potentiate the discriminative effects of THC.

2.4 Tolerance, Physical Dependence and Behavioral Sensitization

2.4.1 Tolerance

The chronic administration of natural or synthetic cannabinoid agonists induces tolerance to most of their pharmacological effects in numerous animal species (Abood and Martin 1992). Tolerance has been shown to develop to the effects of cannabinoids involving antinociception, decreased locomotion, hypothermia and catalepsy, and neuroendocrine effects (Martin 2005), but studies of tolerance to the effects of THC on learning and memory in rats have been contradictory (Delatte et al. 2002; Nava et al. 2001). The development of cannabinoid tolerance is rapid, and a marked decrease of the acute response can sometimes be observed after only the second administration of a cannabinoid agonist (Abood and Martin 1992; Hutcheson et al. 1998). It has been reported that the total number of CB_1-binding sites significantly decreases in several brain areas, including the striatum, cortex, limbic system and cerebellum, during chronic administration of cannabinoids (Rodriguez de Fonseca et al. 1994; Rubino et al. 2000b, c). Also, there are other cellular adaptations observed in some brain regions which play an important role in the induction of synaptic plasticity due to cannabinoid chronic exposure, such as increased activation of the cAMP pathway (Rubino et al. 2000b) and adaptations in the ERK cascade (Rubino et al. 2004, 2005). Together, the downregulation of CB_1 receptors along with the changes in these second messenger systems seems to be responsible for the development of cannabinoid tolerance.

Furthermore, there seems to be a relationship between the status of the CB_1 receptors and the levels of endocannabinoids. In rats tolerant to THC, there are alterations in endocannabinoid content in various brain regions (Martin 2005). Specifically, AEA concentrations were increased in the limbic forebrain and decreased in the striatum, midbrain and diencephalon of THC-tolerant rats (Gonzalez et al. 2004). 2-AG concentrations increased in the cerebellum, brainstem and hippocampus, whereas they decreased only in the striatum. It appears that the most consistent findings with a number of centrally acting drugs of abuse is that chronic administration leads to an elevation in endocannabinoid levels in the limbic system (see Sect. 1.3). This observation is consistent with the notion that endocannabinoids enhance the reinforcing effects of addictive drugs by increasing dopamine release via the inhibition of GABA release in the limbic system (Martin 2005).

Several studies have revealed that cross-tolerance develops for four of the main behavioral/physiological effects of different exogenous CB_1 agonists (analgesia, hypoactivity, catalepsy and hypothermia) (Pertwee et al. 1993). However, there is not always cross-tolerance between AEA and other cannabinoids. For example,

THC and AEA did not show cross-tolerance to hypothermic effects (Pertwee et al. 1993), but did show cross-tolerance to antinociceptive effects (Welch 1997). Cross-tolerance between opioid and cannabinoid compounds has also been revealed. Morphine and THC elicit cross-tolerance to antinociceptive and hypothermic effects in mice (Thorat and Bhargava 1994). On the other hand, AEA-tolerant mice were not cross-tolerant to opioids (Welch 1997). Results such as these probably indicate that tolerance to some effects of AEA involves cannabinoid mechanisms, but tolerance to other effects of AEA does not.

2.4.2 Physical Dependence

Abstinence from cannabis use by chronic users does not produce signs of withdrawal as pronounced as those seen in opioid, ethanol, or barbiturate users. Nonetheless, withdrawal from THC has been reported to induce withdrawal symptoms in both humans (including craving for the drug, decreased appetite, sleep disturbances, anger and aggression (Haney et al. 1999a, b)) and animals (Aceto et al. 1996; Taylor and Fennessy 1982; Verberne et al. 1981). It is likely that the severity of these withdrawal symptoms when use is discontinued is limited by the slow release of THC from its depot in fat tissues, where it is stored due to its highly lipophilic nature. This hypothesis is consistent with the fact that administration of the CB_1 antagonist/inverse agonist rimonabant generally precipitates a pronounced withdrawal syndrome in animals that have been chronically treated with cannabinoids (Aceto et al. 1995; Costa et al. 2000; Hutcheson et al. 1998). There are conflicting reports on the ability of rimonabant to precipitate withdrawal signs in rats chronically treated with AEA, which has a short duration of action (Aceto et al. 1998; Costa et al. 2000).

The behavioral signs of CB_1 antagonist-precipitated cannabinoid withdrawal in rodents include increased grooming, wet-dog shakes, a hunched-back posture, piloerection, body tremors, paw tremors and ptosis. The CB_1 antagonist/inverse agonist rimonabant failed to precipitate behavioral manifestations of abstinence in CB_1 knockout mice given long-term treatment with THC (Ledent et al. 1999), indicating further that somatic signs of abstinence are CB_1-receptor mediated. Microinjection of rimonabant into the cerebellum induced severe manifestations of abstinence in mice dependent on WIN 55, 212-2 (Castane et al. 2004). When the CB_1 antagonist/inverse agonist was administered into the hippocampus and the amygdala, a moderate but significant withdrawal syndrome was also observed. However, no signs of withdrawal were induced when rimonabant was microinjected into the striatum. The cerebellum, and to a lesser extent the hippocampus and the amygdala, participates in the behavioral expression of cannabinoid withdrawal (Castane et al. 2004).

Neurochemical adaptive changes have also been demonstrated during antagonist-precipitated cannabinoid withdrawal in rats and mice, including activation of corticotropin releasing factor (Rodriguez de Fonseca et al. 1997), pronounced increases in the activity of the cAMP pathway in the cerebellum (Hutcheson et al.

1998), and decreases in dopamine transmission in the shell of the nucleus accumbens (Tanda et al. 1999). Some of these signs also occur during withdrawal from other drugs of abuse, such as alcohol (Rossetti et al. 1991), cocaine (Richter et al. 1995) and morphine (Acquas et al. 1991).

Spontaneous cannabinoid withdrawal produced significant time-related alterations in gene transcription (Oliva et al. 2003), such as decreased tyrosine hydroxylase mRNA levels in the ventral tegmental area and increased levels in substantia nigra; increased proenkephalin gene expression in caudate-putamen, nucleus accumbens, olfactory tubercle and piriform cortex; and increased pro-opiomelanocortin gene expression in the arcuate nucleus of the hypothalamus. These alterations induced by spontaneous cannabinoid withdrawal could play a role in the altered vulnerability to other drugs of abuse, as well as in schizoaffective disorders, observed in cannabis users.

2.4.3 Behavioral Sensitization

Behavioral sensitization, an increased response to the drug after repeated exposure, is another adaptive neurobiological alteration that occurs after repeated exposure to drugs. The ability to produce this phenomenon is shared by many drugs abused by humans (e.g., opioids, psychostimulants, nicotine and phencyclidine) and has been proposed to play a role in addiction (Robinson and Berridge 1993, 2001), particularly in drug-seeking behavior persisting long after discontinuation of drug use (De Vries et al. 1998). Repeated exposure to cannabinoid agonists can induce behavioral sensitization (Cadoni et al. 2001; Pontieri et al. 2001b), which is typically observed as an increase in behavioral activity in response to a drug challenge given weeks after the last training injection. However, a recent study (Varvel et al. 2007) was not able to replicate THC-induced behavioral sensitization in rodents under various protocols. Cross-sensitization may occur between cannabinoid agonists and other drugs abused by humans, including heroin (Pontieri et al. 2001a), morphine (Cadoni et al. 2001) and amphetamine (Lamarque et al. 2001).

The adaptive neurobiological changes underlying cannabinoid-induced behavioral sensitization are only beginning to be understood. Altered CB_1 receptor functionality in the striatum and cerebellum of sensitized rats has been observed, as well as lost responsiveness to cannabinoids by the cAMP pathway in the cerebellum (Rubino et al. 2003). In another study (Cadoni et al. 2008), rats pre-exposed to THC showed behavioral sensitization associated with a reduced stimulation of dopamine transmission in the nucleus accumbens shell and an increased stimulation in the nucleus accumbens core in response to THC challenge. Animals pre-treated with morphine showed behavioral sensitization and differential changes in the dopamine response to a THC challenge, with a decreased response in the shell and an increased response in the core. This suggests that THC-induced behavioral sensitization is associated with changes in the responsiveness of dopamine transmission in the nucleus accumbens subdivisions that are similar to those observed with sensitization induced by other drugs of abuse.

3 Opioids

The existence of functional, bidirectional interactions between the endogenous cannabinoid and opioid systems has been demonstrated in numerous studies. Both systems participate in the common circuits involved in the addictive properties of different drugs of abuse. Mu-opioid and CB_1-cannabinoid receptors are both expressed in brain areas involved in reward processes where they share common signaling cascades (Fattore et al. 2005; Maldonado and Rodriguez de Fonseca 2002). The endocannabinoid system is crucial not only for opioid-induced rewarding effects and relapse, but also in the development of physical dependence during chronic opioid administration.

Cross-dependence has been reported between opioid and cannabinoid compounds. In morphine- or methadone-dependent rodents, the opioid antagonist naloxone precipitated a withdrawal syndrome, which was attenuated by THC or AEA (Hine et al. 1975; Lichtman et al. 2001; Vela et al. 1995). Similarly, morphine decreased withdrawal signs in THC-dependent mice undergoing rimonabant-precipitated withdrawal (Lichtman et al. 2001). Furthermore, rimonabant induced behavioral alterations usually associated with opioid withdrawal when given to morphine-dependent rats, and naloxone induced an opioid withdrawal syndrome when given to animals made cannabinoid-dependent by repeated administration of the potent cannabinoid agonist HU210 (Navarro et al. 1998). However, long-term treatment with rimonabant reduced the intensity of naloxone-precipitated withdrawal in morphine-tolerant animals (Rubino et al. 2000a). In CB_1 knockout mice, the severity of naloxone-precipitated morphine withdrawal was robustly attenuated (Ledent et al. 1999). Reciprocally, the expression of cannabinoid withdrawal was decreased in pre-proenkephalin knockout mice compared to wild-type (Valverde et al. 2000). In contrast, rimonabant-precipitated withdrawal in THC-dependent mice was not affected by deletion of mu, kappa, or delta opioid receptors (Ghozland et al. 2002). Another study (Castane et al. 2003) suggested that cooperative actions of both mu and delta receptors were required for the expression of THC dependence.

Studies of rewarding effects of opioids confirm involvement of the endocannabinoid system. In CB_1 knockout mice, morphine did not induce intravenous self-administration (Cossu et al. 2001), but place-conditioning studies show that morphine-induced place preference may or may not develop in these mice dependent on the conditioning paradigm used (Martin et al. 2000; Rice et al. 2002). Rimonabant reduced opioid self-administration and blocked development of heroin-induced conditioned place preference in rodents (De Vries et al. 2003; Navarro et al. 2001). The effects of CB_1 antagonist/inverse agonists like rimonabant appear to be relatively weak when the effort required to obtain heroin is low (fixed-ratio 1 schedules), but become more pronounced when the effort is high (progressive-ratio schedules) (De Vries et al. 2003; Solinas et al. 2003). Furthermore, CB_1 agonists increased the motivation to self-administer heroin under a progressive-ratio schedule (Solinas et al. 2005). On the other hand, opioid antagonists can block cannabinoid-induced place preference or cannabinoid self-administration in rodents and primates (Braida et al. 2001b; Justinova et al. 2004). Deletion of mu-opioid receptors in mice abolished THC place preference, and deletion of kappa-opioid receptors abolished THC place aversion, while unmasking THC place preference (Ghozland et al. 2002). This suggests an opposing activity of mu- and kappa-opioid receptors in modulating reward pathways.

The role of the endocannabinoid system in relapse to opioid use has also been established. Blockade of CB_1 receptors can prevent heroin-induced reinstatement of heroin-seeking behavior after a long period of extinction, and CB_1 agonists can reinstate heroin-seeking behavior in rats (De Vries et al. 2003; Fattore et al. 2003; Solinas et al. 2003). Rimonabant can also block cue-induced heroin seeking in rats (De Vries et al. 2003). On the other hand, heroin reinstated cannabinoid-seeking behavior after a long period of abstinence, and this effect was blocked by rimonabant (Spano et al. 2004). In the same study, naloxone blocked heroin-induced cannabinoid-seeking behavior, which further supports the existence of bidirectional opioid–cannabinoid interactions in the central mechanisms underlying relapse. However, in squirrel monkeys, morphine-induced reinstatement of THC seeking under a second-order schedule was not blocked by rimonabant, and THC-induced reinstatement was not blocked by naltrexone (details in Sect. 2.1.3).

Both opioids' and cannabinoids' rewarding effects are related to their facilitatory effects on mesolimbic dopamine transmission. Heroin or morphine-induced activation of dopamine transmission in the nucleus accumbens does not appear to be mediated by CB_1 receptors, because rimonabant does not block this effect (Caille and Parsons 2003; Tanda et al. 1997) and CB_1 knockout mice show normal accumbal morphine-induced dopamine elevations (Mascia et al. 1999). Naloxone, on the other hand, prevented the cannabinoid-induced dopamine elevations in the same area (Tanda et al. 1997).

4 Alcohol

The endogenous cannabinoid system is involved in both the rewarding effects of alcohol and in relapse to alcohol abuse (Vengeliene et al. 2008). The endocannabinoid system seems to participate in alcohol's rewarding properties by modulating its effects on activation of mesolimbic dopamine transmission. Pharmacological blockade of CB_1 receptors blocks dopamine-releasing effects of alcohol, and alcohol did not increase extracellular levels of dopamine in the nucleus accumbens of CB_1 knockout mice (Cohen et al. 2002; Hungund et al. 2003). Alcohol acutely inhibits endocannabinoid transmission (Ferrer et al. 2007), which in turn leads to above normal endocannabinoid transmission in reward-related brain areas during chronic alcohol administration, as was revealed by the downregulation of CB_1 receptors and by increased levels of AEA and 2-AG (Hungund and Basavarajappa 2004).

Pharmacological manipulations of the CB_1 receptors showed that, generally, CB_1 agonists increase (Colombo et al. 2002) and CB_1 antagonist/inverse agonists decrease rodents' oral alcohol consumption in self-administration studies (Arnone et al. 1997; Cippitelli et al. 2005). Although CB_1 receptor blockade can cause suppression of fluid and food intake (McGregor and Gallate 2004), CB_1 antagonist/inverse agonists were still found to decrease alcohol's rewarding effects when this confounding factor was controlled in place-conditioning procedures by giving alcohol intraperitoneally to bypass the oral route of administration (Gessa et al. 2005; Lallemand and De Witte 2006). Moreover, genetic manipulations of the CB_1 receptor confirmed that rewarding effects of ethanol require CB_1 receptor activation, since knockout mice consumed less alcohol in most studies (Crabbe et al. 2006) and did not develop place preference for an alcohol-paired environment (Thanos et al. 2005).

Exposure to the CB_1 agonists WIN 55, 212-2 or THC promotes relapse to alcohol use in abstinent rats (Lopez-Moreno et al. 2004; McGregor et al. 2005), and the CB_1 antagonist/inverse agonist rimonabant blocks cue-induced relapse to ethanol seeking (Cippitelli et al. 2005). The latter study also showed that in a strain of rats bred for its ethanol preference (alcohol-preferring Marchigian Sardinian – msP rats), there is increased CB_1 receptor mRNA expression in brain areas relevant for the processing of reward and reward-associated behaviors. This suggests that altered function of the CB_1 receptor system may be linked to genetic vulnerability to alcohol misuse. In fact, it has recently been reported (Zuo et al. 2007) in a large case-controlled sample that the human CB_1 receptor, which is encoded by the CNR1 gene, may play a role in the development of alcoholism.

There is also a question of whether increased AEA levels in the brain contribute to sustained high levels of alcohol drinking or facilitate relapse to alcohol seeking. Studies in rodents have yielded a spectrum of results so far. One study showed that chronic alcohol-induced increases in extracellular AEA were due to inhibition of AEA transport, but not FAAH, in cerebellar granular neurons of mice (Basavarajappa et al. 2003). Blockade of AEA transport by AM404 in Wistar rats reduced alcohol self-administration, but did not affect the relapse induced by contextual cues associated with ethanol (Cippitelli et al. 2007). Genetic ablation of FAAH in mice resulted in increased alcohol preference and intake (Blednov et al. 2007). Pharmacological inhibition of FAAH by URB597 produced increased alcohol intake in wild-type mice (Blednov et al. 2007), but had no effect on alcohol intake in Wistar or msP rats (Cippitelli et al. 2008). In the latter study, URB597, like AM404, did not affect relapse to alcohol seeking induced by either cues or stress. The lack of effect of AM404 and URB597 on relapse to alcohol seeking suggests the absence of a primary role of AEA in the regulation of alcohol-ingestive behaviors in the rat.

5 Nicotine

The endocannabinoid system is critically involved in the addictive effects of nicotine. Preclinical evidence clearly implicates CB_1 receptors in nicotine addiction, which has led to clinical trials indicating that CB_1 receptor antagonists (rimonabant) could be useful as therapeutic agents for smoking cessation (Fernandez and Allison 2004). Rimonabant was shown to block nicotine-induced conditioned place preference, nicotine self-administration, cue-induced reinstatement of nicotine seeking, as well as nicotine-induced dopamine release in the nucleus accumbens shell in rats (Cohen et al. 2002; Cohen et al. 2005; De Vries and Schoffelmeer 2005; Le Foll and Goldberg 2004). CB_1 knockout mice did not develop nicotine-induced place preference, but they self-administered nicotine similarly to the wild-type (Castane et al. 2002; Cossu et al. 2001; Merritt et al. 2008). Genetic deletion or pharmacological inhibition of FAAH by URB597 enhanced the expression of nicotine-induced place preference in mice (Merritt et al. 2008). In contrast, in rats pharmacological inhibition of FAAH by URB597 markedly inhibited the development of nicotine-induced place preference, reduced nicotine-induced reinstatement of drug seeking and reduced nicotine-induced dopamine elevations in the nucleus accumbens shell (Scherma et al. 2008b). Also in rats, inhibition of FAAH by URB597 prevented nicotine-induced activation of dopaminergic neurons in the VTA (Pistis et al. 2008). These results point to drugs that inhibit FAAH as potentially useful agents in the treatment of tobacco dependence in humans.

Interactions between nicotine and the endocannabinoid system may underlie the widespread practice of cannabis and tobacco co-administration in humans. For example, in place conditioning procedures, sub-threshold doses of nicotine and THC produced place preference when given in combination (Valjent et al. 2002). Also, nicotine potentiates the discriminative-stimulus effects of low doses of THC, and this effect is mediated in part by the release of AEA (Solinas et al. 2007b). It was further shown that systemic administration of the 7-nicotinic acetylcholine receptor (nACh) antagonist methyllycaconitine significantly reduced not only the discriminative effects of THC and WIN 55, 212-2 and the self-administration of WIN 55, 212-2, but also the ability of THC to increase dopamine levels in the nucleus accumbens shell (Solinas et al. 2007a). These findings suggest that drugs that block 7-nACh receptors can counteract the addictive properties of THC and may be potentially useful agents in the treatment of cannabis abuse in humans.

6 Psychostimulants

The mechanism of action of psychostimulants differs from that of other drugs of abuse. Psychostimulants enhance the activity of dopaminergic neurons by directly acting on the reuptake of monoamines binding to one or multiple monoamine transporters (Rothman and Baumann 2003). There are two primary mechanisms by which psychostimulants affect the dopamine transporter (DAT), but the end result is to inhibit the elimination of dopamine from the synapse and therefore increase the quantity and half-life of synaptic and extrasynaptic dopamine levels (Kalivas 2007). Psychostimulants can be separated into "uptake blockers" (cocaine and methylphenidate) and "releasers" (amphetamines) based on the mechanism of their acute effect on neurotransmitter flux through the DAT. Cocaine binds to DAT, but is not transported into the presynaptic terminal as surrogate dopamine. Amphetamines also bind to DAT, but also translocate into the cell in place of dopamine and enter the dopamine synaptic vesicles. This causes a large buildup of dopamine in the cytosol and reversal of the direction of DAT to release dopamine into the extracellular space. The general separation of drugs into these two classes helps to functionally distinguish the pharmacological profiles of some of the most commonly used psychostimulants. For example, uptake blockers cause little or no persistent dopamine deficits, whereas releasers can cause persistent deficits in monoaminergic neurons (Riddle et al. 2005).

6.1 Cocaine and Methylphenidate

6.1.1 Cocaine

Results of many preclinical studies indicate that CB_1 receptors are not involved in the primary reinforcing effects of cocaine. For example, the ability to self-administer cocaine was unaffected in CB_1 knockout mice, as was development of cocaine-induced place preference (Cossu et al. 2001; Martin et al. 2000). Blockade of CB_1 receptors by rimonabant did not interfere with cocaine self-administration in mice, rats or monkeys (De Vries et al. 2001; Lesscher et al. 2005; Tanda et al. 2000). However, there are also contrasting reports, such as the demonstration that rimonabant can affect acquisition of cocaine-induced conditioned place preference (Chaperon et al. 1998). Another report (Soria et al. 2005) showed reduced acquisition of cocaine self-administration in CB_1 knockout mice and that the maximal effort to obtain cocaine (break-point under a progressive-ratio schedule) was also significantly reduced in CB_1 knockout mice or after CB_1-receptor blockade in wild-type mice. In the same study, acute cocaine administration induced a similar enhancement in extracellular levels of dopamine in the nucleus accumbens of both CB_1 knockout and wild-type mice. This impairment in cocaine self-administration indicates decreased motivation for cocaine-seeking behavior, suggesting a role for CB_1 receptors in consolidation of the cocaine addictive process, but not in its acute effects on mesolimbic dopaminergic transmission (Maldonado et al. 2006).

The endocannabinoid system does appear to be capable of influencing the reinstatement of extinguished cocaine self-administration behavior, since CB_1 agonists can induce reinstatement of cocaine seeking (De Vries et al. 2001; Spano et al. 2004) (see Sect. 1.4 for details). Cocaine, on the other hand, does not reinstate extinguished cannabinoid-seeking behavior (Justinova et al. 2008b; Spano et al. 2004). Recent evidence shows that acute cocaine administration could alter synaptic plasticity in the brain reward system (i.e., nucleus accumbens) by abolishing a retrograde long-term depression (LTD) mediated by endocannabinoids (Fourgeaud et al. 2004). Behavioral sensitization to cocaine is accompanied by a decrease in excitatory drive to the nucleus accumbens (Thomas et al. 2001) and a reduction of basal extracellular glutamate in the nucleus accumbens (Pierce et al.

1996). Thus the abolition of endocannabinoid-mediated LTD in the nucleus accumbens of cocaine-exposed animals might serve as a compensatory mechanism to counterbalance the general decrease in glutamatergic activity measured in response to cocaine (Fourgeaud et al. 2004). Although the endocannabinoid system does not appear to participate in the primary reinforcing effects of cocaine, it is important for maintaining cocaine-seeking behavior, probably by modulating synaptic processes induced by cocaine (Maldonado et al. 2006).

6.1.2 Methylphenidate

Brain dopaminergic and noradrenergic systems play an important role in impulsive behavior, which is manifested at pathological levels in attention-deficit/hyperactivity disorder (ADHD), for which methylphenidate shows therapeutic efficacy. Impulsivity also plays a crucial role in drug addiction, and prolonged drug intake produces disturbances in inhibition of behavior that might contribute to the compulsivity associated with addiction (Jentsch and Taylor 1999). This hypothesis, that drug addiction and impulsivity are strongly interrelated, has been supported by several recent studies in both humans and laboratory animals demonstrating that elevated impulsivity might predispose individuals to initiate or maintain drug seeking and taking (Pattij and Vanderschuren 2008).

The endocannabinoid system, and particularly CB_1 receptors, has been implicated in higher cognitive functions including attention. In healthy volunteers, marijuana and THC have been demonstrated to increase the occurrence of risk-taking behavior in the laboratory and induce impulsive action in a stop signal task, but not delay aversion (McDonald et al. 2003; Ramaekers et al. 2006), which suggests a role for the cannabinoid system in impulsivity. A recent study provided evidence for a differential involvement of the endocannabinoid system in independent measures of impulsivity, as the CB_1 antagonist/inverse agonist rimonabant primarily affected inhibitory control, and did not affect either impulsive choice nor response inhibition, whereas the CB_1 agonist WIN 55, 212-2 only slightly affected response inhibition (Pattij et al. 2007).

6.2 Amphetamine, Methamphetamine and 3,4-Methylendioxymethamphetamine (MDMA)

Dopamine–endocannabinoid interactions have been suggested to be important for the development of amphetamine-induced behavioral sensitization. AEA and 2-AG are differentially modulated by dopamine, via activation of D_1 and D_2 receptors (Patel et al. 2003), which play a significant role in the induction and expression of amphetamine sensitization. Repeated exposure to THC can induce behavioral sensitization not only to cannabinoids, but also to psychostimulants, including amphetamine (Gorriti et al. 1999; Lamarque et al. 2001). In line with this finding is the report that CB_1 knockout mice failed to sensitize to the locomotor stimulant effects of amphetamine (Thiemann et al. 2008). Furthermore, amphetamine-sensitized wild-type animals in that study had decreased levels of AEA and 2-AG in the ventral striatum (which contains the nucleus accumbens). It seems that amphetamine, which directly increases dopamine activity, can trigger a compensatory reduction in cannabinoid levels, most likely via *trans*-synaptic mechanisms within mesolimbic circuitry (van der Stelt and Di Marzo 2003)). However, amphetamine also releases endocannabinoids in rat amygdala, producing LTD by a dopamine-independent mechanism mediated by CB_1 receptors (Huang et al. 2003), and these endocannabinoids participate in the synaptic plasticity produced by amphetamine in mesocorticolimbic structures (Wolf et al. 2004).

Studies investigating involvement of the endocannabinoid system in the reinforcing effects of amphetamines show conflicting results. Amphetamine is self-administered in CB_1 knockout mice (Cossu et al. 2001). On the other hand, the CB_1 antagonist/inverse agonist AM251 decreased and AEA and methanandamide increased methamphetamine self-administration under a fixed-ratio schedule in rats (Vinklerova et al. 2002). Rimonabant was also shown to block methamphetamine- and cue-induced reinstatement of methamphetamine-seeking behavior in rats (Anggadiredja et al. 2004). Studies with MDMA showed contradictory effects as well. Blockade of CB_1 receptors antagonized MDMA-induced place preference (Braida et al. 2005), but increased intracerebroventricular self-administration of MDMA (Braida et al. 2004). The increase in operant responding induced by rimonabant indicates a decreased motivation to self-administer amphetamine and its derivatives, suggesting that the endocannabinoid system influences the mechanisms regulating MDMA's reinforcing effects (Sala and Braida 2005).

It is important to note that, as with alcohol, marijuana, and heroin, a human genetic variant of the cannabinoid CB_1 receptor gene *CNR1* has been associated with susceptibility to cocaine and amphetamine dependence (Ballon et al. 2006; Comings et al. 1997; Zhang et al. 2004).

7 Endocannabinoid System and Treatment of Drug Addiction

As can be seen by the large number of studies in this area in recent years, the role of the endocannabinoid system in drug abuse and addiction is the focus of intense activity. This interest is generated for several important reasons. Endocannabinoids appear to modulate the direct reinforcing effects of many drugs, the ability of these drugs to induce relapse, and perhaps most interestingly, the ability of drug-related cues to induce relapse. The abuse of cannabis itself is a widespread phenomenon, and large numbers of people seek treatment for cannabis dependence each year. Cannabinoid antagonists represent a unique approach to the treatment of substance abuse (including obesity and addiction to both licit and illicit drugs). Along with replacement therapy (e.g., methadone, nicotine replacement), aversion therapy (e.g., Antabuse), and antagonist or mixed agonist therapies that are specific for opioid addiction (e.g., naltrexone and buprenorphine, respectively), manipulations of the endocannabinoid system offer one of the very few kinds of pharmacotherapeutic treatments that have shown promise for treating addiction. Among these treatments, cannabinoid-based therapies may be the only ones with the potential to target addiction and relapse, per se, as opposed to targeting the abuse of a single substance. Unfortunately, the recent rejection of the CB_1 antagonist/inverse agonist rimonabant as an aid to smoking cessation by the FDA indicates that the search for a cannabinoid-related treatment for addiction is just beginning. Recently developed neutral antagonists that in animals appear to lack the unwanted side effects of CB_1 antagonist/inverse agonists such as rimonabant (details in Sect. 4), as well as drugs such as FAAH inhibitors that alter endocannabinoid signaling, are two examples of potentially useful approaches to cannabinoid-related treatment of addiction. As our understanding of the endocannabinoid system rapidly increases, it is hoped that the promise of safe and effective therapies based on this system will soon be realized.

Acknowledgements The preparation of this manuscript was supported in part by the Intramural Research Program of the National Institute on Drug Abuse, National Institutes of Health, Department of Health and Human Services and Maryland Psychiatric Research Center, Department of Psychiatry, University of Maryland School of Medicine, Baltimore, Maryland.

References

Abood ME, Martin BR (1992) Neurobiology of marijuana abuse. Trends Pharmacol Sci 13:201–206

Aceto MD, Scates SM, Lowe JA et al. (1995) Cannabinoid precipitated withdrawal by the selective cannabinoid receptor antagonist, SR 141716A. Eur J Pharmacol 282:R1–R2

Aceto MD, Scates SM, Lowe JA et al. (1996) Dependence on delta 9-tetrahydrocannabinol: studies on precipitated and abrupt withdrawal. J Pharmacol Exp Ther 278:1290–1295

Aceto MD, Scates SM, Razdan RK et al. (1998) Anandamide, an endogenous cannabinoid, has a very low physical dependence potential. J Pharmacol Exp Ther 287:598–605

Acquas E, Carboni E, Di Chiara G (1991) Profound depression of mesolimbic dopamine release after morphine withdrawal in dependent rats. Eur J Pharmacol 193:133–134

Agurell S, Halldin M, Lindgren JE et al. (1986) Pharmacokinetics and metabolism of delta 1-tetrahydrocannabinol and other cannabinoids with emphasis on man. Pharmacol Rev 38:21–43

Alici T, Appel JB (2004) Increasing the selectivity of the discriminative stimulus effects of delta 9-tetrahydrocannabinol: complete substitution with methanandamide. Pharmacol Biochem Behav 79:431–437

American Psychiatric Association (1994) DSM-IV: diagnostic and statistical manual of mental disorders, 4th edn. APA, Washington DC

Anggadiredja K, Nakamichi M, Hiranita T et al. (2004) Endocannabinoid system modulates relapse to methamphetamine seeking: possible mediation by the arachidonic acid cascade. Neuropsychopharmacology 29:1470–1478

Arnold JM, Roberts DC (1997) A critique of fixed and progressive ratio schedules used to examine the neural substrates of drug reinforcement. Pharmacol Biochem Behav 57:441–447

Arnone M, Maruani J, Chaperon F et al. (1997) Selective inhibition of sucrose and ethanol intake by SR 141,716, an antagonist of central cannabinoid (CB1) receptors. Psychopharmacology (Berl) 132:104–106

Arroyo M, Markou A, Robbins TW et al. (1998) Acquisition, maintenance and reinstatement of intravenous cocaine self-administration under a second-order schedule of reinforcement in rats: effects of conditioned cues and continuous access to cocaine. Psychopharmacology (Berl) 140:331–344
Ballon N, Leroy S, Roy C et al. (2006) (AAT)n repeat in the cannabinoid receptor gene (CNR1): association with cocaine addiction in an African–Caribbean population. Pharmacogenomics J 6:126–130
Barrett RL, Wiley JL, Balster RL et al. (1995) Pharmacological specificity of delta 9-tetrahydrocannabinol discrimination in rats. Psychopharmacology (Berl) 118:419–424
Basavarajappa BS, Saito M, Cooper TB et al. (2003) Chronic ethanol inhibits the anandamide transport and increases extracellular anandamide levels in cerebellar granule neurons. Eur J Pharmacol 466:73–83
Bazinet RP, Lee HJ, Felder CC et al. (2005) Rapid high-energy microwave fixation is required to determine the anandamide (N-arachidonoylethanolamine) concentration of rat brain. Neurochem Res 30:597–601
Bequet F, Uzabiaga F, Desbazeille M et al. (2007) CB1 receptor-mediated control of the release of endocannabinoids (as assessed by microdialysis coupled with LC/MS) in the rat hypothalamus. Eur J Neurosci 26:3458–3464
Bergman J, Johanson CE (1985) The reinforcing properties of diazepam under several conditions in the rhesus monkey. Psychopharmacology (Berl) 86:108–113
Blednov YA, Cravatt BF, Boehm SL et al. (2007) Role of endocannabinoids in alcohol consumption and intoxication: studies of mice lacking fatty acid amide hydrolase. Neuropsychopharmacology 32:1570–1582
Bortolato M, Campolongo P, Mangieri RA et al. (2006) Anxiolytic-like properties of the anandamide transport inhibitor AM404. Neuropsychopharmacology 31:2652–2659
Braida D, Pozzi M, Cavallini R et al. (2001a) Conditioned place preference induced by the cannabinoid agonist CP 55,940: interaction with the opioid system. Neuroscience 104:923–926
Braida D, Pozzi M, Parolaro D et al. (2001b) Intracerebral self-administration of the cannabinoid receptor agonist CP 55,940 in the rat: interaction with the opioid system. Eur J Pharmacol 413:227–234
Braida D, Iosue S, Pegorini S et al. (2004) Delta9-tetrahydrocannabinol-induced conditioned place preference and intracerebroventricular self-administration in rats. Eur J Pharmacol 506:63–69
Braida D, Iosue S, Pegorini S et al. (2005) 3, 4 Methylenedioxymethamphetamine-induced conditioned place preference (CPP) is mediated by endocannabinoid system. Pharmacol Res 51:177–182
Branch MN, Dearing ME, Lee DM (1980) Acute and chronic effects of delta 9-tetrahydrocannabinol on complex behavior of squirrel monkeys. Psychopharmacology (Berl) 71:247–256
Brodie MS, Pesold C, Appel SB (1999) Ethanol directly excites dopaminergic ventral tegmental area reward neurons. Alcohol Clin Exp Res 23:1848–1852
Burkey RT, Nation JR (1997) (R)-methanandamide, but not anandamide, substitutes for delta 9-THC in a drug-discrimination procedure. Exp Clin Psychopharmacol 5:195–202
Cadoni C, Pisanu A, Solinas M et al. (2001) Behavioural sensitization after repeated exposure to Delta 9-tetrahydrocannabinol and cross-sensitization with morphine. Psychopharmacology (Berl) 158:259–266
Cadoni C, Valentini V, Di Chiara G (2008) Behavioral sensitization to Delta(9)-tetrahydrocannabinol and cross-sensitization with morphine: differential changes in accumbal shell and core dopamine transmission. J Neurochem 106:1586–1593
Caille S, Parsons LH (2003) SR141716A reduces the reinforcing properties of heroin but not heroin-induced increases in nucleus accumbens dopamine in rats. Eur J Neurosci 18:3145–3149
Caille S, Varez-Jaimes L, Polis I et al. (2007) Specific alterations of extracellular endocannabinoid levels in the ucleus accumbens by ethanol, heroin, and cocaine self-administration. J Neurosci 27:3695–3702

Cami J, Farre M (2003) Drug addiction. N Engl J Med 349:975–986
Carroll ME, Meisch RA (1984) Increased drug-reinforced behavior due to food deprivation. In: Thompson T, Dews PB, Barrett JE (eds) Advances in behavioral pharmacology. Academic Press, New York, pp 47–88
Castane A, Valjent E, Ledent C et al. (2002) Lack of CB1 cannabinoid receptors modifies nicotine behavioural responses, but not nicotine abstinence. Neuropharmacology 43:857–867
Castane A, Robledo P, Matifas A et al. (2003) Cannabinoid withdrawal syndrome is reduced in double mu and delta opioid receptor knockout mice. Eur J Neurosci 17:155–159
Castane A, Maldonado R, Valverde O (2004) Role of different brain structures in the behavioural expression of WIN 55,212–2 withdrawal in mice. Br J Pharmacol 142:1309–1317
CEWG (2007) Epidemiologic trends in drug abuse. Proceedings of the Community Epidemiology Work Group. Highlights and executive summary. National Institute on Drug Abuse, NIH Publication No. 07-6,200, Bethesda, MD
Chaperon F, Thiebot MH (1999) Behavioral effects of cannabinoid agents in animals. Crit Rev Neurobiol 13:243–281
Chaperon F, Soubrie P, Puech AJ et al. (1998) Involvement of central cannabinoid (CB1) receptors in the establishment of place conditioning in rats. Psychopharmacology (Berl) 135:324–332
Cheer JF, Kendall DA, Marsden CA (2000) Cannabinoid receptors and reward in the rat: a conditioned place preference study. Psychopharmacology (Berl) 151:25–30
Chen JP, Paredes W, Li J et al. (1990) Delta 9-tetrahydrocannabinol produces naloxone-blockable enhancement of presynaptic basal dopamine efflux in nucleus accumbens of conscious, freely-moving rats as measured by intracerebral microdialysis. Psychopharmacology (Berl) 102: 156–162
Cheng HY, Laviolette SR, van der Kooy D et al. (2004) DREAM ablation selectively alters THC place aversion and analgesia but leaves intact the motivational and analgesic effects of morphine. Eur J Neurosci 19:3033–3041
Cippitelli A, Bilbao A, Hansson AC et al. (2005) Cannabinoid CB1 receptor antagonism reduces conditioned reinstatement of ethanol-seeking behavior in rats. Eur J Neurosci 21:2243–2251
Cippitelli A, Bilbao A, Gorriti MA et al. (2007) The anandamide transport inhibitor AM404 reduces ethanol self-administration. Eur J Neurosci 26:476–486
Cippitelli A, Cannella N, Braconi S et al. (2008) Increase of brain endocannabinoid anandamide levels by FAAH inhibition and alcohol abuse behaviours in the rat. Psychopharmacology (Berl) 198:449–460
Cohen C, Perrault G, Voltz C et al. (2002) SR141716, a central cannabinoid (CB(1)) receptor antagonist, blocks the motivational and dopamine-releasing effects of nicotine in rats. Behav Pharmacol 13:451–463
Cohen C, Kodas E, Griebel G (2005) CB1 receptor antagonists for the treatment of nicotine addiction. Pharmacol Biochem Behav 81:387–395
Collins RJ, Weeks JR, Cooper MM et al. (1984) Prediction of abuse liability of drugs using IV self-administration by rats. Psychopharmacology (Berl) 82:6–13
Colombo G, Serra S, Brunetti G et al. (2002) Stimulation of voluntary ethanol intake by cannabinoid receptor agonists in ethanol-preferring sP rats. Psychopharmacology (Berl) 159:181–187
Colpaert FC (1999) Drug discrimination in neurobiology. Pharmacol Biochem Behav 64:337–345
Comings DE, Muhleman D, Gade R et al. (1997) Cannabinoid receptor gene (CNR1): association with i.v. drug use. Mol Psychiatry 2:161–168
Cossu G, Ledent C, Fattore L et al. (2001) Cannabinoid CB1 receptor knockout mice fail to self-administer morphine but not other drugs of abuse. Behav Brain Res 118:61–65
Costa B, Giagnoni G, Colleoni M (2000) Precipitated and spontaneous withdrawal in rats tolerant to anandamide. Psychopharmacology (Berl) 149:121–128
Crabbe JC, Phillips TJ, Harris RA et al. (2006) Alcohol-related genes: contributions from studies with genetically engineered mice. Addict Biol 11:195–269
Delatte MS, Winsauer PJ, Moerschbaecher JM (2002) Tolerance to the disruptive effects of Delta (9)-THC on learning in rats. Pharmacol Biochem Behav 74:129–140

De Vries TJ, Schoffelmeer AN (2005) Cannabinoid CB1 receptors control conditioned drug seeking. Trends Pharmacol Sci 26:420–426
De Vries TJ, Schoffelmeer AN, Binnekade R et al. (1998) Drug-induced reinstatement of heroin- and cocaine-seeking behaviour following long-term extinction is associated with expression of behavioural sensitization. Eur J Neurosci 10:3565–3571
De Vries TJ, Shaham Y, Homberg JR et al. (2001) A cannabinoid mechanism in relapse to cocaine seeking. Nat Med 7:1151–1154
De Vries TJ, Homberg JR, Binnekade R et al. (2003) Cannabinoid modulation of the reinforcing and motivational properties of heroin and heroin-associated cues in rats. Psychopharmacology (Berl) 168:164–169
Dewey WL (1986) Cannabinoid pharmacology. Pharmacol Rev 38:151–178
Di Chiara G, Tanda G, Bassareo V et al. (1999) Drug addiction as a disorder of associative learning. Role of nucleus accumbens shell/extended amygdala dopamine. Ann N Y Acad Sci 877:461–485
Di Marzo V, Berrendero F, Bisogno T et al. (2000) Enhancement of anandamide formation in the limbic forebrain and reduction of endocannabinoid contents in the striatum of delta9-tetrahydrocannabinol-tolerant rats. J Neurochem 74:1627–1635
Evenden J, Ko T (2007) The effects of anorexic drugs on free-fed rats responding under a second-order FI15-min (FR10:S) schedule for high incentive foods. Behav Pharmacol 18:61–69
Everitt BJ, Robbins TW (2000) Second-order schedules of drug reinforcement in rats and monkeys: measurement of reinforcing efficacy and drug-seeking behaviour. Psychopharmacology (Berl) 153:17–30
Fattore L, Cossu G, Martellotta CM et al. (2001) Intravenous self-administration of the cannabinoid CB1 receptor agonist WIN 55, 212-2 in rats. Psychopharmacology (Berl) 156:410–416
Fattore L, Spano MS, Cossu G et al. (2003) Cannabinoid mechanism in reinstatement of heroin-seeking after a long period of abstinence in rats. Eur J Neurosci 17:1723–1726
Fattore L, Deiana S, Spano SM et al. (2005) Endocannabinoid system and opioid addiction: behavioural aspects. Pharmacol Biochem Behav 81:343–359
Fattore L, Fadda P, Fratta W (2007) Endocannabinoid regulation of relapse mechanisms. Pharmacol Res 56:418–427
Fernandez JR, Allison DB (2004) Rimonabants sanofi-synthelabo. Curr Opin Invest Drugs 5:430–435
Ferrer B, Bermudez-Silva FJ, Bilbao A et al. (2007) Regulation of brain anandamide by acute administration of ethanol. Biochem J 404:97–104
Fourgeaud L, Mato S, Bouchet D et al. (2004) A single in vivo exposure to cocaine abolishes endocannabinoid-mediated long-term depression in the nucleus accumbens. J Neurosci 24:6939–6945
Fusco FR, Martorana A, Giampa C et al. (2004) Immunolocalization of CB1 receptor in rat striatal neurons: a confocal microscopy study. Synapse 53:159–167
Gessa GL, Serra S, Vacca G et al. (2005) Suppressing effect of the cannabinoid CB1 receptor antagonist, SR147778, on alcohol intake and motivational properties of alcohol in alcohol-preferring sP rats. Alcohol Alcohol 40:46–53
Ghozland S, Matthes HW, Simonin F et al. (2002) Motivational effects of cannabinoids are mediated by mu-opioid and kappa-opioid receptors. J Neurosci 22:1146–1154
Giuffrida A, Parsons LH, Kerr TM et al. (1999) Dopamine activation of endogenous cannabinoid signaling in dorsal striatum. Nat Neurosci 2:358–363
Gobbi G, Bambico FR, Mangieri R et al. (2005) Antidepressant-like activity and modulation of brain monoaminergic transmission by blockade of anandamide hydrolysis. Proc Natl Acad Sci USA 102:18620–18625
Goldberg SR, Hoffmeister F, Schlichting UU et al. (1971) A comparison of pentobarbital and cocaine self-administration in rhesus monkeys: effects of dose and fixed-ratio parameter. J Pharmacol Exp Ther 179:277–283
Goldberg SR, Kelleher RT, Morse WH (1975) Second-order schedules of drug injection. Fed Proc 34:1771–1776

Goldberg SR, Spealman RD, Goldberg DM (1981) Persistent behavior at high rates maintained by intravenous self-administration of nicotine. Science 214:573–575
Gonzalez S, Cascio MG, Fernandez-Ruiz J et al. (2002) Changes in endocannabinoid contents in the brain of rats chronically exposed to nicotine, ethanol or cocaine. Brain Res 954:73–81
Gonzalez S, Schmid PC, Fernandez-Ruiz J et al. (2003) Region-dependent changes in endocannabinoid transmission in the brain of morphine-dependent rats. Addict Biol 8:159–166
Gonzalez S, Fernandez-Ruiz J, Di Marzo V et al. (2004) Behavioral and molecular changes elicited by acute administration of SR141716 to Delta9-tetrahydrocannabinol-tolerant rats: an experimental model of cannabinoid abstinence. Drug Alcohol Depend 74:159–170
Gorriti MA, de Rodriguez FF, Navarro M et al. (1999) Chronic (-)-delta9-tetrahydrocannabinol treatment induces sensitization to the psychomotor effects of amphetamine in rats. Eur J Pharmacol 365:133–142
Haller J, Matyas F, Soproni K et al. (2007) Correlated species differences in the effects of cannabinoid ligands on anxiety and on GABAergic and glutamatergic synaptic transmission. Eur J Neurosci 25:2445–2456
Haney M, Comer SD, Ward AS et al. (1997) Factors influencing marijuana self-administration by humans. Behav Pharmacol 8:101–112
Haney M, Ward AS, Comer SD et al. (1999a) Abstinence symptoms following oral THC administration to humans. Psychopharmacology (Berl) 141:385–394
Haney M, Ward AS, Comer SD et al. (1999b) Abstinence symptoms following smoked marijuana in humans. Psychopharmacology (Berl) 141:395–404
Harris RT, Waters W, McLendon D (1974) Evaluation of reinforcing capability of delta-9-tetrahydrocannabinol in rhesus monkeys. Psychopharmacologia 37:23–29
Hine B, Torrelio M, Gershon S (1975) Attenuation of precipitated abstinence in methadone-dependent rats by delta 9-THC. Psychopharmacol Commun 1:275–283
HODOS W (1961) Progressive ratio as a measure of reward strength. Science 134:943–944
Hohmann AG, Herkenham M (2000) Localization of cannabinoid CB(1) receptor mRNA in neuronal subpopulations of rat striatum: a double-label in situ hybridization study. Synapse 37:71–80
Hollister LE (1986) Health aspects of cannabis. Pharmacol Rev 38:1–20
Huang YC, Wang SJ, Chiou LC et al. (2003) Mediation of amphetamine-induced long-term depression of synaptic transmission by CB1 cannabinoid receptors in the rat amygdala. J Neurosci 23:10311–10320
Huestis MA, Gorelick DA, Heishman SJ et al. (2001) Blockade of effects of smoked marijuana by he CB1-selective cannabinoid receptor antagonist SR141716. Arch Gen Psychiatry 58:322–328
Hungund BL, Basavarajappa BS (2004) Role of endocannabinoids and cannabinoid CB1 receptors in alcohol-related behaviors. Ann N Y Acad Sci 1025:515–527
Hungund BL, Szakall I, Adam A et al. (2003) Cannabinoid CB1 receptor knockout mice exhibit markedly reduced voluntary alcohol consumption and lack alcohol-induced dopamine release in the nucleus accumbens. J Neurochem 84:698–704
Hutcheson DM, Tzavara ET, Smadja C et al. (1998) Behavioural and biochemical evidence for signs of abstinence in mice chronically treated with delta-9-tetrahydrocannabinol. Br J Pharmacol 125:1567–1577
Jacobs EH, Smit AB, De Vries TJ et al. (2003) Neuroadaptive effects of active versus passive drug administration in addiction research. Trends Pharmacol Sci 24:566–573
Jarbe TU, Lamb RJ, Lin S et al. (2001) (R)-methanandamide and Delta 9-THC as discriminative stimuli in rats: tests with the cannabinoid antagonist SR-141,716 and the endogenous ligand anandamide. Psychopharmacology (Berl) 156:369–380
Jarbe TU, Lamb RJ, Liu Q et al. (2006) Discriminative stimulus functions of AM-1,346, a CB1R selective anandamide analog in rats trained with Delta9-THC or (R)-methanandamide (AM-356). Psychopharmacology (Berl) 188:315–323
Jayamanne A, Greenwood R, Mitchell VA et al. (2006) Actions of the FAAH inhibitor URB597 in neuropathic and inflammatory chronic pain models. Br J Pharmacol 147:281–288

Jentsch JD, Taylor JR (1999) Impulsivity resulting from frontostriatal dysfunction in drug abuse: implications for the control of behavior by reward-related stimuli. Psychopharmacology (Berl) 146:373–390

Johanson CE, Balster RL (1978) A summary of the results of a drug self-administration study using substitution procedures in rhesus monkeys. Bull Narc 30:43–54

Justinova Z, Goldberg SR (2004) The abuse potential of the endocannabinoid transport inhibitor AM404: Self-administration by squirrel monkeys. 2005 Symposium on the Cannabinoids, Burlington, Vermont, International Cannabinoid Research Society, page 176

Justinova Z, Tanda G, Redhi GH et al. (2003) Self-administration of Delta(9)-tetrahydrocannabinol (THC) by drug naive squirrel monkeys. Psychopharmacology (Berl) 169:135–140

Justinova Z, Tanda G, Munzar P et al. (2004) The opioid antagonist naltrexone reduces the reinforcing effects of Delta 9 tetrahydrocannabinol (THC) in squirrel monkeys. Psychopharmacology (Berl) 173:186–194

Justinova Z, Goldberg SR, Heishman SJ et al. (2005a) Self-administration of cannabinoids by experimental animals and human marijuana smokers. Pharmacol Biochem Behav 81:285–299

Justinova Z, Solinas M, Tanda G et al. (2005b) The endogenous cannabinoid anandamide and its synthetic analog R(+)-methanandamide are intravenously self-administered by squirrel monkeys. J Neurosci 25:5645–5650

Justinova Z, Bortolato M, Mangieri RA et al. (2007) Lack of abuse liability of the FAAH inhibitor URB597 in squirrel monkeys. FASEB J 21:A409

Justinova Z, Mangieri RA, Bortolato M et al. (2008a) Fatty acid amide hydrolase inhibition heightens anandamide signaling without producing reinforcing effects in primates. Biol Psychiatry 64:930–937

Justinova Z, Munzar P, Panlilio LV et al. (2008a) Blockade of THC-seeking behavior and relapse in monkeys by the cannabinoid CB(1)-receptor antagonist rimonabant. Neuropsychopharmacology 33:2870–2877

Kalivas PW (2007) Cocaine and amphetamine-like psychostimulants: neurocircuitry and glutamate neuroplasticity. Dialogues Clin Neurosci 9:389–397

Kathuria S, Gaetani S, Fegley D et al. (2003) Modulation of anxiety through blockade of anandamide hydrolysis. Nat Med 9:76–81

Kauer JA (2004) Learning mechanisms in addiction: synaptic plasticity in the ventral tegmental area as a result of exposure to drugs of abuse. Annu Rev Physiol 66:447–475

Kaymakcalan S (1973) Tolerance to and dependence on cannabis. Bull Narc 25:39–47

Kofalvi A, Rodrigues RJ, Ledent C et al. (2005) Involvement of cannabinoid receptors in the regulation of neurotransmitter release in the rodent striatum: a combined immunochemical and pharmacological analysis. J Neurosci 25:2874–2884

Koob GF (1992) Drugs of abuse: anatomy, pharmacology and function of reward pathways. Trends Pharmacol Sci 13:177–184

Koob GF, Weiss F (1990) Pharmacology of drug self-administration. Alcohol 7:193–197

Koob GF, Sanna PP, Bloom FE (1998) Neuroscience of addiction. Neuron 21:467–476

Kreek MJ, LaForge KS, Butelman E (2002) Pharmacotherapy of addictions. Nat Rev Drug Discov 1:710–726

Lallemand F, De Witte P (2006) SR147778, a CB1 cannabinoid receptor antagonist, suppresses ethanol preference in chronically alcoholized Wistar rats. Alcohol 39:125–134

Lamarque S, Taghzouti K, Simon H (2001) Chronic treatment with Delta(9)-tetrahydrocannabinol enhances the locomotor response to amphetamine and heroin. Implications for vulnerability to drug addiction. Neuropharmacology 41:118–129

Le Foll B, Goldberg SR (2004) Rimonabant, a CB1 antagonist, blocks nicotine-conditioned place preferences. Neuroreport 15:2139–2143

Le Foll B, Goldberg SR (2005) Cannabinoid CB1 receptor antagonists as promising new medications for drug dependence. J Pharmacol Exp Ther 312:875–883

Le Foll B, Wiggins M, Goldberg SR (2006) Nicotine pre-exposure does not potentiate the locomotor or rewarding effects of Delta-9-tetrahydrocannabinol in rats. Behav Pharmacol 17:195–199

Ledent C, Valverde O, Cossu G et al. (1999) Unresponsiveness to cannabinoids and reduced addictive effects of opiates in CB1 receptor knockout mice. Science 283:401–404
Lepore M, Vorel SR, Lowinson J et al. (1995) Conditioned place preference induced by delta 9-tetrahydrocannabinol: comparison with cocaine, morphine, and food reward. Life Sci 56: 2073–2080
Lesscher HM, Hoogveld E, Burbach JP et al. (2005) Endogenous cannabinoids are not involved in cocaine reinforcement and development of cocaine-induced behavioural sensitization. Eur Neuropsychopharmacol 15:31–37
Lichtman AH, Sheikh SM, Loh HH et al. (2001) Opioid and cannabinoid modulation of precipitated withdrawal in delta(9)-tetrahydrocannabinol and morphine-dependent mice. J Pharmacol Exp Ther 298:1007–1014
Lopez-Moreno JA, Gonzalez-Cuevas G, Rodriguez de Fonseca F et al. (2004) Long-lasting increase of alcohol relapse by the cannabinoid receptor agonist WIN 55, 212–2 during alcohol deprivation. J Neurosci 24:8245–8252
Lupica CR, Riegel AC (2005) Endocannabinoid release from midbrain dopamine neurons: a potential substrate for cannabinoid receptor antagonist treatment of addiction. Neuropharmacology 48:1105–1116
Maldonado R (2002) Study of cannabinoid dependence in animals. Pharmacol Ther 95:153–164
Maldonado R, Rodriguez de Fonseca F (2002) Cannabinoid addiction: behavioral models and neural correlates. J Neurosci 22:3326–3331
Maldonado R, Valverde O, Berrendero F (2006) Involvement of the endocannabinoid system in drug addiction. Trends Neurosci 29:225–232
Mallet PE, Beninger RJ (1998) Delta9-tetrahydrocannabinol, but not the endogenous cannabinoid receptor ligand anandamide, produces conditioned place avoidance. Life Sci 62:2431–2439
Martellotta MC, Cossu G, Fattore L et al. (1998) Self-administration of the cannabinoid receptor agonist WIN 55, 212-2 in drug-naive mice. Neuroscience 85:327–330
Martin BR (2005) Role of lipids and lipid signaling in the development of cannabinoid tolerance. Life Sci 77:1543–1558
Martin BR, Compton DR, Thomas BF et al. (1991) Behavioral, biochemical, and molecular modeling evaluations of cannabinoid analogs. Pharmacol Biochem Behav 40:471–478
Martin M, Ledent C, Parmentier M et al. (2000) Cocaine, but not morphine, induces conditioned place preference and sensitization to locomotor responses in CB1 knockout mice. Eur J Neurosci 12:4038–4046
Mascia MS, Obinu MC, Ledent C et al. (1999) Lack of morphine-induced dopamine release in the nucleus accumbens of cannabinoid CB(1) receptor knockout mice. Eur J Pharmacol 383:R1–R2
Masuzawa M, Nakao S, Miyamoto E et al. (2003) Pentobarbital inhibits ketamine-induced dopamine release in the rat nucleus accumbens: a microdialysis study. Anesth Analg 96:148–152
McDonald J, Schleifer L, Richards JB et al. (2003) Effects of THC on behavioral measures of impulsivity in humans. Neuropsychopharmacology 28:1356–1365
McGregor IS, Gallate JE (2004) Rats on the grog: novel pharmacotherapies for alcohol craving. Addict Behav 29:1341–1357
McGregor IS, Issakidis CN, Prior G (1996) Aversive effects of the synthetic cannabinoid CP 55, 940 in rats. Pharmacol Biochem Behav 53:657–664
McGregor IS, Dam KD, Mallet PE et al. (2005) Delta9-THC reinstates beer- and sucrose-seeking behaviour in abstinent rats: comparison with midazolam, food deprivation and predator odour. Alcohol Alcohol 40:35–45
Melis M, Pistis M, Perra S et al. (2004) Endocannabinoids mediate presynaptic inhibition of glutamatergic transmission in rat ventral tegmental area dopamine neurons through activation of CB1 receptors. J Neurosci 24:53–62
Merritt LL, Martin BR, Walters C et al. (2008) The endogenous cannabinoid system modulates nicotine reward and dependence. J Pharmacol Exp Ther 326:483–492
Mokler DJ, Nelson BD, Harris LS et al. (1986) The role of benzodiazepine receptors in the discriminative stimulus properties of delta-9-tetrahydrocannabinol. Life Sci 38:1581–1589

Nava F, Carta G, Colombo G et al. (2001) Effects of chronic Delta(9)-tetrahydrocannabinol treatment on hippocampal extracellular acetylcholine concentration and alternation performance in the T-maze. Neuropharmacology 41:392–399

Navarro M, Chowen J, Carrera MR et al. (1998) CB1 cannabinoid receptor antagonist-induced opiate withdrawal in morphine-dependent rats. Neuroreport 9:3397–3402

Navarro M, Carrera MR, Fratta W et al. (2001) Functional interaction between opioid and cannabinoid receptors in drug self-administration. J Neurosci 21:5344–5350

O'Brien C (2001) Drug addiction and drug abuse. In: Hardman J, Limbird L, Gilman AG (eds) The pharmacological basis of therapeutics. McGraw-Hill, New York, pp 621–642

Oliva JM, Ortiz S, Palomo T et al. (2003) Behavioural and gene transcription alterations induced by spontaneous cannabinoid withdrawal in mice. J Neurochem 85:94–104

Parker LA, Gillies T (1995) THC-induced place and taste aversions in Lewis and Sprague–Dawley rats. Behav Neurosci 109:71–78

Parker LA, Mcdonald RV (2000) Reinstatement of both a conditioned place preference and a conditioned place aversion with drug primes. Pharmacol Biochem Behav 66:559–561

Patel S, Rademacher DJ, Hillard CJ (2003) Differential regulation of the endocannabinoids anandamide and 2-arachidonylglycerol within the limbic forebrain by dopamine receptor activity. J Pharmacol Exp Ther 306:880–888

Patel S, Carrier EJ, Ho WS et al. (2005) The postmortal accumulation of brain N-arachidonylethanolamine (anandamide) is dependent upon fatty acid amide hydrolase activity. J Lipid Res 46:342–349

Pattij T, Vanderschuren LJ (2008) The neuropharmacology of impulsive behaviour. Trends Pharmacol Sci 29:192–199

Pattij T, Janssen MC, Schepers I et al. (2007) Effects of the cannabinoid CB1 receptor antagonist rimonabant on distinct measures of impulsive behavior in rats. Psychopharmacology (Berl) 193:85–96

Pertwee RG, Stevenson LA, Griffin G (1993) Cross-tolerance between delta-9-tetrahydrocannabinol and the cannabimimetic agents, CP 55, 940, WIN 55, 212- 2 and anandamide. Br J Pharmacol 110:1483–1490

Pickel VM, Chan J, Kash TL et al. (2004) Compartment-specific localization of cannabinoid 1 (CB1) and mu-opioid receptors in rat nucleus accumbens. Neuroscience 127:101–112

Pickel VM, Chan J, Kearn CS et al. (2006) Targeting dopamine D2 and cannabinoid-1 (CB1) receptors in rat nucleus accumbens. J Comp Neurol 495:299–313

Pierce RC, Bell K, Duffy P et al. (1996) Repeated cocaine augments excitatory amino acid transmission in the nucleus accumbens only in rats having developed behavioral sensitization. J Neurosci 16:1550–1560

Pistis M, Pillolla G, Luchicchi A et al. (2008) Inhibition of FAAH blocks the excitatory effects of nicotine on mesolimbic dopamine neurons via CB1 and PPAR-alpha receptors. 18th Annual Symposium on the Cannabinoids Burlington, Vermont, International Cannabinoid Research Society, page 40

Pontieri FE, Tanda G, Di Chiara G (1995) Intravenous cocaine, morphine, and amphetamine preferentially increase extracellular dopamine in the "shell" as compared with the "core" of the rat nucleus accumbens. Proc Natl Acad Sci USA 92:12304–12308

Pontieri FE, Tanda G, Orzi F et al. (1996) Effects of nicotine on the nucleus accumbens and similarity to those of addictive drugs. Nature 382:255–257

Pontieri FE, Monnazzi P, Scontrini A et al. (2001a) Behavioral sensitization to WIN 55,212-2 in rats pretreated with heroin. Brain Res 898:178–180

Pontieri FE, Monnazzi P, Scontrini A et al. (2001b) Behavioral sensitization to heroin by cannabinoid pretreatment in the rat. Eur J Pharmacol 421:R1–R3

Ramaekers JG, Kauert G, van Ruitenbeek P et al. (2006) High-potency marijuana impairs executive function and inhibitory motor control. Neuropsychopharmacology 31:2296–2303

Rice OV, Gordon N, Gifford AN (2002) Conditioned place preference to morphine in cannabinoid CB1 receptor knockout mice. Brain Res 945:135–138

Richter RM, Pich EM, Koob GF et al. (1995) Sensitization of cocaine-stimulated increase in extracellular levels of corticotropin-releasing factor from the rat amygdala after repeated administration as determined by intracranial microdialysis. Neurosci Lett 187:169–172
Riddle EL, Fleckenstein AE, Hanson GR (2005) Role of monoamine transporters in mediating psychostimulant effects. AAPS J 7:E847–E851
Riegel AC, Lupica CR (2004) Independent presynaptic and postsynaptic mechanisms regulate endocannabinoid signaling at multiple synapses in the ventral tegmental area. J Neurosci 24:11070–11078
Robbe D, Alonso G, Duchamp F et al. (2001) Localization and mechanisms of action of cannabinoid receptors at the glutamatergic synapses of the mouse nucleus accumbens. J Neurosci 21:109–116
Robinson TE, Berridge KC (1993) The neural basis of drug craving: an incentive-sensitization theory of addiction. Brain Res Brain Res Rev 18:247–291
Robinson TE, Berridge KC (2001) Incentive-sensitization and addiction. Addiction 96:103–114
Rodriguez de Fonseca F, Gorriti MA, Fernandez-Ruiz JJ et al. (1994) Downregulation of rat brain cannabinoid binding sites after chronic delta 9-tetrahydrocannabinol treatment. Pharmacol Biochem Behav 47:33–40
Rodriguez de Fonseca F, Carrera MR, Navarro M et al. (1997) Activation of corticotropin-releasing factor in the limbic system during cannabinoid withdrawal. Science 276:2050–2054
Rossetti ZL, Melis F, Carboni S et al. (1991) Marked decrease of extraneuronal dopamine after alcohol withdrawal in rats: reversal by MK-801. Eur J Pharmacol 200:371–372
Rothman RB, Baumann MH (2003) Monoamine transporters and psychostimulant drugs. Eur J Pharmacol 479:23–40
Rubino T, Massi P, Vigano D et al. (2000a) Long-term treatment with SR141716A, the CB1 receptor antagonist, influences morphine withdrawal syndrome. Life Sci 66:2213–2219
Rubino T, Vigano' D, Massi P et al. (2000b) Chronic delta-9-tetrahydrocannabinol treatment increases cAMP levels and cAMP-dependent protein kinase activity in some rat brain regions. Neuropharmacology 39:1331–1336
Rubino T, Vigano D, Massi P et al. (2000c) Changes in the cannabinoid receptor binding, G protein coupling, and cyclic AMP cascade in the CNS of rats tolerant to and dependent on the synthetic cannabinoid compound CP55,940. J Neurochem 75:2080–2086
Rubino T, Vigano D, Massi P et al. (2003) Cellular mechanisms of Delta 9-tetrahydrocannabinol behavioural sensitization. Eur J Neurosci 17:325–330
Rubino T, Forlani G, Vigano D et al. (2004) Modulation of extracellular signal-regulated kinases cascade by chronic delta 9-tetrahydrocannabinol treatment. Mol Cell Neurosci 25:355–362
Rubino T, Forlani G, Vigano D et al. (2005) Ras/ERK signalling in cannabinoid tolerance: from behaviour to cellular aspects. J Neurochem 93:984–991
Sala M, Braida D (2005) Endocannabinoids and 3, 4-methylenedioxymethamphetamine (MDMA) interaction. Pharmacol Biochem Behav 81:407–416
Salamone JD, McLaughlin PJ, Sink K et al. (2007) Cannabinoid CB1 receptor inverse agonists and neutral antagonists: effects on food intake, food-reinforced behavior and food aversions. Physiol Behav 91:383–388
Sanudo-Pena MC, Tsou K, Delay ER et al. (1997) Endogenous cannabinoids as an aversive or counter-rewarding system in the rat. Neurosci Lett 223:125–128
Scherma M, Medalie J, Fratta W et al. (2008a) The endogenous cannabinoid anandamide has effects on motivation and anxiety that are revealed by fatty acid amide hydrolase (FAAH) inhibition. Neuropharmacology 54:129–140
Scherma M, Panlilio LV, Fadda P et al. (2008b) Inhibition of anandamide hydrolysis by URB597 reverses abuse-related behavioral and neurochemical effects of nicotine in rats. J Pharmacol Exp Ther 327:482–490
Schindler ACW, Panlilio LV, Goldberg SR (2002) Second-order schedules of drug self-administration in animals. Psychopharmacology (Berl) 163:327–344

Sink KS, McLaughlin PJ, Wood JA et al. (2008) The novel cannabinoid CB1 receptor neutral antagonist AM4113 suppresses food intake and food-reinforced behavior but does not induce signs of nausea in rats. Neuropsychopharmacology 33:946–955

Solinas M, Panlilio LV, Antoniou K et al. (2003) The cannabinoid CB1 antagonist N-piperidinyl-5-(4-chlorophenyl)-1-(2, 4-dichlorophenyl) -4-methylpyrazole-3-carboxamide (SR-141716A) differentially alters the reinforcing effects of heroin under continuous reinforcement, fixed ratio, and progressive ratio schedules of drug self-administration in rats. J Pharmacol Exp Ther 306:93–102

Solinas M, Zangen A, Thiriet N et al. (2004) Beta-endorphin elevations in the ventral tegmental area regulate the discriminative effects of Delta-9-tetrahydrocannabinol. Eur J Neurosci 19:3183–3192

Solinas M, Panlilio LV, Tanda G et al. (2005) Cannabinoid agonists but not inhibitors of endogenous cannabinoid transport or metabolism enhance the reinforcing efficacy of heroin in rats. Neuropsychopharmacology 30:2046–2057

Solinas M, Justinova Z, Goldberg SR et al. (2006a) Anandamide administration alone and after inhibition of fatty acid amide hydrolase (FAAH) increases dopamine levels in the nucleus accumbens shell in rats. J Neurochem 98:408–419

Solinas M, Panlilio LV, Justinova Z et al. (2006b) Using drug-discrimination techniques to study the abuse-related effects of psychoactive drugs in rats. Nat Protoc 1:1194–1206

Solinas M, Scherma M, Fattore L et al. (2007a) Nicotinic alpha 7 receptors as a new target for treatment of cannabis abuse. J Neurosci 27:5615–5620

Solinas M, Scherma M, Tanda G et al. (2007b) Nicotinic facilitation of delta9-tetrahydrocannabinol discrimination involves endogenous anandamide. J Pharmacol Exp Ther 321:1127–1134

Solinas M, Tanda G, Justinova Z et al. (2007c) The endogenous cannabinoid anandamide produces delta-9-tetrahydrocannabinol-like discriminative and neurochemical effects that are enhanced by inhibition of fatty acid amide hydrolase but not by inhibition of anandamide transport. J Pharmacol Exp Ther 321:370–380

Solinas M, Yasar S, Goldberg SR (2007d) Endocannabinoid system involvement in brain reward processes related to drug abuse. Pharmacol Res 56:393–405

Solinas M, Goldberg SR, Piomelli D (2008) The endocannabinoid system in brain reward processes. Br J Pharmacol 154:369–382

Soria G, Mendizabal V, Tourino C et al. (2005) Lack of CB1 cannabinoid receptor impairs cocaine self-administration. Neuropsychopharmacology 30:1670–1680

Spano MS, Fattore L, Cossu G et al. (2004) CB1 receptor agonist and heroin, but not cocaine, reinstate cannabinoid-seeking behaviour in the rat. Br J Pharmacol 143:343–350

Spyraki C, Fibiger HC (1988) A role for the mesolimbic dopamine system in the reinforcing properties of diazepam. Psychopharmacology (Berl) 94:133–137

Substance Abuse and Mental Health Services Administration (2007) Results from the 2005 National Survey on Drug Use and Health: National Findings. Office of Applied Studies, NSDUH Series H-32, DHHS Publication No SMA 07-4,293, Rockville, MD

Takahashi RN, Singer G (1979) Self-administration of delta 9-tetrahydrocannabinol by rats. Pharmacol Biochem Behav 11:737–740

Takahashi RN, Singer G (1980) Effects of body weight levels on cannabis self-injection. Pharmacol Biochem Behav 13:877–881

Tanda G, Goldberg SR (2003) Cannabinoids: reward, dependence, and underlying neurochemical mechanisms-a review of recent preclinical data. Psychopharmacology (Berl) 169:115–134

Tanda G, Pontieri FE, Di Chiara G (1997) Cannabinoid and heroin activation of mesolimbic dopamine transmission by a common mu1 opioid receptor mechanism. Science 276:2048–2050

Tanda G, Loddo P, Di Chiara G (1999) Dependence of mesolimbic dopamine transmission on delta9-tetrahydrocannabinol. Eur J Pharmacol 376:23–26

Tanda G, Munzar P, Goldberg SR (2000) Self-administration behavior is maintained by the psychoactive ingredient of marijuana in squirrel monkeys. Nat Neurosci 3:1073–1074

Taylor DA, Fennessy MR (1982) Time-course of the effects of chronic delta 9-tetrahydrocannabinol on behaviour, body temperature, brain amines and withdrawal-like behaviour in the rat. J Pharm Pharmacol 34:240–245
Thanos PK, Dimitrakakis ES, Rice O et al. (2005) Ethanol self-administration and ethanol conditioned place preference are reduced in mice lacking cannabinoid CB1 receptors. Behav Brain Res 164:206–213
Thiemann G, van der Stelt M, Petrosino S et al. (2008) The role of the CB1 cannabinoid receptor and its endogenous ligands, anandamide and 2-arachidonoylglycerol, in amphetamine-induced behavioural sensitization. Behav Brain Res 187:289–296
Thomas MJ, Beurrier C, Bonci A et al. (2001) Long-term depression in the nucleus accumbens: a neural correlate of behavioral sensitization to cocaine. Nat Neurosci 4:1217–1223
Thorat SN, Bhargava HN (1994) Evidence for a bidirectional cross-tolerance between morphine and delta 9-tetrahydrocannabinol in mice. Eur J Pharmacol 260:5–13
Thornton-Jones ZD, Vickers SP, Clifton PG (2005) The cannabinoid CB1 receptor antagonist SR141716A reduces appetitive and consummatory responses for food. Psychopharmacology (Berl) 179:452–460
Valjent E, Maldonado R (2000) A behavioural model to reveal place preference to delta 9-tetrahydrocannabinol in mice. Psychopharmacology (Berl) 147:436–438
Valjent E, Mitchell JM, Besson MJ et al. (2002) Behavioural and biochemical evidence for interactions between Delta 9-tetrahydrocannabinol and nicotine. Br J Pharmacol 135:564–578
Valverde O, Maldonado R, Valjent E et al. (2000) Cannabinoid withdrawal syndrome is reduced in pre-proenkephalin knock-out mice. J Neurosci 20:9284–9289
van der Laan JW, Eigeman L, Jansen van't Land C (1992) Benzodiazepines preferentially affect mesolimbic dopamine turnover in rats. Eur Neuropsychopharmacol 2:425–431
van der Stelt M, Di Marzo V (2003) The endocannabinoid system in the basal ganglia and in the mesolimbic reward system: implications for neurological and psychiatric disorders. Eur J Pharmacol 480:133–150
Varvel SA, Martin BR, Lichtman AH (2007) Lack of behavioral sensitization after repeated exposure to THC in mice and comparison to methamphetamine. Psychopharmacology (Berl) 193:511–519
Vela G, Ruiz-Gayo M, Fuentes JA (1995) Anandamide decreases naloxone-precipitated withdrawal signs in mice chronically treated with morphine. Neuropharmacology 34:665–668
Vengeliene V, Bilbao A, Molander A et al. (2008) Neuropharmacology of alcohol addiction. Br J Pharmacol 154:299–315
Verberne AJ, Taylor DA, Fennessy MR (1981) Attenuation of delta 9-tetrahydrocannabinol-induced withdrawal-like behaviour by delta 9-tetrahydrocannabinol. Psychopharmacology (Berl) 73:97–98
Vigano D, Grazia CM, Rubino T et al. (2003) Chronic morphine modulates the contents of the endocannabinoid, 2-arachidonoyl glycerol, in rat brain. Neuropsychopharmacology 28:1160–1167
Vigano D, Valenti M, Cascio MG et al. (2004) Changes in endocannabinoid levels in a rat model of behavioural sensitization to morphine. Eur J Neurosci 20:1849–1857
Vinklerova J, Novakova J, Sulcova A (2002) Inhibition of methamphetamine self-administration in rats by cannabinoid receptor antagonist AM 251. J Psychopharmacol 16:139–143
Welch SP (1997) Characterization of anandamide-induced tolerance: comparison to delta 9-THC-induced interactions with dynorphinergic systems. Drug Alcohol Depend 45:39–45
Wiley JL, Martin BR (1999) Effects of SR141716A on diazepam substitution for delta9-tetrahydrocannabinol in rat drug discrimination. Pharmacol Biochem Behav 64:519–522
Wiley JL, Huffman JW, Balster RL et al. (1995a) Pharmacological specificity of the discriminative stimulus effects of delta 9-tetrahydrocannabinol in rhesus monkeys. Drug Alcohol Depend 40:81–86
Wiley JL, Lowe JA, Balster RL et al. (1995b) Antagonism of the discriminative stimulus effects of delta 9-tetrahydrocannabinol in rats and rhesus monkeys. J Pharmacol Exp Ther 275:1–6

Wiley JL, Golden KM, Ryan WJ et al. (1997) Evaluation of cannabimimetic discriminative stimulus effects of anandamide and methylated fluoroanandamide in rhesus monkeys. Pharmacol Biochem Behav 58:1139–1143
Wiley JL, Ryan WJ, Razdan RK et al. (1998) Evaluation of cannabimimetic effects of structural analogs of anandamide in rats. Eur J Pharmacol 355:113–118
Wiley JL, LaVecchia KL, Karp NE et al. (2004) A comparison of the discriminative stimulus effects of Delta(9)-tetrahydrocannabinol and O-1,812, a potent and metabolically stable anandamide analog, in rats. Exp Clin Psychopharmacol 12:173–179
Wise RA (2004) Dopamine, learning and motivation. Nat Rev Neurosci 5:483–494
Wise RA, Bozarth MA (1987) A psychomotor stimulant theory of addiction. Psychol Rev 94: 469–492
Wolf ME, Sun X, Mangiavacchi S et al. (2004) Psychomotor stimulants and neuronal plasticity. Neuropharmacology 47:61–79
Xi ZX, Gilbert JG, Peng XQ et al. (2006) Cannabinoid CB1 receptor antagonist AM251 inhibits cocaine-primed relapse in rats: role of glutamate in the nucleus accumbens. J Neurosci 26:8531–8536
Yokel RA (1987) Intravenous self-administration: response rates, the effects of pharmacological challenges, and drug preference. In: Bozarth MA (ed) Methods of assessing the reinforcing properties of abused drugs. Springer, New York, pp 1–33
Young AM, Herling S (1986) Drugs as reinforcers: studies in laboratory animals. In: Goldberg SR, Stolerman IP (eds) Behavioral analysis of drug dependence. Academic Press, Orlando, pp 9–67
Young AM, Woods JH (1981) Maintenance of behavior by ketamine and related compounds in rhesus monkeys with different self-administration histories. J Pharmacol Exp Ther 218: 720–727
Young AM, Herling S, Woods JH (1981) History of drug exposure as a determinant of drug self-administration. NIDA Res Monogr 37:75–88
Zangen A, Solinas M, Ikemoto S et al. (2006) Two brain sites for cannabinoid reward. J Neurosci 26:4901–4907
Zimmer A, Valjent E, Konig M et al. (2001) Absence of delta -9-tetrahydrocannabinol dysphoric effects in dynorphin-deficient mice. J Neurosci 21:9499–9505
Zhang PW, Ishiguro H, Ohtsuki T et al. (2004) Human cannabinoid receptor 1: 5' exons, candidate regulatory regions, polymorphisms, haplotypes and association with polysubstance abuse. Mol Psychiatry 9:916–931
Zuo L, Kranzler HR, Luo X et al. (2007) CNR1 variation modulates risk for drug and alcohol dependence. Biol Psychiatry 62:616–626

Role of Endocannabinoid Signaling in Anxiety and Depression

Sachin Patel and Cecilia J. Hillard

Contents

Abstract Cannabinoid receptors and their endogenous ligands are located throughout the limbic, or "emotional," brain, where they modulate synaptic neurotransmission. Converging preclinical and clinical data suggest a role for endogenous cannabinoid signaling in the modulation of anxiety and depression. Augmentation

S. Patel and C.J. Hillard (✉)
Department of Pharmacology, Medical College of Wisconsin, 8701 Watertown Plank Road, Milwaukee, WI 53226, USA
e-mail: chillard@mcw.edu

D. Kendall and S. Alexander (eds.), *Behavioral Neurobiology of the Endocannabinoid System*, Current Topics in Behavioral Neurosciences 1,
DOI: 10.1007/978-3-540-88955-7_14,

of endocannabinoid signaling (ECS) has anxiolytic effects, whereas blockade or genetic deletion of CB_1 receptors has anxiogenic properties. Augmentation of ECS also appears to have anti-depressant actions, and in some assays blockade and genetic deletion of CB_1 receptors produces depressive phenotypes. These data provide evidence that ECS serves in an anxiolytic, and possibly anti-depressant, role. These data suggest novel approaches to treatment of affective disorders which could include enhancement of endogenous cannabinoid signaling, and warrant cautious use of CB_1 receptor antagonists in patients with pre-existing affective disorders.

Keywords Cannabis • Fatty acid amide hydrolase • Post-traumatic stress disorder • Marijuana • Anandamide • Cannabinoid

Abbreviations

2AG	2-Arachidonoylglycerol
5-HT	5-Hydroxytryptamine, serotonin
ACC	Anterior cingulate cortex
AEA	Anandamide
BLA	Basolateral amygdala
CCK	Cholecystokinin
CUS	Chronic exposure to an unpredictable and variable set of stressors
ECS	Endocannabinoid signaling
ECT	Electroconvulsive therapy
FAAH	Fatty acid amide hydrolase
HPA	Hypothalamus–pituitary–adrenal
KO	Knockout
PFC	Prefrontal cortex
PTSD	Post-traumatic stress disorder
PVN	Paraventricular nucleus
SSRI	Selective serotonin re-uptake inhibitors

1 Human Studies Suggesting a Role for Endocannabinoid Signaling in Anxiety

Cannabis has been used for centuries for a variety of recreational and medicinal purposes. The primary psychoactive chemical in cannabis, Δ^9-tetrahydrocannabinol (THC), is a partial agonist of the CB_1 cannabinoid receptor (Breivogel et al. 1998). The most commonly cited reasons for continued recreational cannabis use are relaxation and reduction in tension (Reilly et al. 1998; Schofield et al. 2006; Thomas

1993). Paradoxically, the most commonly cited reasons for discontinuation of cannabis use are increased anxiety and panic reactions (Reilly et al. 1998; Szuster et al. 1988). Modulation of anxiety reactions by cannabis appears to be complex in that both dose and environmental context can modulate these effects. Subjects under "experimenter harassment" were more likely to experience anxiety reactions under the influence of cannabis than those in neutral environments (Gregg et al. 1976). Since the subjective effects of cannabis are mediated via the CB_1 receptor (Huestis et al. 2001), these data suggest a role for endocannabinoid signaling (ECS) in the regulation of anxiety.

A CB_1 receptor antagonist, rimonabant (also named Acomplia, SR141716 and SR141716A) has been developed and used in humans for the treatment of obesity, diabetes and dyslipidemia (Van Gaal et al. 2008). Psychiatric adverse effects, including anxiety, were cited as reasons for discontinuation by patients taking rimonabant significantly more than those taking placebo (Van Gaal et al. 2008), although objective measures of anxiety were not significantly increased in patients taking rimonabant (Scheen et al. 2006). A recent meta-analysis pooling data from four large clinical trials indicated that subjects taking rimonabant had a significantly greater increase in anxiety symptoms while taking the drug than patients taking placebo (Christensen et al. 2007). Therefore, human experience with a cannabinoid receptor agonist (THC) and antagonist (rimonabant) support the hypothesis that ECS regulates anxiety in humans and suggest that activation of the CB_1 receptor by endocannabinoids could produce anxiolytic effects.

Support for an inverse relationship between ECS and anxiety in humans also comes from a recent study of serum endocannabinoids in women with depression (Hill et al. 2008). In this study, the severity of anxiety experienced by women with major depression was inversely correlated with serum content of *N*-arachidonylethanolamine (AEA). Although very little is known about the source or potential target of circulating endocannabinoids, these data suggest that some of the somatic manifestations of anxiety could be related to reduced ECS.

2 Animal Studies Indicating a Role for ECS in Anxiety

2.1 *Effects of CB_1 Receptor Blockade and Genetic Deletion on Unconditioned Anxiety Behaviors*

A commonly used and well-validated test of unconditioned anxiety in rodents is the elevated plus-maze. This is an exploration-based test that utilizes the innate fear of open spaces exhibited by rodents. The maze measures the proportion of time rodents spend in well-lit "open" arms, compared to darker "closed" arms. A drug-induced increase in the proportion of time spent in the open arms is suggestive of an anxiolytic effect, whereas an increase in time spent in closed arms is suggestive of an anxiogenic effect. An anxiogenic effect of rimonabant has been demonstrated using an elevated plus-maze test in rats (Navarro et al. 1997) and mice (Arevalo et al. 2001; Patel and Hillard 2006). A second CB_1 receptor antagonist, AM251, a structural analog of rimonabant, also shows anxiogenic effects in rodents in the elevated plus-maze (Haller et al. 2004b; Patel and Hillard 2006). Rimonabant exhibits an anxiogenic profile in the defensive-withdrawal (Navarro et al. 1997) and ultrasonic vocalization tests (McGregor et al. 1996) as well.

In contrast to these findings, other studies have demonstrated either no effect (Bortolato et al. 2006; Kathuria et al. 2003) or an anxiolytic effect of rimonabant (Degroot and Nomikos 2004; Griebel et al. 2005; Rodgers et al. 2003). In the studies in which no effect was seen, relatively low doses of the antagonists were used (Bortolato et al. 2006; Kathuria et al. 2003). Dose-dependent anxiolytic effects of rimonabant were seen in the elevated plus-maze and Vogel conflict test in mice (Griebel et al. 2005). Furthermore, using a design in which rodents were tested twice, rimonabant had no effect in the elevated plus-maze during the first trial, but produced an anxiolytic effect during the second exposure (Rodgers et al. 2003). Interestingly, rimonabant produced anxiolytic effects in CB_1 receptor knockout (KO) mice, leading Haller et al. to suggest its anxiolytic actions are mediated via non-CB_1-dependent mechanisms (Haller et al. 2002). These authors did not observe anxiogenic effects of AM251 in CB_1 receptor KO mice, and concluded that AM251 does not share the non-receptor effect of rimonabant (Haller et al. 2004a). Anxiolytic effects of rimonabant have also been demonstrated in the shock-probe burying test, although this effect could be due to the effect of the drug to enhanced memory function, rather than direct effects of unconditioned anxiety per se (Degroot and Nomikos 2004).

Administration of rimonabant results in activation of brain regions involved in the generation of fear and anxiety. Systemic administration of rimonabant increased Fos expression, a marker of neuronal activity, within the central amygdala, bed nucleus of the stria terminalis, hypothalamus and brainstem (Alonso et al. 1999; Patel et al. 2005b; Rodriguez de Fonseca et al. 1997). These studies further support the hypothesis that ECS is an endogenous anxiolytic system that dampens neuronal activity within brain regions critical for the generation of fear and anxiety responses.

CB_1 receptor KO mice exhibit increased anxiety-like behaviors in the elevated plus-maze (Haller et al. 2002, 2004a, b), and in the light–dark exploration model in young mice only (Maccarrone et al. 2002). Interestingly, these effects appear to be more prominent under environmentally stressful conditions (Haller et al. 2004a; Maccarrone et al. 2002). In particular, in a high light condition, which is considered stressful since rodents are nocturnal and have impaired vision under this condition, CB_1 receptor KO mice exhibit an anxiogenic phenotype; while under low light conditions, this phenotypic difference is absent (Haller et al. 2004a). This finding may explain why some studies have failed to detect an anxiogenic phenotype in CB_1 receptor KO mice (Marsicano et al. 2002). In addition to direct anxiogenic behaviors, CB_1 receptor KO mice display impaired behavioral responses to non-cannabinoid anxiolytics including benzodiazepines and buspirone (Uriguen et al. 2004).

2.2 Effects of Pharmacological and Genetic Augmentation of ECS on Unconditioned Anxiety Behaviors

ECS occurs when synaptic concentrations of the endocannabinoids AEA and/or 2-arachidonoylglycerol (2AG) are increased through either increased synthesis or decreased catabolism. In particular, fatty acid amide hydrolase (FAAH) is a well-characterized enzyme that hydrolyzes and inactivates AEA and other *N*-acylethanolamines (Ho and Hillard 2005). Pharmacologic inhibition or genetic deletion results in significant increases in brain AEA but not 2AG content (Cravatt et al. 1996; Kathuria et al. 2003; Patel et al. 2005a). Systemic administration of a highly efficacious inhibitor of FAAH, URB597, produced anxiolytic effects in the elevated zero-maze (a slight modification of the elevated plus-maze described above) and in the ultrasonic vocalization test in rats (Kathuria et al. 2003). This effect was accompanied by an increase in brain AEA concentrations and blocked by the CB_1 receptor antagonist rimonabant (Kathuria et al. 2003). These data suggest that increased CB_1 receptor signaling by AEA produces anxiolytic behavioral effects that can be enhanced by pharmacological blockade of FAAH. This effect of URB597 has been replicated in mice using the elevated plus-maze (Moreira et al. 2008; Patel and Hillard 2006) and in rats using the light–dark box test (Scherma et al. 2008). FAAH KO mice also exhibit an anxiolytic phenotype in the elevated plus-maze and light–dark box test (Moreira et al. 2008; Naidu et al. 2007); effects that are blocked by pretreatment with rimonabant (Moreira et al. 2008). Taken together, these data support the hypothesis that the ECS in rodents provides an anxiolytic tone that can be enhanced if AEA-mediated signaling is increased. The role of 2AG in this system is not known.

These findings are consistent with data showing that exogenous administration of low doses of direct-acting CB_1 receptor agonists also produce anxiolytic effects in rodents (Patel and Hillard 2006; Scherma et al. 2008). However, unlike direct CB_1 receptor agonists that display anxiogenic effects at higher doses, FAAH inhibitors exhibit only dose-dependent anxiolytic effects without anxiogenic effects at high doses (Kathuria et al. 2003; Patel and Hillard 2006). These data suggest that the spatio-temporal properties of ECS are maintained by FAAH inhibition, in contrast to global CB_1 activation by direct agonists, and that this property of FAAH inhibitors subserves their uniphasic, anxiolytic properties. In other words, global CB_1 receptor activation can result in both decreased and increased anxiety, but the evidence using both inhibition of FAAH and CB_1 receptor antagonism indicate that the anxiogenic "pool" of CB_1 receptors is not endogenously active. We suggested earlier that the functional pools are anatomically distinct (Patel and Hillard 2006), a suggestion that is supported by a recent study using region-selective, virally mediated up-regulation of FAAH. Parolaro and co-workers showed that increasing FAAH expression within the prefrontal cortex (PFC) caused a reduction in AEA concentrations and an increase in anxiety behaviors in the elevated plus-maze (Rubino et al. 2008b). These data confirm a role for ECS in the regulation of anxiety behaviors and suggest that the anatomical site of this ECS function includes the PFC.

However, another explanation for the difference in the effects on anxiety between FAAH inhibition and direct CB_1 receptor agonists is that the inhibition of FAAH increases levels of non-cannabinoid, fatty acid ethanolamides (NAEs) as well as AEA (Cravatt et al. 2001). Since the anxiolytic effects of FAAH inhibitors can be blocked by CB_1 receptor antagonists (Kathuria et al. 2003; Moreira et al. 2008), it can be concluded that CB_1 receptor activation is required for the anti-anxiety efficacy of FAAH inhibition. However, these data do not address the question of whether other NAEs contribute to the efficacy as well. In other words, it is not known whether CB_1 receptor activation is sufficient for the anxiolytic efficacy of FAAH inhibition.

In addition to inhibition of FAAH, inhibitors of endocannabinoid transport have also demonstrated anxiolytic properties. AM404 is an arachidonic acid analog that inhibits uptake of both AEA (Beltramo et al. 1997) and 2AG (Beltramo and Piomelli 2000), inhibits FAAH activity (Jarrahian et al. 2000), and increases brain AEA concentrations (Bortolato et al. 2006). Several studies have demonstrated that systemic administration of AM404 produces anxiolytic effects in the elevated plus-maze, defensive withdrawal test, and social isolation test (Bortolato et al. 2006; Patel and Hillard 2006). These effects are blocked by the CB_1 receptor antagonist rimonabant, consistent with the hypothesis that indirect activation of the ECS can produce anxiolytic effects (Bortolato et al. 2006). However, in another study in which drugs were administered into the periaqueductal gray of rats, AEA produced anxiolytic effects that were enhanced by AM404, but alone AM404 was not anxiolytic (Moreira et al. 2007).

2.3 Effects of CB_1 Receptor Deletion and Pharmacological Blockade on Conditioned Anxiety Behaviors

Conditioned, or "learned," fear is a model for certain types of anxiety disorders including post-traumatic stress disorder (PTSD). In this paradigm, a temporal contingency is established between environmental cues such as an auditory tone or a specific environmental context, i.e., "cage type," and an aversive stimulus such as an electric shock. After single or repeated "paired" presentations of these two stimuli, the environmental cues presented alone can elicit an innate, conditioned fear response such as freezing, and signs of sympathetic nervous system activation. After "conditioned" fear responses to cue presentation are established, presentation of environmental cues in the absence of the aversive stimulus causes a gradual extinction of conditioned fear responses.

Two different conditioning paradigms, context and tone, have been used to examine the role of ECS in the acquisition of conditioned fear responses. Several studies have shown no effect of either CB_1 receptor genetic deletion or pharmacological blockade on the acquisition of contextual or tonal fear conditioning (Marsicano et al. 2002; Suzuki et al. 2004). However, a recent study utilizing a multiple-trial acquisition model found enhanced acquisition of conditioned fear responses in trace and delayed fear conditioning paradigms, which are hippocampus- and amygdala-dependent, respectively (Reich et al. 2008). These data suggest that ECS could impair acquisition of conditioned anxiety responses under specific conditions.

It has been conclusively demonstrated that both pharmacological and genetic inhibition of CB_1 receptors impair the extinction of both contextual and tonal conditioned anxiety responses (Kamprath et al. 2006; Marsicano et al. 2002; Reich et al. 2008; Suzuki et al. 2004). Impaired extinction of aversive associative learning has also been demonstrated using fear-potentiated startle and passive avoidance protocols (Chhatwal et al. 2005), but not an appetitively motivated instrumental responding paradigm (Niyuhire et al. 2007).

Data from a novel paradigm that attempts to separate the associative and non-associative components of conditioned fear responses suggest that impairments in extinction observed in CB_1 receptor KO mice are due to deficits in habituation, the non-associative component of extinction (Kamprath et al. 2006). In this paradigm, presentation of the tone stimulus used in fear conditioning paradigms (preceded by a sensitizing shock) results in freezing behavior that habituates over repeated presentations; this represents a non-associative component of extinction of conditioned fear behavior. Mice lacking CB_1 receptors do not show habituation of these innate fear responses after repeated tone presentation. These authors suggest that the impairments in extinction of conditioned fear behavior observed in CB_1 receptor KO mice and after CB_1 receptor blockade are a result of an impaired "habituation component" of the extinction process (Kamprath et al. 2006). This suggestion is consistent with a growing body of literature supporting a role of the ECS in habituation of the behavioral and endocrine responses to stress (Patel and Hillard 2008).

2.4 *Effects of ECS Augmentation on Conditioned Anxiety Behaviors*

Similarly to unconditioned anxiety, insight into the role of ECS in conditioned anxiety comes from studies in which CB_1 receptor signaling is activated using low doses of agonists. For example, the CB_1 receptor agonist WIN55212-2 impairs acquisition of context-, but not tone-, conditioned anxiety responses (Pamplona and Takahashi 2006) and low doses of WIN55212-2 facilitate extinction of conditioned anxiety responses in a contextual fear-conditioning paradigm (Pamplona et al. 2006). Similarly, the indirect agonist, AM404, impairs extinction of fear-potentiated startle responses (Chhatwal et al. 2005), and FAAH KO mice exhibit enhanced extinction of an aversively motivated, spatial memory task (Varvel et al. 2007). This appears selective for aversively motivated over appetitively motivated learning (Holter et al. 2005).

Taken together, data in animal models of unconditioned and conditioned anxiety support the hypothesis that activation or enhancement of ECS can produce a reduction in anxiety in rodents. This function of the ECS appears to be tonically "on" or easily activated since treatment of rodents in mildly aversive environments with CB_1 receptor antagonists enhances anxiety behaviors. It is likely that changes in CB_1 receptor activation can regulate anxiety in multiple brain regions and through multiple mechanisms (discussed further below). High doses of direct CB_1 receptor agonists can be anxiogenic, which parallels the human experience in which cannabis use can be both anxiolytic and anxiogenic. However, the lack of anxiogenic effects by FAAH inhibitors and the nearly consistent finding that CB_1 receptor blockade is monophasically anxiogenic support the hypothesis that the predominant effect of endogenous CB_1 receptor activation is a reduction in anxiety.

3 Neural Mechanisms Underlying Endocannabinoid Modulation of Anxiety

The neural mechanisms by which ECS affects anxiety are not well understood, yet several mechanisms at the systems, synaptic, and molecular level can be posited based on available data. The majority of available data indicate that ECS has anxiolytic properties in both conditioned and unconditioned anxiety models, and that these effects are more active during states of stress or high arousal (Haller et al. 2004a). The anxiolytic effects of ECS are mimicked by low doses of direct CB_1 receptor agonists (Patel and Hillard 2006); thus data exploiting this phenomenon can be used to increased our understanding of the neural mechanisms subserving the anxiolytic actions of the ECS system.

At the systems level, microinjections of low doses of the direct CB_1 agonist THC into the PFC (Rubino et al. 2008a), ventral hippocampus (Rubino et al. 2008a), and dorsal periaqueductal gray area (Moreira et al. 2007) exert anxiolytic effects in the elevated plus-maze. These effects are blocked by the CB_1 receptor antagonist AM251 (Moreira et al. 2007; Rubino et al. 2008b). Pharmacological inhibition of FAAH within the PFC produces CB_1-receptor-dependent anxiolytic effects, and over-expression of FAAH (which reduces local AEA levels) causes an anxiogenic effect in the elevated plus-maze (Rubino et al. 2008b). In contrast to the PFC and hippocampus, very low doses of THC produce only anxiogenic effects when administered into the basolateral amygdala (BLA); this was also dependent upon CB_1 receptor activation (Rubino et al. 2008a). These data suggest that the PFC and hippocampus are likely anatomical sites of action that subserve the anxiolytic effects of ECS. More specifically, the balance of ECS in favor of an increase in the PFC and/or hippocampus and reduced signaling in the amygdala could be required for maximal anxiolytic effects.

With regard to endocannabinoid facilitation of extinction of conditioned fear responses, direct administration of CB_1 agonists into the lateral amygdala impairs fear memories by blocking reconsolidation in a fear-potentiated startle model (Lin et al. 2006). These data suggest that ECS in the amygdala during presentation of conditioned cues impairs reconsolidation of fear memories, and thus facilitates extinction of conditioned fear responses. Thus, in contrast to unconditioned anxiety responses (which are enhanced by CB_1 receptor activation in the amygdala), impairments in conditioned anxiety responses are observed after amygdalar CB_1 receptor activation. These data suggest a complex and potentially divergent role for amygdalar ECS in the modulation of conditioned vs. unconditioned anxiety behaviors.

At the synaptic level, activation of CB_1 receptors inhibits glutamatergic inputs to principal neurons in the cortex, hippocampus and BLA (Hashimotodani et al. 2007). In addition, CB_1 receptor activation inhibits GABA release from a subpopulation of cholecystokinin (CCK)-expressing interneurons that form perisomatic (and some dendritic) contacts with hippocampal principal neurons; however, this effect is only operative when the firing rates of these interneurons is low (Foldy et al. 2007). Haller and co-workers suggest that the anxiolytic effects of ECS are mediated via inhibition of GABAergic transmission within the hippocampus (Haller et al. 2007). This suggestion is based on data demonstrating an anxiolytic effect of WIN55212-2 in CD-1 mice, in which this compound was significantly more efficacious at inhibiting hippocampal GABAergic than glutamatergic transmission. By contrast, WIN55212-2 produced an anxiogenic effect and affected GABAergic and glutamatergic transmission equally in rats. In addition, AM251 blocked the anxiogenic effect of WIN55212-2 in mice, and blocked the effect of this compound on GABAergic transmission, but not glutamatergic transmission. These data led the authors to conclude that WIN55212-2 produced anxiolytic effects via inhibition of GABAergic transmission within the hippocampus. These pharmacologic studies led to the further suggestion that the anxiogenic effect of WIN55212-2 in rats is mediated by inhibition of glutamatergic transmission. These data provide an interesting hypothesis that requires further experimental evidence; particularly important will be studies using mouse models in which CB_1 receptors on either glutamatergic or GABAergic terminals have been selectively abolished (Monory et al. 2006).

A synaptic mechanism subserving endocannabinoid facilitation of extinction of conditioned fear responses has also been proposed (Lafenetre et al. 2007). These authors incorporate the ability of endocannabinoids to modulate both GABAergic and glutamatergic transmission within the amygdala in their model. They suggest that under basal conditions ECS is not active in the amygdala; a conclusion that is supported by c-Fos studies from our laboratory (Patel et al. 2005b). After tonal fear conditioning, presentation of the tone alone increases ECS in the BLA, which has been demonstrated experimentally (Marsicano et al. 2002). This increase in ECS inhibits GABAergic transmission, which results in dis-inhibition of BLA projection neurons and facilitation of a "no fear" pathway mediated by activation of inhibitory neurons within intercalated cell masses. These neurons provide feed-forward inhibition onto central amygdala neurons, which are output neurons of the amygdala and activate conditioned behavioral and physiological responses. These authors also suggest that ECS signaling could decrease glutamatergic transmission in a "fear" pathway that transmits directly from the BLA to the central amygdala. Such depotentiation of the conditioned "fear" pathway could represent a synaptic mechanism for the habituation component of extinction of conditioned fear. The mechanisms that would segregate ECS into GABAergic and glutamatergic signaling in the "no fear" and "fear" pathways, respectively, remain to be determined.

Although the above data provide anatomical and synaptic insights into the mechanisms subserving the anxiolytic effects of ECS, they do not alone explain the context-dependent effects. Specifically, the anxiogenic effects of CB_1 receptor deletions or blockade are more robust under stress or high arousal (Haller et al. 2004a), suggesting increased ECS counteracts the anxiety produced by environmental stress. These observations suggest that exposure to the fear-evoking or stressful context results in an increase in endocannabinoid release. A potential explanation could involve the neuropeptide CCK, which is expressed by CB_1-receptor-positive, GABAergic interneurons. CCK is released under times of stress and high arousal (Nevo et al. 1996), and activation of CCK_2 receptors appears to result in endocannabinoid release from hippocampal principal neurons, based on the effects of AM251 (Foldy et al. 2007). These endocannabinoids can then activate receptors on GABAergic interneurons to produce anxiolytic effects as suggested above. This hypothesis remains to be experimentally tested.

At the molecular level, anxiolytic effects of low doses of CB_1 receptor agonists are associated with increased CREB expression within the PFC and hippocampus (Rubino et al. 2007). This increase was associated with an increase in ERK activation in the PFC, and a decrease in CAMKII (a kinase that inhibits CREB activation) within the hippocampus. In addition, anxiolytic doses of THC inhibited plus-maze exposure-induced Fos expression with the PFC and amygdala (Rubino et al. 2007). Behaviorally, the anxiolytic effects of low doses of THC are blocked by a mu-opioid receptor antagonist (Berrendero and Maldonado 2002), and a $5HT_{1A}$ serotonin receptor antagonist (Marco et al. 2004); the anxiolytic effects of AM404 are also blocked by a $5\text{-}HT_{1A}$ antagonist (Marco et al. 2004). These data suggest a role for opioid and serotonin receptors in the anxiolytic effects of ECS.

In the case of conditioned fear modulation, roles for ERK and calcineurin have been demonstrated. In response to conditioned tone presentation, CB_1 receptor KO mice exhibit relatively increased freezing behavior as a consequence of impaired extinction (Marsicano et al. 2002). These mice also exhibited decreased tone-induced phosphorylation of ERK and calcineurin expression in the BLA and PFC, while showing increased expression of these two proteins in the central amygdala (Cannich et al. 2004). CB_1 receptor KO mice also showed increased p-AKT in the BLA and dorsal hippocampus in response to conditioned tone presentation compared to wild-type mice (Cannich et al. 2004). It has been shown that ERK signaling in the BLA is required for the acquisition of extinction

(Herry et al. 2006), suggesting that impaired ERK signaling in CB_1 receptor KO mice could contribute to the impaired extinction observed in these mice. In addition, mice lacking forebrain calcineurin exhibit impaired extinction of conditioned fear behaviors (Havekes et al. 2008), supporting a role for this protein in the impaired extinction observed in CB_1 receptor KO mice. These data suggest that ECS could facilitate extinction of conditioned fear via activation of ERK and calcineurin signaling (Davis et al. 2003; Galve-Roperh et al. 2002).

4 Human Studies Suggesting a Role for ECS in Depression

4.1 Cannabis Use and Depression

The thousands of years of human use of the CB_1 receptor agonist, THC, in preparations of *Cannabis sativa* support the hypothesis that there is a relationship between cannabis use and depression. Elevation of mood is one of the commonly cited motivations for the use of cannabis. In a study of young, poly-substance users, 69% of the respondents reported that they used cannabis to "make themselves feel better when down or depressed" (Boys et al. 2001). While this is far less than the 97% who responded that they used cannabis to help relax, it argues that cannabis could exert anti-depressant effects in humans. Several clinical trials in the 1970s designed to determine the anti-depressant efficacy of THC found that it failed to improve symptoms of depression and produced unacceptable adverse effects (Ablon and Goodwin 1974; Kotin et al. 1973). Although it can be argued that these studies were small and did not take into consideration the heterogeneity in depressive illnesses, it is not likely that THC would be broadly useful as an anti-depressant in humans.

A similar hypothesis, that depressed individuals self-administer cannabis because it elevates mood, is not supported by available data (Kandel et al. 1986; Miller-Johnson et al. 1998; Patton et al. 2002). This hypothesis predicts that depressed people use cannabis to elevate mood more frequently than non-depressed users. This prediction was not upheld in a recent study (Arendt et al. 2007); in fact, depressed subjects experienced more depression, aggression and sadness when intoxicated with cannabis than when they were not intoxicated.

There are data to support an alternative hypothesis that cannabis use precipitates depression. For example, cannabis dependence and depression are co-morbid diagnoses more than would be expected by chance (Degenhardt et al. 2003). Furthermore, several prospective studies have found that cannabis use precedes the diagnosis of depression (Bovasso 2001; Patton et al. 2002; Rey and Tennant 2002). Cannabis use was identified in high-school students as a significant, independent predictor of suicidal behaviors after adjustment for depressive symptoms (Chabrol et al. 2008). However, a large (greater than 12,000 participants) longitudinal study did not find that past cannabis use was a significant predictor of depression in adults when baseline differences between users and non-users were carefully controlled (Harder et al. 2006). The authors of this study concluded that the available evidence does not support a causal relationship between cannabis use and depression, but does suggest that a common factor or factors predisposes individuals to both depression and cannabis dependence. In this regard, the hypothesis of a shared genetic predisposition for both cannabis use and depression has received support in the literature. Both cannabis use and dependence (Fu et al. 2002a; Kendler et al. 2000; Lynskey et al. 2002) and depressive/suicidal behaviors (Fu et al. 2002a, b; Statham et al. 1998; Sullivan et al. 2000) are moderately heritable. More importantly, several recent studies have demonstrated that the genetic factors for cannabis dependence and depression/suicidality are moderately correlated (Fu et al. 2002a; Lynskey et al. 2004). Twin studies suggest that shared environmental factors also contribute significantly to the co-morbidity of cannabis dependence and depression (Lynskey et al. 2004).

4.2 Depression and the ECS

The data described above lead to the hypothesis that dysregulation of ECS results in depression. Support for this hypothesis comes from the adverse events profile in humans of the CB_1 receptor antagonist, rimonabant, which demonstrates a small, yet significant, increased likelihood for the development or exacerbation of depression (Van Gaal et al. 2008). The likelihood of depression or mood changes with depressive symptoms increases when patients with pre-existing depressive illness were not excluded from rimonabant treatment (Nissen et al. 2008). These data suggest that endogenous activation of CB_1 receptors serves as a buffer against depression and its elimination or reduction in susceptible individuals can result in depressive symptoms. In another study, the incidence of depression in patients with Parkinson's disease was found to be significantly correlated with polymorphisms in the CB_1 receptor gene (Barrero et al. 2005). There was a trend for the same observation in non-Parkinson patients, but the study was not sufficiently powerful to determine whether CB_1 receptor polymorphisms contribute to the likelihood of developing major depression in the general population.

There have also been some very interesting studies that have investigated the hypothesis that depression changes ECS. Patients with depression who died by suicide had significantly greater CB_1 receptor agonist binding site density and agonist signaling in the dorsolateral PFC than matched controls (Hungund et al. 2004; Vinod et al. 2005). Tissue contents of both AEA and 2AG in the dorsolateral PFC were also increased in alcoholic patients who were depressed compared to alcoholics without depression (Vinod et al. 2005). In a study using immunohistochemical approaches, neuronal CB_1 receptor density in the anterior cingulate cortex (ACC) was not found to be different between patients with major depression and controls (Koethe et al. 2007). However, CB_1 receptor density was significantly decreased in subjects with major depression taking selective serotonin re-uptake inhibitors (SSRIs) compared to patients with major depression who were not being treated with SSRIs, suggesting that the drug therapy reduced CB_1 receptor expression (Koethe et al. 2007). CB_1 receptor density was also decreased in glial cells in the ACC of brains from patients who died with major depression compared to controls (Koethe et al. 2007). This finding is particularly interesting in light of other data suggesting that glial cell function and/or numbers are dysregulated in major depression (Cotter et al. 2001).

Our group has recently published a study in which circulating endocannabinoid concentrations were compared in non-medicated women with major depression and controls (Hill et al. 2008). 2AG contents in the serum were significantly lower in women with major depression than matched controls and were negatively correlated with the length of the current depressive episode. These data, while preliminary, support the possibility that some of the peripheral consequences of depression, such as cardiovascular and metabolic changes, could be related to ECS modulation.

To summarize, the available human data support the general hypothesis that CB_1 receptor activity is involved in the regulation of mood and that pharmacological dysregulation of ECS can alter mood in some individuals. Data suggest that depressed individuals have altered ECS; however, whether changes in ECS precede or follow the development of depression is unknown.

5 Animal Studies Suggesting a Role for ECS in Depression

5.1 Evidence That Alteration of CB_1 Receptor Signaling Results in Anti-Depressant-Like Effects

Immobility assays in rodents have been used extensively as preclinical models of anti-depressant efficacy of various pharmacologic agents. The Porsolt forced swim test is commonly employed; the time that rodents spend in an immobile, floating state is argued to represent a state of behavioral despair and is reduced by monoamine elevating anti-depressants (Porsolt et al. 1978). The highly efficacious CB_1 receptor agonists, HU210 (Hill and Gorzalka 2005b) and WIN55212-2 (Bambico et al. 2007) reduce immobility duration in the forced swim test in male rats at very low doses, consistent with anti-depressant efficacy. These agonist effects are blocked by co-treatment with CB_1 receptor antagonist. Indirect CB_1 receptor agonists, including AM404 (Hill and Gorzalka 2005b) and the FAAH inhibitor, URB597 (Gobbi et al. 2005; Hill et al. 2007b), also exhibit anti-depressant efficacy in the forced swim test. URB597 also has anti-depressant efficacy in a second immobility assay, the mouse tail suspension (Gobbi et al. 2005).

While the direct and indirect agonist data are fairly consistent and support a role for the ECS in the coping response of mice in the forced swim, antagonist data have been inconsistent. In both male and female C57Bl/6N mice, rimonabant had no effect on the duration of immobility and increased struggling during the first exposure to the test (Steiner et al. 2008b). However, these investigators found that chronic treatment with high dose rimonabant significantly decreased immobility (Steiner et al. 2008a). Other studies using acute treatment with antagonists have also reported no effect (Bambico et al. 2007; Gobbi et al. 2005; Gobshtis et al. 2007; Hill and Gorzalka 2005b). On the other hand, several studies have demonstrated that acute treatment with antagonists, usually at high doses, reduces immobility (Shearman et al. 2003). The reasons for the discrepancies in these studies are not clear, but strain/species differences, differences in the parameters examined and differences in the environmental context of the assay (i.e., light vs. dark phase) are all plausible explanations.

Immobility tests comparing KO and wild-type mice have also been used to infer pro-depressant or anti-depressant roles for various proteins or signaling systems (Cryan and Holmes 2005). The duration of immobility of CB_1 receptor KO mice on a CD-1 background is not different from wild-type (Jardinaud et al. 2005). In one study, Steiner and colleagues reported that immobility (floating) was significantly increased in CB_1 receptor KO mice on a C57Bl/6N background compared to wild-type (Steiner et al. 2008b), while a second study from the same laboratory reported no difference in response when KO and wild-type mice were pretreated with a vehicle injection (Steiner et al. 2008a).

Taken together, these data suggest that activation of the CB_1 receptor exogenously can produce an anti-depressant behavioral phenotype in immobility assays; and they provide some support for ECS tone. On the other hand, they suggest that CB_1 receptor activation also contributes to behavioral despair since antagonist treatment can be anti-depressant as well. As for the effects of cannabinoid receptor ligands in anxiety discussed above, it is likely that there are "functional" pools of CB_1 receptors that subserve pro- and anti-depressant behavioral effects.

Most depressive disorders in humans include decreased incentive to seek positive reinforcers or anhedonia as a core symptom (Rush and Weissenburger 1994). This aspect of depression can be modeled using several rodent assays; the most common is the sucrose consumption test. Activation of CB_1 receptors results in a selective increase in the consumption of highly palatable foods, including increased sucrose drinking relative to the drinking of water (Sofia and Knobloch 1976). Inhibition of ECS by antagonists inhibits sucrose consumption in two bottle-choice paradigms (Arnone et al. 1997) and decreases responding reinforced by normal food and sucrose in operant procedures models (Freedland et al. 2001; Perio et al. 2001). CB_1 receptor KO mice also display reduced sucrose intake (Poncelet et al. 2003; Sanchis-Segura et al. 2004). Therefore, there are consistent data that inhibition or removal of the CB_1 receptor in otherwise normal rodents results in a decrease in their motivation to consume sucrose. These data lead to the hypothesis that reduced ECS could contribute to the anhedonia that occurs in depression. In support of this hypothesis, exposure of mice to stress results in a decrease in sucrose consumption that is reversed by direct and indirect CB_1 receptor agonists (Rademacher and Hillard 2007). Interestingly, in this study, rimonabant reduced sucrose consumption in the stressed mice at doses that did not affect sucrose consumption in unstressed mice, consistent with a possible recruitment of ECS in the stressed condition (Rademacher et al. 2008).

5.2 Evidence That Environmental Contexts That Produce Depression-Like Symptoms Alter ECS

Repeated stress has been used to model depressive symptoms in rodents with a reasonable degree of biological and behavioral similarities to humans (Nestler et al. 2002). In particular, chronic exposure to an unpredictable and variable set of stressors (CUS) produces changes in rodents that parallel many aspects of human depression (Willner 2005). Several studies have demonstrated alterations in ECS in rodents exposed to CUS. Hippocampal CB_1 receptor density is reduced in rats exposed to CUS; and perseveration in the water maze induced by CUS is reversed by cannabinoid agonist treatment (Hill et al. 2005a). In another study, CUS was found to reduce body weight and sucrose intake in rats, both of which were reversed by treatment with a FAAH inhibitor (Bortolato et al. 2007). These studies suggest that down-regulation of ECS contributes to the detrimental effects of CUS. This conclusion is supported by the finding that CB_1 receptor KO mice exhibit increased sensitivity to the anhedonic effects of CUS (Martin et al. 2002).

Repeated exposure to the same stressor also recapitulates some of the behavioral effects of depression, including anhedonia. Repeated restraint results in changes in endocannabinoid content in several limbic regions, including a progressive increase in 2AG content within the PFC, amygdala and hypothalamus, as the number of restraint episodes increases (Rademacher et al. 2008). On the other hand, restraint decreases AEA contents in the PFC and amygdala regardless of the number of restraint episodes. These and other data support the hypothesis that repeated exposure to stress alters ECS and that these changes underlie the behavioral alterations induced by stress (Patel and Hillard 2008). Early life stress, which is known to promote the appearance of depression in adulthood, can be mimicked in mice using a 24-h maternal deprivation (Marco et al. 2009). Evidence from Macri and Laviola suggests that early life stress also down-regulates CB_1-receptor-mediated signaling (Macri and Laviola 2004).

In a recent study, Rubino and colleagues demonstrated that chronic THC exposure during adolescence resulted in significantly increased immobility in the forced swim test in females but not males, and significant anhedonia in both males and females (Rubino et al. 2008c). These studies are very interesting, particularly since they bear on the hypothesis that cannabis consumption predisposes humans to depression.

Therefore, an evolving body of evidence supports the hypothesis that altered ECS accompanies the development of depressive-like behaviors in rodents. The specifics of the alteration are not completely clear, but hypofunctional ECS in subcortical regions, particularly the hippocampus and hypothalamus, have been seen in several models.

5.3 Evidence That Anti-Depressant Therapies Alter ECS

While THC itself is not a good anti-depressant in humans, the role of ECS in mood regulation prompts the question of whether altered ECS contributes to the efficacy of other anti-depressant drugs or manipulations. Chronic exposure of rats to desipramine results in a significant increase in CB_1 receptor binding site density in the hippocampus and hypothalamus in non-stressed rats (Hill et al. 2006). Furthermore, the ability of chronic desipramine treatment to inhibit activation of Fos in the paraventricular nucleus (PVN) in response to stress was reversed by CB_1 receptor antagonist treatment. In addition, rimonabant inhibited the weight gain in response to desipramine, but did not affect the ability of desipramine to reduce immobility in the forced swim assay (Gobshtis et al. 2007). These data suggest that the ability of chronic desipramine to inhibit the activation of the hypothalamic–pituitary–adrenal (HPA) axis and increase weight in normal rats is mediated by an increase in ECS, perhaps in the hypothalamus. In contrast to these results, the effect of an acute injection of desipramine to induce immobility was absent in CB_1 receptor KO mice but the dampening effects of desipramine on HPA axis activation were intact (Steiner et al. 2008b). These results also suggest a difference in the mechanisms by which anti-depressants and ECS affect immobility and HPA axis activation, an observation that is discussed further below. The role of ECS in the effects of desipramine is not identical for other anti-depressants. For example, the SSRI citalopram significantly decreases CB_1 -receptor-mediated signaling in the PVN (Hesketh et al. 2008). Electroconvulsive shock treatment (ECT) is the most effective therapeutic option for depression in humans in that it benefits a higher proportion of patients than chemical anti-depressant therapy and requires substantially less time to see benefit (Silverstone and Silverstone 2004). ECT also produces alterations in ECS that can be summarized as an increase in subcortical ECS and a decrease in cortical ECS (Hill et al. 2007a).

Therefore, the treatments for human depression modulate ECS in a regionally specific manner. However, the changes are not consistent with respect to brain region or directionality and more studies are needed to determine which, if any, of these changes are relevant to ECS in depression.

6 Neural Mechanisms Underlying Endocannabinoid Modulation of Depression

The neurobiology of depression is complex; however, a large body of evidence supports the hypothesis that dysregulation of the HPA axis plays a critical role (Hill and Gorzalka 2005a). In particular, HPA axis hyperactivation and reduced feedback inhibition are seen in humans with depression and in animal models of depression. The ability of anti-depressants to suppress HPA axis hyperactivity is coupled to their clinical efficacy (Appelhof et al. 2006). Recent studies strongly suggest that a primary role for ECS is to dampen HPA axis activation by stress and to allow for appropriate stress recovery (Barna et al. 2004; Di et al. 2003; Patel et al. 2004). These findings are consistent with the data obtained in rodents described above that inhibition of ECS is generally pro-depressive while its activation results in an anti-depressant phenotype, and lead to the hypothesis that dampening of the HPA axis is the mechanism by which ECS interacts with depression. However, HPA axis inhibition does not completely explain the effects of ECS to alter coping behaviors in the forced swim assay. For example, desipramine-induced behavioral effects are CB_1 receptor-dependent while its effects on HPA axis activation are not (Steiner et al. 2008a). Recent studies in our laboratory demonstrate that female CB_1 receptor KO mice exhibit normal HPA axis activation by stress but have increased immobility in the forced swim assay compared to wild-type (Roberts and Hillard, unpublished data).

The monoamine hypothesis of depression posits that dysregulation of serotonergic and noradrenergic signaling in the brain contributes to depressive symptoms (Belmaker and Agam 2008). ECS interactions with serotonergic signaling have been demonstrated in many studies. For example, serotonergic neurons have been shown to be involved in many cannabinoid effects, including hypothermia (Malone and Taylor 1998) and sleep (Mendelson and Basile 2001). The effect of WIN55212-2 to reduce immobility in the forced swim test is abolished by the serotonin (5-HT) depleting agent, para-chlorphenylalanine, indicating that this behavior is also 5-HT-mediated (Bambico et al. 2007). Low doses of WIN55212-2 enhance dorsal raphe serotonergic neuronal activity, an effect that is mimicked by the FAAH inhibitor URB597 (Gobbi et al. 2005). This effect appears to be due to ECS activation in the medial PFC since lesions there abolish the WIN55212-2 on raphe firing. Therefore, these studies suggest that activation of 5-HT-mediated signaling in the PFC is involved in the anti-depressant efficacy of activation of ECS. Recent studies have found that both CB_1 receptor blockade (Tzavara et al. 2003) and chronic administration of THC result in increased serotonin levels in the PFC (Sagredo et al. 2006). Chronic administration of another agonist, HU210, results in an enhancement of $5\text{-}HT_{2A}$ behavioral effects and a decrease in $5\text{-}HT_{1A}$ effects (Hill et al. 2005b). On the other hand, the FAAH inhibitor, URB597, increases firing of serotonergic neurons in the dorsal raphe and noradrenergic neurons in the locus coeruleus and ECS has been shown to subserve the regulation of glutamate-induced activation of serotonergic neurons in the raphe (Haj-Dahmane and Shen 2005).

CB_1 receptors are present throughout the limbic system (Herkenham et al. 1990) and can modulate both GABA and glutamate release (Freund et al. 2003). Therefore, it is not surprising that global activation or inhibition of ECS has confusing effects on behavior. A few studies have begun to dissect regional differences in the role of ECS in depression. HU210 injected into the rat hippocampus elicits reduced immobility in the forced swim test while URB597 is not active via this route (McLaughlin et al. 2007). These data, that activation of ECS in the hippocampus exerts anti-depressant effects, are consistent with findings that CUS, which produces depressive-like symptoms, down-regulates hippocampal ECS (Hill et al. 2005a). WIN55212-2 is also an effective anti-depressant when injected into the ventromedial PFC; the effects of indirect agonists and antagonists were not determined (Bambico et al. 2007). The possible role of 5-HT signaling in the PFC effects is discussed above. Interestingly, CUS has been shown to increase CB_1 receptor mRNA expression in the PFC (Bortolato et al. 2007) and human suicides have increased CB_1 receptor density and signaling (Hungund et al. 2004). It will be very interesting to determine the neuronal site of these up-regulated receptors.

7 Clinical Implications for Endocannabinoid-Based Therapeutics for Anxiety and Depressive Disorders

The data reviewed above indicate that ECS has an anxiolytic function. Data from studies of unconditioned anxiety measures suggest that pharmacological augmentation of ECS could represent a novel approach to the treatment of generalized anxiety disorder, and anxiety symptoms associated with depressive disorders. Endocannabinoid augmentation could also be useful in the treatment of PTSD based on the role of ECS in stress response habituation (Patel and Hillard 2008) and enhancement of extinction of conditioned fear and anxiety.

Initial augmentation strategies have focused on inhibition of AEA catabolism by FAAH and endocannabinoid uptake inhibitors. Both of these approaches have been successful in preclinical models. Future drug discovery should be aimed at development of selective inhibitors of 2AG degradation, which could also have anxiolytic properties. It is likely that pharmacological augmentation of ECS will have several advantages over direct CB_1 receptor agonists including less likelihood of precipitating anxiety or panic reactions and less socio-political resistance to widespread clinical use. Lastly, these data suggest that the use of CB_1 receptor antagonists should be minimized in patients with anxiety disorders, due to an increased risk of exacerbating symptoms (Christensen et al. 2007).

The issue of treating depression with ECS-based therapies is far more murky. Human depression is a heterogeneous disease and only a fraction of those treated with conventional therapies have long-term disease remission. There are strong indications (discussed at length above) that ECS dysregulation could contribute to depression in some humans. The challenge to research at this stage is to further our understanding of both depression and ECS in order to elucidate which depressed patients will benefit from ECS-based therapy.

Acknowledgements CJH was supported during the writing of this review by Research for a Healthier Tomorrow, a component of the Advancing a Healthier Wisconsin endowment at the Medical College of Wisconsin and NIH grant R21 DA022439.

References

Ablon SL, Goodwin FK (1974) High frequency of dysphoric reactions to tetrahydrocannabinol among depressed patients. Am J Psychiatry 131:448–453
Alonso R, Voutsinos B, Fournier M et al. (1999) Blockade of cannabinoid receptors by SR141716 selectively increases Fos expression in rat mesocorticolimbic areas via reduced dopamine D2 function. Neuroscience 91:607–620
Appelhof BC, Huyser J, Verweij M et al. (2006) Glucocorticoids and relapse of major depression (dexamethasone/corticotropin-releasing hormone test in relation to relapse of major depression). Biol Psychiatry 59:696–701
Arendt M, Rosenberg R, Fjordback L et al. (2007) Testing the self-medication hypothesis of depression and aggression in cannabis-dependent subjects. Psychol Med 37:935–945
Arevalo C, de Miguel R, Hernandez-Tristan R (2001) Cannabinoid effects on anxiety-related behaviours and hypothalamic neurotransmitters. Pharmacol Biochem Behav 70:123–131
Arnone M, Maruani J, Chaperon F et al. (1997) Selective inhibition of sucrose and ethanol intake by SR 141716, an antagonist of central cannabinoid (CB1) receptors. Psychopharmacology (Berl) 132:104–106
Bambico FR, Katz N, Debonnel G, et al. (2007) Cannabinoids elicit antidepressant-like behavior and activate serotonergic neurons through the medial prefrontal cortex. J Neurosci 27:11700–11711
Barna I, Zelena D, Arszovszki AC et al. (2004) The role of endogenous cannabinoids in the hypothalamo-pituitary-adrenal axis regulation: in vivo and in vitro studies in CB1 receptor knockout mice. Life Sci 75:2959–2970
Barrero FJ, Ampuero I, Morales B et al. (2005) Depression in Parkinson's disease is related to a genetic polymorphism of the cannabinoid receptor gene (CNR1). Pharmacogenomics J 5:135–141
Belmaker RH, Agam G (2008) Major depressive disorder. N Engl J Med 358:55–68
Beltramo M, Piomelli D (2000) Carrier-mediated transport and enzymatic hydrolysis of the endogenous cannabinoid 2-arachidonylglycerol. Neuroreport 11:1231–1235
Beltramo M, Stella N, Calignano A et al. (1997) Functional role of high-affinity anandamide transport, as revealed by selective inhibition. Science 277:1094–1097
Berrendero F, Maldonado R (2002) Involvement of the opioid system in the anxiolytic-like effects induced by Delta(9)-tetrahydrocannabinol. Psychopharmacology (Berl) 163:111–117
Bortolato M, Campolongo P, Mangieri RA et al. (2006) Anxiolytic-like properties of the anandamide transport inhibitor AM404. Neuropsychopharmacology 31:2652–2659
Bortolato M, Mangieri RA, Fu J et al. (2007) Antidepressant-like activity of the fatty acid amide hydrolase inhibitor URB597 in a rat model of chronic mild stress. Biol Psychiatry 62:1103–1110
Bovasso GB (2001) Cannabis abuse as a risk factor for depressive symptoms. Am J Psychiatry 158:2033–2037
Boys A, Marsden J, Strang J (2001) Understanding reasons for drug use amongst young people: a functional perspective. Health Educ Res 16:457–469
Breivogel CS, Selley DE, Childers SR (1998) Cannabinoid receptor agonist efficacy for stimulating [35S]GTPgammaS binding to rat cerebellar membranes correlates with agonist-induced decreases in GDP affinity. J Biol Chem 273:16865–16873
Cannich A, Wotjak CT, Kamprath K et al. (2004) CB1 cannabinoid receptors modulate kinase and phosphatase activity during extinction of conditioned fear in mice. Learn Memory 11:625–632
Chabrol H, Chauchard E, Girabet J (2008) Cannabis use and suicidal behaviours in high-school students. Addict Behav 33:152–155
Chhatwal JP, Davis M, Maguschak KA et al. (2005) Enhancing cannabinoid neurotransmission augments the extinction of conditioned fear. Neuropsychopharmacology 30:516–524
Christensen R, Kristensen PK, Bartels EM et al.(2007) Efficacy and safety of the weight-loss drug rimonabant: a meta-analysis of randomised trials. Lancet 370:1706–1713

Cotter D, Mackay D, Landau S et al. (2001) Reduced glial cell density and neuronal size in the anterior cingulate cortex in major depressive disorder. Arch Gen Psychiatry 58:545–553
Cravatt BF, Giang DK, Mayfield SP et al. (1996) Molecular characterization of an enzyme that degrades neuromodulatory fatty-acid amides. Nature 384:83–87
Cravatt BF, Demarest K, Patricelli MP et al. (2001) Supersensitivity to anandamide and enhanced endogenous cannabinoid signaling in mice lacking fatty acid amide hydrolase. Proc Natl Acad Sci USA 98:9371–9376
Cryan JF, Holmes A (2005) The ascent of mouse: advances in modelling human depression and anxiety. Nat Rev Drug Discov 4:775–790
Davis MI, Ronesi J, Lovinger DM (2003) A predominant role for inhibition of the adenylate cyclase/protein kinase a pathway in ERK activation by cannabinoid receptor 1 in N1E-115 neuroblastoma cells. J Biol Chem 278:48973–48980
Degenhardt L, Hall W, Lynskey M (2003) Exploring the association between cannabis use and depression. Addiction 98:1493–1504
Degroot A, Nomikos GG (2004) Genetic deletion and pharmacological blockade of CB1 receptors modulates anxiety in the shock-probe burying test. Eur J NeuroSci 20:1059–1064
Di S, Malcher-Lopes R, Halmos KC et al. (2003) Nongenomic glucocorticoid inhibition via endocannabinoid release in the hypothalamus: a fast feedback mechanism. J Neurosci 23:4850–4857
Foldy C, Lee SY, Szabadics J et al. (2007) Cell type-specific gating of perisomatic inhibition by cholecystokinin. Nat Neurosci 10:1128–1130
Freedland CS, Sharpe AL, Samson HH et al. (2001) Effects of SR141716A on ethanol and sucrose self-administration. Alcohol Clin Exp Res 25:277–282
Freund TF, Katona I, Piomelli D (2003) Role of endogenous cannabinoids in synaptic signalling. Physiol Rev 83:1017–1066
Fu Q, Heath AC, Bucholz KK et al. (2002a) Shared genetic risk of major depression, alcohol dependence, and marijuana dependence: contribution of antisocial personality disorder in men. Arch Gen Psychiatry 59:1125–1132
Fu Q, Heath AC, Bucholz KK et al. (2002b) A twin study of genetic and environmental influences on suicidality in men. Psychol Med 32:11–24
Galve-Roperh I, Rueda D, Gomez del Pulgar T et al. (2002) Mechanism of extracellular signal-regulated kinase activation by the CB(1) cannabinoid receptor. Mol Pharmacol 62:1385–1392
Gobbi G, Bambico FR, Mangieri R et al. (2005) Antidepressant-like activity and modulation of brain monoaminergic transmission by blockade of anandamide hydrolysis. Proc Natl Acad Sci USA 102:18620–18625
Gobshtis N, Ben-Shabat S, Fride E (2007) Antidepressant-induced undesirable weight gain: prevention with rimonabant without interference with behavioral effectiveness. Eur J Pharmacol 554:155–163
Gregg JM, Small EW, Moore R et al. (1976) Emotional response to intravenous delta9tetrahydrocannabinol during oral surgery. J Oral Surg 34:301–313
Griebel G, Stemmelin J, Scatton B (2005) Effects of the cannabinoid CB1 receptor antagonist rimonabant in models of emotional reactivity in rodents. Biol Psychiatry 57:261–267
Haj-Dahmane S, Shen RY (2005) The wake-promoting peptide orexin-B inhibits glutamatergic transmission to dorsal raphe nucleus serotonin neurons through retrograde endocannabinoid signalling. J Neurosci 25:896–905
Haller J, Bakos N, Szirmay M et al. (2002) The effects of genetic and pharmacological blockade of the CB1 cannabinoid receptor on anxiety. Eur J NeuroSci 16:1395–1398
Haller J, Varga B, Ledent C et al. (2004a) Context-dependent effects of CB1 cannabinoid gene disruption on anxiety-like and social behaviour in mice. Eur J NeuroSci 19:1906–1912
Haller J, Varga B, Ledent C et al. (2004b) CB1 cannabinoid receptors mediate anxiolytic effects: convergent genetic and pharmacological evidence with CB1-specific agents. Behav Pharmacol 15:299–304

Haller J, Matyas F, Soproni K et al. (2007) Correlated species differences in the effects of cannabinoid ligands on anxiety and on GABAergic and glutamatergic synaptic transmission. Eur J NeuroSci 25:2445–2456
Harder VS, Morral AR, Arkes J (2006) Marijuana use and depression among adults: Testing for causal associations. Addiction 101:1463–1472
Hashimotodani Y, Ohno-Shosaku T, Kano M (2007) Endocannabinoids and synaptic function in the CNS. Neuroscientist 13:127–137
Havekes R, Nijholt IM, Visser AK et al. (2008) Transgenic inhibition of neuronal calcineurin activity in the forebrain facilitates fear conditioning, but inhibits the extinction of contextual fear memories. Neurobiol Learn Memory 89:595–598
Herkenham M, Lynn AB, Little MD et al. (1990) Cannabinoid receptor localization in brain. Proc Natl Acad Sci USA 87:1932–1936
Herry C, Trifilieff P, Micheau J et al. (2006) Extinction of auditory fear conditioning requires MAPK/ERK activation in the basolateral amygdala. Eur J NeuroSci 24:261–269
Hesketh SA, Brennan AK, Jessop DS et al. (2008) Effects of chronic treatment with citalopram on cannabinoid and opioid receptor-mediated G-protein coupling in discrete rat brain regions. Psychopharmacology (Berl) 198:29–36
Hill MN, Gorzalka BB (2005a) Is there a role for the endocannabinoid system in the etiology and treatment of melancholic depression? Behav Pharmacol 16:333–352
Hill MN, Gorzalka BB (2005b) Pharmacological enhancement of cannabinoid CB(1) receptor activity elicits an antidepressant-like response in the rat forced swim test. Eur Neuropsychopharmacol 15(6):593–599
Hill MN, Patel S, Carrier EJ et al. (2005a) Downregulation of endocannabinoid signalling in the hippocampus following chronic unpredictable stress. Neuropsychopharmacology 30:508–515
Hill MN, Sun JC, Tse MT et al. (2005b) Altered responsiveness of serotonin receptor subtypes following long-term cannabinoid treatment. Int J Neuropsychopharmacol 9:277–286
Hill MN, Ho WS, Sinopoli KJ et al. (2006) Involvement of the endocannabinoid system in the ability of long-term tricyclic antidepressant treatment to suppress stress-induced activation of the hypothalamic–pituitary–adrenal axis. Neuropsychopharmacology 31:2591–2599
Hill MN, Barr AM, Ho WS et al. (2007a) Electroconvulsive shock treatment differentially modulates cortical and subcortical endocannabinoid activity. J Neurochem 103:47–56
Hill MN, Karacabeyli ES, Gorzalka BB (2007b) Estrogen recruits the endocannabinoid system to modulate emotionality. Psychoneuroendocrinology 32:350–357
Hill MN, Miller GE, Ho WS et al. (2008) Serum endocannabinoid content is altered in females with depressive disorders: a preliminary report. Pharmacopsychiatry 41:48–53
Ho W-SV, Hillard CJ (2005) Modulators of endocannabinoid enzymic hydrolysis and membrane transport. In: Pertwee R (ed) Cannabinoids (handbook of experimental pharmacology). Springer, Freiburg, pp 187–207
Holter SM, Kallnik M, Wurst W et al. (2005) Cannabinoid CB1 receptor is dispensable for memory extinction in an appetitively-motivated learning task. Eur J Pharmacol 510:69–74
Huestis MA, Gorelick DA, Heishman SJ et al. (2001) Blockade of effects of smoked marijuana by the CB1-selective cannabinoid receptor antagonist SR141716. Arch Gen Psychiatry 58:322–328
Hungund BL, Vinod KY, Kassir SA et al. (2004) Upregulation of CB1 receptors and agonist-stimulated [35S]GTPgammaS binding in the prefrontal cortex of depressed suicide victims. Mol Psychiatry 9:184–190
Jardinaud F, Crete D, Canestrelli C et al. (2005) CB1 receptor knockout mice show similar behavioral modifications to wild-type mice when enkephalin catabolism is inhibited. Brain Res 1063:77–83
Jarrahian A, Manna S, Edgemond WS et al. (2000) Structure–activity relationships among N-arachidonylethanolamine (Anandamide) head group analogues for the anandamide transporter. J Neurochem 74:2597–2606

Kamprath K, Marsicano G, Tang J et al. (2006) Cannabinoid CB1 receptor mediates fear extinction via habituation-like processes. J Neurosci 26:6677–6686
Kandel DB, Davies M, Karus D et al. (1986) The consequences in young adulthood of adolescent drug involvement. An overview. Arch Gen Psychiatry 43:746–754
Kathuria S, Gaetani S, Fegley D et al. (2003) Modulation of anxiety through blockade of anandamide hydrolysis. Nat Med 9:76–81
Kendler KS, Karkowski LM, Neale MC et al. (2000) Illicit psychoactive substance use, heavy use, abuse, and dependence in a US population-based sample of male twins. Arch Gen Psychiatry 57:261–269
Koethe D, Llenos IC, Dulay JR et al. (2007) Expression of CB(1) cannabinoid receptor in the anterior cingulate cortex in schizophrenia, bipolar disorder, and major depression. J Neural Transm 114:1055–1063
Kotin J, Post RM, Goodwin FK (1973) Delta9-tetrahydrocannabinol in depressed patients. Arch Gen Psychiatry 28:345–348
Lafenetre P, Chaouloff F, Marsicano G (2007) The endocannabinoid system in the processing of anxiety and fear and how CB1 receptors may modulate fear extinction. Pharmacol Res 56:367–381
Lin HC, Mao SC, Gean PW (2006) Effects of intra-amygdala infusion of CB1 receptor agonists on the reconsolidation of fear-potentiated startle. Learn Memory 13:316–321
Lynskey MT, Heath AC, Nelson EC et al. (2002) Genetic and environmental contributions to cannabis dependence in a national young adult twin sample. Psychol Med 32:195–207
Lynskey MT, Glowinski AL, Todorov AA et al. (2004) Major depressive disorder, suicidal ideation, and suicide attempt in twins discordant for cannabis dependence and early-onset cannabis use. Arch Gen Psychiatry 61:1026–1032
Maccarrone M, Valverde O, Barbaccia ML et al. (2002) Age-related changes of anandamide metabolism in CB1 cannabinoid receptor knockout mice: correlation with behaviour. Eur J NeuroSci 15:1178–1186
Macri S, Laviola G (2004) Single episode of maternal deprivation and adult depressive profile in mice: interaction with cannabinoid exposure during adolescence. Behav Brain Res 154: 231–238
Malone DT, Taylor DA (1998) Modulation of delta9-tetrahydrocannabinol-induced hypothermia by fluoxetine in the rat. Br J Pharmacol 124:1419–1424
Marco EM, Perez-Alvarez L, Borcel E et al. (2004) Involvement of 5-HT1A receptors in behavioural effects of the cannabinoid receptor agonist CP 55, 940 in male rats. Behav Pharmacol 15:21–27
Marco EM, Adriani W, Llorente R et al. (2009) Detrimental psychophysiological effects of early maternal deprivation in adolescent and adult rodents: Altered responses to cannabinoid exposure. Neurosci Biobehav Rev 33(4):498–507
Marsicano G, Wotjak CT, Azad SC et al. (2002) The endogenous cannabinoid system controls extinction of aversive memories. Nature 418:530–534
Martin M, Ledent C, Parmentier M et al. (2002) Involvement of CB1 cannabinoid receptors in emotional behaviour. Psychopharmacology (Berl) 159:379–387
McGregor IS, Dastur FN, McLellan RA et al. (1996) Cannabinoid modulation of rat pup ultrasonic vocalizations. Eur J Pharmacol 313:43–49
McLaughlin RJ, Hill MN, Morrish AC et al. (2007) Local enhancement of cannabinoid CB1 receptor signalling in the dorsal hippocampus elicits an antidepressant-like effect. Behav Pharmacol 18:431–438
Mendelson WB, Basile AS (2001) The hypnotic actions of the fatty acid amide, oleamide. Neuropsychopharmacology 25:S36–S39
Miller-Johnson S, Lochman JE, Coie JD et al. (1998) Comorbidity of conduct and depressive problems at sixth grade: substance use outcomes across adolescence. J Abnorm Child Psychol 26:221–232

Monory K, Massa F, Egertova M et al. (2006) The endocannabinoid system controls key epileptogenic circuits in the hippocampus. Neuron 51:455–466
Moreira FA, Aguiar DC, Guimaraes FS (2007) Anxiolytic-like effect of cannabinoids injected into the rat dorsolateral periaqueductal gray. Neuropharmacology 52:958–965
Moreira FA, Kaiser N, Monory K et al. (2008) Reduced anxiety-like behaviour induced by genetic and pharmacological inhibition of the endocannabinoid-degrading enzyme fatty acid amide hydrolase (FAAH) is mediated by CB1 receptors. Neuropharmacology 54:141–150
Naidu PS, Varvel SA, Ahn K et al. (2007) Evaluation of fatty acid amide hydrolase inhibition in murine models of emotionality. Psychopharmacology (Berl) 192:61–70
Navarro M, Hernandez E, Munoz RM et al. (1997) Acute administration of the CB1 cannabinoid receptor antagonist SR 141716A induces anxiety-like responses in the rat. NeuroReport 8:491–496
Nestler EJ, Gould E, Manji H et al. (2002) Preclinical models: status of basic research in depression. Biol Psychiatry 52:503–528
Nevo I, Becker C, Hamon M et al. (1996) Stress- and yohimbine-induced release of cholecystokinin in the frontal cortex of the freely moving rat: prevention by diazepam but not ondansetron. J Neurochem 66:2041–2049
Nissen SE, Nicholls SJ, Wolski K et al. (2008) Effect of rimonabant on progression of atherosclerosis in patients with abdominal obesity and coronary artery disease: the STRADIVARIUS randomized controlled trial. JAMA 299:1547–1560
Niyuhire F, Varvel SA, Thorpe AJ et al. (2007) The disruptive effects of the CB1 receptor antagonist rimonabant on extinction learning in mice are task-specific. Psychopharmacology (Berl) 191:223–231
Pamplona FA, Takahashi RN (2006) WIN 55212-2 impairs contextual fear conditioning through the activation of CB1 cannabinoid receptors. Neurosci Lett 397:88–92
Pamplona FA, Prediger RD, Pandolfo P et al. (2006) The cannabinoid receptor agonist WIN 55, 212-2 facilitates the extinction of contextual fear memory and spatial memory in rats. Psychopharmacology (Berl) 188:641–649
Patel S, Hillard CJ (2006) Pharmacological evaluation of cannabinoid receptor ligands in a mouse model of anxiety: further evidence for an anxiolytic role for endogenous cannabinoid signalling. J Pharmacol Exp Ther 318:304–311
Patel S, Hillard CJ (2008) Adaptations in endocannabinoid signalling in response to repeated homotypic stress: A novel mechanism for stress habituation. Eur J Neurosci 27(11):2921–2929
Patel S, Roelke CT, Rademacher DJ et al. (2004) Endocannabinoid signalling negatively modulates stress-induced activation of the hypothalamic–pituitary–adrenal axis. Endocrinology 145:5431–5438
Patel S, Carrier EJ, Ho WS et al. (2005a) The postmortal accumulation of brain N-arachidonylethanolamine (anandamide) is dependent upon fatty acid amide hydrolase activity. J Lipid Res 46:342–349
Patel S, Cravatt BF, Hillard CJ (2005b) Synergistic interactions between cannabinoids and environmental stress in the activation of the central amygdala. Neuropsychopharmacology 30:497–507
Patton GC, Coffey C, Carlin JB et al. (2002) Cannabis use and mental health in young people: cohort study. Br Med J 325:1195–1198
Perio A, Barnouin MC, Poncelet M et al. (2001) Activity of SH141716 on post-reinforcement pauses in operant responding for sucrose reward in rats. Behav Pharmacol 12:641–645
Poncelet M, Maruani J, Calassi R et al. (2003) Overeating, alcohol and sucrose consumption decrease in CB1 receptor deleted mice. Neurosci Lett 343:216–218
Porsolt RD, Anton G, Blavet N et al. (1978) Behavioural despair in rats: a new model sensitive to antidepressant treatments. Eur J Pharmacol 47:379–391
Rademacher DJ, Hillard CJ (2007) Interactions between endocannabinoids and stress-induced decreased sensitivity to natural reward. Prog Neuropsychopharmacol Biol Psychiatry 31: 633–641

Rademacher DJ, Meier SE, Shi L et al. (2008) Effects of acute and repeated restraint stress on endocannabinoid content in the amygdala, ventral striatum, and medial prefrontal cortex in mice. Neuropharmacology 54:108–116
Reich CG, Mohammadi MH, Alger BE (2008) Endocannabinoid modulation of fear responses: learning and statedependent performance effects. J Psychopharmacol 22(7):769–777
Reilly D, Didcott P, Swift W et al. (1998) Long-term cannabis use: characteristics of users in an Australian rural area. Addiction 93:837–846
Rey JM, Tennant CC (2002) Cannabis and mental health. Br Med J 325:1183–1184
Rodgers RJ, Haller J, Halasz J et al. (2003) 'One-trial sensitization' to the anxiolytic-like effects of cannabinoid receptor antagonist SR141716A in the mouse elevated plus-maze. Eur J NeuroSci 17:1279–1286
Rodriguez de Fonseca F, Carrera MR, Navarro M et al. (1997) Activation of corticotropin-releasing factor in the limbic system during cannabinoid withdrawal. Science 276:2050–2054
Rubino T, Sala M, Vigano D et al. (2007) Cellular mechanisms underlying the anxiolytic effect of low doses of peripheral delta(9)-tetrahydrocannabinol in rats. Neuropsychopharmacology 32:2036–2045
Rubino T, Guidali C, Vigano D et al. (2008a) CB1 receptor stimulation in specific brain areas differently modulate anxiety-related behaviour. Neuropharmacology 54:151–160
Rubino T, Realini N, Castiglioni C et al. (2008b) Role in anxiety behavior of the endocannabinoid system in the prefrontal cortex. Cereb Cortex 18:1292–1301
Rubino T, Vigano D, Realini N et al. (2008c) Chronic delta(9)-tetrahydrocannabinol during adolescence provokes sex-dependent changes in the emotional profile in adult rats: behavioral and biochemical correlates. Neuropsychopharmacology 33:2760–2771
Rush AJ, Weissenburger JE (1994) Melancholic symptom features and DSM-IV. Am J Psychiatry 151:489–498
Sagredo O, Ramos JA, Fernandez-Ruiz J et al. (2006) Chronic delta(9)-tetrahydrocannabinol administration affects serotonin levels in the rat frontal cortex. Naunyn Schmiedebergs Arch Pharmacol 372:313–317
Sanchis-Segura C, Cline BH, Marsicano G et al. (2004) Reduced sensitivity to reward in CB1 knockout mice. Psychopharmacology (Berl) 176:223–232
Scheen AJ, Finer N, Hollander P et al. (2006) Efficacy and tolerability of rimonabant in overweight or obese patients with type 2 diabetes: a randomised controlled study. Lancet 368:1660–1672
Scherma M, Medalie J, Fratta W et al. (2008) The endogenous cannabinoid anandamide has effects on motivation and anxiety that are revealed by fatty acid amide hydrolase (FAAH) inhibition. Neuropharmacology 54:129–140
Schofield D, Tennant C, Nash L et al. (2006) Reasons for cannabis use in psychosis. Aust NZ J Psychiatry 40:570–574
Shearman LP, Rosko KM, Fleischer R et al. (2003) Antidepressant-like and anorectic effects of the cannabinoid CB1 receptor inverse agonist AM251 in mice. Behav Pharmacol 14:573–582
Silverstone PH, Silverstone T (2004) A review of acute treatments for bipolar depression. Int Clin Psychopharmacol 19:113–124
Sofia RD, Knobloch LC (1976) Comparative effects of various naturally occurring cannabinoids on food, sucrose and water consumption by rats. Pharmacol Biochem Behav 4:591–599
Statham DJ, Heath AC, Madden PA et al. (1998) Suicidal behaviour: an epidemiological and genetic study. Psychol Med 28:839–855
Steiner MA, Marsicano G, Nestler EJ et al. (2008a) Antidepressant-like behavioral effects of impaired cannabinoid receptor type 1 signalling coincide with exaggerated corticosterone secretion in mice. Psychoneuroendocrinology 33:54–67
Steiner MA, Wanisch K, Monory K et al. (2008b) Impaired cannabinoid receptor type 1 signalling interferes with stress-coping behavior in mice. Pharmacogenomics J 8:196–208
Sullivan PF, Neale MC, Kendler KS (2000) Genetic epidemiology of major depression: review and meta-analysis. Am J Psychiatry 157:1552–1562

Suzuki A, Josselyn SA, Frankland PW et al. (2004) Memory reconsolidation and extinction have distinct temporal and biochemical signatures. J Neurosci 24:4787–4795

Szuster RR, Pontius EB, Campos PE (1988) Marijuana sensitivity and panic anxiety. J Clin Psychiatry 49:427–429

Thomas H (1993) Psychiatric symptoms in cannabis users. Br J Psychiatry 163:141–149

Tzavara ET, Davis RJ, Perry KW et al. (2003) The CB1 receptor antagonist SR141716A selectively increases monoaminergic neurotransmission in the medial prefrontal cortex: implications for therapeutic actions. Br J Pharmacol 138:544–553

Uriguen L, Perez-Rial S, Ledent C et al. (2004) Impaired action of anxiolytic drugs in mice deficient in cannabinoid CB1 receptors. Neuropharmacology 46:966–973

Van Gaal L, Pi-Sunyer X, Despres JP et al. (2008) Efficacy and safety of rimonabant for improvement of multiple cardiometabolic risk factors in overweight/obese patients: pooled 1-year data from the rimonabant in obesity (RIO) program. Diabetes Care 31(Suppl 2):S229–S240

Varvel SA, Wise LE, Niyuhire F et al. (2007) Inhibition of fatty-acid amide hydrolase accelerates acquisition and extinction rates in a spatial memory task. Neuropsychopharmacology 32:1032–1041

Vinod KY, Arango V, Xie S et al. (2005) Elevated levels of endocannabinoids and CB1 receptor-mediated G-protein signalling in the prefrontal cortex of alcoholic suicide victims. Biol Psychiatry 57:480–486

Willner P (2005) Chronic mild stress (CMS) revisited: consistency and behavioural-neurobiological concordance in the effects of CMS. Neuropsychobiology 52:90–110

Feeding Disorders and Obesity

Cristina Cervino, Valentina Vicennati, Renato Pasquali, and Uberto Pagotto

Contents

Abstract The ability of the endocannabinoid (EC) system to control appetite, food intake and energy balance has recently received great attention, particularly in the light of the different modes of action underlying these functions. The EC system modulates rewarding properties of food by acting at specific mesolimbic areas in the brain. In the hypothalamus, cannabinoid type 1 receptors (CB1) and ECs are integrated components of the networks controlling appetite and food intake. Interestingly, the EC system has recently been shown to control several metabolic functions by acting on peripheral tissues, such as adipocytes, hepatocytes, the skeletal muscles and the endocrine pancreas. The relevance of the system is further strengthened by the notion that visceral obesity seems to be a condition in which an overactivation of the EC system occurs; therefore, drugs interfering with this

C. Cervino V. Vicennati, R. Pasquali and U. Pagotto (✉)
Endocrinology Unit and C.R.B.A., Department of Clinical Medicine and Gastroenterology, S. Orsola-Malpighi Hospital, Via Massarenti 9, 40138, Bologna, Italy
e-mail: uberto.pagotto@unibo.it

D. Kendall and S. Alexander (eds.), *Behavioral Neurobiology of the Endocannabinoid System*, Current Topics in Behavioral Neurosciences 1,
DOI: 10.1007/978-3-540-88955-7_15, © Springer-Verlag Berlin Heidelberg 2009

overactivation by blocking CB1 receptors are considered valuable candidates for the treatment of obesity and related cardiometabolic risk factors.

Keywords Adipocyte • Cannabinoid type 1 receptor • Endocannabinoids • Hepatocyte • Hypothalamus • Obesity • Rimonabant • Taranabant

Abbreviations

AEA	Anandamide
AMPK	5'-AMP-activated protein kinase
AN	Anorexia nervosa
2AG	2-arachidonoyl-glycerol
BED	Binge-eating disorder
BN	Bulimia nervosa
CART	Cocaine-amphetamine-related transcript
CB1	Cannabinoid receptor type 1
CRH	Corticotropin releasing hormone
GABA	Gamma-aminobutyric acid
EC	Endocannabinoid
HDL	High-density lipoprotein
HOMA-IR	Homeostasis model assessment-insulin resistance
LDL	Low-density lipoprotein
MCH	Melanocortin concentrating hormone
NPY	Neuropeptide Y
PPAR	Peroxisome proliferator-activated receptor
PVN	Paraventricular nucleus
RIO	Rimonabant in obesity
THC	Δ^9-Tetrahydrocannabinol

1 Introduction

In physiology, the notion of the "thrifty genotype", first proposed by Neel (1962), is well known. He argued that certain human genotypes were selected because of their selective advantage over the less "thrifty" ones: in particular the "thrifty genotype" was described as "being exceptionally efficient in the intake and/or utilisation of food". Therefore, during famines, individuals with the thrifty genotype would have an advantage because they relied on more consistent, previously stored energy to maintain homeostasis. In recent years, several genes or systems have been described as members of the family of thrifty genes.

In our opinion, the genetic machinery comprising EC signalling should also be included in this family.

2 Historical Background

Marijuana was used for the treatment of appetite loss and to overcome the sensation of hunger in ancient Indian medical practice, and this represents the first evidence of a therapeutic role of the EC system in feeding disorders (Peters and Nahas 1999). However, it was only with the discovery of Δ^9-tetrahydrocannabinol (THC), the main psychoactive component of marijuana (Gaoni and Mechoulam 1964), and, more importantly, with the identification of the endogenous signalling molecules mimicking the marijuana effect (Piomelli 2003), the so-called endocannabinoids (ECs) such as anandamide (AEA) and 2-arachidonoyl glycerol (2AG), that the molecular basis of the orexigenic stimulus started to become clear.

3 Modes of Action by which Endocannabinoids Promote Energy Storage

3.1 Endocannabinoids Promote Energy Storage in the Brain

By analogy with the hyperphagic effect provided by THC, ECs have also been shown to induce an orexigenic stimulus in a dose-dependent manner when injected into the nucleus accumbens or hypothalamus (Williams and Kirkham 2002). Similar findings were derived from experiments in which the first selective CB1 receptor antagonist, rimonabant, was shown to decrease food intake (Simiand et al. 1998).

ECs are synthesised in the areas of the brain involved in the motivation of eating (Cota et al. 2006) in relation to the feeding state of animals, being increased during inter-meal times, reaching a critical level in order to trigger feeding, and rapidly reducing when food is accessed (Hanus et al. 2003). In rodent hypothalamic brain areas, levels of ECs have been shown to increase during fasting condition and to decrease as the animals are re-fed, returning to normal values in satiated animals (Kirkham et al. 2002).

ECs seem to interact with many different hypothalamic neuropeptides to generate, as a final effect, a strong impulse for food intake (Pagotto et al. 2006). They positively modulate the orexigenic signals provided by neuropeptide Y (NPY) (Gamber et al. 2005), but, on the other hand, they inhibit the anorexigenic action of cocaine-amphetamine-related transcript (CART) (Osei-Hyiaman et al. 2005b). In the paraventricular nucleus (PVN) of the hypothalamus they interact with corticotropin releasing hormone (CRH) (Cota et al. 2003; Hermann and Lutz 2005). Post-synaptically released ECs from the parvocellular neurons have been shown to decrease glutamatergic transmission onto CRH-releasing neurons, acting at pre-synaptic CB1 receptors, finally resulting in an inhibition of CRH release (Di et al. 2003). This mechanism is stimulated by a non-genomic effect of glucocorticoids, suggesting that the well-known effect of glucocorticoids on food intake may partly be caused by the activation of ECs (Di et al. 2003).

A recent study has demonstrated that CB1 receptor stimulation strongly augments the orexin-A-stimulated intracellular pathway, and that this effect can be blocked by the CB1 receptor antagonist rimonabant, suggesting a positive orexigenic role for CB1 in this neural population (Hilairet et al. 2003). Considering the fact that these neurones project to the ventral tegmental area, it is possible that the lateral hypothalamus, and the ECs acting there, represent the functional link between the hypothalamic circuitry controlling consummatory behaviour and limbic structures involved in food reward (Yo et al. 2005). The interactions between ECs, leptin and the orexigenic melanocortin-concentrating hormone (MCH) within the limbic system have recently been characterised in detail. Whereas MCH neurones are inhibited by gamma-aminobutyric acid (GABA)-ergic inputs from the limbic system (Yo et al. 2005), ECs act to reduce GABA release and, thus, might stimulate the excitability of MCH neurones, leading to an increase in food intake. However, this effect appears to be blocked in these neurones via a leptin-mediated inhibition of voltage-gated Ca^{2+} currents, potentially leading to a reduced synthesis and release of ECs and a subsequent reduced excitability of MHC neurones – the overall effect being a reduced orexigenic stimulus (Yo et al. 2005).

The orexigenic cross-talk between EC and ghrelin signalling represents a further intriguing hypothalamic interaction not yet fully explored. Ghrelin stimulation is able to increase hypothalamic EC content, leading, via CB1 receptor activation, to an increase of 5'-AMP-activated protein kinase (AMPK) (Kola et al. 2005). This effect may, in turn, promote appetite (Tucci et al. 2004; Kola et al 2008).

3.2 Endocannabinoids Promote Energy Storage at a Peripheral Level

3.2.1 Adipose Tissue

A key discovery in the emergence of a role of the ECs in the peripheral modulation of metabolic processes was the finding reported independently by two research groups of CB1 receptor expression in rodent white adipocytes (Bensaid et al. 2003; Cota et al. 2003). After these preliminary observations, further studies have shed light on the role of ECs on adipose tissue. Now it is widely accepted that, as in the brain, CB1 activation in the adipose tissue promotes energy storage. ECs stimulate growth and differentiation of pre-adipocytes to the fully mature adipocytes (Gari-Bobo et al. 2006; Bellocchio et al. 2008; Bouaboula et al. 2005; Pagano et al. 2007; Matias et al. 2006) via cross-talk with the peroxisome proliferator-activating receptor-γ (PPAR-γ) (Matias et al. 2006). By inhibiting lipolysis on the one hand (Gasperi et al. 2007; Jbilo et al. 2005), and activating lipoprotein lipase on the other (Cota et al. 2003), and by stimulating de novo lipogenesis, ECs are able to trigger fat accumulation (Osei-Hyiaman et al. 2005a). ECs have also been reported to increase insulin signalling and glucose uptake, to further promote energy storage (Pagano et al. 2007).

Furthermore, CB1 receptor activation modulates important products of adipocyte endocrine activity such as adiponectin and visfatin (Bensaid et al. 2003; Matias et al. 2006; Perwitz et al. 2006). Adiponectin is a circulating adipokine that plays a crucial role in fat and glucose metabolism, and obesity and type 2 diabetes are characterised by reduced adiponectin levels in tissues and blood (Kadowaki and Yamamuchi 2005). Visfatin is an insulin-mimetic growth factor, levels of which are increased in obesity and type 2 diabetes (Tilg and Moschen 2008). In this scenario, the CB1 receptor agonist WIN55,212 has been shown to reduce adiponectin and increase visfatin expression in cultured adipocytes (Perwitz et al. 2006), whereas rimonabant is able to increase adiponectin expression and release from adipocytes in vitro (Bensaid et al. 2003; Gari-Bobo et al. 2006; Matias et al. 2006).

3.2.2 Liver

The CB1 receptor has also been identified in the liver (Osei-Hyiaman et al. 2005a). Its expression is very low in normal hepatocytes, but it is raised in obesity. ECs are able to increase de novo lipogenesis and the expression of the transcription factor sterol regulatory element-binding protein-1c and its targets acetyl-coenzyme A carboxylase-1 and fatty acid synthase; this occurs particularly in association with a high fat diet (Osei-Hyiaman et al., 2005). Intriguingly, mice with selective deletion of CB1 receptors from their hepatocytes develop obesity on a high fat diet, but are protected from diet-induced hepatic steatosis, insulin and leptin resistance and dyslipidaemia (Osei-Hyiaman et al. 2008). All of these findings clearly highlight the role of hepatic CB1 receptors in the regulation of metabolism (Pacher et al. 2006).

3.2.3 Skeletal Muscle

The CB1 receptor is also expressed in skeletal muscles such as the soleus muscle (Pagotto et al. 2006). ECs decrease mRNA expression of several enzymes involved in muscle oxidation, such as AMPK-α1 and -α2, pyruvate dehydrogenase kinase-4, and PPAR-γ coactivator-1 in cultured myotube cells derived from lean and obese subjects. These effects could be reversed by CB1 receptor antagonist treatment, indicating a crucial negative role of ECs on fatty acid and glucose oxidation in skeletal muscle (Liu et al. 2005; Cavuoto et al. 2007).

3.2.4 Endocrine Pancreas

It is only recently that both CB1 and CB2 receptors have been demonstrated in the endocrine pancreas. The CB1 receptor is mainly present in glucagon-containing α-cells, while the CB2 receptor has been detected in both α- and β-cells (Juan-Pico et al. 2006; Bermudez-Silva et al. 2008; Starowicz et al. 2008). The stimulation of CB1 receptors by ECs seems to induce glucose intolerance in rats, and this effect can be blocked by AM251, a specific CB1 receptor antagonist. The mechanism underlying EC-mediated glucose intolerance is probably a reduction of glucose-dependent insulin secretion (Bermudez-Silva et al. 2006).

4 Obesity as a Disease Model of Endocannabinoid Overactivation

Recent data, obtained from both animal and human studies, have identified a tight association between the development of obesity and a simultaneous over-activation of ECs, expressed as raised EC production or increased CB1 receptor expression (Matias and Di Marzo 2007). Di Marzo and Kunos were the first to illustrate such an association. Models of obesity such as *ob/ob* and *db/db* mice, characterised by an impairment of leptinergic signalling, have been shown to have increased (pathological) levels of hypothalamic ECs (Di Marzo et al. 2001), and a single intravenous injection of leptin in these animals was able to reduce the overproduction of ECs (Di Marzo et al. 2001).

In wild-type mice on a high fat diet, an increase in hepatic anandamide (AEA) associated with an increased density of CB1 receptors has been observed (Osei-Hyiaman et al. 2005a). The ECs might, therefore, play a role in several hepatic diseases in which steatotic processes progressively replace the normal liver structure. The high fat diet-induced transformation in hepato-steatosis has not been observed in mice pre-treated with rimonabant or in CB1 receptor knockout mice (Osei-Hyiaman et al. 2005a).

Another target of ECs in obesity may be the skeletal muscle. In fact, as mentioned above, the CB1 receptor is present in murine skeletal muscle and its expression is increased in diet-induced obese mice (Pagotto et al. 2006).

Regarding the endocrine pancreas, it was recently shown that cultured β-cells, under conditions mimicking hyperglycaemia, expressed elevated levels of both AEA and 2AG (Matias et al. 2006), suggesting that, under these conditions, such as during pre-diabetes, type 2 diabetes and obesity, EC production and/or degradation in β-cells, instead of remaining under insulin control, becomes dysregulated.

An increasing body of evidence suggests that an up-regulation of EC signalling may also occur in overweight, obese and hyperglycaemic patients. In obese women without comorbidities, blood levels of either AEA alone or both AEA and 2AG were significantly higher compared with lean subjects (Engeli et al. 2005). Significantly higher levels of 2AG, but not AEA, have been detected in the visceral, but not subcutaneous, fat of obese patients (Matias et al. 2006). Interestingly, increased levels of haematic circulating 2AG are present in visceral obese human patients when compared to subcutaneous obese and lean controls (Bluher et al. 2006). The increase in 2AG was also shown to correlate positively with some important cardiometabolic risk factors, such as body mass index, waist circumference, fasting plasma triglyceride and insulin levels, low high-density lipoprotein (HDL) cholesterol and adiponectin levels (Côtè et al. 2007). However, whether 2AG and/or AEA are important in both obesity and related comorbidities and whether EC over-activation is the effective cause or just a simple consequence of such diseases still remain matters of debate.

5 Endocannabinoids and Central Eating Disorders

There are only a few reports concerning the putative association between alterations in the EC system and eating disorders such as anorexia nervosa. In a recent publication, Monteleone and colleagues reported measurements of plasma levels of AEA, 2AG, and leptin in women with anorexia nervosa (AN), with bulimia nervosa (BN), with binge-eating disorder (BED), and in a group of healthy women. Plasma levels of AEA were significantly enhanced in both anorexic and BED women, but not in bulimic patients. No significant changes occurred in the plasma levels of 2AG in any of the patient groups. Moreover, circulating AEA levels were significantly and inversely correlated with plasma leptin concentrations in both healthy controls and anorexic women. These findings show a derangement in the production of AEA in drug-free symptomatic women with AN or BED. Although the pathophysiological significance of this alteration awaits further clarification, it suggests a possible involvement of AEA in the mediation of the rewarding aspects of the aberrant eating behaviours occurring in AN and BED (Monteleone et al. 2005).

6 How CB1 Receptor Antagonism May Act Against Obesity and Metabolic Complications

A significant body of evidence has recently accumulated suggesting that CB1 blockade may display favourable metabolic effects beyond weight loss alone. In normal rats and in diet-induced obese mice chronically treated with different CB1 receptor antagonists, tolerance to the anorectic effect of the drugs developed in a few days, whereas the reduction of body weight was maintained throughout the treatment period (Colombo et al. 1998; Ravinet Trillou et al. 2003; Hildebrandt et al. 2003). Additional evidence in favour of a peripheral mechanism of action of CB1 blockade came from the CB1 knockout mouse model. These mice showed a significant weight deficit compared with their wild-type littermates (Cota et al. 2003; Osei-Hyiaman et al. 2005a). Pair feeding studies showed that the weight deficit was caused by a reduction in food intake only in young animals, whereas in adult animals it was partially independent of caloric intake (Cota et al. 2003), similar to the dissociation between effects of CB1 receptor antagonists on food intake and body weight. Several studies have been performed in both genetically obese and diet-induced obese mouse models searching for non-appetite-related pathways involved in CB1 receptor antagonist-mediated body weight loss. Dernbach et al. (2005) was the first to note that, in addition to the well-known effect on food intake, administration of rimonabant for 10 days to obese rats induced an increase in energy expenditure, whilst, in the pair-feeding control group, the lowered food intake led to an expected decrease in the same parameter. The low respiratory quotient observed in the rimonabant-treated group led the authors to speculate that a shift to an increase in fat oxidation could be one of the possible mechanisms by which the pharmacological blockade of the CB1 receptor produced such an effect.

Taking advantage of micro-array analysis, Jbilo et al. (2005) were able to screen the expression of a wide panel of genes in adipocytes from diet-induced obese mice after long-term treatment with rimonabant. They found that the transcriptional patterns of treated obese mice were similar to those obtained in the CB1 receptor knockout mice fed with a high-fat diet, supporting a role for CB1 receptors in this process. Functional analysis of these gene modulations indicated that the drug-induced reduction of adipose mass was due to increased energy expenditure, mainly through futile cycling (calcium and substrate) (Jbilo et al. 2005).

Several reports have documented a direct role of ECs in modulation of proteins involved in thermogenesis. It has also recently been shown that treatment of differentiated brown adipocytes with a CB1 receptor agonist decreased the expression of uncoupling protein 1 (Perwitz et al. 2006). The contribution of brown adipose tissue-mediated thermogenesis in the process of energy expenditure in small animals such as rodents is well established, whereas the role of brown adipose tissue is less clear in humans. However, it has been suggested that several physiological and pharmacological stimuli may be capable of *trans*-differentiating white adipocytes into brown adipocytes in humans (Klaus 2004). One could, therefore, speculate that CB1 receptor antagonists may increase numbers of brown adipocytes, leading to an eventual increase in energy expenditure. On the other hand, we recently showed that CB1 receptor blockade provides a potent stimulus to mitochondriogenesis in adipose tissue contributing further to an increase in energy expenditure (Tedesco et al. 2008).

As described above, CB1 receptor blockade may also affect peripheral tissues via elevation of adiponectin levels (Bensaid et al. 2003; Perwitz et al. 2006; Matias et al. 2006). Adiponectin reduces serum hepatic gluconeogenesis and stimulates fatty acid oxidation in skeletal muscles. Adiponectin stimulates skeletal muscle glucose uptake by increasing insulin receptor tyrosine kinase activity, p38 mitogen-activated protein kinase, and the tyrosine phosphorylation of insulin receptor substrate-1. Insulin resistance in skeletal muscle and in the hepatic parenchyma develops as excess triglycerides accumulate; thus, the improvement in insulin action induced by adiponectin may promote clearance of intracellular lipid stores in these tissues. Adiponectin also promotes mitochondrial fatty acid oxidation and, consequently, a reduction in circulating levels of free fatty acids. High concentrations of free fatty acids and insulin resistance associated with visceral obesity are highly detrimental because they inhibit lipoprotein lipase, the enzyme responsible for hydrolysing triglycerides in very-low-density lipoprotein and chylomicrons. Low lipoprotein lipase activity leads to an accumulation of triglycerides and lipoprotein remnant particles that are highly atherogenic. Under these circumstances, HDL cholesterol decreases significantly. Adiponectin also favourably affects this metabolic scenario by increasing PPAR-γ activity in skeletal muscle and liver, leading to an increased synthesis of HDL cholesterol through increased hepatic expression of apoproteins A-I and A-II (Kadowaki and Yamamuchi 2005; Guerre-Millo 2008).

7 Pharmacological Implications in Humans

7.1 Rimonabant

Rimonabant has been evaluated in four multicenter, randomised, placebo-controlled clinical trials: rimonabant in obesity (RIO) – Europe, RIO – North America, RIO – lipids, and RIO – diabetes (Van Gaal et al. 2005; Despres et al. 2005; Pi-Sunyer et al. 2006; Scheen et al. 2006). Patients in these studies were randomised to rimonabant 20 mg/day, rimonabant 5 mg/day, or placebo, given in combination with a hypocaloric diet. The patients were additionally advised to increase their physical activity. In all four studies, treatment with rimonabant (20 mg/day) resulted in significantly greater decreases in body weight and waist circumference than did treatment with placebo.

Treatment with rimonabant (20 mg/day) also produced improvements in a number of cardiovascular and metabolic risk factors that were, in most studies, significantly greater than those achieved with placebo. These included increases in HDL cholesterol, and decreases in triglycerides, low-density lipoprotein (LDL) cholesterol–HDL cholesterol ratio, total cholesterol–HDL cholesterol ratio, fasting glucose, fasting insulin, and homeostasis model assessment-insulin resistance (HOMA-IR).

Additional treatment effects were demonstrated in the RIO-Lipids study, which focused on patients with untreated dyslipidaemia and a high risk of cardiovascular disease (Després et al. 2005). Included in these effects were a shift in the distribution of LDL particles towards a larger size and a decrease in the proportion of small LDL particles in the rimonabant (20 mg/day) group compared with the placebo group ($P<0.001$ and $P=0.002$, respectively), Also, compared with placebo, rimonabant resulted in greater decreases in plasma levels of leptin and C-reactive protein ($P<0.001$ and $P<0.02$), and greater increases in plasma adiponectin ($P<0.001$). This latter effect on adiponectin nicely confirmed the initial findings from in vitro and animal studies.

Results of the RIO-Diabetes (Scheen et al. 2006) and Serenade (Rosenstock et al. 2008) studies extended the findings with rimonabant in overweight or obese non-diabetic patients to overweight or obese patients with type 2 diabetes that was inadequately controlled by metformin or sulphonylureas, or who were naïve-treated.

HbA_{1C} levels were significantly lower in the rimonabant-treated group than in the placebo group in both studies.

It is important to note that the changes in the levels of HDL cholesterol, triglycerides, fasting insulin, HOMA-IR, HbA_{1C}, and adiponectin in the various studies exceeded those expected for weight loss alone. This is consistent with the direct peripheral metabolic effects of rimonabant described earlier.

Regarding the safety of rimonabant for the treatment of obesity, this agent was generally well tolerated in the four randomised, placebo-controlled studies, with the most common adverse events including gastrointestinal disturbances, upper respiratory tract infections, and dizziness. Results of a meta-analysis based on these trials indicated that patients receiving rimonabant, compared with those taking placebo, were 2.5 times more likely to discontinue treatment because of depressive mood disorders and three times more likely to discontinue because of anxiety (Christensen et al. 2007). According to data presented by the manufacturer at an advisory committee meeting of the US Food and Drug Administration (FDA) in June 2007, rates of depression and anxiety for 2742 patients who received rimonabant 20 mg were 3.9% and 5.9%, respectively, compared with 1.7 and 2.1% for 2474 patients who received placebo. Definite suicidal behaviour/ideation, while uncommon, was higher among 3081 patients who received rimonabant (20 mg/day) than among 2214 who received placebo (0.65% vs. 0.36%) (Chew 2007).

7.2 Taranabant

Taranabant, another CB1 receptor inverse agonist, has been studied in a 2-year, phase 3 trial scheduled for completion in the last quarter of 2007. Interim results for this trial are not available. However, a recent shorter clinical trial in which taranabant was tested in humans confirmed the efficacy of CB1 receptor inhibitors in reducing body weight and ameliorating several metabolic conditions. A single-dose administration of 12 mg taranabant caused a small but significant increase in resting energy expenditure and fat oxidation (Addy et al. 2008). The authors concluded that the modest increases in energy expenditure may, nevertheless, exert profound effects on body weight over a period of months. On the other hand, in the same study, Addy et al. were not able to define what proportion of taranabant's effects on resting energy expenditure were mediated centrally via activation of the autonomic nervous system or peripherally by engagement of CB1 receptor distributed in peripheral organs involved in metabolic functions (Addy et al. 2008).

8 Conclusions

The EC system is now recognised as a crucial player in the control of energy balance. It is clear that it influences a large variety of peripheral organs to modulate metabolic processes. Detailed characterisation of each individual contribution and the reciprocal interactions among the organs is necessary in future studies to fully understand the physiological and pathophysiological roles of the EC system. The system appears to be over-activated in conditions such as obesity, and pharmacological blockade of CB1 receptors normalises the imbalance providing an attractive strategy to tackle obesity and associated comorbidities.

References

Addy C, Wright H, Van Laere K et al. (2008) The acyclic CB1R inverse agonist taranabant mediates weight loss by increasing energy expenditure and decreasing caloric intake. Cell Metab 7:68–78

Bellocchio L, Cervino C, Vicennati V et al. (2008) Cannabinoid type 1 receptor: another arrow in the adipocytes' bow. J Neuroendocrinol 20(Suppl 1):130–138

Bensaid M, Gary-Bobo M, Esclangon A et al. (2003) The cannabinoid CB1 receptor antagonist SR141716 increases Acrp30 mRNA expression in adipose tissue of obese *fa/fa* rats and in cultured adipocyte cells. Mol Pharmacol 63:908–914

Bermudez-Silva FJ, Serrano A, Diaz-Molina FJ et al. (2006) Activation of cannabinoid CB1 receptors induces glucose intolerance in rats. Eur J Pharmacol 531:282–284

Bermudez-Silva FJ, Suarez J, Baixeras E et al. (2008) Presence of functional cannabinoid receptors in human endocrine pancreas. Diabetologia 51:476–487

Bluher M, Engeli S, Kloting N et al. (2006) Dysregulation of the peripheral and adipose tissue endocannabinoid system in human abdominal obesity. Diabetes 5:3353–3360

Bouaboula M, Hilairet S, Marchand J et al. (2005) Anandamide induced PPAR gamma transcriptional activation and 3T3–L1 preadipocyte differentiation. Eur J Pharmacol 517:174–181

Cavuoto P, McAincha AJ, Hatzinikolas G et al. (2007) Effects of cannabinoid receptors on skeletal muscle oxidative pathways. Mol Cell Endocrin 267:63–69

Chew P (2007) Clinical safety. Presented to the Endocrinologic and Metabolic Drugs Advisory Committee, FDA, June 13, 2007. Available at www.fda.gov/ohrms/dockets/ac/07/slides/2007-4306s1-05-Sanofi-Aventis Chew.pdf (slides MM-95, MM-109), accessed Dec 15, 2007

Christensen R, Kristensen PK, Bartels EM et al. (2007) Efficacy and safety of the weight-loss drug rimonabant: a meta-analysis of randomised trials. Lancet 370:1706–1713

Colombo G, Agabio R, Diaz G et al. (1998) Appetite suppression and weight gain loss after the cannabinoid antagonist SR141716. Life Sci 63:PL113–PL117

Cota D, Marsicano G, Tschoep M et al. (2003) The endogenous cannabinoid system affects energy balance via central orexigenic drive and peripheral lipogenesis. J Clin Invest 112:423–431

Cota D, Tschoep M, Horvath TL et al. (2006) Cannabinoids, opioids and eating behavior: the molecular face of hedonism? Brain Res Rev 5:85–107

Côtè M, Matias I, Lemieux I et al. (2007) Circulating endocannabinoid levels, abdominal adiposity and related cardiometabolic risk factors in obese men. Int J Obesity 31:692–699

Dernbach H, Haller J, Kaulisch T et al. (2005) Effects of the CB1 antagonist SR141716A (Rimonabant) on energy expenditure and respiratory quotient in diet-induced obese rats. Obesity Rev 6:172–173

Després JP, Golay A, Sjöström L for the Rimonabant in Obesity-Lipids Study Group (2005) Effects of rimonabant on metabolic risk factors in overweight patients with dyslipidemia. N Engl J Med 353:2121–2134

Di Marzo V, Goparaju SK, Wang L et al. (2001) Leptin-regulated endocannabinoids are involved in maintaining food intake. Nature 410:822–825

Di S, Malcher-Lopes R, Halmos KC et al. (2003) Nongenomic glucocorticoid inhibition via endocannabinoid release in the hypothalamus: a fast feedback mechanism. J Neurosci 23:4850–4857

Engeli S, Bohnke J, Feldpausch M et al. (2005) Activation of the peripheral endocannabinoid system in human obesity. Diabetes 54:2838–2843

Gamber KM, Macartur H, Westfall TC (2005) Cannabinoids augment the release of neuropeptide Y in the rat hypotalamus. Neuropharmacology 143:520–523

Gaoni Y, Mechoulam R (1964) Isolation, structure and partial synthesis of an active constituent of hashish. J Am Chem Soc 86:1646–1647

Gari-Bobo M, Elachouri G, Scatton B et al. (2006) The cannabinoid CB1 receptor antagonist rimonabant (SR141716) inhibits cell proliferation and increase markers of adipocyte maturation in cultured mouse 3T3 F442A preadipocytes. Mol Pharmacol 69:471–478

Gasperi V, Fezza F, Pasquariello N et al. (2007) Endocannabinoids in adipocytes during differentiation and their role in glucose uptake. Cell Mol Life Sci 64:219–229

Guerre-Millo M (2008) Adiponectin: an update. Diabetes Metab 34:12–18

Hanus L, Avraham Y, Ben-Shushan D et al. (2003) Short term fasting and prolonged semistarvation have opposite effects on 2-AG levels in the mouse brain. Brain Res 983:144–151

Hermann H, Lutz B (2005) Coexpression of cannabinoid receptor type 1 with the corticotrophin releasing hormone receptor type-1 in distinct region of mouse adult forebrain. Neurosci Lett 375:13–18

Hilairet S, Bouaboula M, Carriere D et al. (2003) Hypersensitization of the Orexin 1 receptor by the CB1 receptor: evidence for crosstalk blocked by the specific CB1 antagonist, SR141716. J Biol Chem 278:23731–23737

Hildebrandt AL, Kelly-Sullivan DM, Black SC (2003) Anti-obesity effects of chronic CB1 receptor antagonist treatment in diet induced obese mice. Eur J Pharmacol 462:125–132

Jbilo O, Ravinet-Trillou C, Arnone M et al. (2005) The CB1 receptor antagonist rimonabant reverses the diet-induced obesity phenotype through the regulation of lipolysis and energy balance. FASEB J 19:1567–1569

Juan-Pico P, Fuentes E, Bermudez-Silva FJ et al. (2006) Cannabinoid receptor regulates Ca(2+) signals and insulin secretion in pancreatic beta-cells. Cell Calcium 39:155–162

Kadowaki T, Yamamuchi T (2005) Adiponectin and adiponectin receptors. Endocrin Rev 26:439–451

Kirkham TC, Williams CM, Fezza F et al. (2002) Endocannabinoid levels in rat limbic forebrain and hypothalamus in relation to fasting, feeding and satiation: stimulation of eating by 2-arachidonoyl glycerol. Br J Pharmacol 136:550–557

Klaus C (2004) Adipose tissue as a regulator of energy balance. Curr Drug Targets 5:241–250

Kola B, Hubina E, Tucci SA et al. (2005) Cannabinoids and ghrelin have both central and peripheral metabolic effects via AMP-activated protein kinase. J Biol Chem 280:25196–25202

Kola B, Farkas I, Christ-Crain M et al. (2008) The orexigenic effect of ghrelin is mediated through central activation of the endogenous cannabinoid system. PLoS ONE Mar 12; 3:e1797 PMID: 18335063

Liu YL, Connoley IP, Wilson CA et al. (2005) Effects of the cannabinoid CB1 receptor antagonist SR141716 on oxygen consumption and soleus muscle glucose uptake in Lep(ob)/Lep(ob) mice. Int J Obesity 29:183–187

Matias I, Di Marzo V (2007) Endocannabinoids and the control of energy balance. Trends Endocrinol Metab 18:27–37

Matias I, Gonthier MP, Orlando P et al. (2006) Regulation, function and dysregulation of endocannabinoids in models of adipose and β-pancreatic cells and in obesity and hyperglicemia. J Clin Endocrinol Metab 91:3171–3180

Monteleone P, Matias I, Martiadis V et al. (2005) Blood levels of the endocannabinoid anandamide are increased in anorexia nervosa and in binge-eating disorder, but not in bulimia nervosa. Neuropsychopharmacology 30:1216–1221

Neel JV (1962) Diabetes mellitus: A "thrifty" genotype rendered detrimental by "progress". Am Hum Genet 14:353–362

Osei-Hyiaman D, De Petrillo M, Pacher P et al. (2005a) Endocannabinoids activation at hepatic CB1 receptors stimulates fatty acid synthesis and contributes to diet induced obesity. J Clin Invest 115:1298–1305

Osei-Hyiaman D, Depetrillo M, Harvey-White J et al. (2005b) Cocaine- and amphetamine-related transcript is involved in the orexigenic effect of anandamide. Neuroendocrinology 81:273–282
Osei-Hyiaman D, Liu J, Godlewski G et al. (2008) Hepatic CB1 receptor is required for development of diet-induced steatosis, dyslipidemia, and insulin and leptin resistance in mice. J Clin Invest 118:3160–3169
Pacher P, Batkai S, Kunos G (2006) The endocannabinoid system as an emerging target of pharmacotherapy. Pharmacol Rev 58:389–462
Pagano C, Pilon C, Calcagno A et al. (2007) The endogenous cannabinoid system stimulates glucose uptake in human fat cells via phosphatidylinositol 3-kinase and calcium-dependent mechanisms. J Clin Endocrinol Metab 92:4810–4819
Pagotto U, Marsicano G, Cota D et al. (2006) The emerging role of endocannabinoid system in endocrine regulation and energy balance. Endocr Rev 27:73–100
Perwitz N, Fasshauer M, Klein J (2006) Cannabinoid receptor signalling directly inhibits thermogenesis and alters expression of adiponectin and visfatin. Horm Metab Res 38:356–358
Peters H, Nahas GG (1999) A brief history of four millennia (B.C.2000–A.D. 1974). In: Nahas GG, Sutin K, Harvey D, Agurell S (eds) Marihuana and medicine. Humana Press Inc, Totowa (NJ), pp 3–7
Piomelli D (2003) The molecular logic of endocannabinoid signalling. Nat Rev Neurosci 4:873–884
Pi-Sunyer FX, Aronne LJ, Heshmati HM, Devin J, Rosenstock J for the RIO-North America Study Group (2006) Effect of rimonabant, a cannabinoid-1 receptor blocker, on weight and cardiometabolic risk factors in overweight or obese patients: RIO-North America: a randomized controlled trial. JAMA 295:761–775
Ravinet Trillou C, Arnone M, Delgorge C et al. (2003) Anti-obesity effect of SR141716, a CB1 receptor antagonist, in diet induced obese mice. Am J Physiol Regul Integr Comp Physiol 284: R345–R353
Rosenstock J, Hollander P, Chevalier S, Iranmanesh A for the SERENADE study group (2008) The SERENADE trial: effects of monotherapy with rimonabant, the first selective CB1 receptor antagonist, on glycemic control, body weight and lipid profile in drug-naive type 2 diabetes. Diabetes Care Aug 4. PMID: 18678611
Scheen AJ, Finer N, Hollander P, Jensen MD, Van Gaal LF for the RIO-Diabetes Study Group (2006) Efficacy and tolerability of rimonabant in overweight or obese patients with type 2 diabetes: a randomised controlled study. Lancet 368:1660–1672
Simiand J, Keane M, Keane PE et al. (1998) SR 141716, a CB1 cannabinoid receptor antagonist, selectively reduces sweet food intake in marmoset. Behav Pharmacol 9:179–181
Starowicz KA, Cristino L, Matias I et al. (2008) Endocannabinoid dysregulation in the pancreas and adipose tissue of mice fed with a high-fat diet. Obesity 16:553–565
Tedesco L, Valerio A, Cervino C et al. (2008) Cannabinoid type 1 receptor blockade promotes mitochondrial biogenesis through endothelial nitric oxide synthase expression in white adipocytes. Diabetes 57:2028–2036
Tilg H, Moschen AR (2008) Role of adiponectin and PBEF/visfatin as regulators of inflammation: involvement in obesity-associated diseases. Clin Sci (Lond) 114:275–288
Tucci SA, Rogers EK, Korbonits M et al. (2004) The cannabinoid receptor antagonist SR141716 blocks the orexigenic effect of intrahypothalamic ghrelin. Br J Pharmacol 143:520–523
Van Gaal LF, Rissanen AM, Scheen AJ, Ziegler O, Rössner S for the RIO-Europe Study Group (2005) Effects of the cannabinoid-1 receptor blocker rimonabant on weight reduction and cardiovascular risk factors in overweight patients: 1-year experience from the RIO-Europe study. Lancet 365:1389–1397
Williams CM, Kirkham TC (2002) Observational analysis of feeding induced by Δ^9-THC and anandamide. Physiol Behav 76:241–250
Yo JH, Chen YJ, Talmage DA et al. (2005) Integration of leptin signalling in an appetite-related neural circuit. Neuron 48:1055–1066

Schizophrenia

F. Markus Leweke

Contents

Abstract Since the discovery of the endocannabinoid system, a growing body of psychiatric research has emerged focusing on the potential role of this system in schizophrenia. On the basis of earlier epidemiological studies and results from animal models, endocannabinoids and their relation to symptoms are considered in clinical studies as well as in *post-mortem* analyses of cannabinoid CB_1 receptor densities. A possible neurobiological mechanism for the deleterious influence of cannabis use in schizophrenia is discussed, involving the disruption of endogenous cannabinoid signalling and function.

Keywords Schizophrenia • Endocannabinoids • Δ^9-THC • Animal modes • CSF

F. Markus Leweke
Department of Psychiatry and Psychotherapy, Central Institute of Mental Health, 68159, Mannheim, Germany

D. Kendall and S. Alexander (eds.), *Behavioral Neurobiology of the Endocannabinoid System*, Current Topics in Behavioral Neurosciences 1,
DOI: 10.1007/978-3-540-88955-7_16,

Abbreviations

ACC Anterior cingulate cortex
CSF Cerebrospinal fluid
PCC Posterior cingulate cortex

1 Introduction

Investigations of the psychotomimetic properties of herbal cannabis preparations and certain purified phytocannabinoid compounds – and thus, indirectly, of manipulating the endogenous cannabinoid system – started as early as 1845, when Moreau de Tours observed the effects of cannabis in an experimental setting and described the occurrence of psychotic symptoms in healthy individuals after cannabis administration. In 1932, Beringer et al. described specific thought disturbances, perceptual alterations and delusions in healthy subjects after administration of a standardised extract from *Cannabis sativa*. These psychopathological findings were quite similar to the acute symptoms of schizophrenia. After more than 60 years of declining interest in this issue, further studies in healthy volunteers were initiated. It was now possible to use synthetic Δ^9-tetrahydrocannabinol (Δ^9-THC) instead of a natural preparation of *C. sativa* with its abundance of different cannabinoid compounds. These studies have observed psychotic symptoms and alterations of perception (Leweke et al. 1999b) and emotional processing (Leweke et al. 1998) after oral administration of Δ^9-THC. Furthermore, transient schizophreniform symptoms and cognitive disturbances were caused by both orally and intravenously administered Δ^9-THC (D'Souza et al. 2004; Koethe et al. 2006).

These findings were in agreement with the impressions of clinicians, over many decades, that frequent cannabis use increases the risk of schizophrenia and that schizophrenic patients are more likely than healthy individuals to abuse cannabis. Both observations have been verified over recent years. First, it has been demonstrated in a number of epidemiological studies that frequent cannabis use is associated with a greater risk of suffering psychotic symptoms or developing schizophrenia (Arseneault et al. 2002; Henquet et al. 2005; van Os et al. 2002; Zammit et al. 2002). This was first described by Andreasson and colleagues (1987), who reported an increased risk of developing schizophrenia correlated with a higher frequency of cannabis use in a cohort of Swedish conscripts. Even after correction for certain confounding variables, a twofold increased risk for schizophrenia was estimated following more than twenty episodes of cannabis use (Andreasson et al. 1989). These findings have been a source of controversy for more than a decade, but a number of more recent epidemiological studies applying different strategies also suggest an overall twofold increased risk of suffering schizophrenia in the wake of frequent cannabis use (for review see Arseneault et al. 2004), confirming the initial Swedish results. Additionally, cannabis users tend to be significantly younger when developing schizophrenia and to suffer earlier from negative symptoms of the disease, both representing negative prognostic factors in schizophrenia (Veen et al. 2004). Second, schizophrenic patients are more likely than healthy individuals to abuse cannabis (Kovasznay et al. 1997). Several models have been proposed to explain the etiological relationship between substance abuse and psychosis, but no single model is able to explain all co-morbidity adequately. However, there is no substantial epidemiological evidence that cannabis abuse serves as a kind of self-medication for schizophrenic patients, while greater support has been found in self-reporting studies for an "alleviation of dysphoria" model, in which patients see substance misuse as a means of alleviating unpleasant affective states (Gregg et al. 2007). Furthermore, in schizophrenic patients, the abuse of cannabis seems to trigger psychotic symptoms and may worsen the outcome of the disease (e.g. Linszen et al. 1994). Whereas cannabis use may precipitate the development of psychosis in vulnerable people, the hypothesis that cannabis use causes schizophrenia is not supported (Leweke et al. 2007a).

Taken together, there exists a large body of evidence that there is a relevant association between acute and chronic cannabis, or more specifically Δ^9-THC, intake and behavioural, cognitive and psychotic symptoms or the development of schizophrenia. It is known that Δ^9-THC, the main psychoactive compound of herbal cannabis preparations, binds to and activates CB_1 receptors and has been suggested to disrupt the physiological role of endogenous cannabinoids, its primary target system. While, over the last decade, the endogenous cannabinoid system has become a major theme of interest in a variety of fields such as pain modulation, neurotransmitter systems, energy metabolism and immune functions, the system has also been hypothesised to be involved in the pathophysiology of schizophrenia (Emrich et al. 1997).

2 Neurobiology of the Endocannabinoid System in Schizophrenia

2.1 Animal Studies

Hypotheses concerning a disturbance of endocannabinoid functions in schizophrenia have been developed on the basis of valid animal models of cannabinoid-associated schizophrenia-like symptoms in rodents, and researchers have tested various cannabinoid compounds on a variety of animal models for schizophrenia. For example, animal models reflecting certain aspects of schizophreniform symptoms have been developed using acute or chronic treatment of rodents with the cannabinoid receptor agonist WIN55212-2 (Schneider and Koch 2003, 2005; Schneider et al. 2005). It is noteworthy that most of these models target the developing brain, suggesting its heightened susceptibility to the effects of exogenous cannabinoids both during the perinatal period through maternal cannabis use and in young adolescent users (Schneider 2008). A number of studies have demonstrated a subtle rather than gross effect of cannabis upon later brain function, including the development of schizophreniform psychotic symptoms (for review see Sundram 2006).

Animal experiments suggested improvement of psychotic symptoms by CB receptor antagonism: rimonabant (SR141716A), a selective CB_1 receptor antagonist, was able to reduce the hyperactivity induced in gerbils by various stimulant drugs known to produce schizophrenic-like symptoms (Poncelet et al. 1999). Interestingly, in a clinical study with acute schizophrenic patients, the blockade of the CB_1 receptor by rimonabant did not show corresponding effects and the psychotic symptoms did not improve (Meltzer et al. 2004). To date, however, there are insufficient clinical data to support the hypothesis that CB_1 receptor antagonists could work as antipsychotic drugs, although there is evidence from basic research involving animal studies. Further animal studies showed that administration of Δ^9-THC increased dopaminergic activity in the mesolimbic dopaminergic system, and (indirectly) acetylcholine release in the hippocampus and prefrontal cortex, whereas Δ^9-THC (twice daily for 7 or 14 days) caused a persistent and selective reduction in medial prefrontal cortical dopamine turnover (Pisanu et al. 2006; Verrico et al. 2003). Interestingly cannabidiol, the non-psychotropic main compound in *C. sativa*, revealed antipsychotic properties in a study with rats (Zuardi et al. 1991). Cannabidiol is now being tested in clinical trials with promising preliminary results concerning both psychopathology and cognitive improvement (Leweke et al. 2005).

2.2 *Human Post-Mortem Studies*

With regard to endocannabinoid receptors, Dean et al. (2001) reported increased binding of the cannabinoid receptor agonist [^{3}H]-CP55940 to CB_1 receptors in the dorsolateral prefrontal cortex of schizophrenic patients as compared to controls using quantitative autoradiography, but showed no relationship with recent cannabis use. Zavitsanou et al. (2004) examined the distribution and density of CB_1 receptors in post-mortem anterior cingulate cortex (ACC) from schizophrenic patients by radioligand binding of the antagonist radioligand [^{3}H]-rimonabant. The CB_1 receptors displayed a homogeneous distribution among the layers of the ACC, and a significant increase of 64% in [^{3}H]-rimonabant-specific binding to CB_1 receptors was found in schizophrenic patients compared to controls. Recently, Newell et al. (2006) reported elevated binding of [^{3}H]-CP55940 to CB_1 receptors in the posterior cingulate cortex (PCC) in schizophrenic patients compared to controls. A 25% increase in CB_1 receptor binding was found in the superficial layers (I, II), which was not related to cannabis use. No difference was found in the deeper layers of the PCC. In addition, Koethe et al. (2007) analysed the expression of the CB_1 receptors in ACC at the protein level using immunohistochemistry. Five patients suffering from schizophrenia and fifteen controls were included in aquantitative post-mortem study. Densities of neurons and glial cells immunopositive for CB_1 receptors were quantified. No evidence of an increased or decreased density of CB_1 receptor-immunopositive cells in schizophrenia was found. In this study, confounding variables such as post-mortem time, fixation time, use of illicit drugs, alcohol abuse or sex did not significantly influence the parameters measured. Such post-mortem studies do suffer from the limitation that the patients are mainly chronic cases treated with various medications over decades, which might have an impact on the expression of CB_1 receptors. These findings suggest some, albeit not particularly dominant, role of endocannabinoid receptors in the pathophysiology of schizophrenia.

Most recently, a post-mortem study was able to show that, in the dorsolateral prefrontal cortex, levels of both CB_1 receptor mRNA and protein were lower in subjects with schizophrenia than in well-matched controls. In a parallel study, mRNA levels in antipsychotic-exposed monkeys were analysed and it was found that they were unchanged compared to those in untreated animals. This finding limits the possibility that the reported post-mortem differences between schizophrenic patients and controls are due to antipsychotic treatment (Eggan et al. 2008).

2.3 *Clinical Studies*

Administration of intravenous Δ^9-THC to pharmacologically stabilised schizophrenic patients led to a transient worsening of psychotic symptoms and of cognitive function (D'Souza et al. 2005). Interestingly, another major non-psychotomimetic compound from *C. sativa*, cannabidiol, has been demonstrated to ameliorate psychotic symptoms and perceptual alterations induced by the synthetic Δ^9-THC analogue nabilone (Leweke et al. 2000), a finding in agreement with animal studies suggesting antipsychotic properties of cannabidiol (Zuardi et al. 1991).

In this context, pioneering clinical studies of the endocannabinoid system measured changes in the levels of anandamide, the most intensely investigated endogenous ligand of the CB_1 receptor, as well as its structural analogues palmitoylethanolamide and oleoylethanolamide in cerebrospinal fluid (CSF) of acutely psychotic schizophrenic patients and healthy controls (Leweke et al. 1999a). In this study, significantly elevated levels of both anandamide and palmitoylethanolamide were observed in the CSF of schizophrenic patients, suggesting up- or dysregulation of the endocannabinoid system in acute schizophrenia. This latter finding was replicated with a larger sample of schizophrenic patients revealing more detailed data on the role of the endocannabinoid system in the disorder (Giuffrida et al. 2004). Not only was a significant elevation of anandamide in the CSF of first-onset, antipsychotic-naïve schizophrenic patients observed, but it was also demonstrated – for the first time for any neurotransmitter investigated so far – that anandamide levels in CSF were significantly and inversely correlated to psychotic symptoms and to negative symptoms in particular (Fig. 1). In addition, neither anandamide nor palmitoylethanolamide or oleoylethanolamide levels were affected in serum, indicating the up-regulation of anandamide to be exclusive to the central nervous

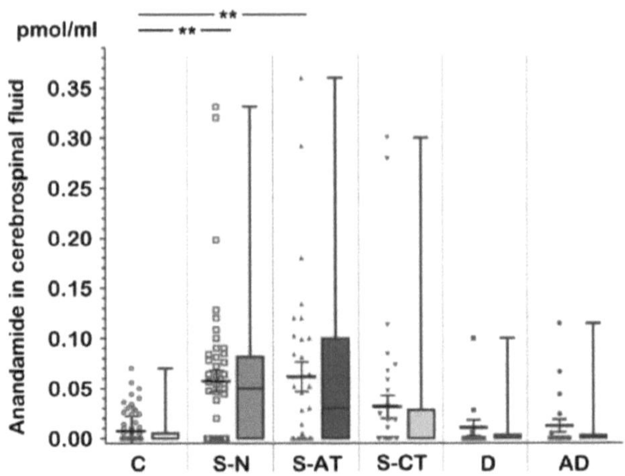

Fig. 1 Anandamide levels are elevated in the CSF of antipsychotic-free first-episode schizophrenic patients. CSF anandamide in healthy volunteers (C); antipsychotic-free schizophrenics with paranoid schizophrenia (S-N); acute schizophrenics (paranoid type) treated with 'atypical' (S-AT) or 'typical' (S-CT) antipsychotic drugs; and patients affected by dementia (D) or affective disorders (AD) without psychotic symptoms. Single values are given with mean ±SEM as well as corresponding boxplots illustrating median, range, and quartiles for each group. Statistically significant differences between groups are shown (*$P \leq 0.01$; **$P \leq 0.001$)

system and, since patients suffering affective disorders or dementia showed no such changes, to be specific to schizophrenia. Interestingly, patients treated with antipsychotic drugs predominantly blocking dopamine D_2 receptors showed much lower levels of anandamide in CSF that were not statistically different from those in healthy volunteers. This was not the case for combined serotonin $5HT_{2A}$ receptor and D_2 receptor blocking antipsychotics (so called second-generation antipsychotics), pointing to a D_2 receptor-mediated increase of anandamide in CSF that had been previously observed in relation to motor control in rodents (Giuffrida et al. 1999).

2.4 *Cannabis Administration*

At present we are only beginning to understand why cannabis abuse may have deleterious effects on the course of schizophrenia, both in its early stages and during the later course of the disease. Most recently, a study from our group investigated the influence of previous, more frequent, cannabis use in first-episode antipsychotic-naïve schizophrenia on anandamide levels in CSF and serum (Leweke et al. 2007a). This analysis revealed significantly elevated levels of anandamide in the CSF of acute schizophrenic patients with a frequency of lifetime cannabis use of less than five times. There was an even stronger significant inverse correlation between anandamide in CSF and psychotic negative symptoms compared to the entire sample of patients. Interestingly, those patients with a frequency of lifetime cannabis use

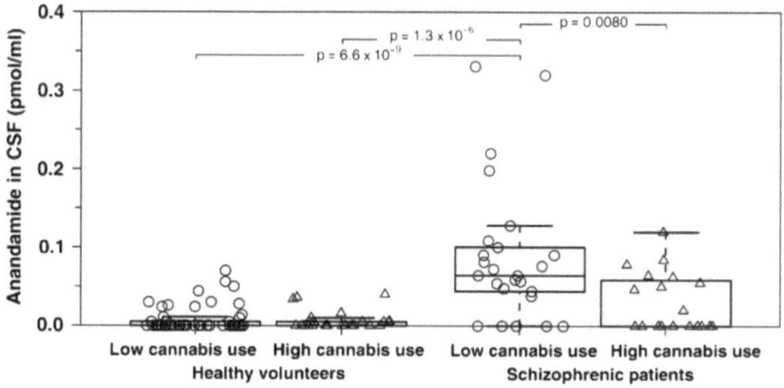

Fig. 2 Box-whiskers-plots (box shows 25th, 50th and 75th percentile of the empirical distribution; whiskers extend to smallest and largest value excluding outliers) of anandamide levels in cerebrospinal fluid (CSF) of schizophrenic patients and healthy volunteers. Left panel, anandamide levels in CSF of healthy volunteers with lowfrequency cannabis use ($\leq$ 5 times in life; column with circles, left panel; n = 55) or with high-frequency cannabis use (< 20 and >50 times in life; column with triangles, left panel; n = 26). Right panel, anandamide levels in CSF of acute antipsychoticnaïve patients suffering from paranoid schizophrenia or schizophreniform psychosis with $\leq$ 5 times of cannabis use in life (column with circles, right panel; n = 25) or >20 times of cannabis use in life (column with triangles, right panel; n = 19).

of more than 20 times showed significantly lower levels of anandamide in CSF than schizophrenic patients who had used cannabis less than five times in total (Fig. 2). In addition, CSF anandamide levels from schizophrenic patients who frequently used cannabis did not significantly differ from matched controls.

3 Conclusions and Model

Accordingly, a model of dopamine/endocannabinoid interaction in acute schizophrenia was proposed in which over-activation of dopamine D_2 receptors is associated with an increased release of anandamide, counterbalancing dopamine-mediated psychotic symptoms by strengthening the endogenous adaptive feedback loop (Fig. 3). This model suggests an adaptation of endocannabinoid function over a longer period of time in response to a slowly, potentially stepwise, increasing level of dopaminergic neurotransmission. While the endocannabinoid system in this hypothetical model may fail to fully counterbalance dopaminergic over-excitation during the initial course of the illness, those patients able to raise levels of anandamide higher than others suffer fewer symptoms (Leweke et al. 2007a). This view is further supported by recent data from patients at risk of psychosis (initial prodromal states of psychosis) showing significantly elevated levels of anandamide in CSF already at this stage of their illness. In addition, those patients with higher levels of anandamide in CSF are less likely to develop frank psychosis during an observational period of at least 42 months (Leweke et al. 2007b).

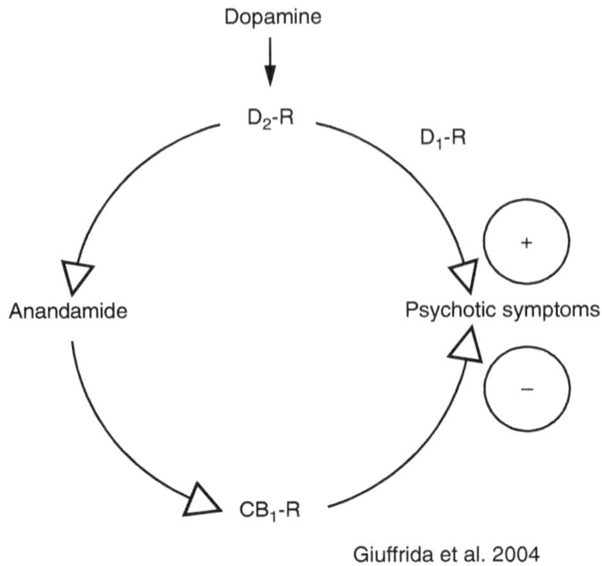

Fig. 3 Model of dopamine/endocannabinoid ineraction in acute schizophrenia. Increased anandamide release, associated with over-activation of dopamine D2 receptors, counterbalances dopamine-mediated psychotic symptoms.

Giuffrida et al. (1999, 2004) suggested a model based upon a dopamine/endocannabinoid interaction in which the activation of CB_1 receptors by anandamide serves as a feedback loop for D_2 receptor-mediated motor control in rodents (Giuffrida et al. 1999) as well as psychotic symptoms in humans (Giuffrida et al. 2004) (Fig. 2). Based on this model, chronic and more frequent administration of Δ^9-THC in schizophrenic patients may disrupt anandamide release and may, thereby, weaken the proposed inhibitory feedback loop on dopamine-mediated processes (Fig. 2). This model is further supported by the fact that schizophrenic patients treated primarily with dopamine D_2-antagonist antipsychotics show markedly lower anandamide levels in CSF than antipsychotic-naïve or those patients treated with $5HT_{2A}$-/weaker D_2-antagonistic antipsychotics (Giuffrida et al. 2004). For the first time, a potential neurobiological mechanism for the negative influence of more frequent use of Δ^9-THC-containing cannabis preparations on schizophrenia symptoms and outcome is provided.

Over recent years, our understanding of the pathophysiolgical role of the endocannabinoid system in schizophrenia has been expanded far beyond expectations. The pre-clinical and clinical data support the contention that this highly expressed regulatory neurotransmitter system is deeply involved in the underlying neurobiological processes in schizophrenia. However, many open questions and controversial results still remain and further research is required to clarify and extend the findings in this area. Successful outcomes will not only contribute to our understanding of this complex disease but will also open new avenues for the treatment of affected patients.

Acknowledgements This research was funded by the Stanley Medical Research Institute (01-315 and 03-NV-003 to FML) and the Koeln Fortune Program (108-2000 to FML). I gratefully acknowledge the contributions of Drs. L. Kranaster, D. Koethe, C.W. Gerth, C. Hoyer, B.M. Nolden, C. Mauss, D. Schreiber, F. Pahlisch, and C. Jöpen to this research.

References

Andreasson S, Allebeck P, Engstrom A et al. (1987) Cannabis and schizophrenia. A longitudinal study of Swedish conscripts. Lancet 2:1483–1486

Andreasson S, Allebeck P, Rydberg U (1989) Schizophrenia in users and nonusers of cannabis. A longitudinal study in Stockholm County. Acta Psychiat Scand 79:505–510

Arseneault L, Cannon M, Poulton R et al. (2002) Cannabis use in adolescence and risk for adult psychosis: longitudinal prospective study. Br Med J 325:1212–1213

Arseneault L, Cannon M, Witton J et al. (2004) Causal association between cannabis and psychosis: examination of the evidence. Br J Psychiatry 184:110–117

Beringer K, von Baeyer W, Marx H (1932) Zur Klinik des Haschischrausches.Nervenarzt 5:37

Dean B, Sundram S, Bradbury R et al. (2001) Studies on [3H]CP-55940 binding in the human central nervous system: regional specific changes in density of cannabinoid-1 receptors associated with schizophrenia and cannabis use. Neuroscience 103:9–15

D'Souza CD, Perry E, MacDougall L et al. (2004) The psychotomimetic effects of intravenous delta-9-tetrahydrocannabinol in healthy individuals: implications for psychosis. Neuropsychopharmacology 29:1558–1572

D'Souza DC, Abi-Saab WM, Madonick S et al. (2005) Delta-9-tetrahydrocannabinol effects in schizophrenia: implications for cognition, psychosis, and addiction. Biol Psychiatry 57:594–608

Eggan SM, Hashimoto T, Lewis DA (2008) Reduced cortical cannabinoid 1 receptor messenger RNA and protein expression in schizophrenia. Arch Gen Psychiatry 65:772–784

Emrich HM, Leweke FM, Schneider U (1997) Towards a cannabinoid hypothesis of schizophrenia: cognitive impairments due to a dysregulation of the endogenous cannabinoid system. Pharmacol Biochem Behav 56:803–807

Giuffrida A, Parsons LH, Kerr TM et al. (1999) Dopamine activation of endogenous cannabinoid signaling in dorsal striatum. Nat Neurosci 2:358–363

Giuffrida A, Leweke FM, Gerth CW et al. (2004) Cerebrospinal anandamide levels are elevated in acute schizophrenia and are inversely correlated with psychotic symptoms. Neuropsychopharmacology 29:2108–2114

Gregg L, Barrowclough C, Haddock G (2007) Reasons for increased substance use in psychosis. Clin Psychol Rev 27:494–510

Henquet C, Krabbendam L, Spauwen J et al. (2005) Prospective cohort study of cannabis use, predisposition for psychosis, and psychotic symptoms in young people. Br Med J 330:11–16

Koethe D, Gerth CW, Neatby MA et al. (2006) Disturbances of visual information processing in early states of psychosis and experimental delta-9-tetrahydrocannabinol altered states of consciousness. Schizophr Res 88:142–150

Koethe D, Llenos IC, Dulay JR et al. (2007) Expression of CB_1 cannabinoid receptor in the anterior cingulate cortex in schizophrenia, bipolar disorder, and major depression. J Neural Transm 114:1055–1063

Koethe D, Giuffrida A, Schreiber D, Hellmich M, Schultze-Lutter F, Ruhrmann S, Klosterkötter J, Piomelli D, Leweke, FM (2009) Anandamide elevation in cerebrospinal fluid in initial prodromal states of psychosis. Br J Psychiatry 194:371–372

Kovasznay B, Fleischer J, Tanenberg Karant M et al. (1997) Substance use disorder and the early course of illness in schizophrenia and affective psychosis. Schizophr Bull 23:195–201

Leweke FM, Kampmann C, Radwan M et al. (1998) The effects of tetrahydrocannabinol on the recognition of emotionally charged words: an analysis using event-related brain potentials. Neuropsychobiology 37:104–111

Leweke FM, Giuffrida A, Wurster U et al. (1999a) Elevated endogenous cannabinoids in schizophrenia. Neuroreport 10:1665–1669

Leweke FM, Schneider U, Thies M et al. (1999b) Effects of synthetic Δ^9-tetrahydrocannabinol on binocular depth inversion of natural and artificial objects in man. Psychopharmacology 142:230–235

Leweke FM, Schneider U, Radwan M et al. (2000) Different effects of nabilone and cannabidiol on binocular depth inversion in man. Pharmacol Biochem Behav 66:175–181

Leweke FM, Koethe D, Gerth CW et al. (2005) The endocannabinoid modulator cannabidiol as an antipsychotic. Results from the first controlled clinical trial in acute schizophrenia. Biol Psychiat 57:S135

Leweke FM, Giuffrida A, Koethe D et al. (2007a) Anandamide levels in cerebrospinal fluid of first-episode schizophrenic patients: impact of cannabis use. Schizophr Res 94(1–3):29–36

Linszen DH, Dingemans PM, Lenior ME (1994) Cannabis abuse and the course of recent-onset schizophrenic disorders. Arch Gen Psychiatry 51:273–279

Meltzer HY, Arvanitis L, Bauer D et al. (2004) Placebo-controlled evaluation of four novel compounds for the treatment of schizophrenia and schizoaffective disorder. Am J Psychiatry 161:975–984

Moreau de Tours J-J (1845) Du hachisch et de l'aliénation mentale, Études psychologiques. Paris, Fortin et Masson

Newell KA, Deng C, Huang XF (2006) Increased cannabinoid receptor density in the posterior cingulate cortex in schizophrenia. Exp Brain Res 172:556–560

Pisanu A, Acquas E, Fenu S, Di Chiara G (2006) Modulation of Delta(9)-THC-induced increase of cortical and hippocampal acetylcholine release by micro opioid and D(1) dopamine receptors. Neuropharmacology 50:661–670

Poncelet M, Barnouin M-C, Breliere J-C et al. (1999) Blockade of cannabinoid (CB1) receptors by SR 141716 selectively antagonizes drug-induced reinstatement of exploratory behaviour in gerbils. Psychopharmacology 144:144–150

Schneider M (2008) Puberty as a highly vulnerable developmental period for the consequences of cannabis exposure. Addict Biol 13:253–263

Schneider M, Koch M (2003) Chronic pubertal, but not adult chronic cannabinoid treatment impairs sensorimotor gating, recognition memory, and the performance in a progressive ratio task in adult rats. Neuropsychopharmacology 28:1760–1769

Schneider M, Koch M (2005) Deficient social and play behavior in juvenile and adult rats after neonatal cortical lesion: effects of chronic pubertal cannabinoid treatment. Neuropsychopharmacology 30:944–957

Schneider M, Drews E, Koch M (2005) Behavioral effects in adult rats of chronic prepubertal treatment with the cannabinoid receptor agonist WIN 55, 212-2. Behav Pharmacol 16:447–454

Sundram S (2006) Cannabis and neurodevelopment: implications for psychiatric disorders. Hum Psychopharmacol 21:245–254

van Os J, Bak M, Hanssen M et al. (2002) Cannabis use and psychosis: a longitudinal population-based study. Am J Epidemiol 156:319–327

Veen ND, Selten JP, van der Tweel I et al. (2004) Cannabis use and age at onset of schizophrenia. Am J Psychiatry 161:501–506

Verrico CD, Jentsch JD, Dazzi L, Roth RH, Taylor JR (2003) Repeated, intermittent delta(9)-tetrahydrocannabinol administration to rats impairs acquisition and performance of a test of visuospatial divided attention. Neuropsychopharmacology 29:522–529

Zammit S, Allebeck P, Andreasson S et al. (2002) Self reported cannabis use as a risk factor for schizophrenia in Swedish conscripts of 1969: historical cohort study. Br Med J 325:1199–1203

Zavitsanou K, Garrick T, Huang XF (2004) Selective antagonist [3H]SR141716A binding to cannabinoid CB1 receptors is increased in the anterior cingulate cortex in schizophrenia. Prog Neuropsychopharmacol Biol Psychiatry 28:355–360

Zuardi AW, Rodrigues JA, Cunha JM (1991) Effects of cannabidiol in animal models predictive of antipsychotic activity. Psychopharmacology 104:260–264

Tourette's Syndrome

Kirsten R. Müller-Vahl

Contents

Abstract Tourette's syndrome (TS) is a chronic disorder characterized by motor and vocal tics and a variety of associated behaviour disorders. Because current therapy is often unsatisfactory, there is expanding interest in new therapeutic strategies that are more effective, cause less side effects and ameliorate not only tics but also behavioural problems. From anecdotal reports and preliminary controlled studies it is suggested that – at least in a subgroup of patients – cannabinoids are effective in the treatment of TS. While most patients report beneficial effects

K. R. Müller-Vahl
Clinic of Psychiatry, Social Psychiatry and Psychotherapy, Hannover Medical School, Carl-Neuberg-Str. 1, D-30625, Hannover, Germany
e-mail: mueller-vahl.kirsten@mh-hannover.de

D. Kendall and S. Alexander (eds.), *Behavioral Neurobiology of the Endocannabinoid System*, Current Topics in Behavioral Neurosciences 1,
DOI: 10.1007/978-3-540-88955-7_17,

when smoking marijuana (*Cannabis sativa L.*), available clinical trials have been performed using oral Δ^9-tetrahydrocannabinol (THC). In otherwise treatment-resistant TS patients, therefore, therapy with THC should not be left unattempted. To date, it is unknown whether other drugs that interact with the endocannabinoid receptor system might be more effective in the treatment of TS than smoked marijuana or pure THC. Since it has been suggested that abnormalities within the endocannabinoid receptor system might underlie TS pathophysiology, it would be of interest to investigate the effect of substances that for example bind more selectively to the central cannabinoid receptor or inhibit the uptake or the degradation of different endocannabinoids.

Keywords Tourette's syndrome • Tic • Attention obsessive compulsive disorder • OCD

Abbreviations

2AG	2-Arachidonoylglycerol
ADHD	Attention deficit hyperactivity disorder
GCIS	Global clinical impression scale
GP	Globus pallidus
MRI	Magnetic resonance imaging
NL	Neuroleptics
PD	Parkinson's disease
PET	Positron emission tomography
OCB	Obsessive compulsive behaviour
SSCP	Single-strand conformation polymorphism
SSRI	Selective serotonin-reuptake inhibitors
STSS	Shapiro Tourette-syndrome severity scale
TS	Tourette's syndrome
TSGS	Tourette's syndrome global scale
TSSL	Tourette syndrome symptom list
YGTSS	Yale global tic severity scale

1 Tourette's Syndrome

1.1 The Clinical Picture of Tourette's Syndrome

Tourette's syndrome (TS) is defined as a childhood-onset chronic neuropsychiatric disorder characterized by multiple motor and one or more vocal tics (The Tourette

Syndrome Classification Study Group 1993). Tics are sudden, repetitive, stereotyped movements or phonic productions that predominantly involve facial, shoulder or upper limb muscles. Salient features of tics are premonitory urges preceding the tics and the ability to suppress the tics for a short period of time. Beside such simple tics, complex tics can occur including copro- and echophenomena. In the majority of patients behavioural problems are associated such as attention deficit hyperactivity disorder (ADHD), obsessive compulsive behaviour (OCB), self injurious behaviour, depression, anxiety disorder, rage, learning disorders, conduct disorder, oppositional deficient disorder, and addiction. Most typically, tics start between the age of 6–8 years, reach their maximum between the age of 10–14 years, and decrease spontaneously in the further course of the disease (Robertson 2000; Singer 2000).

1.2 The Aetiology of Tourette's Syndrome

The neurobiology of TS is still unknown. Findings from in vivo neuroimaging studies provided evidence that different parallel circuits that connect frontal association areas with the basal ganglia are pathophysiologically involved (Gerard and Peterson 2003). It is thought that these loops are involved in the selection, programming, initiation, and control of movement (Alexander et al. 1990). Furthermore, it has been suggested that abnormal function of basal ganglia circuits with abnormal excessive activity of multiple discrete sets of striatal neurons can produce tics (Mink 2001). However, recent findings from magnetic resonance imaging (MRI) studies provided evidence that TS is primarily caused by anomalous frontal lobe association and projection fibre bundles resulting in both basal ganglia function abnormalities and disinhibition of the cingulate gyrus (Müller-Vahl 2006).

Most neurotransmitters involved in frontal-subcortical circuits have been suggested to play a role in the pathobiology of TS, including the dopaminergic, GABAergic, glutamatergic, cholinergic, serotonergic, noradrenergic, opiod, second messenger, and cannabinoid receptor systems (Singer and Wendtlandt 2001; Müller-Vahl et al. 1998). Although multiple clinical and laboratory studies favour an involvement of the dopaminergic system, to date no characteristic dopaminergic dysfunction has been consistently identified. Therefore, it has been speculated that dysfunctions in other transmitter systems might underlie TS pathology and changes in the dopaminergic system might be secondary to these defects (Singer and Wendtlandt 2001).

1.3 Treatment of Tourette's Syndrome

In 1961, haloperidol was proven to be effective in the treatment of tics in patients suffering from TS. Since then, dopamine receptor blocking drugs (neuroleptics, NL) such as haloperidol, pimozide, sulpiride, risperidone, tiapride, and other typical and atypical NL are considered the most effective agents in the treatment of tics. However, treatment with neuroleptic drugs is often unsatisfactory due to low efficacy or significant side effects (sedation, drowsiness, impaired motivation, weight gain, depression, akathisia, and acute dystonic reactions). Therefore, NL are recommended particularly in those patients who are significantly impaired and/or suffer from severe tics. Alternatively, only a limited number of substances can be used in the treatment of tics including clonidine, an α-adrenoceptor agonist, and dopamine receptor agonists such as pergolide. To date, there is no therapy known that is not only effective in the treatment of tics, but also improves associated behavioural disorders. Therefore, selective serotonin-reuptake inhibitors (SSRI) are recommended for the treatment of associated OCB, and psychostimulants such as methylphenidate are the treatment of choice in patients suffering from additional ADHD. In patients with severe and complex symptoms combined treatment with several drugs is often inevitable (Müller-Vahl 2002).

1.4 Future Perspectives in the Treatment of Tourette's Syndrome

At present, therapy of TS often remains unsatisfactory. There is no drug known that is curative. All available drugs are associated with potentially disabling adverse effects. Although there is general agreement that available drug therapy should be limited to those patients who are significantly impaired by their symptoms, it is well known that not only severe, but also mild, tics can be functionally disabling. In the treatment of TS, therefore, new therapeutic strategies are desirable that (1) are more effective in the treatment of tics, (2) cause less adverse effects, and (3) improve not only tics but also associated behavioural disorders such as ADHD and OCB.

Against this background, many TS patients seek alternative or complementary medicine including special diets and nutritional supplements (Mantel et al. 2004; Müller-Vahl et al. 2008) as well as legal and illegal drugs such as nicotine, alcohol and *Cannabis sativa* (Müller-Vahl et al. 1997a, b). Based on such self-monitoring, further investigations were stimulated on the therapeutic use of cannabinoids in the treatment of TS.

2 Treatment of Tourette's Syndrome with Cannabinoids

2.1 Anecdotal Reports

In 1988 Sandyk and Awerbuch and in 1993 Hemming and Yellowlees for the first time suggested that the use of smoked marijuana (*Cannabis sativa*) might be useful in the treatment of TS. Sandyk and Awerbuch (1988) reported on three 15–39-year-old male patients who experienced an improvement not only of their tics and the preceding urge to tic, but also of several associated behavioural problems such as self-mutilatory behaviour, attention span, and hypersexuality when smoking 1/2 to 2 marijuana cigarettes per day. Five years later, Hemming and Yellowlees (1993), in addition, described a single case of a 36-year-old man suffering from TS who reported that he had been symptom-free for more than 1 year when taking one "cone" of marijuana per night.

These initial case reports were corroborated by results obtained from a retrospective survey that has been performed at a specialized TS outpatient clinic (Clinic of Psychiatry, Hannover Medical School) in 1998 (Müller-Vahl et al. 1998). Using a standardized questionnaire, 64 consecutive adult TS patients were interviewed about the use and the potential effect of cannabinoids on their symptoms. Of 17 patients reporting prior use of marijuana, 14 (82%) experienced a reduction or complete remission of motor and vocal tics and/or an amelioration of premonitory urges, OCB, and ADHD. None of these patients reported serious side effects or a deterioration of symptoms when smoking marijuana. Beneficial effects were noted not only in drug-free patients, but also in patients with ongoing treatment.

2.2 *Uncontrolled Single Case Studies*

Because in Germany use of marijuana is illegal and the cannabis herb is not licenced for clinical use, consecutive clinical trials investigating the therapeutic effect of cannabinoids in TS were performed using Δ^9-tetrahydrocannabinol (THC), the most psychoactive ingredient of *Cannabis sativa*.

In an uncontrolled single case study beneficial effects of a single dose treatment with 10 mg THC orally were reported in a 25-year-old male patient who suffered from TS in association with ADHD, OCB, anxiety, lack of impulse control, and self injurious behaviour (Müller-Vahl et al. 1999). For several years he had used marijuana (2–3 g per day) illegally and reported a marked improvement of his tics and behavioural problems when smoking marijuana. In this prospective single case study, for the first time, valid and reliable rating scales were used to assess the clinical effect of THC in TS. At the time of investigation, the patient was unmedicated and had stopped smoking marijuana 3 days before. Using the tic section of the Tourette's Syndrome Global Scale (TSGS) (Leckman et al. 1988), the total tic severity score was 41 before treatment and was reduced to 7 two hours after THC treatment. Both motor and vocal tics improved and coprolalia disappeared. The improvement began 30 min after treatment and lasted for about 7 h. No adverse effects occurred. Measuring cognitive functions, neuropsychological tests showed improved signal detection, sustained attention, and reaction time after treatment. The patient himself noted an improvement of motor and vocal tics of about 70%. Furthermore, he felt an amelioration in attention, impulse control, OCB, and premonitory feeling.

In another single case study, THC in combination with a neuroleptic medication was described as superior to THC or NL alone with respect to the treatment of tics (Müller-Vahl et al. 2002a). In this 24-year-old female suffering from extreme motor and vocal tics, treatment with 10 mg/day THC plus 1200 mg/day amisulpride (an atypical neuroleptic drug) was found to be the most effective treatment. However, due to NL-induced side effects such as galactorrhoea, weight gain, and sedation, later on she decided to discontinue pharmacotherapy. Nonetheless, from this clinical observation it is suggested that THC might augment anti-tic effects of dopamine receptor blocking drugs. These results are in line with animal studies in rats demonstrating that hypokinesia induced by the dopamine receptor antagonist haloperidol significantly increases after co-administration of THC (Moss et al. 1984). It, therefore, has been suggested that combined treatment with cannabinoids and NL might be of therapeutic value in hyperkinetic movement disorders such as TS (Moss et al. 1989).

2.3 Controlled Single-Dose Trial

Based on these initial case reports, a randomized double-blind placebo-controlled crossover single-dose trial of THC in TS was performed (Müller-Vahl et al. 2002b). In this study 12 adult patients (11 men, 1 woman, mean age = 34 ± 13 (SD) years, range 18–66 years) were included. Patients were randomly assigned a single dose of oral THC first or a single dose of visually identical placebo first on two days separated by a 4-week washout phase before they were crossed over to receive the other treatment. According to their body weight, sex, age and prior use of marijuana, patients were treated with 5, 7.5 or 10 mg THC. Both self (Tourette Syndrome Symptom List (TSSL) (Leckman et al. 1988)) and examiner rating scales (Shapiro Tourette-Syndrome Severity Scale (STSS) (Shapiro et al. 1988), Yale Global Tic Severity Scale (YGTSS) (Harcherik et al. 1984) and TSGS (Leckman et al. 1988)) were used to determine the effect of THC. Using the TSSL, there was a significant global tic improvement after THC compared with placebo ($p = 0.015$). Examiner ratings demonstrated a significant improvement ($p = 0.015$) for complex motor tics (TSGS). Using the TSSL, in addition, there was a significant improvement of OCB ($p = 0.041$). Including only those patients who had received either 7.5 or 10.0 mg THC ($n = 8$), data became more robust suggesting that higher dosages are more effective. On the THC treatment day, 10 of 12 patients experienced a global improvement (mean +35% ± 28.0, range 20–90%). In contrast, on the placebo day only three patients reported a global improvement (mean of +7% ± 13.7, range 10–40%). No serious adverse reactions occurred. Five patients experienced transient mild side effects lasting for 1–6 h. Four of them reported headache, nausea, dizziness, hot flush, tiredness, poor powers of concentration, and cheerfulness. One patient who was treated with 10 mg THC experienced dizziness, anxiety, tremble, sensitivity to noise and light, dry mouth, and ataxia lasting for about half an hour.

In addition, a variety of neuropsychological tests was performed to investigate the influence of a single-dose treatment of THC on neuropsychological performance (Müller-Vahl et al. 2001). No detrimental effect of THC was found on short-term verbal and visual memory, recognition, verbal learning, intelligence, information processing, vigilance, reaction time, sustained attention and divided attention. In healthy cannabis users there is evidence that cannabis use causes cognitive impairments that correlate with frequency and duration of cannabis use (Solowij et al. 1995; Block and Ghoneim 1993). Since it has been suggested that the central cannabinoid system might be involved in the pathophysiology of TS (Müller-Vahl et al. 1998, 1999), it can be hypothesized that the effect of THC on neuropsychological performance may be different in patients suffering from TS compared to healthy users.

Furthermore, treatment with THC did not result in a deterioration of depression, somatization, interpersonal sensitivity, anxiety, anger–hostility, paranoid ideation, and psychoticism. Using the Symptom Checklist 90-R (SCL-90-R) (Derogatis et al. 1973; Derogatis 1977), data provided evidence for a deterioration of OCB and a trend towards an increase in phobic anxiety. However, limitations of the SCL-90-R in measuring OCB are known. From other studies, in contrast, it is suggested that cannabinoids may even improve OCB (Müller-Vahl et al. 1998, 1999). The increase in phobic anxiety is probably due to the study design, because the dosage could not be administered slowly.

2.4 Six-Week Randomized Trial

Based on these encouraging results, a randomized, double-blind, placebo-controlled study was performed to confirm these findings (Müller-Vahl et al. 2003a, b). In this study 24 adult patients (19 men, 5 women, mean age = 33 ± 11 (SD) years, range 18–68 years) with TS were included. Patients were treated over a period of 6 weeks with 5–10 mg THC. The dosage was titrated to the target dosage of10 mg THC. Starting at 2.5 mg/day, the dosage was increased by increments of 2.5 mg/day every 4 days. The study consisted of six visits (visit 1 = baseline, visits 2–4 during treatment, visits 5 and 6 after withdrawal). At each visit tic severity was measured using different examiner rating scales (Global Clinical Impression Scale, GCIS) (Leckman et al. 1988), STSS, YGTSS, and a videotape-based rating scale (Goetz et al. 1987) as well as a self rating (TSSL).

Using the GCIS at visits 3 and 4 there was a significant difference ($p < 0.05$) between the THC and placebo group. At visit 4, in addition, a significant difference between both groups was found when using the STSS ($p = 0.033$), the subscore "motor global scale" of the YGTSS ($p = 0.040$) and the videotape-based rating scale ($p = 0.030$). The TSSL demonstrated a significant difference ($p < 0.05$) between the placebo and THC group at 10 treatment days (between day 16 and 41). ANOVA also demonstrated a significant difference between both groups ($p = 0.037$). Several other measures, in addition, demonstrated a trend towards a significant difference ($p < 0.1$) at visits 2, 3, and 4, respectively, either in global tic scores or in several subscores.

Seven patients dropped out of the study or had to be excluded, but only one due to side effects. No serious adverse effects occurred. Five patients in the THC group reported mild side effects (tiredness, dry mouth, dizziness, and muzziness) and three patients in the placebo group (tiredness, dizziness, anxiety, and depression). One patient in the THC group stopped medication at day 4 (first day at dose 5 mg) due to side effects like anxiety and restlessness.

In addition, the influence of a 6-week THC treatment on neuropsychological performance was investigated (Müller-Vahl et al. 2003). To measure cognitive functions the following tests were used: (1) German version of the Auditory Verbal Learning Test (VLMT) (Helmstaedter and Durwen 1990), (2) Benton-Visual-Retention-Test (BVRT) (Benton 1945), (3) Divided Attention (TAP) (Zimmermann and Fimm 1989), and (4) multiple choice vocabulary test (Mehrfachwahl-Wortschatztest, MWT-B) (Merz et al. 1975). Neither during medication, nor immediately after medication was stopped, nor 5–6 weeks after withdrawal, were any detrimental effects seen on learning curve, interference, recall and recognition of word lists, immediate visual memory span, and divided attention. Measuring immediate verbal memory span, there was even a trend towards a significant improvement during and after treatment. Furthermore, no significant influence on OCB, anxiety, depression and "the current emotional state" was found [unpublished data].

3 Adverse Effects

Based on the available data it can be concluded that in most TS patients treatment with THC causes only mild adverse reactions. Overall, adverse effects were comparable to those seen in other groups of patients including headache, dry mouth, nausea, dizziness, muzziness, hot flush, tiredness, poor powers of concentration, and cheerfulness. Previous studies were afflicted with a low drop-out rate due to side effects. Only rarely more significant adverse effects were observed such as anxiety, tremble, ataxia, and restlessness. In contrast to other studies, neuropsychological tests did not demonstrate detrimental effects on cognition in TS patients. Since it has been suggested that changes in the cannabinoid receptor system might be involved in the pathophysiology of TS, it can be speculated that in TS patients cognitive functions are less impaired (or even improved) by THC compared to healthy users.

In general, cannabinoids are contraindicated in patients suffering from a psychotic illness and significant cardiac disorder. THC should be used with caution in patients with a history of substance abuse. Patients receiving treatment with THC should be warned not to drive or operate machinery until it is established that they are able to tolerate the drug. In addition, THC should not be used in pregnant and breast-feeding women and children because there is evidence that frequent cannabis use in young people is associated with increased rates of psychotic symptoms, depression and anxiety (Fergusson et al. 2003; Patton et al. 2002). However, in a small study in eight children (3–13 years) suffering from hematologic cancers, treatment with Δ^8-THC was well tolerated (Abrahamov et al. 1995). The authors suggested that in children side effects may occur less frequently because the central cannabinoid CB_1 receptor system is not fully developed.

4 Central Cannabinoid Receptor (*CNR1*) Gene in Tourette's Syndrome

Based on the beneficial effects of cannabinoids in the treatment of TS, it has been hypothesized that the central cannabinoid (CB_1) receptor system is involved in the pathophysiology of TS. Therefore, the central cannabinoid receptor gene (*CNR1*) encoding the CB_1 was considered as a candidate gene for TS and systematically screened by single-strand conformation polymorphism (SSCP) analysis and sequencing. However, investigating 40 TS patients and 81 healthy controls and, in addition, two subsequent cohorts of 56 TS patients and 55 controls, and 64 patients and 66 controls, there was no evidence suggesting that TS is caused by genetic variations of the *CNR1* gene (Gadzicki et al. 2004).

5 In Vivo Imaging of Central Cannabinoid CB_1 Receptors in TS Using [^{123}I]AM281 and SPECT

In vivo neuroimaging using positron emission tomography (PET) and single photon emission computed tomography (SPECT) to investigate different aspects of the cannabinoid CB_1 receptor system is in a very preliminary state. Although some ligands that are suitable for measuring specific binding to CB_1 receptors in vivo in humans are already forthcoming, there is only a single study available investigating central cannabinoid CB_1 receptors in TS using the CB_1 antagonist [^{123}I]AM281 (*N*-(morpholin-4-yl)-1-(2,4-dichlorophenyl)-5-(4-[^{123}I]iodophenyl)-4-methyl-1H-pyrazole-3-carboxamide) and single photon emission computed tomography (SPECT) (Berding et al. 2004). [^{123}I]AM281 was employed in six TS patients before and after THC treatment and specific over nonspecific partition coefficients V3" were calculated. Although mean V3" did not change significantly after THC treatment, V3" clearly declined in the only patient with a marked clinical response after THC treatment. Results from this first study, therefore, suggest that specific binding of [^{123}I]AM281 to CB_1 receptors can be detected in patients using SPECT. Because in this study a control group is lacking, the question as to whether CB_1 receptor binding sites are pathologically changed in TS patients as measured by [^{123}I]AM281 and SPECT remains unanswered.

6 Possible Explanations for Beneficial Effects of Cannabinoids in TS

In TS, positive effects of THC in the treatment of tics may be explained by different mechanisms. In the CNS, the highest densities of CB_1 receptors were found in the basal ganglia, cerebellum, and hippocampus (Herkenham et al. 1990; Glass et al. 1997). Within the basal ganglia, CB_1 receptors are particularly prominent in the globus pallidus (GP) and substantia nigra pars reticulata – the indirect and direct output pathways (Herkenham et al. 1990). In TS there is evidence for an involvement of both the basal ganglia and the limbic system. This might account for the effects of cannabinoids on tics and behavioural problems in TS.

There are several lines of evidence suggesting a complex interaction between the CB_1 receptor system and the dopaminergic system, which is suggested to be overactive in TS patients. In rats it has been demonstrated that the release of the endocannabinid anandamide was eight-fold increased in the dorsal striatum after administration of a D_2-like dopamine receptor agonist (Giuffrida et al. 1999). This response could be prevented by administration of a D_2-like receptor antagonist. Pretreatment with the cannabinoid antagonist rimonabant enhanced the stimulation of motor behaviour elicited by a D_2-like dopamine receptor agonist, while administration of rimonabant alone had no effect on motor activity. It therefore can be speculated that the endocannabinoid system may act as an inhibitory feedback mechanism countering dopamine stimulation of motor activity (Giuffrida et al. 1999). In addition, it has been demonstrated that anandamide increases the release of dopamine both in the striatum (Cadogan et al. 1997) and in the mesolimbic system (Gessa et al. 1998). Treatments with the dopamine D_2 receptor antagonist haloperidol and sulpiride resulted in significantly increased cannabinoid receptor mRNA levels in the caudate-putamen. Therefore, it has been suggested that the expression of the cannabinoid receptor gene in the striatum is under the negative control of dopamine receptor-mediated events (Mailleux and Vanderhaeghen 1993).

Furthermore, it has been demonstrated that constitutive hyperdopaminergia in dopamine transporter (DAT) knockout (KO) mice, an animal model linked with hyperdopaminergia, is associated with a significant decrease of striatal anandamide levels (Tzavara et al. 2006). These results further support that hyperdopaminergia leads to alterations of the endocannabinoid system and suggest that normalization of decreased anandamide levels might constitute an alternative therapeutic strategy for disorders associated with hyperdopaminergia such as TS (Tzavara et al. 2006).

In the reserpine-treated rat, an animal model for Parkinson's disease, a seven-fold increase in the levels of the endocannabinoid 2-arachidonoylglycerol (2AG) was observed in the GP. Administration of a dopamine D_2 receptor agonist increased locomotion accompanied by reduced 2AG and anandamide levels in the GP (Di Marzo et al. 2000). In humans, it has been shown that nabilone, a classical synthetic THC analogue, ameliorates levodopa-induced dyskinesia in PD (Sieradzan et al. 2001). Therefore, it can be speculated that THC inhibits dopaminergic activity in motor-control centres and, through this, reduces tics in TS.

On the other hand several other neurotransmitters involved in frontal-subcortical circuits have been suggested to play a role in the pathobiology of TS including the GABAergic, glutamatergic, cholinergic, serotonergic, noradrenergic, opiod, and second messenger systems. There is experimental evidence that the activity of most of these transmitters – both excitatory neurotransmitters such as glutamate and inhibitory transmitters such as GABA and glycine – is affected by cannabinoids as well. Therefore, beneficial effects of THC in TS might also be explained by a modulation of one or several of these neurotransmitter systems.

7 Conclusions and Perspective

Available results from a limited number of case reports and preliminary studies consistently provide evidence for beneficial effects of *Cannabis sativa* and THC, respectively, in the treatment of tics and possibly behavioural problems (OCB, attention span, impulsivity, autoaggression) in TS patients. Based on the available data, it can be speculated that even low dosages (5–10 mg) are effective in this group of patients. In most TS patients treated with THC observed adverse effects were mild. Overall, adverse reactions were comparable to those seen in other groups of patients. However, in TS patients no detrimental effects of THC on neuropsychological tests were observed. Since it has been suggested that changes in the cannabinoid receptor system might be involved in the pathophysiology of TS, it can be speculated that the effect of exogenous cannabinoids on the endocannabinoid CB_1 receptor system might be different compared to healthy people. Such a hypothesis might explain why cannabinoids may induce different effects in different groups of patients.

In many cases, TS is associated with comorbid ADHD. The two main behavioural features of this disorder are impaired attention and an impulsive–hyperactive behavioural trait. From case reports it is suggested that impaired attention in TS patients may improve after smoking marijuana (Sandyk and Awerbuch 1988; Müller-Vahl et al. 1998) or the intake of oral THC (Müller-Vahl et al. 1999, 2003a, b). These clinical observations are in line with results from an animal model of ADHD (spontaneously hypertensive rat) suggesting that enhanced impulsivity is associated with a reduced cortical density of cannabinoid CB_1 receptors. In these rats impulsivity could be normalized with acute administration of a cannabinoid receptor agonist (WIN55212-2) (Adriani et al. 2003). In addition, there is a single uncontrolled case study available reporting a 28-year-old male suffering from ADHD who demonstrated a significant improvement of his driving-related performance after the oral intake of THC (Strohbeck-Kühner et al. 2007). The authors, therefore concluded that "... in persons with ADHD THC may have atypical and even performance-enhancing effects".

To date it is unknown whether herbal cannabis, cannabis extracts, other cannabinoid receptor agonists that bind more selectively to the central cannabinoid CB_1 receptor, or agents that interfere with the inactivation of endocannabinoids by inhibiting the uptake or the degradation might be superior to pure THC in the treatment of tics. There is some evidence that THC might augment the anti-tic effect of neuroleptic drugs. Further studies would be desirable investigating the effect of different drugs that interact with the endocannabinoid receptor system on different clinical features in patients suffering from TS.

References

Abrahamov A, Abrahamov A, Mechoulam R (1995) An efficient new cannabinoid antiemetic in pediatric oncology. Life Sci 56:2097–2102

Adriani W, Caprioli A, Granstrem O et al. (2003) The spontaneously hypertensive-rat as an animal model of ADHD: evidence for impulsive and non-impulsive subpopulations. Neurosci Biobehav Rev 2:639–651

Alexander GE, Crutcher MD, DeLong MR (1990) Basal ganglia-thalamocortical circuits: parallel substrates for motor, oculomotor, "prefrontal" and "limbic" functions. Prog Brain Res 85:119–146

Benton AL (1945) A visual retention test for clinical use. Arch Neurol Psychol 59:273–291

Berding G, Müller-Vahl K, Schneider U et al. (2004) [123I]AM281 single-photon emission computed tomography imaging of central cannabinoid CB1 receptors before and after Delta9-tetrahydrocannabinol therapy and whole-body scanning for assessment of radiation dose in tourette patients. Biol Psychiatry 55:904–915

Block RI, Ghoneim MM (1993) Effects of chronic marijuana use on human cognition. Psychopharmacology (Berl) 110:219–228

Cadogan AK, Alexander SP, Boyd EA et al. (1997) Influence of cannabinoids on electrically evoked dopamine release and cyclic AMP generation in the rat striatum. J Neurochem 69:1131–1137

Derogatis LR (1977) SCL 90. Administration, scoring and procedures. Manual-I for the R(evised) version and other instruments of the psychopathology rating scale series. John Hopkins University School of Medicine

Derogatis LR, Lipman RS, Covi L (1973) SCL-90: an outpatient psychiatric rating scale–preliminary report. Psychopharmacol Bull 9:13–28

Di Marzo V, Hill MP, Bisogno T et al. (2000) Enhanced levels of endogenous cannabinoids in the globus pallidus are associated with a reduction in movement in an animal model of Parkinson's disease. FASEB J 14:1432–1438

Fergusson DM, Horwood LJ, Swain-Campbell NR (2003) Cannabis dependence and psychotic symptoms in young people. Psychol Med 33:15–21

Gadzicki D, Müller-Vahl KR, Heller D et al. (2004) Tourette syndrome is not caused by mutations in the central cannabinoid receptor (CNR1) gene. Am J Med Genet B Neuropsychiatr Genet 127:97–103

Gerard E, Peterson BS (2003) Developmental processes and brain imaging studies in Tourette syndrome. J Psychosom Res 55:13–22

Gessa GL, Melis M, Muntoni AL et al. (1998) Cannabinoids activate mesolimbic dopamine neurons by an action on cannabinoid CB1 receptors. Eur J Pharmacol 341:39–44

Giuffrida A, Parsons LH, Kerr TM et al. (1999) Dopamine activation of endogenous cannabinoid signaling in dorsal striatum. Nat Neurosci 2:358–363

Glass M, Dragunow M, Faull RL (1997) Cannabinoid receptors in the human brain: a detailed anatomical and quantitative autoradiographic study in the fetal, neonatal and adult human brain. Neuroscience 77:299–318

Goetz CG, Tanner CM, Wilson RS et al. (1987) A rating scale for Gilles de la Tourette's syndrome: description, reliability, and validity data. Neurology 37:1542–1544

Harcherik DF, Leckman JF, Detlor J et al. (1984) A new instrument for clinical studies of Tourette's syndrome. Am J Acad Child Psychiatry 23:153–160

Helmstaedter C, Durwen HF (1990) VLMT. Verbaler Lern- und Merkfähigkeitstest. Schweiz Arch Neurol Psychiatr 141:21–30

Hemming M, Yellowlees PM (1993) Effective treatment of Tourette's syndrome with marijuana. J Psychopharmacol 7:389–391
Herkenham M, Lynn AB, Little MD et al. (1990) Cannabinoid receptor localization in brain. Proc Natl Acad Sci USA 87:1932–1936
Leckman JF, Towbin KE, Ort SI et al. (1988) Clinical assessment of tic disorder severity. In: Cohen DJ, Bruun RD, Leckman JF (eds) Tourette's syndrome and tic disorders. Wiley, New York
Mailleux P, Vanderhaeghen JJ (1993) Dopaminergic regulation of cannabinoid receptor mRNA levels in the rat caudate-putamen: an in situ hybridization study. J Neurochem 61:1705–1712
Mantel BJ, Meyers A, Tran QY et al. (2004) Nutritional supplements and complementary/alternative medicine in Tourette syndrome. J Child Adolesc Psychopharmacol 14:582–589
Merz J, Lehrl S, Galster JV et al. (1975) MWT-B – ein Intelligenzkurztest. Psychiatr Neurol Med Psychol 27:423–428
Mink JM (2001) Neurobiology of basal ganglia circuit in Tourette Syndrome: faulty inhibition of unwanted motor patterns? Adv Neurol 85:113–122
Moss DE, Montgomery SP, Salo AA et al. (1984) Tetrahydrocannabinol effects on extrapyramidal motor behaviors in an animal model of Parkinson's disease. In: Agurell S, Dewey WL, Willette RE (eds) The cannabinoids: chemical, pharmacologic, and therapeutic aspects. Academic Press, New York
Moss DE, Manderscheid PZ, Montgomery SP et al. (1989) Nicotine and cannabinoids as adjuncts to neuroleptics in the treatment of Tourette syndrome and other motor disorders. Life Sci 44:1521–1525
Müller-Vahl KR (2002) The treatment of Tourette's syndrome: current opinions. Expert Opin Pharmacother 3:899–914
Müller-Vahl KR (2006) Involvement of fronto-striatal pathways in Tourette syndrome with and without comorbid ADHD and OCD. Eur Arch Psychiatry Clin Neurosci 256:II19
Müller-Vahl KR, Kolbe H, Dengler R (1997a) Alcohol withdrawal and Tourette's syndrome. Neurology 48:1478–1479
Müller-Vahl KR, Kolbe H, Dengler R (1997b) Gilles de la Tourette syndrome. Effect of nicotine, alcohol and marihuana on clinical symptoms. Nervenarzt 68:985–989
Müller-Vahl KR, Kolbe H, Schneider U et al. (1998) Cannabinoids: Possible role in pathophysiology of Gilles de la Tourette-syndrome. Acta Psychiat Scand 98:502–506
Müller-Vahl KR, Schneider U, Kolbe H et al. (1999) Treatment of Tourette-syndrome with delta-9-Tetrahydrocannabinol. Am J Psychiatry 156:395
Müller-Vahl KR, Koblenz A, Jobges M et al. (2001) Influence of delta-9-Tetrahydrocannabinol (Δ^9-THC) treatment of Tourette-syndrome on neuropsychological testing. Pharmacopsychiatry 34:19–24
Müller-Vahl KR, Schneider U, Emrich HM (2002a) Combined treatment of Tourette-syndrome with Δ^9-THC and dopamine receptor antagonists. J Cannabis Ther 2:145–154
Müller-Vahl KR, Schneider U, Koblenz A et al. (2002b) Treatment of Tourette-syndrome with Δ^9-Tetrahydrocannabinol (THC). Pharmacopsychiatry 35:57–61
Müller-Vahl KR, Prevedel H, Theloe K et al. (2003a) Treatment of Tourette-syndrome with Δ^9-Tetrahydrocannabinol (Δ^9-THC): no influence on neuropsychological performance. Neuropsychopharmacology 28:384–388
Müller-Vahl KR, Schneider U, Prevedel H et al. (2003b) Δ^9-Tetrahydrocannabinol (THC) is effective in the treatment of tics in Tourette-Syndrome: a 6-week randomized trial. J Clin Psychiatry 64:459–465
Müller-Vahl KR, Buddensiek N, Geomelas M et al. (2008) The influence of different food and drink on tics in Tourette syndrome. Acta Paediatr 97:442–446
Patton GC, Coffey C, Carlin JB et al. (2002) Cannabis use and mental health in young people: cohort study. BMJ 325:1195–1198
Robertson MM (2000) Tourette syndrome, associated conditions and the complexities of treatment. Brain 123:425–462

Sandyk R, Awerbuch G (1988) Marijuana and Tourette's syndrome. J Clin Psychopharmacol 8:444–445
Shapiro AK, Shapiro ES, Young JG et al. (1988) Signs, symptoms, and clinical course. In: Shapiro AK, Shapiro ES, Young JG, Feinberg TE (eds) Gilles de la Tourette syndrome, 2nd edn. Raven Press, New York
Sieradzan KA, Fox SH, Hill M et al. (2001) Cannabinoids reduce levodopa-induced dyskinesia in Parkinson's disease: a pilot study. Neurology 57:2108–2111
Singer HS (2000) Current issues in Tourette syndrome. Mov Disord 15:1051–1063
Singer HS, Wendtlandt JT (2001) Neurochemistry and synaptic neurotransmission in Tourette syndrome. Adv Neurol 85:163–178
Solowij N, Michie PT, Fox AM (1995) Differential impairments of selective attention due to frequency and duration of cannabis use. Biol Psychiatry 37:731–739
Strohbeck-Kühner P, Skopp G, Mattern R (2007) Fitness to drive in spite (because) of THC. Arch Kriminol 220:11–19
The Tourette syndrome classification study group (1993) Definitions and classification of tic disorders. Arch Neurol 50:1013–1016
Tzavara ET, Li DL, Moutsimilli L et al. (2006) Endocannabinoids activate transient receptor potential vanilloid 1 receptors to reduce hyperdopaminergia-related hyperactivity: therapeutic implications. Biol Psychiatry 59:508–515
Zimmermann P, Fimm B (1989) Neuropsychologische Testbatterie zur Erfassung von Aufmerksamkeitsdefiziten. Psychologisches Institut der Universität Freiburg

Index

MIX
Papier aus verantwortungsvollen Quellen
Paper from responsible sources
FSC® C105338

If you have any concerns about our products,
you can contact us on
ProductSafety@springernature.com

In case Publisher is established outside the EU,
the EU authorized representative is:
Springer Nature Customer Service Center GmbH
Europaplatz 3, 69115 Heidelberg, Germany

Printed by Libri Plureos GmbH
in Hamburg, Germany